Kausik Chattopadhyay •
Subhash C. Basu
Editors

Biochemical and Biophysical Roles of Cell Surface Molecules

Springer

Editors
Kausik Chattopadhyay
Centre for Protein Science, Design and Engineering, Department of Biological Sciences
Indian Institute of Science Education and Research (IISER) Mohali
Mohali, Punjab, India

Subhash C. Basu
Department of Chemistry and Biochemistry
University of Notre Dame
Notre Dame, IN, USA

ISSN 0065-2598 ISSN 2214-8019 (electronic)
Advances in Experimental Medicine and Biology
ISBN 978-981-13-3064-3 ISBN 978-981-13-3065-0 (eBook)
https://doi.org/10.1007/978-981-13-3065-0

Library of Congress Control Number: 2018966549

This Springer imprint is published by the registered company Springer Nature Singapore Pte Ltd.
The registered company address is: 152 Beach Road, #21-01/04 Gateway East, Singapore 189721, Singapore

Preface

This book is an edited collection of the scientific contributions from the 11th International Symposium on Cell Surface Macromolecules (11th ISCSM). To value the importance of the book, one needs to know the genesis of the ISCSM. It was at Ronneby, Sweden, the home for the 1983 IGO (International Glycoconjugate Organization) Meeting, where a handful of glycobiologists, Dr. Sailen Mookerjee (Canada), Dr. Michell Monsigny (France), Dr. Debkumar Basu (India), Dr. Guido Tettamanti (Italy), Dr. Sandro Sonnino (Italy), Dr. Herbert Wiegandt (Germany), Dr. Hans Vliegenthart (the Netherlands), Dr. Lars Svennerholm (Sweden), Dr. Subhash C. Basu (USA), Dr. Manju Basu (USA), Dr. Yu-T. Li (USA), Dr. Su-C. Li (USA), and Dr. Saul Roseman (USA), conceived the idea to hold an international meeting on "cell surface macromolecules." The first meeting of the ISCSM was held in New Delhi, India, in 1987. Late Professor Dr. Bimal K. Bachhawat served as a chair of the meeting, and it was supported by an actively engaged executive committee. The next three meetings (in 1990, 1993, and 1996) were also held in New Delhi, India. Since 1999, the ISCSM meetings are being held in different cities in India. Dr. Avadhesha Surolia, Dr. Chitra Mandal, Dr. Amitabha Chattopadhyay, Dr. Perumana Sudhakaran, and Dr. Abhijit Chakrabarti served as the organizers of these ISCSM meetings. Since the first meeting of the ISCSM, this symposium series has provided a unique platform to discuss and present some of the prominent advances made in the area of research on cell surface macromolecules.

Cell membrane plays one of the most critical roles in the functioning of the cellular life forms. By acting as a semipermeability barrier, it compartmentalizes the cellular system from the extracellular environment, while, at the same time, it allows regulated exchange of materials and critical biological information in and out of the cells. Phospholipid bilayer architecture acts as the central scaffold of the cellular plasma membranes. In addition to this core structure, plasma membranes are decorated with a wide array of cell surface macromolecules, like lipids, proteins, glycolipids, and glycoproteins, which act to impart specialized biological functions. Disproportionate distribution of such membrane constituents contributes to the loss of membrane integrity and the disrupted membrane function. Integrity of the cell surface architecture is critical to maintain efficient and optimal intercellular and cell-to-matrix interactions, thus playing a crucial role in diverse biological functions that include cell survival and homeostasis, apoptosis, differentiation, embryonic development, and growth. Accordingly, alteration of the cell membrane/surface and cell surface molecules may lead to diverse pathological conditions. Vast array of cell surface molecules are now known to take part in the cellular processes. Some of the prominent examples include members of the cholesterol family (including steroid hormones), the glycolipid family (sphingolipids), the glycoprotein family (both N-linked and O-linked), and the cell surface protein receptors (like G protein-coupled receptors). Elucidating the structure-function relationship of these cell surface molecules is crucial to understand their roles in the context of the physiological and pathophysiological conditions. Study of the cell surface molecules is challenging owing to their distinct physicochemical properties that equip them to prefer amphipathic membrane environment, rather than the typical aqueous environment. Biochemical, biophysical, cell biology, molecular biology, and computational techniques are routinely used to study the cell surface molecules and their functionalities. In addition, the advent of the new techniques involving crystallography-based structure determination of the membrane proteins, as well as cryo-electron microscopy-based structure determination of the macromolecular complexes, has opened new avenues to extract high-resolution structural information regarding the membrane proteins and the cell surface receptors.

The 11th ISCSM was held in Mohali, India, during February 2017. This volume titled *Biochemical and Biophysical Roles of Cell Surface Molecules* presents contributions from this symposium, covering current topics on the multiple facets of the functional implications of the cell membranes, membrane components, and other cell surface molecules. The volume is organized into six subsections representing some of the prominent topics related to the studies on cell surface and cell surface macromolecules. All the contributed chapters are either comprehensive critical reviews, overviews, or original research articles on the biochemical and biophysical roles of some of the important cell surface molecules that play essential functions in the biological processes. These include contributions on biochemical and biophysical aspects of the membrane components and cell surface molecules (membrane lipids, membrane proteins like GPCRs, and the membrane-associated glycome that includes glycoproteins, glycolipids, and proteoglycans) and their implications for the gene regulations, and other cell biological functions such as apoptosis, cancer homeostasis, immune and neuronal responses, pore formation on the membranes. Overall, this volume aims to provide an in-depth description of our present understanding, current trends, and future perspectives in the field of cell surface biology.

We express our sincere thanks and gratitude to all the authors for their valuable contributions. We would like to thank all the reviewers for their critical evaluation of the manuscripts. We would also like to express our sincere gratitude to Springer Book Editorial Office, Ms. Madhurima Kahali, and Mr. Daniel Jagadisan for their support in publishing this volume in the Advances in Experimental Medicine and Biology series.

Mohali, India — Kausik Chattopadhyay
Notre Dame, IN, USA — Subhash C. Basu

Contents

About the Editors

Dr. Kausik Chattopadhyay is an associate professor in the Department of Biological Sciences at the Indian Institute of Science Education and Research Mohali (IISER Mohali). He received his PhD degree in biochemistry from the Calcutta University in 2004 and subsequently completed his postdoctoral research with Prof. Stanley G. Nathenson in the Department of Microbiology and Immunology at the Albert Einstein College of Medicine, Bronx, New York, USA. He joined IISER Mohali as an assistant professor in 2009 and became an associate professor in 2014. His research interests focus on understanding the structure-function mechanisms of membrane-damaging, bacterial, pore-forming toxins, and their implications for the host-pathogen interaction processes and immunity. He received the prestigious National BioScience Award for Career Development for the year 2014, awarded by the Department of Biotechnology, Ministry of Science & Technology, Government of India for his work in this area.

Dr. Subhash C. Basu is an emeritus professor at the University of Notre Dame (UND), Department of Chemistry and Biochemistry, and also the president of the CDDRF (Cancer Drug Delivery Research Foundation). He received his PhD in biochemistry in 1966 from the University of Michigan and was a Helen Hay Whitney Research Fellow (1966–1970) under Professor Saul Roseman at the Johns Hopkins University. He joined both the teaching and research faculty at the UND in 1970 and became a full professor and chairman of the Biochemistry, Biophysics and Molecular Biology Program in 1983. He was awarded a DSc in biochemistry from the University of Calcutta, India, in 1976. He received the Jacob Javits Neuroscience Research Award from NIH (1989–1998), and the S. C. Roy Medal of Achievement Award from the University of Calcutta, India, in 1990. He was elected to the prestigious Johns Hopkins Honor Society in 1995. He has been an elected fellow of the American Association for the Advancement of Sciences (FAAAS) since 1988. He received the "NRI-2015 Award" Trophy in the International Professional category from the respected *India Times* newspaper in April 2015.

Part I

Membrane Organization

Status of Membrane Asymmetry in Erythrocytes: Role of Spectrin

1

Sauvik Sarkar, Dipayan Bose, Rajendra P. Giri, Mrinmay K. Mukhopadhyay, and Abhijit Chakrabarti

Abstract

Spectrin-based proteinaceous membrane skeletal network has been found to be implicated in membrane disorders like hereditary spherocytosis (HS). HS greatly affects eryptosis via loss of membrane asymmetry which is seen to be the case in haemoglobin disorders like thalassemia and sickle cell disease as well. The biological implications of the status of membrane asymmetry are strongly correlated to spectrin interactions with aminophospholipids, e.g. PE and PS. Fluorescence and X-ray reflectivity (XRR) measurements of spectrin interactions with small unilamellar vesicles (SUVs) and cushioned bilayers of phospholipids, respectively, were studied. Both the XRR and fluorescence measurements led to the characterization of spectrin orientation on the surface of lipid bilayer of phosphatidylcholine (PC) and PC/aminophospholipid mixed membrane systems showing formation of a uniform layer of spectrin on top of the mixed phospholipid bilayer. Fluorescence studies show that spectrin interacts with PC and phosphatidylethanolamine (PE)/phosphatidylserine (PS) membranes with binding dissociation constants (K_d) in the nanomolar range indicating the role of spectrin in the maintenance of the overall membrane asymmetry of erythrocytes.

Keywords

Spectrin · Membrane asymmetry · Lipid-protein interaction · Membrane skeleton

Abbreviations

DM	Dimyristoyl
HE	Hereditary elliptocytosis
HS	Hereditary spherocytosis
PC	Phosphatidylcholine
PE	Phosphatidylethanolamine
PG	Phosphoglycerate
PI	Phosphatidylinositol
PS	Phosphatidylserine
SUV	Small unilamellar vesicles

1.1 Introduction

The asymmetric distribution of lipids across the inner and outer leaflets of the plasma membrane and the presence of lipid rafts are crucial to the functioning of proteins that are embedded in the

S. Sarkar · D. Bose · A. Chakrabarti (✉)
Crystallography & Molecular Biology Division, Saha Institute of Nuclear Physics, HBNI, Kolkata, India
e-mail: abhijit.chakrabarti@saha.ac.in

R. P. Giri · M. K. Mukhopadhyay
Surface Physics & Material Science Division, Saha Institute of Nuclear Physics, HBNI, Kolkata, India

K. Chattopadhyay, S. C. Basu (eds.), *Biochemical and Biophysical Roles of Cell Surface Molecules*, Advances in Experimental Medicine and Biology 1112, https://doi.org/10.1007/978-981-13-3065-0_1

membrane or bound to it (Alessandrini and Facci 2011; Fadeel and Xue 2009). This lipid asymmetry is a key factor in the role of the plasma membrane in various physiological functions such as cell signalling, transport processes, cell–cell interactions, cellular adhesion, etc. However, the mechanical properties of cell membranes arise from the dense proteinaceous meshwork – the membrane skeleton – that remains bound to the cytosolic surface of the inner leaflet rather than the lipid composition of the membrane (Grzybek et al. 2006; Mitra et al. 2015).

Evolutionary complexity, i.e. integration into multicellular tissues and organs, and motility give rise to significant mechanical stress on a cell and as such membrane skeletons which help in maintaining the shape and integrity of the membrane become indispensable (Machnicka et al. 2014). The main protein component of this skeletal network is spectrin which forms the meshwork through interactions with other membrane skeletal proteins and direct interactions to the phospholipid bilayer.

Spectrin was first isolated from red blood cells by Marchesi and Steers in 1968 (Grzybek et al. 2006). Later on it was discovered in non-erythroid cells as well. It is a ~100 nm long, filamentous, wormlike, flexible protein, made up of 2 subunits – α (280 kDa) and β (247 kDa). The subunits are antiparallelly arranged to form the heterodimer. Individual spectrin subunits are mostly made up of repeating domains called 'spectrin repeat' domains – each such domain is typically a 106-amino acid-long triple-helical coiled-coil motif. α-Spectrin contains 22 such repeats, and β-spectrin contains 17 of these domains. Other than these domains, spectrin contains some specialized structural domains, viz. SH3 domain, actin-binding domain, ankyrin-binding domain, calponin homology (CH) domain, pleckstrin homology (PH) domain, etc.

Even before the discovery of ankyrin and other membrane skeletal proteins, spectrin has been known to interact with lipids (Sweet and Zull 1970; Juliano et al. 1971). Spectrin interaction with various phospholipids has been studied in great detail by various methods such as measurement of surface pressure of lipid mono- and bilayers, fluorescence-based studies, measuring enthalpy change to study phase transition, etc. Experiments on phospholipid interaction of spectrin using liposomes of defined compositions led to the discovery that spectrin interacts with many different lipids like phosphatidylcholine (PC), phosphatidylserine (PS), phosphatidylethanolamine (PE) and phosphatidylinositol (PI). Studies with anionic phospholipids like PS showed that spectrin binding causes enthalpic changes associated with a change in phase transition temperature of these lipid bilayers and moreover, this interaction was found to be governed by pH, the optima being at pH 5.5 (Mombers et al. 1980; Grzybek et al. 2006). Specificity of spectrin subunits for phospholipids has been shown in favour of the β-spectrin over α-subunit (Grzybek et al. 2006).

Spectrin-lipid interaction has been studied by using natural membranes as well. To ascertain whether spectrin directly interacts with phospholipids or through protein receptors, natural membranes were treated with NaOH and proteases so as to deplete the natural membranes from proteinaceous receptors. Spectrin was found to interact with these NaOH and protease-treated natural membranes with K_d values in nanomolar range suggesting direct spectrin-lipid interaction (Diakowski et al. 2003).

Recently spectrin was shown to be present in detergent-resistant-membrane (DRM) fragments. These DRM fragments are rich in cholesterol and glycosphingolipids and are treated as representatives of lipid rafts from cell membranes. Though the mechanism of spectrin attachment to these membrane components is not clearly understood, the probability of attachment through direct spectrin-lipid interaction cannot be excluded (Nebl et al. 2002; Machnicka et al. 2014).

In general the phospholipid-binding ability of spectrin has been attributed to its repeat motifs over the decades (DeWolf et al. 1997). But in recent years, the existence of specialized lipid-binding domains has been discovered in spectrin. They include pleckstrin homology (PH) domain with high specificity for phosphatidylinositol-bisphosphate (PIP_2) (Hyvonen et al. 1995; Saraste and Hyvonen 1995) and other structural elements specific for PS and PE. PE binding is shown to be mediated by the ankyrin-binding domain, located at the N-terminal part of β-spectrin (Kennedy et al. 1991). Conformational changes are induced in the 14th repeat of β-spectrin at the ankyrin-binding domain upon binding of PE (Machnicka et al. 2014).

Although spectrin-lipid interaction has been studied in great detail, the direct physiological relevance of such an interaction is poorly understood (Machnicka et al. 2014). Spectrin-based skeleton gives mechanical strength to the plasma membrane, acts as a scaffold for activation and normal functioning of different receptors and stabilizes membrane microdomains or raft-like domains. The mechanisms of these functions attributed to spectrin-lipid interaction are not clearly understood. However, apart from these functions, spectrin-lipid binding can be strongly correlated to the maintenance of lipid asymmetry in biological membranes. It has been observed that in hereditary red blood cell disorders like hereditary spherocytosis (HS) and hereditary elliptocytosis (HE), spectrin is mutated and loses its membrane-binding ability, and concurrently membrane asymmetry is lost (Gallagher 2005; Basu et al. 2010; Basu and Chakrabarti 2015). In HS, vertical linkage between membrane skeleton and lipid bilayer is lost, thereby leading to membrane loss by blebbing. This loss of lipid is correlated with the decrease in the membrane surface area, and erythrocyte deformability is also compromised. HE characterized by the presence of elliptically shaped RBCs in the peripheral blood smear is also associated with mutations in α- and β-spectrin. In haemoglobin disorders like thalassemia and sickle cell disease also, eryptosis is greatly increased through the loss of membrane asymmetry (Datta et al. 2006; Chakrabarti et al. 2008).

We have summarized our results on spectrin-lipid interactions with small unilamellar vesicles and cushioned bilayers of different phospholipid compositions using fluorescence spectroscopy and X-ray reflectivity (XRR) techniques and report our findings leading to a clearer picture on the molecular association of the spectrin-based membrane skeleton with the inner leaflet of the cell surface membrane shown in a cartoon diagram at the end.

1.2 Methods

1.2.1 Fluorescence Measurements of Spectrin-Phospholipid SUV Interactions

Small unilamellar vesicles of different phospholipid compositions were prepared by sonicating hydrated lipid films formed by dissolving appropriate amount of lipids in 2:1 chloroform: methanol followed by the evaporation of solvent under thin stream of nitrogen and overnight drying in vacuum desiccators. This method produces small unilamellar vesicles in the size range of 300–450 Å as seen by electron micrography (Ray and Chakrabarti 2004). Tryptophan fluorescence quenching of erythroid spectrin was studied by the successive addition of small aliquots of lipid SUVs. The excitation wavelength was 295 nm, and emission spectrum was recorded from 310 nm to 400 nm. Excitation and emission slits both were 5 nm. Baseline subtractions were done by adding identical aliquots of lipids to the buffer. Binding parameters (binding dissociation constant, K_d; maximum extent of quenching, Q_{max} and the ratio of the number of dimeric spectrin molecules per single SUV particle, spectrin/SUV) of spectrin-phospholipid interactions were estimated following published procedures (Ray and Chakrabarti 2004; Mitra et al. 2015).

1.2.2 Polymer-Cushioned Lipid Bilayer Deposition and X-ray Reflectivity Measurement

The silicon wafers with one side polished were treated with a UV/O3 cleaner prior to the film deposition to make them hydrophilic. Freshly cleaned silicon wafers were immersed in a 2% (v/v) toluene solution of (3-Aminopropyl) triethoxysilane (APTES) for 15 min and dried under nitrogen flow. Next, a methanol solution of polyacrylic acid (PAA) was spin coated on top of the APTES layer. Samples were baked in each step to gain better stability. Further description of the substrate preparation for the polymer deposition has been elaborated earlier (El-khouri et al. 2011). Polymer-cushioned silicon wafers were then used further for the deposition of DMPC bilayer at constant surface pressure of 30 mN/m from the Langmuir monolayer. The first layer was deposited using the Langmuir-Blodgett (LB) technique, and the next layer was deposited from the same monolayer using Langmuir-Schaefer (LS) method, and finally the deposited bilayer sample was immersed gently inside the water subphase. Afterwards, the sample was transferred into a specially designed water cell which was used for the X-ray reflectivity measurement from the cushioned bilayer. Proteins are added to the subphase from a hole atop the cell.

The XRR measurement was carried out using the X-ray synchrotron source at the Indian beamline at Photon Factory, KEK, Japan, at an X-ray wavelength of 0.854 Å. A highly collimated beam was allowed to illuminate the sample, and the specular beam reflected from the sample was collected as a function of the wave vector transfer ($q_z = 4\pi \text{Sin}\ \theta/\lambda$). The XRR measurement was performed in the presence of spectrin in the subphase at a temperature of 30 °C which corresponds to the bilayer fluid phase. The obtained profile was normalized with respect to the highest intensity and plotted as a function of q_z. We have used Parratt's formalism (Parratt 1954; Basu and Sanyal 2002) to fit the experimentally obtained profile by assuming the simplest model of the film structure and dividing the film into a number of fine slices having sufficient electron density. They were contrasted along the film depth to obtain the electron density profile (EDPs) and was further analyzed to gain insight into the structure of the film (Giri et al. 2017).

1.3 Results and Discussions

The binding parameters of spectrin interaction with phospholipid SUVs of different compositions have been extensively studied and reported from our lab, as summarized in Table 1.1 (Ray and Chakrabarti 2004; Mitra et al. 2015), using lipid-induced quenching of spectrin fluorescence. Spectrin has been shown to laminate DMPC/DMPE bilayer in a PE surface density-dependent manner. Increasing DMPE content up to 80% in DMPC bilayer leads to spectrin lamination of the bilayer surface with an increasing K_d value below the phase transition temperature, i.e. in gel phase. However, in the case of pure DMPE vesicles, appreciably lower extent of quenching of tryptophan fluorescence, much stronger binding affinity and a large number of spectrin dimers bound to lipid vesicles were seen. Electron microscopy revealed spectrin dimers sticking out from pure DMPE vesicles indicating that spectrin interacts with DMPE through one of its ends (Ray and Chakrabarti 2004). In addition, it is also now well established that in the presence of up to 20% cholesterol in DMPC/DMPE bilayer in the gel phase at 15 °C (Table 1.1), spectrin interacts much strongly due to the coexistence of both the gel and fluid phases (induction of l_o phase) (Mitra et al. 2015).

One of the remarkable features of spectrin structure is that tryptophan (Trp) residues are positionally conserved in the spectrin repeat domains. Trp residues are present at the 45th position in 20 spectrin repeat domains out of 22 in α-spectrin and in of all the 17 repeats of β-spectrin. Altogether there are 42 Trps in α-subunit and 35 in β-subunit. The tryptophan residues present in the spectrin repeat domains represent almost 90% of the total Trps present in spectrin heterodimer. This kind of regular arrangement of Trp residues in each repeat motif of spectrin makes it a convenient intrinsic fluorescence reporter to study different functions including phospholipid-binding ability (Ray and Chakrabarti 2004). The maximum extent of quenching, Q_{max}, thus could be directly correlated with the membrane orientation of dimeric spectrin in the phospholipid bilayer. The association of lesser number of spectrin molecules with one SUV along with increasing Q_{max} indicated larger portions of the threadlike spectrin to wrap around the phospholipid vesicles laminating the membrane surface with multiple points of contacts indicated by the Trp residues uniformly distributed across the entire length of the

Table 1.1 Binding parameters associated with spectrin-PC/PE interactions at 50 °C above the main phase transition temperature (T_m)

Membrane systems	K_d (nM)	Q_{max} (%)	Spectrin/ SUV	Reference
DMPC (100%)	45 ± 7	43.7	7	Ray and Chakrabarti (2004)
DMPC:DMPE (60:40)	158 ± 20	81.4	3	
DMPC:DMPE (50:50)	209 ± 20	99.7	2	
DMPC:DMPE (20:80)	174 ± 10	61.3	2	
DMPE (100%)	2.6 ± 0.7	16.1	126	
DMPC:DMPE (75:25)	94 ± 10 (196 ± 27)	35.1	6	Mitra et al. (2015)
DMPC:DMPE (75:25) + 20% cholesterol	74 ± 15 (18 ± 5)	32.3	5	
DOPC (100%)	113 ± 21	58.2	4	
DOPC:DOPE (75:25)	42 ± 6	35.7	4	
DMPC:DMPS (90:10)	116 ± 10	21.8	4	Unpublished data

Error estimates are the standard error of mean obtained from 5 to 8 independent experiments.
Values in parenthesis are at 15 °C below T_m

molecule in each of the repeat domains. This is best reflected in the membrane systems containing increasing amounts of aminophospholipid, in DMPC/DMPE membranes (Table 1.1).

We've also adopted a newer way of studying spectrin-phospholipid interactions through X-ray reflectivity measurements. The obtained experimental XRR profile along with the model fitting from the polymer-cushioned DMPC bilayer in presence of spectrin has been shown in Fig. 1.1. The roughness convoluted electron density profile (EDP) extracted from the fitting along with the box EDP has also been shown in the inset indicating similar orientation of spectrin molecules consistent with fluorescence experiments. The cartoon diagram in the inset of Fig. 1.1 depicts the molecular arrangement in the film. From the EDP, a thin spectrin layer in addition to a smeared DMPC bilayer of total thickness ~6.2 nm is evident. In addition, a polymer layer of thickness ~20.5 nm, a self-assembled APTES-PAA junction, an APTES layer and a thin SiO_x layers have also been observed.

The electron density of the bulk polymer is nearly similar to that of water, while the averaged electron density of the bilayer region has been observed to be slightly lesser than that. From the analysis of the experimental reflectivity profile and also from the lower amplitude in the Kiessig oscillations, it is clear that the film possesses higher surface and interface roughness at the top. That might be due to the lower electron density observed near the top surface of the polymer-brush region that resulted in a larger surface roughness at the lipid bilayer–polymer film interface. The previous report shows that the growth of the Langmuir-Blodgett films are conformal to the surface topology of the underneath substrate (Basu and Sanyal 2002). Thus, the roughness of the lipid bilayer–polymer-film interface is expected to propagate into the bilayer film causing a relatively poor electron density in the bilayer region. The quantitative comparison shows that the averaged electron density in the polymer film is ~0.33 $e\text{Å}^{-3}$, while that of the spectrin-added lipid bilayer is ~0.31 $e\text{Å}^{-3}$.

The currently accepted model of the RBC membrane skeleton was described by Salomao et al. in 2008. The model states that spectrin meshwork is attached to the cell membrane with the help of two multiprotein complexes, the Band3-ankyrin-based complex and the Band3-4.1R-GPC-actin-based complex. The Band3 complex attaches to the spectrin-based membrane skeleton near the dimerization domain of spectrin with the help of ankyrin. The second Band3-4.1R-GPC-actin complex attaches to the spectrin at the tetramerization domain. According to this model, the spectrin-based membrane skeleton attachment to the membrane lipid bilayer is shown to be mediated only with the help of

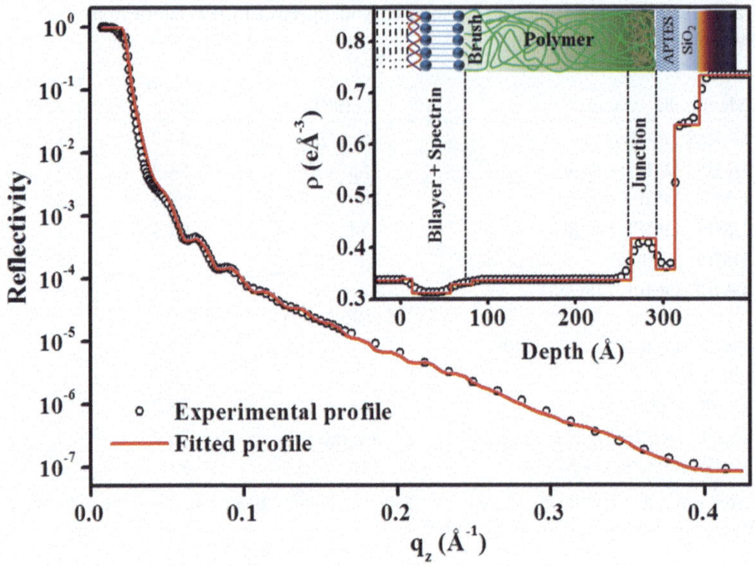

Fig. 1.1 Experimental XRR profile along with the fitted profile obtained from the polymer-cushioned DMPC bilayer deposited at 30 mN/m of surface pressure and at 30 °C of subphase temperature in the presence of spectrin. Inset shows the extracted electron density profile (EDP), and the cartoon represents the corresponding molecular arrangement in the film

these two multiprotein complexes, and the presence of direct interaction of spectrin and other skeletal proteins are completely ignored. However, it is well established that spectrin can directly interact with the phospholipid bilayer. In addition to spectrin-lipid interactions, other membrane skeletal proteins like 4.1R, actin, ankyrin, p55 or MPP1 also interact with different phospholipids.

Protein 4.1R is one of the important components of the erythroid membrane skeleton and is responsible for the organization of skeletal and transmembrane components at the junctional complexes. Experiments from various labs proves that protein 4.1 R, purified from erythrocytes, interacts with phospholipid vesicles especially to those containing phosphatidylserine, in pH and ionic strength-dependent manner. Treatment of erythrocyte inside-out vesicles with PS decarboxylases abolishes the binding, indicating its preferential binding to phosphatidylserine. In addition, this binding can be inhibited by Ca^{+2} ions. Studies with lipid monolayers showed that 4.1R can penetrate the lipid monolayers composed of PC and PS, whereas it can cause permeability changes in PS-containing vesicles but not in PC vesicles (An et al. 2001, 2006; Salomao et al. 2008; Gauthier et al. 2011).

Ankyrin is another well-known peripheral membrane protein found in cells. It is one of the essential members of the multiprotein complexes of membrane skeleton. Although its various functions have been studied in great detail, the membrane lipid-binding ability of ankyrin has not been studied substantially. Only a few reports show that it can bind to hydrophobic ligands as well as to different phospholipids, but the affinity of ankyrin-lipid interaction is much less than that of spectrin-phospholipid interaction (Sikorski et al. 2000).

Actin, a major contributor to the mechanical resilience of the plasma membrane, is present in junctional complexes along with spectrin, and both the forms of actin, F- and G-actin, can bind to uncharged as well as charged lipid vesicles. Experiments on lipid-binding ability of actin show that it interacts with PS and phosphoglycerate (PG) vesicles through electrostatic interactions in the presence of divalent cations in mM range. The cation-driven actin alignment on the membrane bilayer has also been shown to change membrane microviscosity and permeability of the bilayer. Actin can also interact with DMPC membrane, and this interaction depends on the length of the actin filament. G- and F-actin have also been shown to bind actin-depleted liver membrane. All these results indicate that actin can interact with the membrane bilayer in a divalent cation-dependent manner (Gicquaud 1993; Gicquaud and Wong 1994; Takenawa and Itoh 2001; Liu and Fletcher 2006).

MPP1 or p55, another major component of the membrane skeleton, has been implicated in the regulation of physicochemical behaviours of bilayer. It modulates the physicochemical properties of cytoskeleton-depleted giant plasma membrane vesicles (GPMV). It has been reported that p55 modulates the fluidity and induces phase separation in the bilayer. It is known to promote the stabilization of l_o domain in a palmitoylation-dependent manner. Upon palmitoylation, the membrane skeleton binding property of p55 decreases, and then it binds to the preformed raft components, thereby stabilizing the membrane microdomains. Later on Listowski and coworkers showed that it interacts with cholesterol through CRAC and CRAC-like motifs on its surface in a palmitoylation-independent manner with binding dissociation constants in the μM to nM range (Edidin 2006; Lach et al. 2012; Biernatowska et al. 2013; Listowski et al. 2015; Podkalicka et al. 2015).

1.4 Conclusion

Taken everything together, it becomes clear that spectrin and other membrane skeletal proteins have moderately strong ability to interact with phospholipids and aminophospholipids in particular. It is also clear that the phospholipid interactions are sensitive to phase transition temperature, surface densities of the lipids and the presence of other modulatory lipids such as cholesterol. On the basis of the available data, we propose a newer model of the erythrocyte membrane skeleton that not only appreciates the role of the two multiprotein complexes but also emphasizes on the direct phospholipid interactions of the membrane skeletal proteins, particularly spectrin, for the stabilization and normal functioning of the membrane and in the maintenance of the overall membrane asymmetry, as shown in Fig. 1.2.

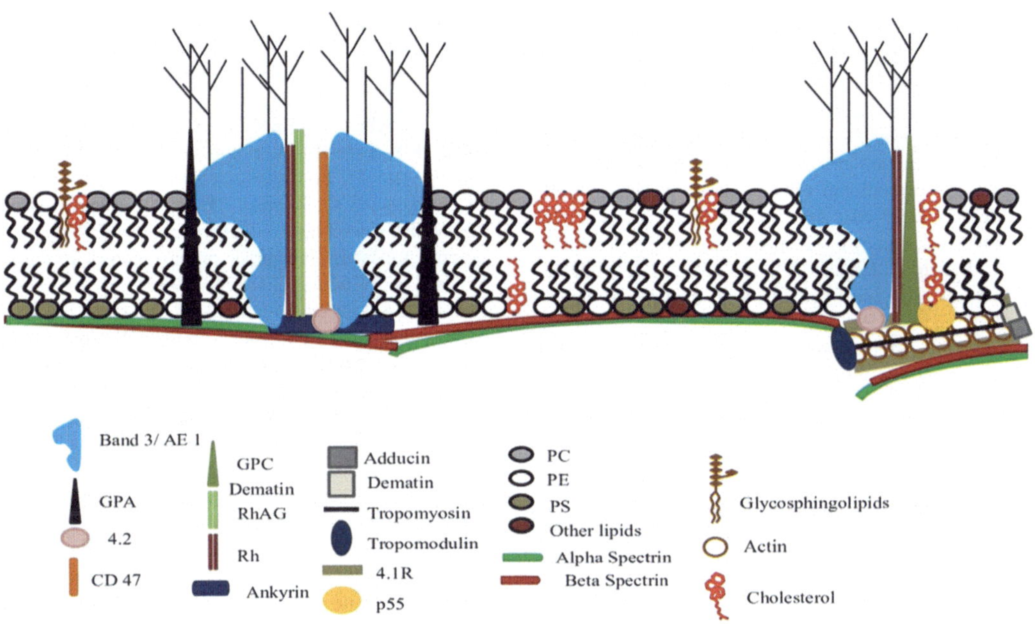

Fig. 1.2 Proposed model of erythrocyte membrane skeleton acknowledging the direct lipid interactions of the component proteins

References

Alessandrini A, Facci P (2011) Unraveling lipid/protein interaction in model lipid bilayers by atomic force microscopy. J Mol Recognit 24:387–396

An XL, Takakuwa Y, Manno S, Han BG, Gascard P, Mohandas N (2001) Structural and functional characterization of protein 4.1R-phosphatidylserine interaction: potential role in 4.1R sorting within cells. J Biol Chem 276:35778–35785

An X, Zhang X, Debnath G, Baines AJ, Mohandas N (2006) Phosphatidylinositol-4,5-biphosphate (PIP2) differentially regulates the interaction of human erythrocyte protein 4.1 (4.1R) with membrane proteins. Biochemistry 45:5725–5732

Basu A, Chakrabarti A (2015) Defects in erythrocyte membrane skeletal architecture. Adv Exp Med Biol 842:41–59

Basu JK, Sanyal MK (2002) Ordering and growth of Langmuir–Blodgett films: x-ray scattering studies. Phys Rep 363:1–84

Basu S, Banerjee D, Chandra S, Chakrabarti A (2010) Eryptosis in hereditary spherocytosis and thalassemia: role of glycoconjugates. Glycoconj J 27:717–722

Biernatowska A, Podkalicka J, Majkowski M, Hryniewicz-Jankowska A, Augoff K, Kozak K, Korzeniewski J, Sikorski AF (2013) The role of MPP1/p55 and its palmitoylation in resting state raft organization in HEL cells. Biochim Biophys Acta 1833:1876–1884

Chakrabarti A, Datta P, Bhattacharya D, Basu S, Saha S (2008) Oxidative crosslinking, spectrin and membrane interactions of hemoglobin mixtures in HbEbeta-thalassemia. Hematology 13:361–368

Datta P, Basu S, Chakravarty SB, Chakravarty A, Banerjee D, Chandra S, Chakrabarti A (2006) Enhanced oxidative cross-linking of hemoglobin E with spectrin and loss of erythrocyte membrane asymmetry in hemoglobin Ebeta-thalassemia. Blood Cells Mol Dis 37:77–81

DeWolf C, McCauley P, Sikorski AF, Winlove CP, Bailey AI, Kahana E, Pinder JC, Gratzer WB (1997) Interaction of dystrophin fragments with model membranes. Biophys J 72:2599–2604

Diakowski W, Szopa J, Sikorski AF (2003) Occurrence of lipid receptors inferred from brain and erythrocyte spectrins binding NaOH-extracted and protease-treated neuronal and erythrocyte membranes. Biochim Biophys Acta 1611:115–122

Edidin M (2006) Switching sides: the actin/membrane lipid connection. Biophys J 91:3963

El-khouri RJ, Bricarello DA, Watkins EB, Kim CY, Miller CE, Patten TE, Parikh AN, Kuhl TL (2011) pH responsive polymer cushions for probing membrane environment interactions. Nano Lett 11:2169–2172

Fadeel B, Xue D (2009) The ins and outs of phospholipid asymmetry in the plasma membrane: roles in health and disease. Crit Rev Biochem Mol Biol 44 (5):264–277

Gallagher PG (2005) Red cell membrane disorders. Hematology Am Soc Hematol Educ Program 2005:13–18

Gauthier E, Guo X, Mohandas N, An X (2011) Phosphorylation-dependent perturbations of the 4.1R-associated multiprotein complex of the erythrocyte membrane. Biochemistry 50:4561–4567

Gicquaud C (1993) Actin conformation is drastically altered by direct interaction with membrane lipids: a differential scanning calorimetry study. Biochemistry 32:11873–11877

Gicquaud C, Wong P (1994) Mechanism of interaction between actin and membrane lipids: a pressure-tuning infrared spectroscopy study. Biochem J 303. (Pt 3:769–774

Giri RP, Chakrabarti A, Mukhopadhyay MK (2017) Cholesterol-induced structural changes in saturated phospholipid model membranes revealed through x-ray scattering technique. J Phys Chem B 121:4081–4090

Grzybek M, Chorzalska A, Bok E, Hryniewicz-Jankowska A, Czogalla A, Diakowski W, Sikorski AF (2006) Spectrin-phospholipid interactions. Existence of multiple kinds of binding sites? Chem Phys Lipids 141:133–141

Hyvonen M, Macias MJ, Nilges M, Oschkinat H, Saraste M, Wilmanns M (1995) Structure of the binding site for inositol phosphates in a PH domain. EMBO J 14:4676–4685

Juliano RL, Kimelberg HK, Papahadjopoulos D (1971) Synergistic effects of a membrane protein (spectrin) and Ca 2+ on the Na + permeability of phospholipid vesicles. Biochim Biophys Acta 241:894–905

Kennedy SP, Warren SL, Forget BG, Morrow JS (1991) Ankyrin binds to the 15th repetitive unit of erythroid and nonerythroid beta-spectrin. J Cell Biol 115:267–277

Lach A, Grzybek M, Heger E, Korycka J, Wolny M, Kubiak J, Kolondra A, Boguslawska DM, Augoff K, Majkowski M, Podkalicka J, Kaczor J, Stefanko A, Kuliczkowski K, Sikorski AF (2012) Palmitoylation of MPP1 (membrane-palmitoylated protein 1)/p55 is crucial for lateral membrane organization in erythroid cells. J Biol Chem 287:18974–18984

Listowski MA, Leluk J, Kraszewski S, Sikorski AF (2015) Cholesterol interaction with the MAGUK protein family member, MPP1, via CRAC and CRAC-like motifs: an in silico docking analysis. PLoS One 10:e0133141

Liu AP, Fletcher DA (2006) Actin polymerization serves as a membrane domain switch in model lipid bilayers. Biophys J 91:4064–4070

Machnicka B, Czogalla A, Hryniewicz-Jankowska A, Boguslawska DM, Grochowalska R, Heger E, Sikorski AF (2014) Spectrins: a structural platform for stabilization and activation of membrane channels, receptors and transporters. Biochim Biophys Acta 1838:620–634

Mitra M, Patra M, Chakrabarti A (2015) Fluorescence study of the effect of cholesterol on spectrin-aminophospholipid interactions. Eur Biophys J 44:635–645

Mombers C, de Gier J, Demel RA, van Deenen LL (1980) Spectrin-phospholipid interaction. A monolayer study. Biochim Biophys Acta 603:52–62

Nebl T, Pestonjamasp KN, Leszyk JD, Crowley JL, Oh SW, Luna EJ (2002) Proteomic analysis of a detergent-resistant membrane skeleton from neutrophil plasma membranes. J Biol Chem 277:43399–43409

Parratt LG (1954) Surface studies of solids by total reflection of x-rays. Phys Rev 95:359–369

Podkalicka J, Biernatowska A, Majkowski M, Grzybek M, Sikorski AF (2015) MPP1 as a factor regulating phase separation in Giant plasma membrane-derived vesicles. Biophys J 108:2201–2211

Ray S, Chakrabarti A (2004) Membrane interaction of erythroid spectrin: surface-density-dependent high-affinity binding to phosphatidylethanolamine. Mol Membr Biol 21:93–100

Salomao M, Zhang X, Yang Y, Lee S, Hartwig JH, Chasis JA, Mohandas N, An X (2008) Protein 4.1R-dependent multiprotein complex: new insights into the structural organization of the red blood cell membrane. Proc Natl Acad Sci USA 105:8026–8031

Saraste M, Hyvonen M (1995) Pleckstrin homology domains: a fact file. Curr Opin Struct Biol 5:403–408

Sikorski AF, Hanus-Lorenz B, Jezierski A, Dluzewski AR (2000) Interaction of membrane skeletal proteins with membrane lipid domain. Acta Biochim Pol 47:565–578

Sweet C, Zull JE (1970) Interaction of the erythrocyte--membrane protein, spectrin, with model membrane systems. Biochem Biophys Res Commun 41:135–141

Takenawa T, Itoh T (2001) Phosphoinositides, key molecules for regulation of actin cytoskeletal organization and membrane traffic from the plasma membrane. Biochim Biophys Acta 1533:190–206

Multiple Roles, Multiple Adaptors: Dynein During Cell Cycle

2

Devashish Dwivedi and Mahak Sharma

Abstract

Dynein is an essential protein complex present in most eukaryotes that regulate biological processes ranging from ciliary beating, intracellular transport, to cell division. Elucidating the detailed mechanism of dynein function has been a challenging task owing to its large molecular weight and high complexity of the motor. With the advent of technologies in the last two decades, studies have uncovered a wealth of information about the structural, biochemical, and cell biological roles of this motor protein. Cytoplasmic dynein associates with dynactin through adaptor proteins to mediate retrograde transport of vesicles, mRNA, proteins, and organelles on the microtubule tracts. In a mitotic cell, dynein has multiple localizations, such as at the nuclear envelope, kinetochores, mitotic spindle and spindle poles, and cell cortex. In line with this, dynein regulates multiple events during the cell cycle, such as centrosome separation, nuclear envelope breakdown, spindle assembly checkpoint inactivation, chromosome segregation, and spindle positioning. Here, we provide an overview of dynein structure and function with focus on the roles played by this motor during different stages of the cell cycle. Further, we review in detail the role of dynactin and dynein adaptors that regulate both recruitment and activity of dynein during the cell cycle.

Keywords

Dynein · Dynactin · Adaptor · Kinetochore · Hook2 · Cell cycle · Mitotic spindle

2.1 Introduction to Dynein Motor

Dynein (derived from the Greek word *dyne* meaning "force") was originally identified as an ATPase required for axoneme motility in the ciliated protozoan *Tetrahymena pyriformis* (Gibbons and Rowe 1965). Cytoplasmic dynein was discovered years later by Paschal and co-workers in 1987 and was described as a microtubule-associated ATPase related to the ciliary dynein, producing force in direction opposite to the known motor protein, kinesin (Paschal et al. 1987; Paschal and Vallee 1987). Notably until the late twentieth century, little was known about this microtubule-based retrograde motor protein, owing to its large molecular weight and high complexity of the motor. Three different forms of dynein are known today. Cytoplasmic dynein 1 performs retrograde transport of proteins, mRNA, vesicles, and organelles in

D. Dwivedi (✉) · M. Sharma (✉)
Department of Biological Sciences, Indian Institute of Science Education and Research (IISER), Mohali, Punjab, India
e-mail: devashishdwivedi@iisermohali.ac.in; msharma@iisermohali.ac.in

K. Chattopadhyay, S. C. Basu (eds.), *Biochemical and Biophysical Roles of Cell Surface Molecules*, Advances in Experimental Medicine and Biology 1112, https://doi.org/10.1007/978-981-13-3065-0_2

interphase and mitotic cells; cytoplasmic dynein 2 is required for transport inside the cilia; and 14 different axonemal dyneins perform movement of the cilia (Yagi 2009). Dynein regulates several physiological processes including ciliary motion, vesicular transport, spatiotemporal distribution of organelles, cell division, etc.

Dynein differs greatly from the kinesin motor proteins in its structure and function although both the motors are fueled by the energy derived from ATP hydrolysis. While kinesins are structurally similar to the G-proteins, the catalytic subunit of dynein belongs to the AAA+ ATPase superfamily (Neuwald et al. 1999; Vale and Milligan 2000). Unlike kinesin motors that move in head-over-head manner in anterograde direction with a step size of 8 nm, dynein takes the 8.3-nm-long uncoordinated steps with its two heads moving independent of each other in the retrograde direction (DeWitt et al. 2012; Mallik et al. 2004; Rank and Rayment 2013). While carrying a cargo a single molecule of kinesin produces a force of 6 pN, on the other hand, a single molecule of dynein produces a force of only 1 pN (DeWitt et al. 2012; Mallik et al. 2004; Visscher et al. 1999). A number of studies suggest that kinesins coordinate among each other to regulate cargo motility but do not show productive cooperation, i.e., presence of multiple kinesin(s) on the same cargo does not generate larger force (Jamison et al. 2010, 2012; Rank and Rayment 2013). On the contrary, dynein shows extensive productive cooperation, and the presence of multiple dynein molecules on a cargo results in higher force generation and long-range robust transport (Belyy et al. 2016; Rai et al. 2013; Torisawa et al. 2014; Vallee et al. 2012).

2.2 Structure of Dynein

The dynein heavy chain (DHC), which is the largest subunit (500 kDa) of this complex, contains two functional moieties, i.e., tail and the motor regions (Fig. 2.1a). The tail region of DHC is required for its dimerization and binding to other subunits, including dynein intermediate chain (DIC), light intermediate chain (LIC), and multiple light chains (LC) that link the motor to different cargos. DIC interacts with multiple dynein regulators like dynactin and NudE/Ndel (McKenney et al. 2011; Nyarko et al. 2012; Schliwa and Woehlke 2003; Wang et al. 2013). The motor region of the DHC contains six AAA+ ATPase domains (named as AAA1–6) that are arranged in a series. AAA1 domain is the major site for ATP hydrolysis which makes it the minimal region required for motility (Kon et al. 2004). Notably, AAA2 domain lacks the key residues required for ATP hydrolysis, while AAA3 and AAA4 domains retain ATP hydrolysis activity. Mutations preventing ATP binding or hydrolysis in AAA3 domain severely impact dynein motility as compared to AAA4, indicating nucleotide binding and hydrolysis by AAA3 domain play a more crucial role as compared to AAA4 (Cho et al. 2008; DeWitt et al. 2014; Kon et al. 2004). Although the rate of ATP hydrolysis by AAA3 domain is one magnitude lower than AAA1 domain, binding of AAA3 domain to ATP is required for robust retrograde transport (DeWitt et al. 2014). Interestingly, several studies indicate that every step of dynein costs only one molecule of ATP, suggesting that AAA3 and AAA4 domains do not hydrolyze ATP at each step (DeWitt et al. 2014; Mallik et al. 2004; Reck-Peterson et al. 2006; Toba et al. 2006). Lastly, AAA5 and AAA6 domains function to transmit conformational changes through AAA+ ring instead of enabling ATP hydrolysis (Schmidt et al. 2015). The motor region of DHC also contains a microtubule-binding domain (MTBD), a small α-helical region present at the end of AAA4 domain. MTBD binds to the cleft between α- and β-tubulin monomers and regulates the microtubule association of dynein (Carter et al. 2008, 2011; Gee et al. 1997; Redwine et al. 2012). AAA4 and AAA5 domain extensions termed as stalk and buttress, respectively, allosterically coordinate the ATP binding of AAA1 and microtubule association of the MTBD (Carter et al. 2008; Gibbons et al. 2005; Kon et al. 2009; Schmidt et al. 2015) (Fig. 2.1a). At the N-terminal end of AAA+ ring, a linker region comprising of four α-helical segments regulates the AAA1-triggered dynein motility by alternating

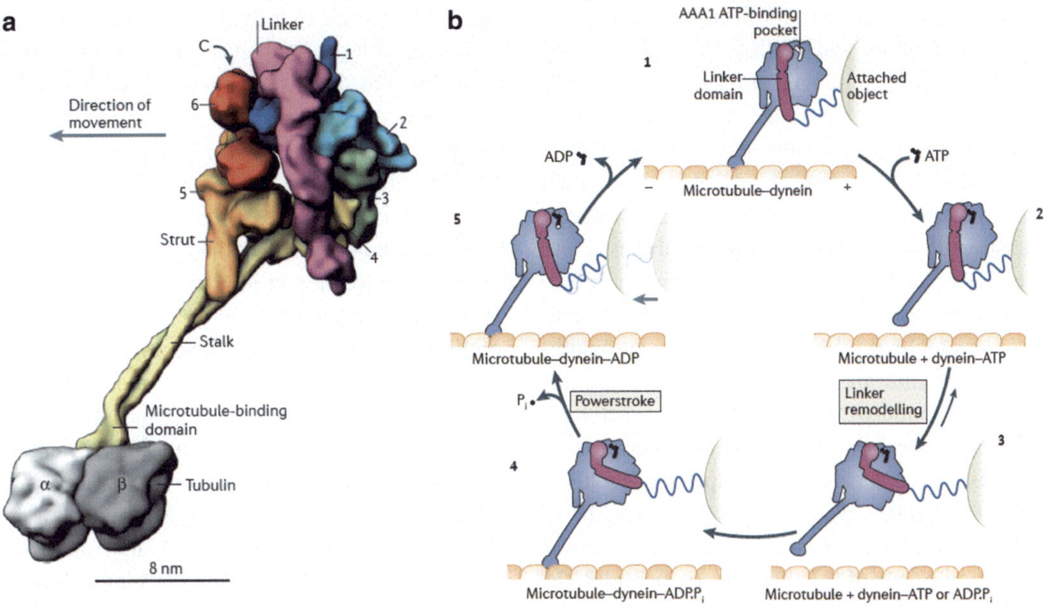

Fig. 2.1 Mechanochemical cycle of dynein motor protein. (**a**) 3D model depicting dynein heavy chain bound to tubulin monomer (other dynein subunits are not shown for simplicity). The six AAA+ ATPase domains are numbered accordingly. Subdomains are shown in surface representation. (**b**) Model showing mechanochemical cycle of dynein. Plus and minus signs indicate microtubule polarity. Dynein moves toward the minus end of the microtubules. Stretching of the spring is shown to indicate force production. Hydrolysis of ATP at AAA1 domain results in dissociation of dynein's MTBD from microtubules (1) resulting in change in conformation of linker from bent to straight (2). This extends the range of MTBD to bind a new site along the microtubules (3) and is coupled with ATP hydrolysis at AAA1 domain. Post ATP hydrolysis, MTBD of dynein associates with a new site on the microtubules, and this association is strengthened by the release of the inorganic phosphate (4), restoring back original conformation of the linker region. Finally, release of ADP from the AAA1 domain restarts the mechanochemical cycle of dynein (5). This figure is adapted and modified from Roberts et al. (2013) and has been reproduced with permission from Springer Nature

between bent (pre-power stroke) and straight (power stroke) confirmations (Kon et al. 2009; Roberts et al. 2012; Schmidt et al. 2012, 2015) (Fig. 2.1b).The C-terminal region after the AAA6 domain contacts the AAA5 and AAA6 domains on the side opposite to the linker region (Carter et al. 2011; Roberts et al. 2009). However, its functional role is debatable as this region is reduced to a small α-helix in *S. cerevisiae*, while its removal in *D. discoideum* dynein reduces its in vitro motility (Carter et al. 2011; Numata et al. 2011; Schmidt et al. 2012, 2015). In contrast, a similar truncation in mammalian dynein results in generation of a greater force by the motor (Nicholas et al. 2015). The C-terminal region appears to regulate the force produced by dynein; however, its actual role is poorly understood and requires further investigation. A key aspect of understanding the dynein function is the repeated ATP hydrolysis by AAA + domain that results in a cycle of multiple conformational changes within the DHC motor domain as revealed by biophysical and structural studies (Bhabha et al. 2014; Carter et al. 2011; DeWitt et al. 2012, 2014; Kon et al. 2012; Roberts et al. 2012; Schmidt et al. 2012, 2015). These changes result in cyclic oscillations in the state from bent to straight of the N-terminal linker region, leading to dynein motility (Fig. 2.1b) (Roberts et al. 2013).

2.3 Structure and Function of Dynactin

Pioneering studies on purification of dynein reported a set of polypeptides consistently co-purifying with dynein which was later found to constitute another multisubunit complex, dynactin (Collins and Vallee 1989; Karki and Holzbaur 1995). Dynactin was originally described as a cofactor activating the dynein-mediated vesicle motility (Gill et al. 1991). Subsequent studies have revealed that dynactin links dynein to various cargos, activates dynein, and strengthens its binding to microtubules (Lenz et al. 2006; McKenney et al. 2014; Schroer 2004). This is evident by the neurodegenerative genetic disorders including hereditary motor neuropathy and Perry syndrome whereby point mutations in dynactin subunit-p150^{glued} disrupt its microtubule association, abrogating retrograde transport in neurons (Farrer et al. 2009; Lipka et al. 2013; Puls et al. 2003). Similar to dynein, vertebrate dynactin is a large ~1.2 MDa multisubunit complex comprising of 23 distinct polypeptides. In agreement with previous studies, recent cryo-EM data at near atomic resolution of dynactin bound to dynein has revealed that dynactin comprises of a thin long arm and a filament connected by a "shoulder" complex (Chowdhury et al. 2015; Imai et al. 2014; Urnavicius et al. 2015). p150 ^{glued}, the largest subunit of dynactin, dimerizes and forms the thin arm with its coiled-coils (CC1A, CC1B, and CC2) and CAP-GLY domains present at the N-terminus of each monomer. The shoulder complex containing p50/dynamitin and p24 subunits link the other end of the thin arm to the 40-nm-long filament made up of concatenated Arp1 units followed by β-actin and Arp11 subunits (Urnavicius et al. 2015). Arp11 subunit binds the capping proteins, p25, p27, and p62, forming the pointed end, while CapZα and CapZβ capping proteins form what is known as the barbed end (Chowdhury et al. 2015; Urnavicius et al. 2015). Subunits of the pointed end complex (namely, p25 and p27) are involved in tethering dynein to the cargo (Yeh et al. 2012; Zhang et al. 2011). Pointed-end complex is absent in *S. cerevisiae*, indicating the role of other unknown proteins in regulating dynein recruitment to cargo (Schroer 2004).

Recent structural insights into the microtubule-bound dynein-dynactin complex have revealed the stoichiometry of this complex where a single molecule of dynactin acts as a scaffold to recruit two dyneins on microtubules for collective force production and faster motility (Grotjahn et al. 2018; Urnavicius et al. 2018). In vertebrates, this association is tightly regulated by multiple adaptor proteins (McKenney et al. 2014). In vitro studies have shown that mammalian dynein is non-processive and quickly falls of the microtubule tracts in the absence of these adaptors (McKenney et al. 2014; Schlager et al. 2014). Adaptor proteins form a stable ternary complex linking dynein and dynactin and activate dynein by changing the confirmation of p150^{glued} from an auto-inhibited state to a processive state (Lee et al. 2018; McKenney et al. 2014; Olenick et al. 2016; Schlager et al. 2014; Schroeder and Vale 2016; Zhang et al. 2017). Multiple dynein adaptors have been discovered that activate dynein in vitro, providing possible hint to how a single retrograde motor regulates a plethora of biological processes (Redwine et al. 2017a). However, the in vivo spatial and temporal significance of these protein-protein interactions is yet to be understood for many dynein-dynactin-adaptor tripartite complexes.

In this review, we focus on the role of cytoplasmic dynein 1 (hereafter referred as "dynein") during different stages of cell cycle with emphasis on the role played by adaptor proteins in regulating the dynein activity during cell division.

2.4 Role of Dynein in Regulating Cell Cycle

Cell cycle is the series of events occurring in a eukaryotic cell that ultimately result in its division. In addition to routine metabolism, a cell must replicate its contents, especially the genome, to divide into two daughter cells. The complete process of cell cycle consists of two phases: interphase and mitotic (M)-phase. Interphase is the longest stage of cell cycle constituting of two gap phases (G1 and G2) separated by a S-phase when the entire genome is replicated, while the M-phase constitutes the steps involved in segregation of genome and cytosol. To ensure proper replication and segregation of genome, the cells pass through different checkpoints. Errors during cell division are often associated with chromosome abnormality defects and cancer (Potapova and Gorbsky 2017). Thus, all the steps in cell cycle are under tight regulation. Different motor proteins regulate these events by either of the three ways: transport of mitotic cargos, cross bridging and sliding microtubules relative to the adjacent microtubules, and regulating microtubule growth and shrinkage (Sharp et al. 2000b). In light of the earlier studies, it is now well appreciated that dynein plays a crucial role in regulating different stages of the cell cycle and this will be discussed later. Targeted by different adaptors, dynein localizes to kinetochores, centrosomes, cell cortex, nuclear envelope, and spindle microtubules (Fig. 2.2), applying force for the movement of nucleus, centrosomes, chromosomes, and even the entire spindle (Kardon and Vale 2009; Sharp et al. 2000a).

2.5 Dynein at the Nuclear Envelope

As the cell prepares to divide, one of the first events regulated by dynein is separation of the duplicated centrosomes during the late G2 phase of the cell cycle (Gonczy et al. 1999; Robinson et al. 1999). This process is also driven by a kinesin, Eg5, which is the major force producer in this process (Tanenbaum and Medema 2010). In addition, dynein also tethers centrosomes to the nuclear envelope (NE), and this NE-associated dynein has been shown to mediate centrosome separation during prophase, independent of the Eg5 activity (Raaijmakers et al. 2012; van Heesbeen et al. 2013). Dynein is recruited to the nuclear pores by associating with its adaptor bicaudal-D2 (BicD2) that tethers to the nuclear pore via RAN binding protein 2 (RANBP2) (Splinter et al. 2010). An additional mechanism of dynein retention at the NE is by binding of NudE/Nde1 to centromere protein F (CENP-F) and possibly to the nuclear pore protein, Nup133 (Bolhy et al. 2011). Previous studies have shown that dynein homogenously localizes to NE during the late G2 phase of the cell cycle (Busson et al. 1998; Salina et al. 2002) that raises the question of how a directional force on centrosomes can be generated for their separation. Studies in the one-cell *C. elegans* embryo have revealed how dynein at the nuclear envelope and at the cell cortex generates this directional force for centrosome separation (De Simone and Gonczy 2017; De Simone et al. 2016). Whereas the initial centrosome position and steric interactions between microtubules and centrosome are required for anisotropic organization of the nuclear dynein forces, dynein at the cell cortex harnesses the directional actomyosin flow for centrosome separation (De Simone and Gonczy 2017; De Simone et al. 2016). Future studies on dynein localization at the NE by techniques including single-molecule super-resolution microscopy will shed light on how dynein regulates centrosome separation. Another ascribed role of dynein is to regulate nuclear

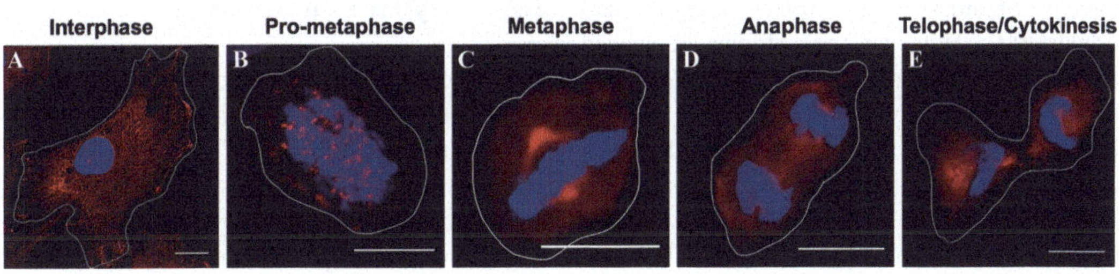

Fig. 2.2 Localization of dynein during different stages of cell cycle. Asynchronous HeLa cells were fixed and stained with anti-DIC antibody (red) to mark dynein, and the nucleus was stained using DAPI (blue). Cells in different stages of cell cycle were imaged in this experiment to highlight dynein localization. Dynein localizes to centrosome, nuclear envelope, and microtubules during the interphase (**a**); kinetochores during prometaphase (**b**); mitotic spindles during metaphase (**c**); weak staining at the interpolar microtubules during anaphase (**d**); and cytokinetic bridge during telophase (**e**). Scale bars, 10 μm

envelope breakdown (NEBD), as interfering with dynein function impedes the NEBD (Salina et al. 2002). A plausible mechanism to explain these observations might be generation of pulling forces by the NE-associated dynein moving on the centrosomal microtubule track, thereby tearing the nuclear membrane (Tanenbaum et al. 2010). Indeed, recent studies have shown the role of microtubules in mechanical shearing of NE, resulting in nuclear envelope breakdown (Beaudouin et al. 2002; Salina et al. 2002).

2.6 Dynein at Kinetochores

Microtubules attach to chromosomes on large proteinaceous structures at the centromere termed as "kinetochores (KT)." Dynein is recruited to KT prior to their microtubule attachments, indicating the role of dynein in capturing microtubules at KT. Recruitment of dynein to KT occurs through the Rod/Zwilch/ZW10 (RZZ) complex and the dynein/dynactin adaptor, Spindly. The RZZ complex and Spindly can also independently interact with dynein and dynactin (Karess 2005; Starr et al. 1998). In addition to dynein-dynactin recruitment at KT, the RZZ complex also recruits the mitotic checkpoint proteins Mad1 and Mad2 on KT to ensure proper microtubule-KT attachments (Buffin et al. 2005; Kops et al. 2005). Recent studies have also shown the role of two additional proteins, CENP-F and NudE/Nde1/Lis1, in dynein recruitment to KT, as depletion of these proteins impaired KT localization of dynein (Stehman et al. 2007; Vergnolle and Taylor 2007). However, more studies are needed to elucidate whether CENP-F-NudE/Nde1/Lis1-dependent dynein-dynactin recruitment requires RZZ complex or it is an alternative pathway of dynein recruitment to the KT. Dynactin is also implicated in proper microtubule-KT attachments by promoting recruitment of polo-like kinase-1 (Plk1) to KT, which recruits mitotic checkpoint protein, Mad1, to KTs for checkpoint signaling (Yeh et al. 2013). Disruption of dynein activity by introducing loss-of-function mutations in DHC or injecting antibodies against dynein or dynein depletion has revealed defects in chromosome alignment together with a delay in anaphase onset in these cells (Raaijmakers et al. 2013; Schmidt et al. 2005; Sharp et al. 2000a; Yang et al. 2007). These studies suggest a crucial role of dynein in chromosome alignment and segregation. However, a lack of these approaches in targeting specific dynein population limits the possibility of assigning these functions to any specific pool of dynein.

Once all KTs are properly attached to the mitotic spindle during metaphase, the SAC needs to be inactivated to allow chromosome segregation and initiate the onset of anaphase. Although different groups have suggested multiple roles of dynein at KT, the most evident function of KT-dynein appears to be silencing of the spindle assemble checkpoint and promoting chromosome segregation. Dynein, along with dynactin and spindly, inactivates the SAC by stripping the SAC components (Mad1, Mad2, Bub1, BubR1) and outer KT components, such as the RZZ complex, from KT and transporting them to the spindle poles (Basto et al. 2004; Chan et al. 2009; Gassmann et al. 2008). Depletion of either dynactin, spindly, or RZZ complex results in extremely long metaphase arrest and a severe delay in mitotic exit (Raaijmakers et al. 2013). The role of dynein at KT needs further exploration wherein identifying mutations that specifically perturb dynein recruitment to KT will be a useful tool to unravel the significance of this localization.

2.7 Dynein on Spindles

During metaphase, three different populations of microtubules exist: astral (microtubules between centrosomes and cell cortex), K-fibers (microtubule bundles between centrosomes and KT), and polar microtubules (antiparallel microtubules originating from opposite centrosomes). The formation of a bipolar spindle is governed by sliding of these antiparallel polar microtubules. Dynein, along with the kinesin motor, Eg5, is the key regulator of this microtubule sliding (Ferenz et al. 2009; Kashina et al. 1996). It has been proposed that dynein-dynactin interact with Eg5 on spindles and transport the antagonistic motor toward spindle poles. This dynein-mediated transportation restricts the microtubule sliding activity of Eg5 near the KT and the microtubule cross-linking activity of Eg5 near the spindle poles to regulate spindle length, orientation, and focusing (Uteng et al. 2008). Dynein is also implicated in spindle focusing at the two poles as disruption of dynein has been linked to spindle focusing defects (Goshima et al. 2005; Maiato et al. 2004; Morales-Mulia and Scholey 2005; Verde et al. 1991). The defects in spindle pole focusing appear to be linked specifically to dynein but not dynactin, as dynactin-depleted cells had normally focused spindle at the poles (Raaijmakers et al. 2013). Spindle positioning, a key determinant of the position of cleavage furrow and consequently cell fate, is regulated by pulling forces generated on astral microtubules by cortical dynein (Gonczy 2008; Nguyen-Ngoc et al. 2007; Schmidt et al. 2005). In yeast, dynein anchor protein, Mcp5, binds to the cell membranes and regulate cortical localization of dynein (Thankachan et al. 2017). In higher eukaryotes, a similar function is performed by the nuclear protein NuMA (nuclear mitotic apparatus) that directly binds to dynein-dynactin and phospholipids at the plasma membrane (Kotak et al. 2014). During anaphase, NuMA localizes to the cell cortex, juxtapositioned to the spindle midzone. NuMA binds to dynein-dynactin and controls the localization of dynein to the cell cortex, thus regulating spindle positioning (Kotak et al. 2012). Further during anaphase as the mitotic spindle elongates, a structure known as central spindle is formed by bundling of the interpolar microtubules between the two separating chromosomes. Previous studies have shown that dynactin subunit $p150^{glued}$ localizes at the central spindles (Delcros et al. 2006; Reboutier et al. 2013). $p150^{glued}$ interacts with the centralspindlin complex subunit, MKLP1/Kif23, and promotes its recruitment to central spindles, which in turn is required for cleavage furrow ingression (Delcros et al. 2006). Indeed, whether the central spindle pool of $p150^{glued}$ is associated with dynein and function of dynein at the central spindles needs further investigation.

2.8 Dynein Function During Cytokinesis

Anaphase onset is coupled with simultaneous formation of cleavage furrow that keeps ingressing, ultimately resulting in cytokinesis and formation of two separate daughter cells. Earlier studies have shown localization of dynein and dynactin to cleavage furrow and later to midbody during cytokinesis; however, the functional significance of these observations is not known (Campbell et al. 1998; Karki et al. 1998). During abscission, dynein is thought to regulate trafficking of Rab8 vesicles from Golgi complex to the midbody for the completion of cytokinesis (Kaplan and Reiner 2011). However, the approach used in this study involved inhibition of dynein by a chemical compound, EHNA, which is a potent adenosine deaminase inhibitor and does not selectively inhibit dynein but has additional effects of depletion of the cellular ATP pool (Burton et al. 2010; Nakajima et al. 2015; Penningroth et al. 1982; Schliwa et al. 1984; Siaw et al. 1980).

It is now well appreciated that dynein function is required at every stage of the cell cycle (Fig. 2.3). Since the phenotypes of dynein depletion or inhibition are too prominent in the early stages of cell cycle, investigating dynein function (s) after metaphase has been a technical challenge. Not surprisingly, information on the role of dynein post-anaphase onset is very limited and needs further investigation. Development of new tools that can rapidly and reversibly inactivate dynein function at specific stages of cell cycle would be an important breakthrough in gaining comprehensive knowledge of the roles performed by dynein and its accessory partners during the late stages of the cell cycle.

2.9 Function of Dynein Adaptors during Cell Division

Previous studies have uncovered that adaptor proteins associate with dynein and provide a scaffold for stably linking dynein and dynactin (McKenney et al. 2014). In addition, these adaptor

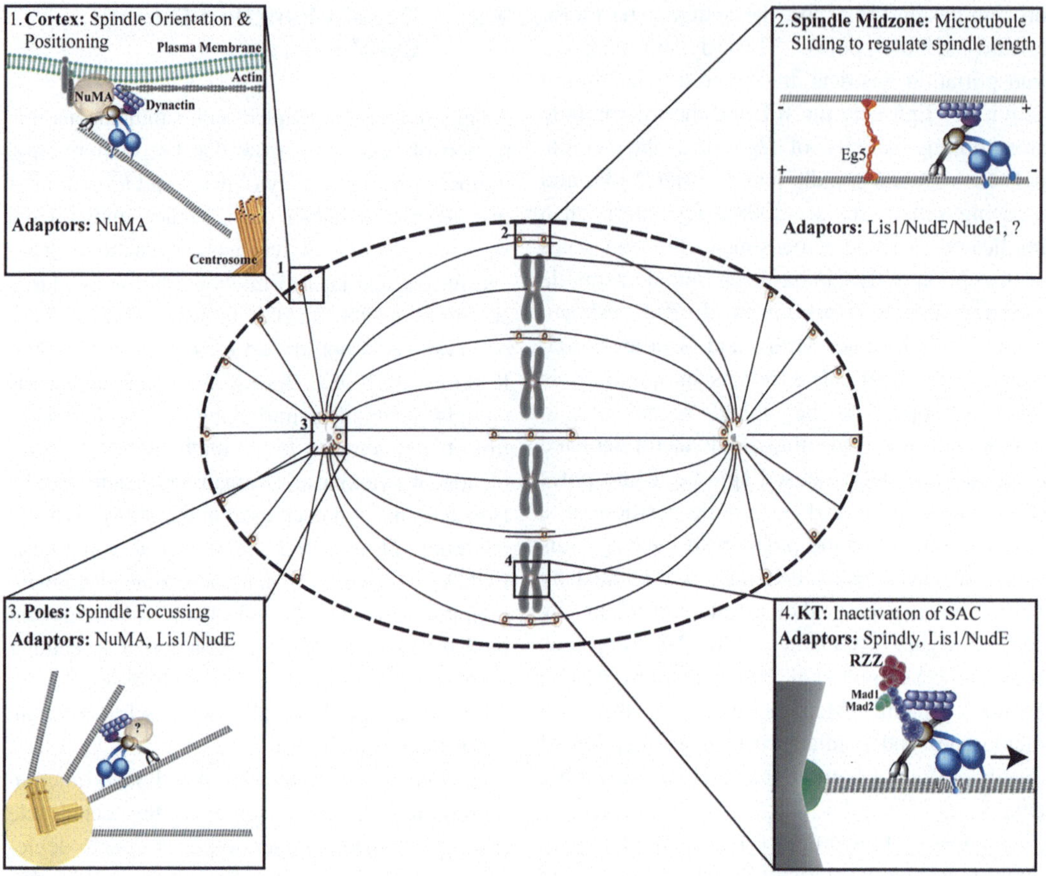

Fig. 2.3 Functions of dynein adaptors during different stages of the cell cycle. Schematic representing dynein localization and function in a mitotic cell and the plausible roles of different dynein adaptors at these sites. The actual adaptors in few scenarios, such as during spindle focusing, are not known

proteins activate dynein by altering its mechanochemical properties and removing the autoinhibited state of p150^{glued}, resulting in robust long-range retrograde transport (McKenney et al. 2014; Olenick et al. 2016; Schroeder and Vale 2016). In the subsequent sections, we will cover the literature on different dynein adaptors that impact cell division, possibly by temporally regulating dynein function during different stages of the cell cycle.

2.9.1 Lis1-NudE/Nde1

Lis1 was initially reported as the gene mutated in the neuron migration disease known as "lissencephaly," characterized by a smooth cerebral surface, cognitive defects, and seizures (Reiner et al. 1993). Lis1 is a dosage-sensitive gene as its deletion or overexpression results in lissencephaly (Bi et al. 2009), microcephaly, and delay in neurodevelopment (Lockrow et al. 2011). Lis1 is reported to participate in multiple protein-protein interactions, which might be the reason behind the dosage sensitivity of this gene (Wynshaw-Boris 2007). One of these interactions of Lis1 was reported with the coiled-coil proteins NudE and Nde1, which are dynein accessory proteins implicated in neurodevelopment (Bradshaw Nicholas et al. 2013). Studies involving genetic screens probing nuclear positioning and migration in filamentous fungi (*A. nidulans*) first linked Lis1 and NudE/Nde1 to dynein (Willins et al. 1997; Xiang et al. 1995), which was later confirmed by in vitro protein-protein interaction experiments (Faulkner et al. 2000; Sasaki et al. 2000; Smith et al. 2000). The interaction between Lis1-NudE/Nde1 and dynein is conserved across evolution and is required for coordinating dynein cargo binding and motor activity (McKenney et al. 2010; Murdoch et al. 2011; Tai et al. 2002; Ye et al. 2011).

The N-terminus of Lis1 contains a LH (Lis-homology) domain and two coiled-coil domains important for its dimerization. This is followed by seven WD-repeat constituting β-propeller regions that directly bind to the AAA3 and AAA4 regions of DHC, making Lis1 the only dynein regulator that directly binds to the dynein motor domain (Huang et al. 2012; Toropova et al. 2014). Similar to Lis1, NudE and Nde1 are capable of forming functional dimers and interacting with DIC and Lis1, thereby tethering Lis1 to dynein (Huang et al. 2012; Wang et al. 2013; Wang and Zheng 2011; Żyłkiewicz et al. 2011). Single-molecule motility assays and bead-based optical trap assays confirmed that Lis1 controls the velocity of dynein by achieving prolonged force-producing state and enhancing the microtubule association of dynein without affecting the intrinsic affinity of dynein to microtubules (Huang et al. 2012; McKenney et al. 2010). These evidences indicate Lis1 acts similar to a clutch by uncoupling microtubule binding and release of dynein with cycles of ATP hydrolysis by the AAA+ region of dynein motor domain (DeSantis et al. 2017).

During interphase the Lis1/NudE/Nde1-dynein complex regulates spatiotemporal distribution of subcellular organelles like Golgi complex, endosomes, and lysosomes (Egan et al. 2012; Shao et al. 2013). In addition, Lis1-NudE/Nde1 regulates chromosome alignment and segregation during metaphase and anaphase, respectively. Further by localizing dynein to the mitotic spindle and central spindle microtubules, Lis1 also regulates microtubule-chromosome attachment, spindle morphology and integrity, and centrosome number (Aumais et al. 2003; Faulkner et al. 2000; Moon et al. 2014; Raaijmakers et al. 2013). Consistent with this, depletion of Lis1/NudE/Nde1 resulted in misaligned chromosomes with missing microtubule-KT attachments, leading to significant delay in metaphase to anaphase transition. Lis1 depletion also led to lesser recruitment of dynein at KT, resulting in chromosome missegregation. Loss of Lis1 was associated with emergence of lagging chromosomes in telophase. However, the duration between NE breakdown and metaphase was not significantly affected upon loss of Lis1 (Moon et al. 2014). Lis1 depletion also resulted in abnormal amplification of centrosomes, leading to formation of more than two centrosomes per cells that cluster in groups to form bipolar spindle during metaphase. Furthermore, gene knockout of Lis1 led to abnormal size and shape of the pericentriolar matrix, leading to asymmetry in the spindle poles and spindle positioning defects (Moon et al. 2014; Zimdahl et al. 2014). Loss of Lis1 in planarian *S. mediterranea* and in fruit fly *D. melanogaster* hematopoietic stem cells has been shown to affect asymmetric cell division, leading to increase in their mitotic index and accelerated differentiation, finally leading to depletion of the tissue-specific stem cell pool (Cowles et al. 2012; Zimdahl et al. 2014).

2.9.2 Spindly

Spindly/Ccdc99 is a dynein adaptor conserved only in metazoans. Spindly was identified during RNAi screen of *Drosophila* S2 cells as a key player in localization of dynein to KT (Griffis et al. 2007). The N-terminus of Spindly contains four coiled-coil regions crucial for dynein-dynactin association and four consensus repeats with undetermined function at its C-terminus (Griffis et al. 2007). The complex of KT proteins, Rod/ZW10/Zwilch (known as RZZ complex), recruits Spindly to KT (Chan et al. 2009; Gassmann et al. 2010). Spindly with its farnesylated C-terminal associates with the Zwilch subunit of the RZZ complex (Holland et al. 2015; Moudgil and Chan 2015; Moudgil et al. 2015).The N-terminal CC1 box and Spindly motifs present between the second and the third coiled-coil domains associate with dynein and dynactin, respectively (Gama et al. 2017; Gassmann et al. 2010; Mosalaganti et al. 2017).

This association results in the formation of a tripartite stable complex between dynein-dynactin and Spindly attached to KT through the RZZ complex (Barisic and Geley 2011; Barisic et al. 2010). More recently, Spindly, like other dynein adaptors (such as Bicd2 and Hook3), has been shown to promote dynein-dynactin interaction and activate dynein motility in vitro (McKenney et al. 2014).

Once microtubules are properly attached to KT during metaphase, Spindly regulates the transport of SAC proteins (Mad1, Mad2, Bub1, and BubR1) and the RZZ complex from KT to the spindle poles, thereby silencing the SAC and promoting anaphase onset (Barisic and Geley 2011; Silva et al. 2015, 2017; Vallee et al. 2006). In addition, Spindly regulates stable microtubule-KT attachments independent of dynein-dynactin (Gassmann et al. 2010). Multiple studies have confirmed metaphase arrest coupled with a delay in mitotic exit in cultured cells upon Spindly depletion (Barisic et al. 2010; Silva et al. 2015, 2017). In addition, Spindly has recently been shown to regulate migration of the ovarian border cell cluster in *Drosophila*, revealing Spindly function in postmitotic cells for the first time (Clemente et al. 2017). Although in vitro experiments support activation of dynein by Spindly (McKenney et al. 2014), how proper KT attachment of microtubules regulates dynein activation and whether this is Spindly-dependent remain unknown.

2.9.3 NuMA

NuMA (nuclear mitotic apparatus) was first identified and named by Lydersen and Pettijohn for its localization to nucleus and mitotic spindle poles during interphase and mitosis, respectively (Lydersen and Pettijohn 1980). In early years, multiple groups discovered and rediscovered NuMA (and named centrophilin, SPN, and SP-H) independently which was later confirmed to be the same protein (Compton et al. 1992; Kallajoki et al. 1993; Lydersen and Pettijohn 1980; Tousson et al. 1991; Yang et al. 1992). NuMA contain two N-terminal globular heads and a C-terminal tail, which are separated by a relatively long coiled-coil region. Both the globular heads contain multiple DNA-binding S/TP motifs, while the C-terminal tail contains nuclear localization signal (NLS) and a stretch of 100 residues that directly associate with microtubules and promote microtubule bundling (Gueth-Hallonet et al. 1996; Haren and Merdes 2002; Luderus et al. 1994; Seldin et al. 2016).

During metaphase, Cdk1-mediated phosphorylation of NuMA negatively regulates its cortical localization. Inactivation of CDK1 at anaphase onset results in dephosphorylation and increase in cortical NuMA, which in turn recruits dynein at the cortex thus ensuring correct spindle positioning (Kotak et al. 2013; Kotak and Gonczy 2013; Sparks et al. 1995). NuMA is recruited to the minus ends of the microtubules at the mitotic spindle and, subsequently, recruits dynactin to regulate dynein activity during mitosis (Bosveld et al. 2017; Hueschen et al. 2017). As a result, loss of NuMA leads to dissociation of centrosomes from mitotic spindles, resulting in frayed spindles with positioning defects similar to the defects observed in dynein-depleted cells (Kotak et al. 2012; Merdes et al. 1996). An additional phosphorylation of NuMA by Aurora-A during anaphase relocates it from the spindle poles to the cell cortex, promoting dynein recruitment to cell cortex that is required for spindle positioning (Gallini et al. 2016; Jin et al. 2017; Kotak et al. 2016; Schmidt et al. 2017). Unlike other dynein adaptors that form homodimers, NuMA oligomerizes into higher-order structures (Harborth et al. 1999; Saredi et al. 1997). This raises the possibility that it could recruit multiple dynein molecules to generate more force required for focused mitotic spindle formation in the presence of high load and friction, such as upon microtubule cross-linking and coupling of chromosomes. Recruitment of a team of motors for efficient transport of large cellular cargos by dynein has already been shown in previous studies (Fu and Holzbaur 2014; Rai et al. 2013). Future structural studies can provide further insights into how NuMA regulates the force production by dynein on mitotic spindles and cell cortex.

2.9.4 Other Dynein Adaptors Involved in Cell Cycle

A number of studies have now established the role of dynein adaptors Lis1, NuMA, and Spindly in regulating different stages of cell cycle, and removal of these proteins has been linked to defects in cell cycle progression (Kotak and Gonczy 2013; Raaijmakers et al. 2013). Nevertheless, cell cycle-specific functions of other dynein adaptors have also been noted that need further exploration in the future. Bicaudal-D2 (BicD2), one of the best characterized dynein adaptors, is required for linking dynein to endosomes and Golgi complex during interphase (Chowdhury et al. 2015; McKenney et al. 2011, 2014; Schlager et al. 2014; Urnavicius et al. 2015). BicD2 also mediates dynein-dependent RNA localization (Splinter et al. 2010). In addition, BicD2 binds to RANBP2, a nucleoporin, and recruits dynein to the NE during G2 phase and facilitates centrosome separation and breakdown of the NE (Splinter et al. 2010). Similarly, a KT kinesin CENP-F along with Lis1/NudE has been shown to regulate localization of dynein and dynactin at nuclear pores and KT during late G2 phase and prometaphase, respectively (Bolhy et al. 2011; Raaijmakers et al. 2013). Two newly enlisted dynein adaptors, Ninein and Ninein-like protein (Nlp), are known to regulate centrosome maturation, spindle organization, and cytokinesis (Lin et al. 2006; Ou et al. 2002; Redwine et al. 2017b; Wang and Zhan 2007). How these adaptors regulate dynein function during cell cycle is an important question to be determined in the near future. Members of the Hook protein family have also been recently characterized as dynein adaptors that promote dynein-dynactin interaction and activate in vitro motility of dynein (Bielska et al. 2014; McKenney et al. 2014; Olenick et al. 2016; Schroeder and Vale 2016; Zhang et al. 2014). Mammals express three hook paralogs with distinct subcellular localization (Walenta et al. 2001). While Hook1 and Hook3 localize to endosomes and Golgi membranes, respectively, Hook2 localizes to the centrosome. Our own findings indicate that Hook2 acts as a dynein-dynactin adaptor that facilitates cytokinesis by regulating dynein-dynactin recruitment to the central spindles (Dwivedi et al. unpublished). Although recent studies have identified different adaptors that activate dynein motility, information on the cellular functions of these adaptors is scanty, and their role in dynein regulation during cell cycle remains unexplored (McKenney et al. 2014; Olenick et al. 2016; Redwine et al. 2017b; Schroeder and Vale 2016).

2.10 Conclusion and Future Prospects

A multitude of cell biology, biochemical, and structural studies have provided detailed information on the structure, function, and regulation of dynein. Mutations in dynein or its accessory partners impair retrograde transport leading to human disorders, including neurodegenerative diseases (Aridor and Hannan 2000, 2002). However, whether these mutations also affect cell cycle progression in actively dividing cells has not been explored. Dynein at KT regulates anaphase onset by inactivating mitotic checkpoint. Identifying the key residues in DHC or dynein adaptors that abrogate KT localization of dynein can address much debatable role of dynein at KT during prometaphase. It is now well appreciated that dynein regulates multiple events during cell cycle. However, unavailability of tools that instantly inhibit dynein activity limits most of the studies to the early stages of cell cycle. Ciliobrevins that are small-molecule dynein inhibitors have been reported to disrupt dynein-dependent retrograde organelle transport, but they suffer with the drawback of having mild activity and low potency thereby limiting their use (Firestone et al. 2012; Roossien et al. 2015). Molecular species that instantly, but reversibly, inhibit dynein activity might prove very helpful in determining dynein function during later stages of the cell cycle. Similar approaches have been used for uncovering function of the kinase Plk1 during cytokinesis, albeit it is a key regulator of prometaphase to metaphase transition (Petronczki et al. 2007). Recently, dynapyrazoles and dynarrestin that inhibit dynein activity at much lower concentrations as compared to ciliobrevins have been developed using chemical structure-based analysis (Hoing et al. 2018; Steinman et al. 2017). These inhibitors might prove to be useful tools to explore newer functions of dynein, especially during later stages of cell cycle, provided they act instantaneously. Finally, we are now beginning to understand how various events during cell cycle that require variable degrees of force are regulated by dynein.

Acknowledgments D.D. acknowledges financial support from the Council of Scientific and Industrial Research (CSIR)-University Grants Commission (UGC) and IISER Mohali. M.S. acknowledges financial support from the Wellcome Trust/Department of Biotechnology (DBT) India Alliance [grant number IA/I/12/1/500523] and IISER Mohali. M.S. is a recipient of the Wellcome Trust/ DBT India Alliance Intermediate Fellowship and SERB Women Excellence Award.

Contributions D.D. reviewed the literature and wrote the manuscript. M.S. helped in literature review and in writing and editing the manuscript. The authors declare no competing financial interests.

References

Aridor M, Hannan LA (2000) Traffic jam: a compendium of human diseases that affect intracellular transport processes. Traffic 1:836–851

Aridor M, Hannan LA (2002) Traffic jams II: an update of diseases of intracellular transport. Traffic 3:781–790

Aumais JP, Williams SN, Luo W, Nishino M, Caldwell KA, Caldwell GA, Lin S-H, Yu-Lee L-Y (2003) Role for NudC, a dynein-associated nuclear movement protein, in mitosis and cytokinesis. J Cell Sci 116:1991

Barisic M, Geley S (2011) Spindly switch controls anaphase: spindly and RZZ functions in chromosome attachment and mitotic checkpoint control. Cell Cycle 10:449–456

Barisic M, Sohm B, Mikolcevic P, Wandke C, Rauch V, Ringer T, Hess M, Bonn G, Geley S (2010) Spindly/ CCDC99 is required for efficient chromosome Congression and mitotic checkpoint regulation. Mol Biol Cell 21:1968–1981

Basto R, Scaerou F, Mische S, Wojcik E, Lefebvre C, Gomes R, Hays T, Karess R (2004) In vivo dynamics of the rough deal checkpoint protein during Drosophila mitosis. Curr Biol 14:56–61

Beaudouin J, Gerlich D, Daigle N, Eils R, Ellenberg J (2002) Nuclear envelope breakdown proceeds by microtubule-induced tearing of the lamina. Cell 108:83–96

Belyy V, Schlager MA, Foster H, Reimer AE, Carter AP, Yildiz A (2016) The mammalian dynein–dynactin complex is a strong opponent to kinesin in a tug-of-war competition. Nat Cell Biol 18:1018

Bhabha G, Cheng HC, Zhang N, Moeller A, Liao M, Speir JA, Cheng Y, Vale RD (2014) Allosteric communication in the dynein motor domain. Cell 159:857–868

Bi W, Sapir T, Shchelochkov OA, Zhang F, Withers MA, Hunter JV, Levy T, Shinder V, Peiffer DA, Gunderson KL et al (2009) Increased LIS1 expression affects human and mouse brain development. Nat Genet 41:168–177

Bielska E, Schuster M, Roger Y, Berepiki A, Soanes DM, Talbot NJ, Steinberg G (2014) Hook is an adapter that coordinates kinesin-3 and dynein cargo attachment on early endosomes. J Cell Biol 204:989

Bolhy S, Bouhlel I, Dultz E, Nayak T, Zuccolo M, Gatti X, Vallee R, Ellenberg J, Doye V (2011) A Nup133-dependent NPC-anchored network tethers centrosomes to the nuclear envelope in prophase. J Cell Biol 192:855–871

Bosveld F, Ainslie A, Bellaïche Y (2017) Sequential activities of dynein, mud and asp in centrosome–spindle coupling maintain centrosome number upon mitosis. J Cell Sci 130:3557

Bradshaw Nicholas J, Hennah W, Soares Dinesh C (2013) NDE1 and NDEL1: twin neurodevelopmental proteins with similar 'nature' but different 'nurture'. In: BioMolecular concepts, p 447

Buffin E, Lefebvre C, Huang J, Gagou ME, Karess RE (2005) Recruitment of Mad2 to the kinetochore requires the Rod/Zw10 complex. Curr Biol 15:856–861

Burton P, Adams DR, Abraham A, Allcock RW, Jiang Z, McCahill A, Gilmour J, McAbney J, Kaupisch A, Kane NM et al (2010) Erythro-9-(2-hydroxy-3-nonyl) adenine (EHNA) blocks differentiation and maintains the expression of pluripotency markers in human embryonic stem cells. Biochem J 432:575

Busson S, Dujardin D, Moreau A, Dompierre J, De Mey JR (1998) Dynein and dynactin are localized to astral microtubules and at cortical sites in mitotic epithelial cells. Curr Biol 8:541–544

Campbell KS, Cooper S, Dessing M, Yates S, Buder A (1998) Interaction of p59fyn kinase with the dynein light chain, Tctex-1, and colocalization during cytokinesis. J Immunol 161:1728–1737

Carter AP, Garbarino JE, Wilson-Kubalek EM, Shipley WE, Cho C, Milligan RA, Vale RD, Gibbons IR (2008) Structure and functional role of Dynein's microtubule-binding domain. Science 322:1691

Carter AP, Cho C, Jin L, Vale RD (2011) Crystal structure of the dynein motor domain. Science 331:1159

Chan YW, Fava LL, Uldschmid A, Schmitz MHA, Gerlich DW, Nigg EA, Santamaria A (2009) Mitotic control of

kinetochore-associated dynein and spindle orientation by human spindly. J Cell Biol 185:859
Cho C, Reck-Peterson SL, Vale RD (2008) Regulatory ATPase sites of cytoplasmic dynein affect processivity and force generation. J Biol Chem 283:25839–25845
Chowdhury S, Ketcham SA, Schroer TA, Lander GC (2015) Structural organization of the dynein–dynactin complex bound to microtubules. Nat Struct Mol Biol 22:345
Clemente GD, Hannaford MR, Januschke J, Griffis ER, Muller H-AJ (2017) Requirement of the dynein-adaptor spindly for mitotic and post-mitotic functions in drosophila. J Dev Biol 2018 Mar 30 6(2): pii: E9. https://doi.org/10.3390/jdb6020009
Collins CA, Vallee RB (1989) Preparation of microtubules from rat liver and testis: cytoplasmic dynein is a major microtubule associated protein. Cell Motil Cytoskeleton 14:491–500
Compton DA, Szilak I, Cleveland DW (1992) Primary structure of NuMA, an intranuclear protein that defines a novel pathway for segregation of proteins at mitosis. J Cell Biol 116:1395–1408
Cowles MW, Hubert A, Zayas RM (2012) A Lissencephaly-1 homologue is essential for mitotic progression in the planarian Schmidtea mediterranea. Dev Dyn 241:901–910
De Simone A, Gonczy P (2017) Computer simulations reveal mechanisms that organize nuclear dynein forces to separate centrosomes. Mol Biol Cell 28:3165–3170
De Simone A, Nedelec F, Gonczy P (2016) Dynein transmits polarized Actomyosin cortical flows to promote centrosome separation. Cell Rep 14:2250–2262
Delcros J-G, Prigent C, Giet R (2006) Dynactin targets Pavarotti-KLP to the central spindle during anaphase and facilitates cytokinesis in Drosophila S2 cells. J Cell Sci 119:4431
DeSantis ME, Cianfrocco MA, Htet ZM, Tran PT, Reck-Peterson SL, Leschziner AE (2017) Lis1 has two opposing modes of regulating cytoplasmic dynein. Cell 170:1197–1208. e1112
DeWitt MA, Chang AY, Combs PA, Yildiz A (2012) Cytoplasmic dynein moves through uncoordinated stepping of the AAA+ ring domains. Science 335:221
DeWitt MA, Cypranowska CA, Cleary FB, Belyy V, Yildiz A (2014) The AAA3 domain of cytoplasmic dynein acts as a switch to facilitate microtubule release. Nat Struct Mol Biol 22:73
Egan MJ, Tan K, Reck-Peterson SL (2012) Lis1 is an initiation factor for dynein-driven organelle transport. J Cell Biol 197:971
Farrer MJ, Hulihan MM, Kachergus JM, Dachsel JC, Stoessl AJ, Grantier LL, Calne S, Calne DB, Lechevalier B, Chapon F et al (2009) DCTN1 mutations in Perry syndrome. Nat Genet 41:163–165
Faulkner NE, Dujardin DL, Tai C-Y, Vaughan KT, O'Connell CB, Wang YL, Vallee RB (2000) A role for the lissencephaly gene LIS1 in mitosis and cytoplasmic dynein function. Nat Cell Biol 2:784
Ferenz NP, Paul R, Fagerstrom C, Mogilner A, Wadsworth P (2009) Dynein antagonizes eg5 by crosslinking and sliding antiparallel microtubules. Curr Biol 19:1833–1838
Firestone AJ, Weinger JS, Maldonado M, Barlan K, Langston LD, O'Donnell M, Gelfand VI, Kapoor TM, Chen JK (2012) Small-molecule inhibitors of the AAA+ ATPase motor cytoplasmic dynein. Nature 484:125–129
Fu MM, Holzbaur EL (2014) Integrated regulation of motor-driven organelle transport by scaffolding proteins. Trends Cell Biol 24:564–574
Gallini S, Carminati M, De Mattia F, Pirovano L, Martini E, Oldani A, Asteriti IA, Guarguaglini G, Mapelli M (2016) NuMA phosphorylation by Aurora-A Orchestrates spindle orientation. Curr Biol 26:458–469
Gama JB, Pereira C, Simões PA, Celestino R, Reis RM, Barbosa DJ, Pires HR, Carvalho C, Amorim J, Carvalho AX et al (2017) Molecular mechanism of dynein recruitment to kinetochores by the Rod–Zw10–Zwilch complex and spindly. J Cell Biol 216:943
Gassmann R, Essex A, Hu J-S, Maddox PS, Motegi F, Sugimoto A, O'Rourke SM, Bowerman B, McLeod I, Yates JR et al (2008) A new mechanism controlling kinetochore–microtubule interactions revealed by comparison of two dynein-targeting components: SPDL-1 and the Rod/Zwilch/Zw10 complex. Genes Dev 22:2385–2399
Gassmann R, Holland AJ, Varma D, Wan X, Çivril F, Cleveland DW, Oegema K, Salmon ED, Desai A (2010) Removal of spindly from microtubule-attached kinetochores controls spindle checkpoint silencing in human cells. Genes Dev 24:957–971
Gee MA, Heuser JE, Vallee RB (1997) An extended microtubule-binding structure within the dynein motor domain. Nature 390:636–639
Gibbons IR, Rowe AJ (1965) Dynein: a protein with adenosine Triphosphatase activity from cilia. Science 149:424
Gibbons IR, Garbarino JE, Tan CE, Reck-Peterson SL, Vale RD, Carter AP (2005) The affinity of the dynein microtubule-binding domain is modulated by the conformation of its coiled-coil stalk. J Biol Chem 280:23960–23965
Gill SR, Schroer TA, Szilak I, Steuer ER, Sheetz MP, Cleveland DW (1991) Dynactin, a conserved, ubiquitously expressed component of an activator of vesicle motility mediated by cytoplasmic dynein. J Cell Biol 115:1639–1650
Gonczy P (2008) Mechanisms of asymmetric cell division: flies and worms pave the way. Nat Rev Mol Cell Biol 9:355–366
Gonczy P, Pichler S, Kirkham M, Hyman AA (1999) Cytoplasmic dynein is required for distinct aspects of MTOC positioning, including centrosome separation, in the one cell stage Caenorhabditis elegans embryo. J Cell Biol 147:135–150

Goshima G, Nedelec F, Vale RD (2005) Mechanisms for focusing mitotic spindle poles by minus end-directed motor proteins. J Cell Biol 171:229–240

Griffis ER, Stuurman N, Vale RD (2007) Spindly, a novel protein essential for silencing the spindle assembly checkpoint, recruits dynein to the kinetochore. J Cell Biol 177:1005

Grotjahn DA, Chowdhury S, Xu Y, McKenney RJ, Schroer TA, Lander GC (2018) Cryo-electron tomography reveals that dynactin recruits a team of dyneins for processive motility. Nat Struct Mol Biol 25:203–207

Gueth-Hallonet C, Weber K, Osborn M (1996) NuMA: a bipartite nuclear location signal and other functional properties of the tail domain. Exp Cell Res 225:207–218

Harborth J, Wang J, Gueth-Hallonet C, Weber K, Osborn M (1999) Self assembly of NuMA: multiarm oligomers as structural units of a nuclear lattice. EMBO J 18:1689–1700

Haren L, Merdes A (2002) Direct binding of NuMA to tubulin is mediated by a novel sequence motif in the tail domain that bundles and stabilizes microtubules. J Cell Sci 115:1815–1824

Hoing S, Yeh TY, Baumann M, Martinez NE, Habenberger P, Kremer L, Drexler HCA, Kuchler P, Reinhardt P, Choidas A et al (2018) Dynarrestin, a novel inhibitor of cytoplasmic dynein. Cell Chem Biol 25(4):357–369

Holland AJ, Reis RM, Niessen S, Pereira C, Andres DA, Spielmann HP, Cleveland DW, Desai A, Gassmann R (2015) Preventing farnesylation of the dynein adaptor spindly contributes to the mitotic defects caused by farnesyltransferase inhibitors. Mol Biol Cell 26:1845–1856

Huang J, Roberts AJ, Leschziner AE, Reck-Peterson SL (2012) Lis1 acts as a "clutch" between the ATPase and microtubule-binding domains of the dynein motor. Cell 150:975–986

Hueschen CL, Kenny SJ, Xu K, Dumont S (2017) NuMA recruits dynein activity to microtubule minus-ends at mitosis. elife 2017 Nov 29 6. pii: e29328. https://doi.org/10.7554/eLife.29328

Imai H, Narita A, Maeda Y, Schroer TA (2014) Dynactin 3D structure: implications for assembly and dynein binding. J Mol Biol 426:3262–3271

Jamison DK, Driver JW, Rogers AR, Constantinou PE, Diehl MR (2010) Two kinesins transport cargo primarily via the action of one motor: implications for intracellular transport. Biophys J 99:2967–2977

Jamison DK, Driver JW, Diehl MR (2012) Cooperative responses of multiple kinesins to variable and constant loads. J Biol Chem 287:3357–3365

Jin M, Pomp O, Shinoda T, Toba S, Torisawa T, Furuta Ky, Oiwa K, Yasunaga T, Kitagawa D, Matsumura S et al (2017) Katanin p80, NuMA and cytoplasmic dynein cooperate to control microtubule dynamics. Sci Rep 7:39902

Kallajoki M, Harborth J, Weber K, Osborn M (1993) Microinjection of a monoclonal antibody against SPN antigen, now identified by peptide sequences as the NuMA protein, induces micronuclei in PtK2 cells. J Cell Sci 104(Pt 1):139–150

Kaplan A, Reiner O (2011) Linking cytoplasmic dynein and transport of Rab8 vesicles to the midbody during cytokinesis by the doublecortin domain-containing 5 protein. J Cell Sci 124:3989

Kardon JR, Vale RD (2009) Regulators of the cytoplasmic dynein motor. Nat Rev Mol Cell Biol 10:854

Karess R (2005) Rod-Zw10-Zwilch: a key player in the spindle checkpoint. Trends Cell Biol 15:386–392

Karki S, Holzbaur EL (1995) Affinity chromatography demonstrates a direct binding between cytoplasmic dynein and the dynactin complex. J Biol Chem 270:28806–28811

Karki S, LaMonte B, Holzbaur EL (1998) Characterization of the p22 subunit of dynactin reveals the localization of cytoplasmic dynein and dynactin to the midbody of dividing cells. J Cell Biol 142:1023–1034

Kashina AS, Baskin RJ, Cole DG, Wedaman KP, Saxton WM, Scholey JM (1996) A bipolar kinesin. Nature 379:270–272

Kon T, Nishiura M, Ohkura R, Toyoshima YY, Sutoh K (2004) Distinct functions of nucleotide-binding/hydrolysis sites in the four AAA modules of cytoplasmic dynein. Biochemistry 43:11266–11274

Kon T, Imamula K, Roberts AJ, Ohkura R, Knight PJ, Gibbons IR, Burgess SA, Sutoh K (2009) Helix sliding in the stalk coiled coil of dynein couples ATPase and microtubule binding. Nat Struct Mol Biol 16:325

Kon T, Oyama T, Shimo-Kon R, Imamula K, Shima T, Sutoh K, Kurisu G (2012) The 2.8 Å crystal structure of the dynein motor domain. Nature 484:345

Kops GJ, Kim Y, Weaver BA, Mao Y, McLeod I, Yates JR 3rd, Tagaya M, Cleveland DW (2005) ZW10 links mitotic checkpoint signaling to the structural kinetochore. J Cell Biol 169:49–60

Kotak S, Gonczy P (2013) Mechanisms of spindle positioning: cortical force generators in the limelight. Curr Opin Cell Biol 25:741–748

Kotak S, Busso C, Gönczy P (2012) Cortical dynein is critical for proper spindle positioning in human cells. J Cell Biol 199:97

Kotak S, Busso C, Gönczy P (2013) NuMA phosphorylation by CDK1 couples mitotic progression with cortical dynein function. EMBO J 32:2517

Kotak S, Busso C, Gönczy P (2014) NuMA interacts with phosphoinositides and links the mitotic spindle with the plasma membrane. EMBO J 33:1815

Kotak S, Afshar K, Busso C, Gonczy P (2016) Aurora A kinase regulates proper spindle positioning in C. elegans and in human cells. J Cell Sci 129:3015–3025

Lee IG, Olenick MA, Boczkowska M, Franzini-Armstrong C, Holzbaur ELF, Dominguez R (2018) A conserved interaction of the dynein light intermediate chain with dynein-dynactin effectors necessary for processivity. Nat Commun 9:986

Lenz JH, Schuchardt I, Straube A, Steinberg G (2006) A dynein loading zone for retrograde endosome motility at microtubule plus-ends. EMBO J 25:2275–2286

Lin C-C, Cheng T-S, Hsu C-M, Wu C-H, Chang L-S, Shen Z-S, Yeh H-M, Chang L-K, Howng S-L, Hong Y-R (2006) Characterization and functional aspects of human Ninein isoforms that regulated by Centrosomal targeting signals and evidence for docking sites to direct gamma-tubulin. Cell Cycle 5:2517–2527

Lipka J, Kuijpers M, Jaworski J, Hoogenraad CC (2013) Mutations in cytoplasmic dynein and its regulators cause malformations of cortical development and neurodegenerative diseases. Biochem Soc Trans 41:1605–1612

Lockrow JP, Holden KR, Dwivedi A, Matheus MG, Lyons MJ (2011) LIS1 duplication: expanding the phenotype. J Child Neurol 27:791–795

Luderus ME, den Blaauwen JL, de Smit OJ, Compton DA, van Driel R (1994) Binding of matrix attachment regions to Lamin polymers involves single stranded regions and the minor groove. Mol Cell Biol 14:6297–6305

Lydersen BK, Pettijohn DE (1980) Human-specific nuclear protein that associates with the polar region of the mitotic apparatus: distribution in a human/hamster hybrid cell. Cell 22:489–499

Maiato H, Rieder CL, Khodjakov A (2004) Kinetochore-driven formation of kinetochore fibers contributes to spindle assembly during animal mitosis. J Cell Biol 167:831–840

Mallik R, Carter BC, Lex SA, King SJ, Gross SP (2004) Cytoplasmic dynein functions as a gear in response to load. Nature 427:649–652

McKenney RJ, Vershinin M, Kunwar A, Vallee RB, Gross SP (2010) LIS1 and NudE induce a persistent dynein force-producing state. Cell 141:304–314

McKenney RJ, Weil SJ, Scherer J, Vallee RB (2011) Mutually exclusive cytoplasmic dynein regulation by NudE-Lis1 and dynactin. J Biol Chem 286:39615–39622

McKenney RJ, Huynh W, Tanenbaum ME, Bhabha G, Vale RD (2014) Activation of cytoplasmic dynein motility by dynactin-cargo adapter complexes. Science 345:337–341

Merdes A, Ramyar K, Vechio JD, Cleveland DW (1996) A complex of NuMA and cytoplasmic dynein is essential for mitotic spindle assembly. Cell 87:447–458

Moon HM, Youn YH, Pemble H, Yingling J, Wittmann T, Wynshaw-Boris A (2014) LIS1 controls mitosis and mitotic spindle organization via the LIS1–NDEL1–dynein complex. Hum Mol Genet 23:449–466

Morales-Mulia S, Scholey JM (2005) Spindle pole organization in Drosophila S2 cells by dynein, abnormal spindle protein (Asp), and KLP10A. Mol Biol Cell 16:3176–3186

Mosalaganti S, Keller J, Altenfeld A, Winzker M, Rombaut P, Saur M, Petrovic A, Wehenkel A, Wohlgemuth S, Müller F et al (2017) Structure of the RZZ complex and molecular basis of its interaction with spindly. J Cell Biol 216:961

Moudgil DK, Chan GKT (2015) Lipids beyond membranes; farnesylation targets spindly to kinetochores. Cell Cycle 14:2185–2186

Moudgil DK, Westcott N, Famulski JK, Patel K, Macdonald D, Hang H, Chan GKT (2015) A novel role of farnesylation in targeting a mitotic checkpoint protein, human spindly, to kinetochores. J Cell Biol 208:881

Murdoch H, Vadrevu S, Prinz A, Dunlop AJ, Klussmann E, Bolger GB, Norman JC, Houslay MD (2011) Interaction between LIS1 and PDE4, and its role in cytoplasmic dynein function. J Cell Sci 124:2253

Nakajima Y, Kanno T, Nagaya T, Kuribayashi K, Nakano T, Gotoh A, Nishizaki T (2015) Adenosine deaminase inhibitor EHNA exhibits a potent anticancer effect against malignant pleural mesothelioma. Cell Physiol Biochem 35:51–60

Neuwald AF, Aravind L, Spouge JL, Koonin EV (1999) AAA+: A class of chaperone-like ATPases associated with the assembly, operation, and disassembly of protein complexes. Genome Res 9:27–43

Nguyen-Ngoc T, Afshar K, Gonczy P (2007) Coupling of cortical dynein and G alpha proteins mediates spindle positioning in Caenorhabditis elegans. Nat Cell Biol 9:1294–1302

Nicholas MP, Hook P, Brenner S, Wynne CL, Vallee RB, Gennerich A (2015) Control of cytoplasmic dynein force production and processivity by its C-terminal domain. Nat Commun 6:6206

Numata N, Shima T, Ohkura R, Kon T, Sutoh K (2011) C-sequence of the Dictyostelium cytoplasmic dynein participates in processivity modulation. FEBS Lett 585:1185–1190

Nyarko A, Song Y, Barbar E (2012) Intrinsic disorder in dynein intermediate chain modulates its interactions with NudE and dynactin. J Biol Chem 287:24884–24893

Olenick MA, Tokito M, Boczkowska M, Dominguez R, Holzbaur ELF (2016) Hook adaptors induce unidirectional processive motility by enhancing the dynein-dynactin interaction. J Biol Chem 2016 Aug 26 291(35): 18239–18251. https://doi.org/10.1074/jbc.M116.738211. Epub 2016 Jun 30

Ou YY, Mack GJ, Zhang M, Rattner JB (2002) CEP110 and ninein are located in a specific domain of the centrosome associated with centrosome maturation. J Cell Sci 115:1825

Paschal BM, Vallee RB (1987) Retrograde transport by the microtubule-associated protein MAP 1C. Nature 330:181–183

Paschal BM, Shpetner HS, Vallee RB (1987) MAP 1C is a microtubule-activated ATPase which translocates microtubules in vitro and has dynein-like properties. J Cell Biol 105:1273

Penningroth SM, Cheung A, Bouchard P, Gagnon C, Bardin CW (1982) Dynein ATPase is inhibited selectively in vitro by erythro-9-[3-2-(hydroxynonyl)]adenine. Biochem Biophys Res Commun 104:234–240

Petronczki M, Glotzer M, Kraut N, Peters J-M (2007) Polo-like kinase 1 triggers the initiation of cytokinesis in human cells by promoting recruitment of the RhoGEF Ect2 to the central spindle. Dev Cell 12:713–725

Potapova T, Gorbsky GJ (2017) The consequences of chromosome segregation errors in mitosis and meiosis. Biology (Basel) 6

Puls I, Jonnakuty C, LaMonte BH, Holzbaur EL, Tokito M, Mann E, Floeter MK, Bidus K, Drayna D, Oh SJ et al (2003) Mutant dynactin in motor neuron disease. Nat Genet 33:455–456

Raaijmakers JA, van Heesbeen RG, Meaders JL, Geers EF, Fernandez-Garcia B, Medema RH, Tanenbaum ME (2012) Nuclear envelope-associated dynein drives prophase centrosome separation and enables Eg5-independent bipolar spindle formation. EMBO J 31:4179–4190

Raaijmakers JA, Tanenbaum ME, Medema RH (2013) Systematic dissection of dynein regulators in mitosis. J Cell Biol 201:201–215

Rai AK, Rai A, Ramaiya AJ, Jha R, Mallik R (2013) Molecular adaptations allow dynein to generate large collective forces inside cells. Cell 152:172–182

Rank KC, Rayment I (2013) Functional asymmetry in kinesin and dynein dimers. Biol Cell 105:1–13

Reboutier D, Troadec MB, Cremet JY, Chauvin L, Guen V, Salaun P, Prigent C (2013) Aurora A is involved in central spindle assembly through phosphorylation of Ser 19 in P150Glued. J Cell Biol 201:65–79

Reck-Peterson SL, Yildiz A, Carter AP, Gennerich A, Zhang N, Vale RD (2006) Single-molecule analysis of dynein processivity and stepping behavior. Cell 126:335–348

Redwine WB, Hernandez-Lopez R, Zou S, Huang J, Reck-Peterson SL, Leschziner AE (2012) Structural basis for microtubule binding and release by dynein. Science 337:1532–1536

Redwine WB, DeSantis ME, Hollyer I, Htet ZM, Tran PT, Swanson SK, Florens L, Washburn MP, Reck-Peterson SL (2017a) The human cytoplasmic dynein interactome reveals novel activators of motility. elife 6

Redwine WB, DeSantis ME, Hollyer I, Htet ZM, Tran PT, Swanson SK, Florens L, Washburn MP, Reck-Peterson SL (2017b) The human cytoplasmic dynein interactome reveals novel activators of motility. elife 6:e28257

Reiner O, Carrozzo R, Shen Y, Wehnert M, Faustinella F, Dobyns WB, Caskey CT, Ledbetter DH (1993) Isolation of a Miller-Dicker lissencephaly gene containing G protein [beta]-subunit-like repeats. Nature 364:717–721

Roberts AJ, Numata N, Walker ML, Kato YS, Malkova B, Kon T, Ohkura R, Arisaka F, Knight PJ, Sutoh K et al (2009) AAA+ Ring and linker swing mechanism in the dynein motor. Cell 136:485–495

Roberts AJ, Malkova B, Walker ML, Sakakibara H, Numata N, Kon T, Ohkura R, Edwards TA, Knight PJ, Sutoh K et al (2012) ATP-driven remodeling of the linker domain in the dynein motor. Structure 20:1670–1680

Roberts AJ, Kon T, Knight PJ, Sutoh K, Burgess SA (2013) Functions and mechanics of dynein motor proteins. Nat Rev Mol Cell Biol 14:713

Robinson JT, Wojcik EJ, Sanders MA, McGrail M, Hays TS (1999) Cytoplasmic dynein is required for the nuclear attachment and migration of centrosomes during mitosis in Drosophila. J Cell Biol 146:597–608

Roossien D, Miller K, Gallo G (2015) Ciliobrevins as tools for studying dynein motor function. Front Cell Neurosci 9:252

Salina D, Bodoor K, Eckley DM, Schroer TA, Rattner JB, Burke B (2002) Cytoplasmic dynein as a facilitator of nuclear envelope breakdown. Cell 108:97–107

Saredi A, Howard L, Compton DA (1997) Phosphorylation regulates the assembly of NuMA in a mammalian mitotic extract. J Cell Sci 110(Pt 11):1287–1297

Sasaki S, Shionoya A, Ishida M, Gambello MJ, Yingling J, Wynshaw-Boris A, Hirotsune S (2000) A LIS1/NUDEL/cytoplasmic dynein heavy chain complex in the developing and adult nervous system. Neuron 28:681–696

Schlager MA, Serra-Marques A, Grigoriev I, Gumy LF, Esteves da Silva M, Wulf PS, Akhmanova A, Hoogenraad CC (2014) Bicaudal d family adaptor proteins control the velocity of dynein-based movements. Cell Rep 8:1248–1256

Schliwa M, Woehlke G (2003) Molecular motors. Nature 422:759–765

Schliwa M, Ezzell RM, Euteneuer U (1984) erythro-9-[3-(2-Hydroxynonyl)]adenine is an effective inhibitor of cell motility and actin assembly. Proc Natl Acad Sci USA 81:6044–6048

Schmidt DJ, Rose DJ, Saxton WM, Strome S (2005) Functional analysis of cytoplasmic dynein heavy chain in Caenorhabditis elegans with fast-acting temperature-sensitive mutations. Mol Biol Cell 16:1200–1212

Schmidt H, Gleave ES, Carter AP (2012) Insights into dynein motor domain function from a 3.3-Å crystal structure. Nat Struct Mol Biol 19:492

Schmidt H, Zalyte R, Urnavicius L, Carter AP (2015) Structure of human cytoplasmic dynein-2 primed for its power stroke. Nature 518:435–438

Schmidt R, Fielmich L-E, Grigoriev I, Katrukha EA, Akhmanova A, van den Heuvel S (2017) Two populations of cytoplasmic dynein contribute to spindle positioning in C. elegans embryos. J Cell Biol 216:2777

Schroeder CM, Vale RD (2016) Assembly and activation of dynein-dynactin by the cargo adaptor protein Hook3. J Cell Biol 214:309–318

Schroer TA (2004) Dynactin. Annu Rev Cell Dev Biol 20:759–779
Seldin L, Muroyama A, Lechler T (2016) NuMA-microtubule interactions are critical for spindle orientation and the morphogenesis of diverse epidermal structures. elife 5:e12504
Shao C-Y, Zhu J, Xie Y-J, Wang Z, Wang Y-N, Wang Y, Su L-D, Zhou L, Zhou T-H, Shen Y (2013) Distinct functions of nuclear distribution proteins LIS1, Ndel1 and NudCL in regulating axonal mitochondrial transport. Traffic 14:785–797
Sharp DJ, Rogers GC, Scholey JM (2000a) Cytoplasmic dynein is required for poleward chromosome movement during mitosis in Drosophila embryos. Nat Cell Biol 2:922–930
Sharp DJ, Rogers GC, Scholey JM (2000b) Microtubule motors in mitosis. Nature 407:41–47
Siaw MF, Mitchell BS, Koller CA, Coleman MS, Hutton JJ (1980) ATP depletion as a consequence of adenosine deaminase inhibition in man. Proc Natl Acad Sci USA 77:6157–6161
Silva PMA, Tavares ÁA, Bousbaa H (2015) Co silencing of human Bub3 and dynein highlights an antagonistic relationship in regulating kinetochore–microtubule attachments. FEBS Lett 589:3588–3594
Silva PMA, Ribeiro N, Lima RT, Andrade C, Diogo V, Teixeira J, Florindo C, Tavares Á, Vasconcelos MH, Bousbaa H (2017) Suppression of spindly delays mitotic exit and exacerbates cell death response of cancer cells treated with low doses of paclitaxel. Cancer Lett 394:33–42
Smith DS, Niethammer M, Ayala R, Zhou Y, Gambello MJ, Wynshaw-Boris A, Tsai L-H (2000) Regulation of cytoplasmic dynein behaviour and microtubule organization by mammalian Lis1. Nat Cell Biol 2:767
Sparks CA, Fey EG, Vidair CA, Doxsey SJ (1995) Phosphorylation of NUMA occurs during nuclear breakdown and not mitotic spindle assembly. J Cell Sci 108(Pt 11):3389–3396
Splinter D, Tanenbaum ME, Lindqvist A, Jaarsma D, Flotho A, Yu KL, Grigoriev I, Engelsma D, Haasdijk ED, Keijzer N et al (2010) Bicaudal D2, dynein, and Kinesin-1 associate with nuclear pore complexes and regulate centrosome and nuclear positioning during mitotic entry. PLoS Biol 8:e1000350
Starr DA, Williams BC, Hays TS, Goldberg ML (1998) ZW10 helps recruit dynactin and dynein to the kinetochore. J Cell Biol 142:763–774
Stehman SA, Chen Y, McKenney RJ, Vallee RB (2007) NudE and NudEL are required for mitotic progression and are involved in dynein recruitment to kinetochores. J Cell Biol 178:583–594
Steinman JB, Santarossa CC, Miller RM, Yu LS, Serpinskaya AS, Furukawa H, Morimoto S, Tanaka Y, Nishitani M, Asano M et al (2017) Chemical structure-guided design of dynapyrazoles, cell-permeable dynein inhibitors with a unique mode of action. elife 6:e25174
Tai C-Y, Dujardin DL, Faulkner NE, Vallee RB (2002) Role of dynein, dynactin, and CLIP-170 interactions in LIS1 kinetochore function. J Cell Biol 156:959–968
Tanenbaum ME, Medema RH (2010) Mechanisms of centrosome separation and bipolar spindle assembly. Dev Cell 19:797–806
Tanenbaum ME, Akhmanova A, Medema RH (2010) Dynein at the nuclear envelope. EMBO Rep 11:649
Thankachan JM, Nuthalapati SS, Addanki Tirumala N, Ananthanarayanan V (2017) Fission yeast myosin I facilitates PI(4,5)P2-mediated anchoring of cytoplasmic dynein to the cortex. Proc Natl Acad Sci USA 114: E2672–E2681
Toba S, Watanabe TM, Yamaguchi-Okimoto L, Toyoshima YY, Higuchi H (2006) Overlapping hand-over-hand mechanism of single molecular motility of cytoplasmic dynein. Proc Natl Acad Sci USA 103:5741–5745
Torisawa T, Ichikawa M, Furuta A, Saito K, Oiwa K, Kojima H, Toyoshima YY, Furuta Ky (2014) Autoinhibition and cooperative activation mechanisms of cytoplasmic dynein. Nat Cell Biol 16:1118–1124
Toropova K, Zou S, Roberts AJ, Redwine WB, Goodman BS, Reck-Peterson SL, Leschziner AE (2014) Lis1 regulates dynein by sterically blocking its mechanochemical cycle. elife 3:e03372
Tousson A, Zeng C, Brinkley BR, Valdivia MM (1991) Centrophilin: a novel mitotic spindle protein involved in microtubule nucleation. J Cell Biol 112:427–440
Urnavicius L, Zhang K, Diamant AG, Motz C, Schlager MA, Yu M, Patel NA, Robinson CV, Carter AP (2015) The structure of the dynactin complex and its interaction with dynein. Science 347:1441–1446
Urnavicius L, Lau CK, Elshenawy MM, Morales-Rios E, Motz C, Yildiz A, Carter AP (2018) Cryo-EM shows how dynactin recruits two dyneins for faster movement. Nature 554:202–206
Uteng M, Hentrich C, Miura K, Bieling P, Surrey T (2008) Poleward transport of Eg5 by dynein–dynactin in Xenopus laevis egg extract spindles. J Cell Biol 182:715–726
Vale RD, Milligan RA (2000) The way things move: looking under the Hood of molecular motor proteins. Science 288:88
Vallee RB, Varma D, Dujardin DL (2006) ZW10 function in mitotic checkpoint control, dynein targeting, and membrane trafficking: is dynein the unifying theme? Cell Cycle 5:2447–2451
Vallee RB, McKenney RJ, Ori-McKenney KM (2012) Multiple modes of cytoplasmic dynein regulation. Nat Cell Biol 14:224
van Heesbeen RGHP, Raaijmakers JA, Tanenbaum ME, Medema RH (2013) Nuclear envelope-associated dynein cooperates with Eg5 to drive prophase centrosome separation. Commun Integr Biol 6:e23841
Verde F, Berrez JM, Antony C, Karsenti E (1991) Taxol-induced microtubule asters in mitotic extracts of Xenopus eggs: requirement for phosphorylated factors and cytoplasmic dynein. J Cell Biol 112:1177–1187

Vergnolle MA, Taylor SS (2007) Cenp-F links kinetochores to Ndel1/Nde1/Lis1/dynein microtubule motor complexes. Curr Biol 17:1173–1179
Visscher K, Schnitzer MJ, Block SM (1999) Single kinesin molecules studied with a molecular force clamp. Nature 400:184–189
Walenta JH, Didier AJ, Liu X, Krämer H (2001) The Golgi-associated Hook3 protein is a member of a novel family of microtubule-binding proteins. J Cell Biol 152:923
Wang Y, Zhan Q (2007) Cell cycle-dependent expression of Centrosomal Ninein-like protein in human cells is regulated by the anaphase-promoting complex. J Biol Chem 282:17712–17719
Wang S, Zheng Y (2011) Identification of a novel dynein binding domain in Nudel essential for spindle pole Organization in Xenopus egg Extract. J Biol Chem 286:587–593
Wang S, Ketcham SA, Schön A, Goodman B, Wang Y, Yates J, Freire E, Schroer TA, Zheng Y (2013) Nudel/NudE and Lis1 promote dynein and dynactin interaction in the context of spindle morphogenesis. Mol Biol Cell 24:3522–3533
Willins DA, Liu B, Xiang X, Morris NR (1997) Mutations in the heavy chain of cytoplasmic dynein suppress the nudF nuclear migration mutation of aspergillus nidulans. Mol Gen Genet 255:194–200
Wynshaw-Boris A (2007) Lissencephaly and LIS1: insights into the molecular mechanisms of neuronal migration and development. Clin Genet 72:296–304
Xiang X, Osmani AH, Osmani SA, Xin M, Morris NR (1995) NudF, a nuclear migration gene in aspergillus nidulans, is similar to the human LIS-1 gene required for neuronal migration. Mol Biol Cell 6:297–310
Yagi T (2009) Chapter 1 – Bioinformatic approaches to dynein heavy chain classification. In: King SM, Pazour GJ (eds) Methods in cell biology. Academic Press, New York, pp 1–9
Yang CH, Lambie EJ, Snyder M (1992) NuMA: an unusually long coiled-coil related protein in the mammalian nucleus. J Cell Biol 116:1303–1317
Yang Z, Tulu US, Wadsworth P, Rieder CL (2007) Kinetochore dynein is required for chromosome motion and congression independent of the spindle checkpoint. Curr Biol 17:973–980
Ye S, Fowler TW, Pavlos NJ, Ng PY, Liang K, Feng Y, Zheng M, Kurten R, Manolagas SC, Zhao H (2011) LIS1 regulates osteoclast formation and function through its interactions with dynein/dynactin and Plekhm1. PLoS One 6:e27285
Yeh TY, Quintyne NJ, Scipioni BR, Eckley DM, Schroer TA (2012) Dynactin's pointed-end complex is a cargo-targeting module. Mol Biol Cell 23:3827–3837
Yeh TY, Kowalska AK, Scipioni BR, Cheong FKY, Zheng M, Derewenda U, Derewenda ZS, Schroer TA (2013) Dynactin helps target polo-like kinase 1 to kinetochores via its left-handed beta-helical p27 subunit. EMBO J 32:1023
Zhang J, Yao X, Fischer L, Abenza JF, Penalva MA, Xiang X (2011) The p25 subunit of the dynactin complex is required for dynein-early endosome interaction. J Cell Biol 193:1245–1255
Zhang J, Qiu R, Arst HN Jr, Penalva MA, Xiang X (2014) HookA is a novel dynein-early endosome linker critical for cargo movement in vivo. J Cell Biol 204:1009–1026
Zhang K, Foster HE, Rondelet A, Lacey SE, Bahi-Buisson N, Bird AW, Carter AP (2017) Cryo-EM reveals how human cytoplasmic dynein is auto-inhibited and activated. Cell 169:1303–1314. e1318
Zimdahl B, Ito T, Blevins A, Bajaj J, Konuma T, Weeks J, Koechlein CS, Kwon HY, Arami O, Rizzieri D et al (2014) Lis1 regulates asymmetric division in hematopoietic stem cells and in leukemia. Nat Genet 46:245–252
Żyłkiewicz E, Kijańska M, Choi W-C, Derewenda U, Derewenda ZS, Stukenberg PT (2011) The N-terminal coiled-coil of Ndel1 is a regulated scaffold that recruits LIS1 to dynein. J Cell Biol 192:433–445

3 Changes in the Nuclear Envelope in Laminopathies

Subarna Dutta, Maitree Bhattacharyya, and Kaushik Sengupta

Abstract

Double-membrane-bound nucleus is the major organelle of every metazoan cell, which controls various nuclear processes like chromatin maintenance, DNA replication, transcription and nucleoskeleton-cytoskeleton coupling. Nuclear homeostasis depends on the integrity of nuclear membrane and associated proteins. Lamins, underlying the inner nuclear membrane (INM), play a crucial role in maintaining nuclear homeostasis. In this review, we have focussed on the disruption of nuclear homeostasis due to lamin A/C mutation which produces a plethora of diseases, termed as laminopathies.

Keywords

Laminopathies · Lamins · Linker of nucleoskeleton and cytoskeleton · Emerin · Lamin B receptor · Nuclear pore complex

S. Dutta
Biophysics & Structural Genomics Division, Saha Institute of Nuclear Physics, Kolkata, West Bengal, India

Department of Biochemistry, University of Calcutta, Kolkata, West Bengal, India

M. Bhattacharyya
Department of Biochemistry, University of Calcutta, Kolkata, West Bengal, India

K. Sengupta (✉)
Biophysics & Structural Genomics Division, Saha Institute of Nuclear Physics, Kolkata, West Bengal, India

3.1 Introduction

The nucleus is the major organelle of every eukaryotic cell and is responsible for maintaining cellular homeostasis. It is a double-membrane-bound structure which comprises of integral membrane proteins as well as nuclear lamins and confines densely packed chromatin. Over a hundred years since the discovery of nuclear envelop, the importance of the subject has increased dramatically with the discovery of numerous envelopathies. The double membranes are continuous with the endoplasmic reticulum (ER), and both act as site of attachment for the NPCs. Inner nuclear membrane is closely associated with a mesh or lamina which helps in tethering the chromosome at the periphery. This lamina is composed of A- and B-type lamins. The mutation of lamins particularly lamin A/C has led to the discovery of 14 different diseases collectively called laminopathies. For the past few decades, there has been a major impetus in getting deeper insights into lamin biology and hence changes in nuclear architecture in the light of laminopathies. In this review we have elucidated the pathological effects of laminopathies on the nuclear envelope and the membrane proteins associated there in.

K. Chattopadhyay, S. C. Basu (eds.), *Biochemical and Biophysical Roles of Cell Surface Molecules*, Advances in Experimental Medicine and Biology 1112, https://doi.org/10.1007/978-981-13-3065-0_3

3.2 Lamins

Underneath the inner nuclear membrane, there is a meshwork-like structure or "lamina" which imparts the structural rigidity to the nucleus. Lamins are major constituents of this meshwork. Lamins are present in all metazoans except plants. In mammals lamins are of two types A and B. A type includes lamins A and C which are alternate splice variants of the LMNA gene, whereas major B types are lamins B1 and B2 encoded by LMNB1 and LMNB2 genes, respectively. B-type lamins are expressed from embryonic stage while lamin A/C expressed in postembryonic stage and developmentally regulated (Rober et al. 1989). A- and B-type lamins with the exception of lamin C possess a –CAAX box at the C-terminus which marks the site for farnesylation required for membrane anchorage (Rusinol and Sinensky 2006). At first, the cysteine residue in CAAX is farnesylated followed by proteolytic digestion after the cysteine residue. Finally, the enzyme carboxymethyltransferase mediates carboxymethylation of the new C-terminal residue. A second proteolytic occurs in lamin A mediated by the zinc metalloproteinase Zmpste24 leading to the removal of another 15 amino acids. Thus, mature lamin A lacks the farnesyl group (Dechat et al. 2008), whereas B-type lamins retain the group and hence remain associated with the inner nuclear membrane.

The farnesylation of lamin A might recruit lamin A to the nuclear periphery by dint of hydrophobic interaction with the nuclear envelope (Hennekes and Nigg 1994). However, several reports using mice and/or cells expressing mutant forms of lamin A indicate that farnesylation is not prerequisite for recruitment to the nuclear envelope (Davies et al. 2010, 2011). This is evident from the fact that the non-farnesylated form of lamin A is expressed at the nuclear periphery (Davies et al. 2010, 2011).

Over a couple of decades, more than 400 mutations have been found in LMNA gene that are known to cause a plethora of diseases collectively called laminopathies (http://www.umd.be/LMNA/). On the contrary, two diseases are known to be caused by B-type lamins, of which one results from LMNB1 duplications and corresponding overexpression of lamin B1 (Padiath et al. 2006) and the other results from polymorphisms of LMNB2 (Hegele et al. 2006). A list of the laminopathies is given in Table 3.1.

These diseases primarily affect the muscle tissue in muscular dystrophies, adipocytes in lipodystrophy, peripheral neurons in neuropathies and multiple organs in premature ageing syndrome. In summary laminopathies which are caused by mutations in lamin A/C show tissue-specific phenotypes. But the commonality of laminopathies lies in the fact that nuclear envelope herniation or blebbing associated with chromatin aberration, loss of B-type lamins and mislocalization of NPCs, emerin and other inner nuclear membrane structural proteins thereby tend to weaken the nucleus in terms of its rigidity (Sullivan et al. 1999; Vigouroux et al. 2001). A cartoon showing the localization of major proteins in plasma and nuclear membrane as well as the interconnection is depicted in Fig. 3.1. There is also some evidence showing abnormal nuclear envelope dynamics due to misregulation of lamin B1 (Vergnes et al. 2004; Vargas et al. 2012). Nuclei from the liver of mice lacking lamin A/C or lamin B−/− in muscle tissue of flies showed the nuclear envelope disruption. Immunohistochemistry analysis of laminopathic patients also showed clearer sign of nuclear envelope rupturing. Ultrastructure analysis of nucleus from a laminopathic cardiomyocytes showed changed nuclear shape along with chromatin disorganization and appearance of sarcoplasmic organelles within nuclear matrix (Fidzianska et al. 2008; Gupta et al. 2010). Furthermore, improper post-translational modification also exerts unfavourable effect on nuclear shape. Y646F mutant of lamin A (which can undergo prenylation but not second proteolytic cleavage) showed abnormal nuclear morphology. It has also been shown that unprenylated prelamin A is not toxic to human embryonic kidney 293 cells. The toxicity of prenylated prelamin A may be due to its association and/or accumulation at the nuclear pore

Table 3.1 Laminopathies and affected tissues

Gene	Disease	Tissue affected
LMNA	Emery-Dreifuss muscular dystrophy (EDMD)	Striated muscle
LMNA	Limb-girdle muscular dystrophy	Striated muscle
LMNA	Congenital muscular dystrophy	Striated muscle
LMNA	Dilated cardiomyopathy (DCM) with/without conduction defect (CD)	Striated muscle
LMNA	Familial partial lipodystrophy type 2 (Dunnigan-type familial partial lipodystrophy)	Adipose tissue
LMNA	Atypical lipodystrophy	Adipose tissue
LMNA	Mandibuloacral dysplasia with type A lipodystrophy	Adipose tissue
LMNA	Hutchinson-Gilford progeria syndrome (premature ageing)	Multiple organs
LMNA	Atypical Werner syndrome	Multiple organs
LMNA	Charcot-Marie-tooth disease	Peripheral nerve
LMNB1	Autosomal dominant adult-onset leukodystrophy	Central nervous system
LMNB2	Acquired partial lipodystrophy (Barraquer-Simons syndrome)	Adipose tissue

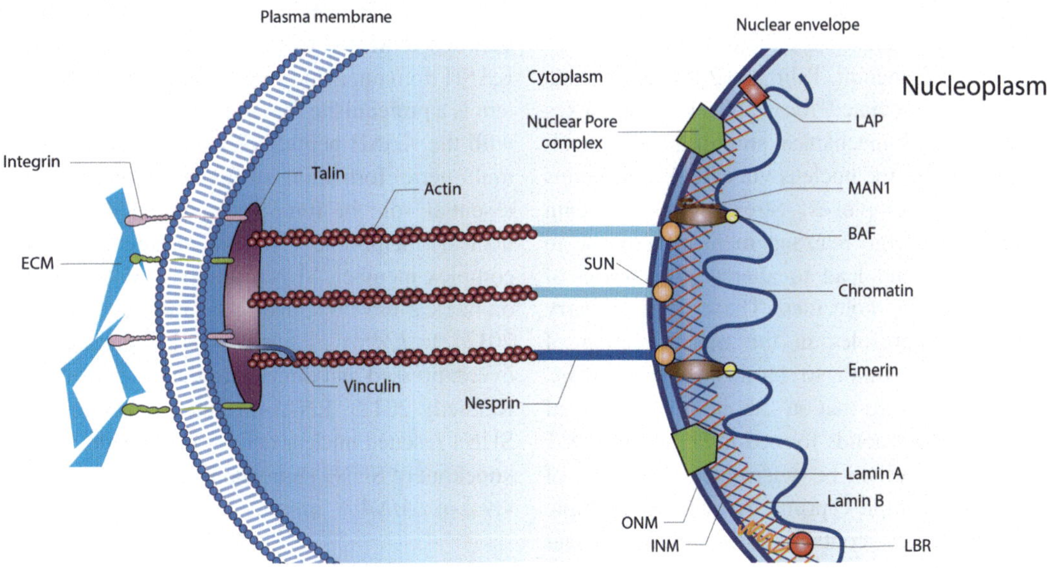

Fig. 3.1 Localization of major proteins in plasma membrane and nuclear membrane
Every nucleus has double-layered membrane, namely, outer nuclear membrane (ONM) and inner nuclear membrane (INM). ONM proteins make connection from nucleus to cytoplasm. INM proteins bridge nuclear envelope with chromatin thereby modulating gene expression and chromatin states. Nuclear lamina underneath the inner nuclear membrane is composed of lamin proteins. INM proteins like LBR, MAN 1, Emerin and LAP 2 interact with lamins and collectively act as nuclear shape modulator. Lamins A and B form 10 nm fibre which resides as a complex meshwork-like structure underneath INM. SUN proteins of INM interact with Nesprin proteins of ONM to form LINC complexes. Cytoplasmic actin by the help of focal adhesion molecules (Vinculin, Paxillin, Talin) via integrin molecules keeps connection with extracellular matrix

complex which could be partially reversed by farnesyl transferase inhibitors (Pan et al. 2007). Thus, nuclear envelope disruption could be an important contributor to the pathophysiology of laminopathies.

3.3 Linker of Nucleoskeleton and Cytoskeleton (LINC)

The additional factors which regulate nuclear structure integrity are proper organization of cytoplasmic actin and active tension on nuclear envelope which is maintained by proper signalling between extracellular matrix molecules (Houben et al. 2007) and the nucleus via LINC (linker of nucleoskeleton and cytoskeleton) complexes. At the cell membrane, focal adhesion (FA) complexes tethering the actin cytoskeleton collect the extracellular mechanical cues and pass down to the nucleus for effector function.

Paxillin, Vinculin, Talin and Zyxin are plasma membrane-associated focal adhesion complexes which transmit mechanical stimuli from extracellular matrix to the nucleus via cytoplasmic actins and LINC complexes. Stress fibres or actin filaments undergo rearrangement in response to shear stress and lead to altered distribution of focal adhesion molecules. These proteins carry out significant roles in the transformation of extracellular stimuli to biomechanical signals, but precise information about the types of complexes and bonds formed in between the FA molecules is yet to be determined. Members of the LINC complex primarily include multiple spectrin repeat containing Nesprins (nuclear envelope spectrin repeat proteins) and SUN proteins. Nesprins have a KASH domain. KASH domain (klarsicht/ANC-1/Syne-1 homology) or KLS domain is a highly hydrophobic domain of 60 amino acid residue which helps it to embed into nuclear envelope. LINC complex proteins help in transmitting the extracellular mechanical cue across the nucleus via the nuclear membrane as depicted in Fig. 3.1. Interestingly, major murine KASH domain proteins were termed as Syne-1 and Syne-2. Nesprin 1 and 2 are very crucial players for nuclear envelope integrity (Zhang et al. 2007a). A deletion of N-terminal domain of Syne-2/Nesprin 2 has also been shown to disrupt nuclear architecture (Luke et al. 2008). Missense mutations in Nesprins 1 and 2 also lead to EDMD-like phenotypes in some cases. Moreover, mutations in Nesprin 1 are shown to be associated with DCM-CD. Minor isoforms like Nesprins 3 and 4 encoded by SYNE-3 and SYNE-4, respectively, are less characterized with respect to disease-causing mutations, although there are reports of hearing loss associated with Nesprin 4 mutations.

On the other hand, deletion of KASH domain of Syne-1 severely hampers both synaptic and nonsynaptic nuclear anchorage (Grady et al. 2005; Zhang et al. 2007b). SUN proteins are inner nuclear membrane-associated proteins which interact with lamins and NPCs via N-terminal domain, whereas the C-terminal domain inserts itself into the perinuclear space between INM and ONM and crosstalks with the KASH domain. Multimerization of the SUN protein is a prerequisite for facilitating the interaction with the KASH peptide (Zhou et al. 2012). This multimeric formation of SUN might play an essential role in force transduction across the nuclear membrane. Defects in another LINC complex member, SUN1, have a profound effect on nuclear envelope disruption (Tapley and Starr 2013). In *LMNA*$^{-/-}$ mice, SUN1 is dramatically overexpressed and directed to the Golgi (Suh and Kennedy 2012). RNAi-mediated knockdown of SUN1 rescued nuclear defects in cell culture, and knockout of SUN1 significantly extended the survival of *LMNA*$^{-/-}$ mice.

3.4 Emerin

Another lamin-associated unclear membrane protein of cardinal importance is emerin which has a significant role in maintaining nuclear stability. Emerin is a component of LEM 2 (LAP2-Emerin-MAN 1) family of proteins present in the nuclear envelope. Emerin protein is comprised of N-terminal serine-rich hydrophilic region (1–221) which helps it to insert in the membrane. This domain is directly responsible for its interactions with SUN proteins, actins, lamins and DNA-binding protein like BAF (barrier-to-autointegration factor). Interestingly, residues 170–220 determine the homotypic associations of emerin. Emerin remains directly or indirectly connected with lamin A, chromatin and LINC complexes. Mutation in emerin has a vital role for X-linked Emery-Dreifuss muscular dystrophy. Emerin-deficient cells show abnormal-shaped nucleus and improper mechanotransduction (Lammerding et al. 2005; Rowat et al. 2006). There is another report which has shown that lamin A/C–emerin interaction is crucial for its localization on nuclear envelope (Vaughan et al. 2001). Interaction of inner nuclear membrane protein LEM 2 with lamins is also crucial for maintaining normal nuclear shape (Ulbert et al. 2006). As emerin tends to mislocalize to the endoplasmic reticulum in EDMD, its role at the nuclear envelope is very vital for the pathogenesis in EDMD. In autosomal EDMD caused by mutation of LMNA, there is fluctuation of emerin distribution between nuclear envelope and endoplasmic reticulum (Sullivan et al. 1999).

3.5 Lamin B Receptor (LBR)

Lamin B receptor (LBR) is one of the most well-characterized integral membrane proteins of INM and is believed to interact with lamin B (Zwerger et al. 2008). It comprises of multiple transmembrane hydrophobic domain and bears structural similarity to sterol reductase enzymes (Holmer et al. 1998). LBR has a bipartite NLS signal and an N-terminal nucleoplasmic domain rich in basic sequences. This N-terminal domain is crucial for the incorporation of LBR into the nuclear envelop. LBR has a propensity to oligomerize in vitro and form insoluble aggregates, a property characteristic of integral membrane proteins. This self-association is modulated by phosphorylation and hyperphosphorylation of LBR by cdk 1 and is observed at the beginning of mitosis (Nikolakaki et al. 1997). It has been shown that LBR is involved in peripheral heterochromatin tethering vis lamin A/C in early stages of development (Solovei et al. 2013). Mutations throughout LBR protein cause Pelger-Huët anomaly while the same in the C-terminal hydrophobic transmembrane domains lead to Greenberg skeletal dysplasia in human being (Clayton et al. 2010; Hoffmann et al. 2002). The latter is primarily associated with perturbed sterol reductase activity. Therefore one of the ways by which LBR maintains nuclear architecture is by dint of its sterol reductase activity which in turn is necessary to nuclear lipid rafts or microdomains. This activity is visualized at the onset of mitosis in animal cells when the nuclear envelope ruptures. Secondly, heterochromatin tethering to the INM is another important role of LBR in nuclear homeostasis. A point mutation in lamin A has been reported to result in LBR relocation from the nuclear envelope to the endoplasmic reticulum (Stegh et al. 2000), but lamin A null cells appear to have normal retention of LBR in the nuclear envelope (Stewart et al. 2007). These evidences prove a more indirect coupling between LBR and lamin A/C. There is increasing evidence that gene silencing occurs at the nuclear envelope (Gaines et al. 2008) promoted, in part, by integral proteins of the nuclear envelope. It is tempting to speculate that increased LBR content may augment repression of the lamin A/C genes and thus facilitate sequestration of the epigenetically repressed lamin A/C genes to the peripheral nuclear heterochromatin.

3.6 Nuclear Pore Complex (NPC)

Nuclear pore complexes provide another link of communication between the cytoplasm and nuclear compartment. These are selective permeable barrier which allow small molecules to pass through by slow diffusion where larger macromolecules are gated by nuclear localization signals (NLS) which are recognized by specific transporters to be ferried across the NPCs. Nuclear pore complexes are approximately 125 MDa structure embedded on nuclear envelope by the help of lamins (Eibauer et al. 2015). Nuclear pore complexes are formed of 30 different nucleoporins (Nups). Its structure can be defined by a central framework which carries a transmembrane part and nuclear basket that protrudes in the cytoplasm and nucleoplasm. NPCs are composed of scaffold-like structures which constitute the membrane annulus. Numerous FG repeats in the channel make up the semipermeable barrier. These complexes are considered to be somewhat mobile in yeasts but are firmly anchored to the lamina in higher complex metazoan cells (Daigle et al. 2001). The association of NPCs to the LINC was shown to be mediated by Nup 153, a nucleoporin, and SUN1 where deletion of SUN1 led to clustering of NPCs. On a similar note, Nup 153 and Nup 53 were shown to interact with B-type lamins (Hawryluk-Gara et al. 2005; Smythe et al. 2000). Kelley et al. (2015) have shown that improper nuclear compartmentalization mediated by Hutchinson-Gilford progeria syndrome patients with dominant negative form of lamin A resulted in disruption of nuclear transport (Kelley et al. 2015). NPC assembly requires at least 450 NUP protomers aided by integral inner nuclear membrane proteins like Lap2α, Emerin, MAN1 (Yewdell et al. 2011) and SUN (Sad 1, UNC 84) domain proteins (Talamas and Hetzer 2011). HGPS is associated with clustering of nuclear pore complex among other phenotypes (Goldman et al. 2004). Furthermore, the interaction of Nup 88 and Nup 153 was shown to be disrupted in the presence of mutation in Ig-fold domain of lamin A/C causing EDMD and FPLD (Lussi et al. 2011; Al-Haboubi et al. 2011).

3.7 Conclusion and Perspectives

The nuclear envelope in the metazoan cell is studded with at least 60 well-characterized NET (nuclear envelope transmembrane) with diverse functions but committed to maintaining nuclear integrity and homeostasis. Some of these envelope proteins are connected to the cytoskeleton via numerous proteins. In this way the cell can be visualized as a complex circuitry which is fine-tuned to respond to external cues via effectors from the nucleus. This is epitomized in Fig. 3.1 which is a simplified cartoon depicting some of these proteins discussed in the review. The field of envelopathies, in particular laminopathies, has unravelled some of the crosstalk mechanisms between these proteins and tried to establish the pathophysiology of the disease progressions. However, we are at the dawn of our knowledge of molecular signalling pathways underlying laminopathies. Understanding this would significantly help in the design and synthesis of antagonists to combat laminopathies thereby improving public health.

References

Al-Haboubi T, Shumaker DK, Koser J, Wehnert M, Fahrenkrog B (2011) Distinct association of the nuclear pore protein Nup153 with A- and B-type lamins. Nucleus 2:500–509

Clayton P, Fischer B, Mann A, Mansour S, Rossier E, Veen M, Lang C, Baasanjav S, Kieslich M, Brossuleit K et al (2010) Mutations causing Greenberg dysplasia but not Pelger anomaly uncouple enzymatic from structural functions of a nuclear membrane protein. Nucleus 1:354–366

Daigle N, Beaudouin J, Hartnell L, Imreh G, Hallberg E, Lippincott-Schwartz J, Ellenberg J (2001) Nuclear pore complexes form immobile networks and have a very low turnover in live mammalian cells. J Cell Biol 154:71–84

Davies BS, Barnes RH 2nd, Tu Y, Ren S, Andres DA, Spielmann HP, Lammerding J, Wang Y, Young SG, Fong LG (2010) An accumulation of non-farnesylated prelamin A causes cardiomyopathy but not progeria. Hum Mol Genet 19:2682–2694

Davies BS, Coffinier C, Yang SH, Barnes RH 2nd, Jung HJ, Young SG, Fong LG (2011) Investigating the purpose of prelamin A processing. Nucleus 2:4–9

Dechat T, Pfleghaar K, Sengupta K, Shimi T, Shumaker DK, Solimando L, Goldman RD (2008) Nuclear lamins: major factors in the structural organization and function of the nucleus and chromatin. Genes Dev 22:832–853

Eibauer M, Pellanda M, Turgay Y, Dubrovsky A, Wild A, Medalia O (2015) Structure and gating of the nuclear pore complex. Nat Commun 6:7532

Fidzianska A, Bilinska ZT, Tesson F, Wagner T, Walski M, Grzybowski J, Ruzyllo W, Hausmanowa-Petrusewicz I (2008) Obliteration of cardiomyocyte nuclear architecture in a patient with LMNA gene mutation. J Neurol Sci 271:91–96

Gaines P, Tien CW, Olins AL, Olins DE, Shultz LD, Carney L, Berliner N (2008) Mouse neutrophils lacking Lamin B-receptor expression exhibit aberrant development and lack critical functional responses. Exp Hematol 36:965–976

Goldman RD, Shumaker DK, Erdos MR, Eriksson M, Goldman AE, Gordon LB, Gruenbaum Y, Khuon S, Mendez M, Varga R et al (2004) Accumulation of mutant Lamin A causes progressive changes in nuclear architecture in Hutchinson-Gilford progeria syndrome. Proc Natl Acad Sci USA 101:8963–8968

Grady RM, Starr DA, Ackerman GL, Sanes JR, Han M (2005) Syne proteins anchor muscle nuclei at the neuromuscular junction. Proc Natl Acad Sci USA 102:4359–4364

Gupta P, Bilinska ZT, Sylvius N, Boudreau E, Veinot JP, Labib S, Bolongo PM, Hamza A, Jackson T, Ploski R et al (2010) Genetic and ultrastructural studies in dilated cardiomyopathy patients: a large deletion in the Lamin A/C gene is associated with cardiomyocyte nuclear envelope disruption. Basic Res Cardiol 105:365–377

Hawryluk-Gara LA, Shibuya EK, Wozniak RW (2005) Vertebrate Nup53 interacts with the nuclear lamina and is required for the assembly of a Nup93-containing complex. Mol Biol Cell 16:2382–2394

Hegele RA, Cao H, Liu DM, Costain GA, Charlton-Menys V, Rodger NW, Durrington PN (2006) Sequencing of the reannotated LMNB2 gene reveals novel mutations in patients with acquired partial lipodystrophy. Am J Hum Genet 79:383–389

Hennekes H, Nigg EA (1994) The role of isoprenylation in membrane attachment of nuclear lamins. A single point mutation prevents proteolytic cleavage of the Lamin A precursor and confers membrane binding properties. J Cell Sci 107(Pt 4):1019–1029

Hoffmann K, Dreger CK, Olins AL, Olins DE, Shultz LD, Lucke B, Karl H, Kaps R, Muller D, Vaya A et al (2002) Mutations in the gene encoding the Lamin B receptor produce an altered nuclear morphology in granulocytes (Pelger-Huet anomaly). Nat Genet 31:410–414

Holmer L, Pezhman A, Worman HJ (1998) The human Lamin B receptor/sterol reductase multigene family. Genomics 54:469–476

Houben F, Ramaekers FC, Snoeckx LH, Broers JL (2007) Role of nuclear lamina-cytoskeleton interactions in the maintenance of cellular strength. Biochim Biophys Acta 1773:675–686

Kelley K, Knockenhauer KE, Kabachinski G, Schwartz TU (2015) Atomic structure of the Y complex of the nuclear pore. Nat Struct Mol Biol 22:425–431

Lammerding J, Hsiao J, Schulze PC, Kozlov S, Stewart CL, Lee RT (2005) Abnormal nuclear shape and impaired mechanotransduction in emerin-deficient cells. J Cell Biol 170:781–791

Luke Y, Zaim H, Karakesisoglou I, Jaeger VM, Sellin L, Lu W, Schneider M, Neumann S, Beijer A, Munck M et al (2008) Nesprin-2 Giant (NUANCE) maintains nuclear envelope architecture and composition in skin. J Cell Sci 121:1887–1898

Lussi YC, Hugi I, Laurell E, Kutay U, Fahrenkrog B (2011) The nucleoporin Nup88 is interacting with nuclear Lamin A. Mol Biol Cell 22:1080–1090

Nikolakaki E, Meier J, Simos G, Georgatos SD, Giannakouros T (1997) Mitotic phosphorylation of the Lamin B receptor by a serine/arginine kinase and p34(cdc2). J Biol Chem 272:6208–6213

Padiath QS, Saigoh K, Schiffmann R, Asahara H, Yamada T, Koeppen A, Hogan K, Ptacek LJ, Fu YH (2006) Lamin B1 duplications cause autosomal dominant leukodystrophy. Nat Genet 38:1114–1123

Pan Y, Garg A, Agarwal AK (2007) Mislocalization of prelamin A Tyr646Phe mutant to the nuclear pore complex in human embryonic kidney 293 cells. Biochem Biophys Res Commun 355:78–84

Rober RA, Weber K, Osborn M (1989) Differential timing of nuclear Lamin A/C expression in the various organs of the mouse embryo and the young animal: a developmental study. Development 105:365–378

Rowat AC, Lammerding J, Ipsen JH (2006) Mechanical properties of the cell nucleus and the effect of emerin deficiency. Biophys J 91:4649–4664

Rusinol AE, Sinensky MS (2006) Farnesylated lamins, progeroid syndromes and farnesyl transferase inhibitors. J Cell Sci 119:3265–3272

Smythe C, Jenkins HE, Hutchison CJ (2000) Incorporation of the nuclear pore basket protein nup153 into nuclear pore structures is dependent upon lamina assembly: evidence from cell-free extracts of Xenopus eggs. EMBO J 19:3918–3931

Solovei I, Wang AS, Thanisch K, Schmidt CS, Krebs S, Zwerger M, Cohen TV, Devys D, Foisner R, Peichl L et al (2013) LBR and Lamin A/C sequentially tether peripheral heterochromatin and inversely regulate differentiation. Cell 152:584–598

Stegh AH, Herrmann H, Lampel S, Weisenberger D, Andra K, Seper M, Wiche G, Krammer PH, Peter ME (2000) Identification of the cytolinker plectin as a major early in vivo substrate for caspase 8 during CD95- and tumor necrosis factor receptor-mediated apoptosis. Mol Cell Biol 20:5665–5679

Stewart CL, Kozlov S, Fong LG, Young SG (2007) Mouse models of the laminopathies. Exp Cell Res 313:2144–2156

Suh Y, Kennedy BK (2012) Dialing down SUN1 for laminopathies. Cell 149:509–510

Sullivan T, Escalante-Alcalde D, Bhatt H, Anver M, Bhat N, Nagashima K, Stewart CL, Burke B (1999) Loss of A-type Lamin expression compromises nuclear envelope integrity leading to muscular dystrophy. J Cell Biol 147:913–920

Talamas JA, Hetzer MW (2011) POM121 and Sun1 play a role in early steps of interphase NPC assembly. J Cell Biol 194:27–37

Tapley EC, Starr DA (2013) Connecting the nucleus to the cytoskeleton by SUN-KASH bridges across the nuclear envelope. Curr Opin Cell Biol 25:57–62

Ulbert S, Antonin W, Platani M, Mattaj IW (2006) The inner nuclear membrane protein Lem2 is critical for normal nuclear envelope morphology. FEBS Lett 580:6435–6441

Vargas JD, Hatch EM, Anderson DJ, Hetzer MW (2012) Transient nuclear envelope rupturing during interphase in human cancer cells. Nucleus 3:88–100

Vaughan A, Alvarez-Reyes M, Bridger JM, Broers JL, Ramaekers FC, Wehnert M, Morris GE, Whitfield WGF, Hutchison CJ (2001) Both emerin and Lamin C depend on Lamin A for localization at the nuclear envelope. J Cell Sci 114:2577–2590

Vergnes L, Peterfy M, Bergo MO, Young SG, Reue K (2004) Lamin B1 is required for mouse development and nuclear integrity. Proc Natl Acad Sci USA 101:10428–10433

Vigouroux C, Auclair M, Dubosclard E, Pouchelet M, Capeau J, Courvalin JC, Buendia B (2001) Nuclear envelope disorganization in fibroblasts from lipodystrophic patients with heterozygous R482Q/W mutations in the Lamin A/C gene. J Cell Sci 114:4459–4468

Yewdell WT, Colombi P, Makhnevych T, Lusk CP (2011) Lumenal interactions in nuclear pore complex assembly and stability. Mol Biol Cell 22:1375–1388

Zhang Q, Bethmann C, Worth NF, Davies JD, Wasner C, Feuer A, Ragnauth CD, Yi Q, Mellad JA, Warren DT et al (2007a) Nesprin-1 and -2 are involved in the pathogenesis of Emery Dreifuss muscular dystrophy and are critical for nuclear envelope integrity. Hum Mol Genet 16:2816–2833

Zhang X, Xu R, Zhu B, Yang X, Ding X, Duan S, Xu T, Zhuang Y, Han M (2007b) Syne-1 and Syne-2 play crucial roles in myonuclear anchorage and motor neuron innervation. Development 134:901–908

Zhou Z, Du X, Cai Z, Song X, Zhang H, Mizuno T, Suzuki E, Yee MR, Berezov A, Murali R et al (2012) Structure of Sad1-UNC84 homology (SUN) domain defines features of molecular bridge in nuclear envelope. J Biol Chem 287:5317–5326

Zwerger M, Herrmann H, Gaines P, Olins AL, Olins DE (2008) Granulocytic nuclear differentiation of Lamin B receptor-deficient mouse EPRO cells. Exp Hematol 36:977–987

Part II

Biophysics of Membrane Proteins and Cell Surface Macromolecules

4 The Effect of Nanoparticles on the Cluster Size Distributions of Activated EGFR Measured with Photobleaching Image Correlation Spectroscopy

Chiara Paviolo, James W. M. Chon, and Andrew H. A. Clayton

Abstract

The epidermal growth factor receptor (EGFR) is an important cell surface receptor in normal physiology and disease. Recent work has shown that EGF-gold nanoparticle conjugates can influence cell behaviour, but the underlying mechanism at the receptor quaternary structural level remains poorly understood.

In the present work, the cluster density and cluster size of activated (phosphorylated) EGFR clusters in HeLa cells were determined with photobleaching image correlation spectroscopy. EGFR activation was probed via immunofluorescence-detected phosphorylation of tyrosines (pY-mAb) located in the kinase domain of EGFR (Y845) and at the EGFR cytoplasmic tail (Y1173). Cell activation was probed via nuclear extracellular-regulated kinase (ERK) phosphorylation. The cluster size of activated EGFR was 1.3–2.4 pY-mAb/cluster in unstimulated HeLa cells. EGF or nanorod treatment led to an increase in EGFR oligomers containing multiple phosphotyrosines (>2 phosphotyrosines per EGFR oligomer, average cluster size range = 3–5 pY-mAb/cluster) which paralleled increases in nuclear p-ERK. In contrast, EGF-nanorods decreased the contribution from higher-order phospho-clusters and decreased nuclear p-ERK relative to the nanorod control. These studies provide direct evidence that targeted nanotechnology can manipulate receptor organization and lead to changes in receptor activation and subsequent signalling processes.

Keywords

Gold nanoparticles · Epidermal growth factor receptor · Phosphorylation · Higher-order oligomers · Y845 · Y1173 · MAPK · Image correlation spectroscopy

4.1 Introduction

The epidermal growth factor receptor (EGFR, also known as ErbB1 or HER-1) is a member of the EGFR/ErbB/HER family of type I transmembrane tyrosine kinase receptors, which also includes ErbB2/HER-2/Neu, ErbB3/HER-3 and ErbB4/HER-4. EGFRs play an essential role in organ development by regulating both the differentiation and growth of cells and tissues. However, EGFR overexpression and dysfunction occur in malignant development and uncontrolled cell proliferation of many solid tumours (Flynn et al. 2009).

EGFR consists of a 612 kDa ectodomain (which mediates ligand binding and

C. Paviolo · J. W. M. Chon (✉) · A. H. A. Clayton (✉)
Centre for Micro-Photonics, Faculty of Science, Engineering and Technology, Swinburne University of Technology, Hawthorn, VIC, Australia
e-mail: jchon@swin.edu.au; aclayton@swin.edu.au

K. Chattopadhyay, S. C. Basu (eds.), *Biochemical and Biophysical Roles of Cell Surface Molecules*, Advances in Experimental Medicine and Biology 1112, https://doi.org/10.1007/978-981-13-3065-0_4

dimerization), a short transmembrane domain (that anchors the receptor to the cell membrane), an intracellular juxta-membrane domain which regulates various functional aspects of EGFR (such as ligand internalization and downregulation of the receptor), a tyrosine kinase domain essential for the functional activation of the receptor and a carboxyl-terminal tail (C-tail) containing several tyrosine (Tyr) residues (Flynn et al. 2009; Kovacs et al. 2015). Different members of the peptide growth factors, such as the epidermal growth factor (EGF), serve as agonists for the EGFR. Ligand binding causes homo- or hetero-dimerization of the receptor, leading to activation of the intracellular tyrosine kinase domain and autophosphorylation of the Tyr residues in the kinase domain and C-tail (Wilson et al. 2009). These phosphorylated Tyr sites serve as docking sites for effector proteins that transduce the EGF signals to generate particular biological responses through the activation of specific intracellular pathways (e.g. the mitogen-activated protein kinase cascade, MAPK) (Kolch and Pitt 2010).

In addition to dimers, the clustering of EGFR in higher-order oligomers or assemblies has attracted increased interest (Clayton et al. 2005, 2008; Kozer et al. 2013, 2014; Needham et al. 2016; Huang et al. 2016). Recent work has shown that EGF-induced higher-order oligomers control EGFR phosphorylation (Clayton et al. 2008; Kozer et al. 2013; Needham et al. 2016; Huang et al. 2016) and assembly of signalling complexes (Kozer et al. 2014; Needham et al. 2016). Moreover, EGFR pre-clustering, mediated by the tumour cell membrane environment, has also been suggested to be an important determinant of tumorigenicity (Clayton et al. 2007; Wang et al. 2014). Consequently, there is a pressing need to measure and control EGFR clustering.

One strategy is to perturb the EGFR clustering using a nanotechnology approach. Indeed, perturbing and monitoring cell receptor conformation and activation with the aid of nanoplatforms are an emerging field of research that has gained more and more attention over the past recent years (Paviolo et al. 2015; Delcassian et al. 2013; Peckys et al. 2013; Crow et al. 2011; Dreaden et al. 2012). Amongst all of the available structures, gold nanoparticles (Au NPs) hold great potential as nano-perturbation agents for cell surface macromolecules. Aside from their excellent optical properties, Au NPs provide relatively favourable biocompatibility and can be produced over a wide range of diameters, aspect ratios and surface functionalization options (Dreaden et al. 2012). Recently, Paviolo et al. showed the possibility to use EGF-conjugated Au NPs as molecular spacers to control EGFR clustering and proliferation in HeLa cervical cancer cells (Paviolo et al. 2015). EGF-conjugated particles of diameter < 50 nm could increase proliferation of HeLa cells, but a 100 nm EGF-nanorod conjugate inhibited proliferation (Paviolo et al. 2015). Although NPs proved to influence cell behaviour, little is known about how NPs can affect the spatial organization of EGFR clusters (i.e. dimers versus oligomers), EGFR phosphorylation levels and the functional consequences on the signalling cascades at the level of single cells.

In this manuscript, we addressed these questions by measuring the influence of EGF-conjugated gold nanorods (EGF-NRs) on clustering of activated receptors and downstream signalling. Photobleaching image correlation spectroscopy provided us with a quantitative measure of changes in activated receptor organization in terms of cluster densities and cluster sizes.

4.2 Material and Methods

EGF-NR Functionalization EGF and Au NR functionalization were as previously described (Paviolo et al. 2015).

Cell Culture, Staining and Imaging Human cervical adenocarcinoma (HeLa) cells (American Type Culture Collection, USA) were grown in Dulbecco's Modified Eagle Medium (DMEM; Sigma Aldrich, AU) containing 10% (v/v) foetal bovine serum (FBS; Life Technologies, AU), 1% (v/v) L-glutamine (Life Technologies, AU), 1% (v/v) penicillin/streptomycin (Life Technologies, AU) and 0.5% (v/v) amphotericin B (Life

Technologies, AU). When 70% confluent, 1.5×10^4 cells·cm^{-2} were seeded in a μ-slide chamber (Dksh, AU) with 2.6 g·dL^{-1} of bovine serum albumin (BSA – Sigma Aldrich, AU) and serum-starved for 24 h. Samples were then treated with 24 nM EGF-T, NR and EGF-NRs for 15 min in a humidified atmosphere (95% (v/v) air, 5% (v/v) carbon dioxide) at 37 °C.

For EGFR phosphorylation imaging and analysis, after stimulation, cells were immediately fixed in ice-cold ethanol-acetic acid [1:1] for 60 s at room temperature and then washed three times with PBS for 5 min. Unreacted binding sites were then blocked with 5% (w/v) BSA for 60 min, followed by an overnight incubation with anti-EGFR-phospho Y845 (1:50 in 1% BSA in PBS; Abcam, AU) and anti-EGFR-phospho Y1173 (1:100 in 1% BSA in PBS; Abcam, AU) at 4 °C. On the following day, cells were incubated in the dark with a fluorescein isothiocyanate (FITC)-conjugated antibody (1:1000 in 1% BSA in PBS; Abcam, AU) for 3 h. Each labelling stage was followed by two rounds of washing for 5 min with PBS to remove excess unbound antibody.

For confocal imaging of ppErk1/2, cells were fixed in formalin for 10 min, followed by three rounds of washing with PBS for 5 min. Samples were then made permeable with Triton X-100 (0.1% v/v in PBS; Sigma Aldrich, AU) for 30 min and again washed three times with PBS for 5 min. Unreacted binding sites were then blocked with 5% (w/v) BSA for 60 min, followed by an overnight incubation with Anti-Erk1 (pT202/pY204) + Anti-Erk2 (pT185/pY187) (1:200 in 1% BSA in PBS; Abcam, AU) at 4 °C. On the following day, cells were incubated in the dark with a FITC-conjugated antibody (1:1000 in 1% BSA in PBS; Abcam, AU) for 3 h. After that, cell nuclei were marked with PI (500 nM in PBS; Invitrogen, AU) for 15 min at room temperature. Each labelling stage was followed by three rounds of washing for 5 min with PBS to remove excess unbound antibody. Three independent measurements ($n = 3$) were performed for each type of stimulation.

Confocal Microscopy Confocal images were acquired with a 100× oil immersion objective (NA = 1.35) on an inverted confocal microscope (FluoView FV1000; Olympus, AU). An argon-ion laser (488 nm) was used to excite the FITC-labelled antibody, while PI was stimulated with a green helium-neon laser (543 nm). For both fluorophores, the recommended filter sets were used for imaging. All of the confocal parameters (laser power, sampling speed and confocality) were kept constant for each experimental part. DIC images were also acquired during the ppErk1/2 analysis. Post-processing of the data was performed with Fiji/ImageJ software and custom-coded programmes developed with MATLAB. Single-cell analysis on cytoplasmic and nuclear activation of ERK (i.e. ppErk1/2) was in differential interference contrast mode (DIC). Cell borders were located with DIC, while nuclear regions were identified with PI staining (Fig. 4.4b).

Photobleaching Image Correlation Spectroscopy Sixty sequential images for each cell were acquired in a time series with the confocal microscope, using a resolution of 640 × 640 pixels and a 2 μs·pixel^{-1} sampling speed. Pixel resolution varied between the samples, being on average approximately 0.06 μm·pixel^{-1}. Images were initially cropped in different region of interests (ROIs) and corrected for background noise, followed by a rescaling to 128 × 128 pixels using a bilinear average process. The autocorrelation analysis was performed using the built-in FD math function in ImageJ. The autocorrelation function was then normalized by the square of the average intensity, the number of pixels in the ROI and the point-spread function of the microscope. The cluster density (CD) of fluorescent clusters can be calculated with the corrected value of the autocorrelation function at zero-lag g(0,0) using Eq. 4.1:

$$g(0,0) = 1/\langle CD \rangle \tag{4.1}$$

Plots of CD as a function of fractional fluorescent intensity remaining (p) were fit to Eq. 4.2:

$$\mathrm{CD}(p)_{\{J\}} = \frac{\mathrm{CD}Jp}{1 + (J - 1)p} \tag{4.2}$$

where CD is the cluster density (at $p = 1$) in units of clusters per square micron and J is the average cluster size (or average brightness) in units of number of fluorophores per cluster. Data were fit in the range of $p = 0.2 - 1$.

Our analysis delivers the cluster size in units of number of fluorophores per cluster. To convert to oligomeric state, we need to know the number of fluorophores per antibody. To convert the cluster size values to the number of antibodies per cluster, cluster size values (in units of fluorophores per cluster) were divided by the number of fluorophores per antibody, estimated to be 1.5 fluorophores/antibody from pbICS experiments of antibodies deposited on a glass slide.

Photobleaching data from several sets of cells (typically 20 cells per treatment condition) were analysed to obtain population mean CD (<CD>), population mean J (<J>), average density of antibodies (<mAb>) and densities of antibodies in lower-order phospho-cluster and higher-order phospho-clusters.

The population mean CD ($p = 1$, units: clusters per square micron) is defined as.

$$< \mathrm{CD} > = \sum\nolimits_1^n \mathrm{CD}_n / n \tag{4.3}$$

The population mean J (units: number of antibodies per cluster) is defined as

$$< J > = \sum\nolimits_1^n J_n^2 \mathrm{CD}_n / \sum\nolimits_1^n J_n \mathrm{CD}_n \tag{4.4}$$

The average density of antibodies (units: antibodies per square micron) is given as

$$< \mathrm{mAb} > = < \mathrm{CD} >< J > \tag{4.5}$$

Data was also binned to reflect populations with cluster size of 1–2 antibodies per cluster and populations with >2 antibodies per cluster. These data were represented as histograms with total density of antibodies on the vertical axis and 1–2 and > 2 antibodies per cluster categories on the horizontal axis.

4.3 Determination of the Average Cluster Size of EGFR

The pbICS experiments deliver cluster densities and cluster sizes of the phosphorylated receptors, as defined by binding of a phospho-specific immunoglobulin. The binding of the immunoglobulin is not 100%, however, so a correction for partial binding needs to be made to extract average cluster sizes of EGFR.

To determine the average cluster size of EGFR we assume that the antibody binding is sub-stoichiometric (due to inactive receptors or limited antibody accessibility).

The average fraction of bound antibody is given by:

$$\text{Fraction bound} = < \mathrm{mAb} > / < \mathrm{EGFR} > \tag{4.6}$$

where <mAb> is the average density of detected antibodies (Eq. 4.5) and <EGFR> is the average density of EGFR on HeLa cells. We used a value for <EGFR> of 50 EGFRs/square micron based on the literature value for the expression level of HeLa cells (50,000 receptors/cell) (Berkers et al. 1991) and typical cell surface area of 1000 square microns.

For an average EGFR cluster size of J_{EGFR}, fractional antibody binding will create a new cluster size distribution given by the binomial formula.

For example, if we assume $J_{\mathrm{EGFR}} = 2$ and use a fraction antibody bound of 0.5, the distribution of detected antibody-bound clusters is:

$$\text{Zero antibody bound}: 1(1-\mathrm{f})(1-\mathrm{f}) = 1(1/2)(1/2) = 1/4$$

$$\text{One antibody bound}: 2(\mathrm{f})(1-\mathrm{f}) = 2(1/2)(1/2) = 1/2$$

$$\text{Two antibodies bound}: 1(\mathrm{f})(\mathrm{f}) = 1/4$$

The average cluster size of the detected antibodies in clusters is then given by Eq. (4.4) as

$$< J >= (1^*0.5 + 4^*0.25)/(1^*0.5 + 2^*0.25) = 1.5/1 = 1.5 \text{ antibodies per cluster.}$$

We begin by assuming a value for J_{EGFR} and use experiment to guide the value for fraction antibody bound (Eq. 4.6) and then evaluate J_{theory} using the binomial formula and Eq. (4.4). The value of J observed experimentally was then compared with J_{theory}. This process was repeated iteratively for each experimental $<J>$ and fraction antibody binding to estimate $<J_{EGFR}>$.

We note that a model involving an average cluster size is of course approximate. More sophisticated approaches, which take into multiple oligomeric states or multivalency of nanoparticles, await further experimental and theoretical developments.

Statistics Results were expressed as mean and standard error of the mean. The variance of each group of data was compared with a two-tailed F-test following an F-distribution under the null hypothesis (data sets have equal variances). A two-tailed *t*-test following a Student's *t* distribution under null hypothesis (data sets with equal means) was used to compare the different groups of data. During the data analysis, a probability lower than 0.05 ($p < 0.05$) was considered to be statistically significant.

4.4 Results and Discussion

4.4.1 Preactivated EGFR Oligomers and EGF-Activated EGFR Oligomers in HeLa Cells

Figure 4.1a shows a schematic of EGFR monomer structure illustrating the different receptor sections and intracellular EGFR tyrosine residues. In our study, we targeted phosphorylation of Y845 and Y1173 to gain information on different phosphorylation sites within the receptor. Y845 is located in the activation loop of the kinase domain and phosphorylation of Y845 is responsible for stabilizing the active kinase (by destabilizing an (inactive) disordered conformation of the kinase) (Shaw et al. 2014). Phosphorylation of EGFR Y845 in EGFR (which occurs via EGFR activation and/or by phosphorylation by c-Src) is also linked to EGFR signal amplification, mitogenesis and cell proliferation (Tice et al. 1999; Shaw et al. 2014). Y1173 is located in the c-terminal portion of the EGFR cytoplasmic domain, distal from the kinase domain. Phosphorylation of Y1173 is associated with binding of SH2-domain proteins including PLCγ and SHIP.

HeLa cells were stimulated with 24 nM EGF-T for 15 min at 37 °C and labelled for anti-EGFR-phosphotyrosine Y845 (pY845) and anti-EGFR-phosphotyrosine Y1173 (pY1173). Cell-by-cell information on EGFR phosphorylation was obtained with quantitative confocal immunofluorescence microscopy. A typical image of HeLa cells immunolabelled with pY845 is displayed in Fig. 4.1b. In agreement with Oyarizin et al. (Oyarizin et al. 2014), robust increases in EGFR phosphorylation were detected following stimulation with EGF.

To evaluate the number of phosphor-tyrosines per EGFR cluster, we used photobleaching image correlation spectroscopy (pbICS). This approach relies on the principle that the survival probability of clusters during photobleaching is dependent on the total number of fluorophores per cluster. Consequently, a plot of the cluster density as a function of remaining fluorescent intensity provides an estimation of the cluster size distribution within the analysed area (Ciccotosto et al. 2013; Lajevardipour et al. 2015).

From the analysis of several sets of cells (see Materials and Methods), we extracted the population average cluster density (CD, clusters/square micron), the population average cluster size (or brightness, *J* in mAb/cluster) and the average density of molecules (CDJ, mAb/square micron). These parameters are collected in Tables 4.1 and 4.2. We also produced histograms of the cluster size distributions by grouping the data into two categories: (i) density of molecules containing 1–2

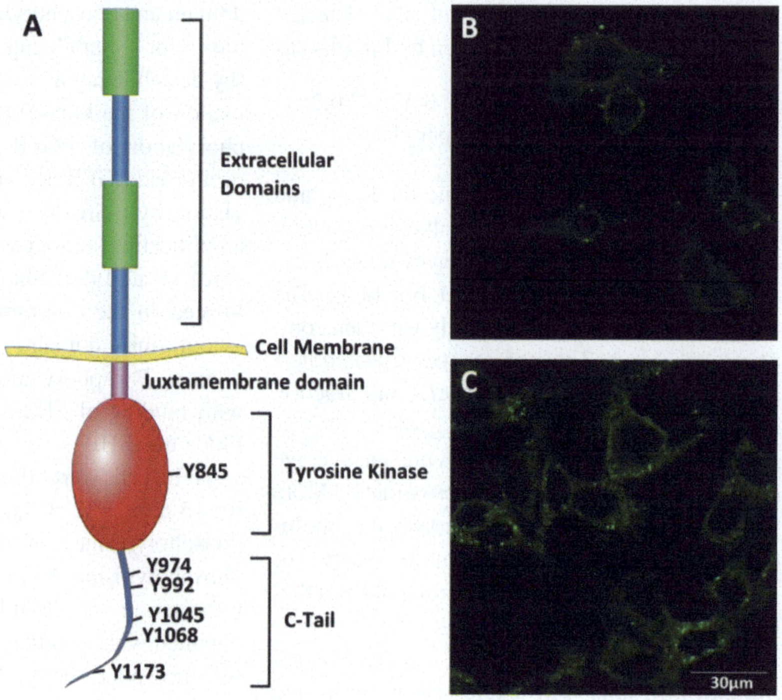

Fig. 4.1 Tyrosine phosphorylation on the EGFR. (**a**) Schematic representation of EGFR monomer showing the different receptor domains and intracellular Tyr residues. (**b**) Typical image of activated EGFR in HeLa cells after stimulation with 24 nM of EGF-T for 15 min at 37 °C. (**c**) Same as (**b**) but after stimulation with 48 nM EGF-T. Cells were fixed and labelled for EGFR pY845 before confocal imaging. Scale bar denotes a length of 30 μm

antibodies per cluster and (ii) density of molecules containing >2 antibodies per cluster. These histograms are presented in Figs. 4.2 and 4.3.

Table 4.1 summarizes the population-averaged pbICS data for EGFR pY1173, while Fig. 4.2a depicts the cluster size distribution of EGFR pY1173 before and after stimulation with 24 nM EGF-T. In the absence of EGF, EGFR oligomers bound to pY1173 mAb dispersed at an average of 16 ± 3 clusters per square micron and contained 1.3 ± 0.3 mAb/cluster. All of the detected clusters contained 1–2 antibodies per cluster (Fig. 4.2a). Based on the average density of detected antibodies (Table 4.1 CDJ, 21 mAbs/square micron) and the EGFR cell surface expression of 50,000 EGFRs/ HeLa cell, we estimate that the fraction of EGFRs bound to pY1173 mAb is 0.42. Therefore, we estimated the average cluster size of the EGFR as 2 EGFRs per cluster, i.e. pre-clustered to some degree. Stimulation with 24 nM EGF-T increased the proportion EGFR oligomers containing multiple phosphotyrosines (from 0% to 71%, Table 4.1 and Fig. 4.2a) and increased the average phospho-cluster size from 1.3 ± 0.3 to 4.6 ± 1 mAbs/cluster (Table 4.1). A *t*-test revealed that the increase in average phospho-cluster size was statistically significant (mean1 = 1.3, SEM1 = 0.3; mean2 = 4.6, SEM2 = 1; $N1 = N2 = 20$, $p = 0.0031$). Making the same calculation as previously, we estimate that 0.9 of EGFR bound to pY1173 mAb in the EGF-stimulated cells. Consequently, the estimated EGFR cluster size for the EGF-stimulated cells is about 5 EGFRs per cluster.

The pbICS results for the pY845-detected EGFR revealed qualitatively different behaviour, as shown in Table 4.2 and Fig. 4.2b. In the serum-starved cells, the pY845-detected EGFR clusters dispersed as mainly lower-order phospho-clusters containing 1–2 phosphotyrosines per cluster with a minor population of phospho-clusters containing greater than 2 phosphotyrosines per cluster. The average phospho-cluster size was 2.4 ± 0.3 mAb/cluster (Table 4.2). Based on the number of detected antibodies (15 mAbs/square micron) and the fractional antibody bound to EGFR of 0.3 mAb/EGFR, we estimate that

Table 4.1 pbICS[a] parameters for phosphorylated pY1173 EGFR in cervical cancer cells

Treatment	CD[b] (±SEM[c])	J[d] (±SEM)	mAb[e](±SEM)	Oligomer[f] (%)	N[g]
Control	16 ± 3	1.3 ± 0.3	21 ± 4	0	20
EGF	10 ± 3	4.6 ± 1	46 ± 9	71	20
NR control	9 ± 3	4.3 ± 1	44 ± 9	100	20
NR-EGF	2 ± 0.3	5.8 ± 1	14 ± 2	90	20

[a]Confocal microscopy: excitation, 488 nm; emission 500–530 nm
[b]CD, calculated average density of clusters, clusters per square micron
[c]SEM, standard error of the mean
[d]J, average cluster size, antibodies per cluster
[e]mAb, total density of detected antibodies, antibodies per square micron
[f]Oligomer percentage, ratio of no. of antibodies in higher-order clusters/no. of antibodies in total
[g]N, number of independent photobleaching experiments

Table 4.2 pbICS[a] parameters for phosphorylated pY845 EGFR in cervical cancer cells

Treatment	CD[b] (±SEM[c])	J[d] (±SEM)	mAb[e] (±SEM)	Oligomer[f] (%)	N[g]
Control	7 ± 2	2.4 ± 0.3	15 ± 5	11	20
EGF	6 ± 2	3.2 ± 1	19 ± 6	46	20
NR control	3 ± 1	8.3 ± 1	24 ± 8	95	20
NR-EGF	3 ± 1	5.2 ± 1	16 ± 5	81	20

[a]Confocal microscopy: excitation, 488 nm; emission 500–530 nm
[b]CD, calculated average density of clusters, clusters per square micron
[c]SEM, standard error of the mean
[d]J, average cluster size, antibodies per cluster
[e]mAb, total density of detected antibodies, antibodies per square micron
[f]Oligomer percentage, ratio of no. of antibodies in higher-order clusters/no. of antibodies in total
[g]N, number of independent photobleaching experiments

pY845 is binding to higher-order EGFR clusters with average size of 6 EGFRs/cluster. Stimulation with EGF increased the proportion of EGFR clusters containing multiple antibodies (from 11% to 46%, Table 4.2 and Fig. 4.2b) and increased the average phospho-cluster size from 2.4 ± 0.3 to 3.2 ± 1 antibodies/cluster. A *t*-test revealed that the increase in mean phospho-cluster size was not statistically significant (mean1 = 2.4, SEM1 = 0.3; mean2 = 3.2, SEM2 = 1; $N1 = N2 = 20$, $p = 0.4483$). Making the same calculation as for untreated cells, the fraction of antibody binding increased slightly from 0.3 to 0.4, and the estimated cluster size of EGF-treated EGFR remained at 6 EGFRs/cluster.

The pbICS results can be compared with a single-molecule study of activated EGFR on HeLa cells (Inchinose et al. 2004). The average cluster size deduced from our pbICS measurements with EGF-treated (24 nM) receptors was 3.2 ± 1 (pY845) activated EGFR/cluster and 4.6 ± 1 (pY1173) activated EGFR/cluster. The average of the pY845 and pY1173 cluster sizes is 3.9 ± 1 activated EGFRs per cluster, and this value agrees reasonably well (ca 20%) with the value of 3.1 activated EGFRs per cluster from single-molecule measurements on HeLa cells by Sako's laboratory (Inchinose et al. 2004). Sako's experiments used a lower concentration of EGF, which might explain the lower cluster size value obtained.

The extrapolated (i.e. corrected for sub-stoichiometric labelling) EGFR cluster sizes obtained from the pbICS measurements can also be compared with those derived from scanning near-field optical microscopy experiments. Abulrob (Abulrob et al. 2010) found a modal population of 6 EGFR per cluster corresponding to the EGF-treated cells, which agrees remarkably well with 5–6 EGFR/cluster using our pbICS approach.

The level of EGFR phosphorylation estimated for pY1173 from antibody immunofluorescence and pbICS approaches the maximum of 100%

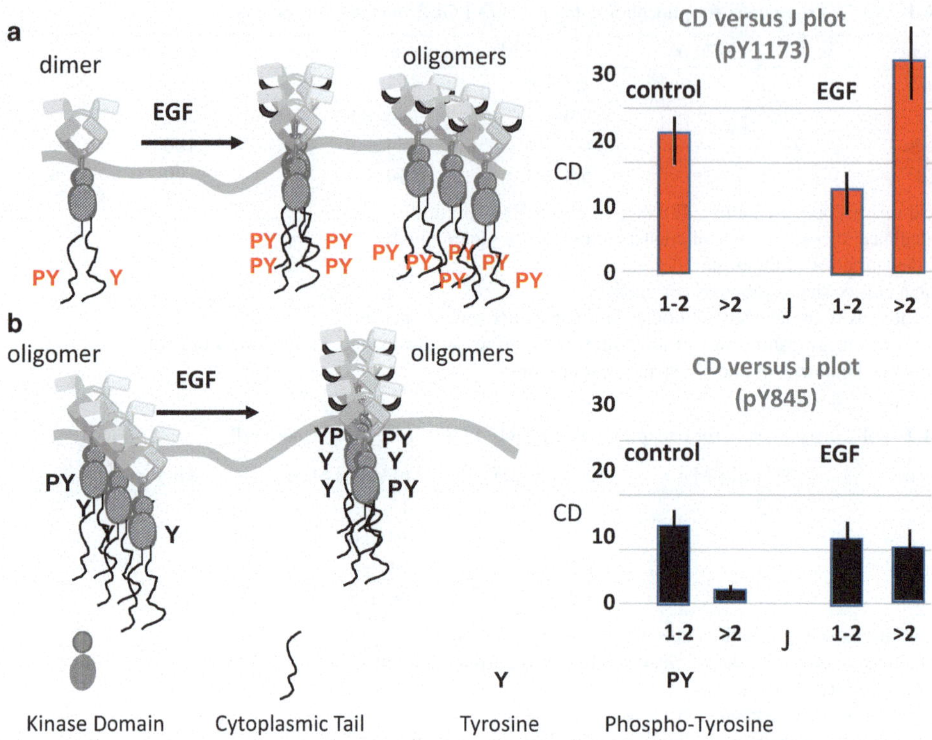

Fig. 4.2 Basal- and EGF-induced activation of EGFR produces lower-order and higher-order activated phospho-clusters on HeLa cells (**a**) pbICS on mAb-labelled pY1173 detects partially activated EGFR dimers in the absence of ligand. EGF-T increases higher-order oligomers containing multiple pY1173. (**b**) pbICS on mAb-labelled pY845 detects partially activated EGFR oligomers in the absence of ligand. EGF-T increases the activation of the preformed oligomers. Red bars denote total density of detected pY1173 (in units of number of mAb/square micron), and black bars denote concentration of pY845 (in units of number of mAb/square micron). 1–2 refers to EGFR oligomers containing 1 or 2 anti-phosphotyrosine mAbs per oligomer, and > 2 refers to EGFR oligomers containing greater than 2 anti-phosphotyrosine mAbs per oligomer

under our conditions of EGF treatment, as expected. This indicates that any observed sub-stoichiometric binding is a result of reduced EGFR phosphorylation and not reduced mAb concentration. The lower level of pY845 as compared to pY1173 in the EGF-treated cells is consistent with the lower level of phosphorylation reported for phosphotyrosines proximal to the kinase domain as compared with those proximal to the tail (Huang et al. 2016).

Taken together, the pbICS experiments appear to be complementary to other biophysical approaches on HeLa cells. We next investigate the effect of nanoparticle and nanoparticle-EGF stimulation on the distribution of activated EGFRs.

4.4.2 Activation of EGFR with NRs and EGF-NRs

It is known that some nanoparticles can stimulate EGFR phosphorylation at a similar level to growth factor stimulation and lead to downstream signalling (Rauch et al. 2012). To address this possibility, the cluster size distributions of EGFR pY845 and EGFR pY1173 after treatment with non-conjugated (i.e. BSA-coated) Au NRs were also determined. Overall, EGFR treated with NRs exhibited phosphorylation at both Y845 and Y1173 residues. When compared to soluble 24 nM EGF-treated EGFR, the pY1173 on NR-treated EGFR showed a comparable overall level of phosphorylation at the cytoplasmic tail site (total pY1173 density of 44 antibodies per

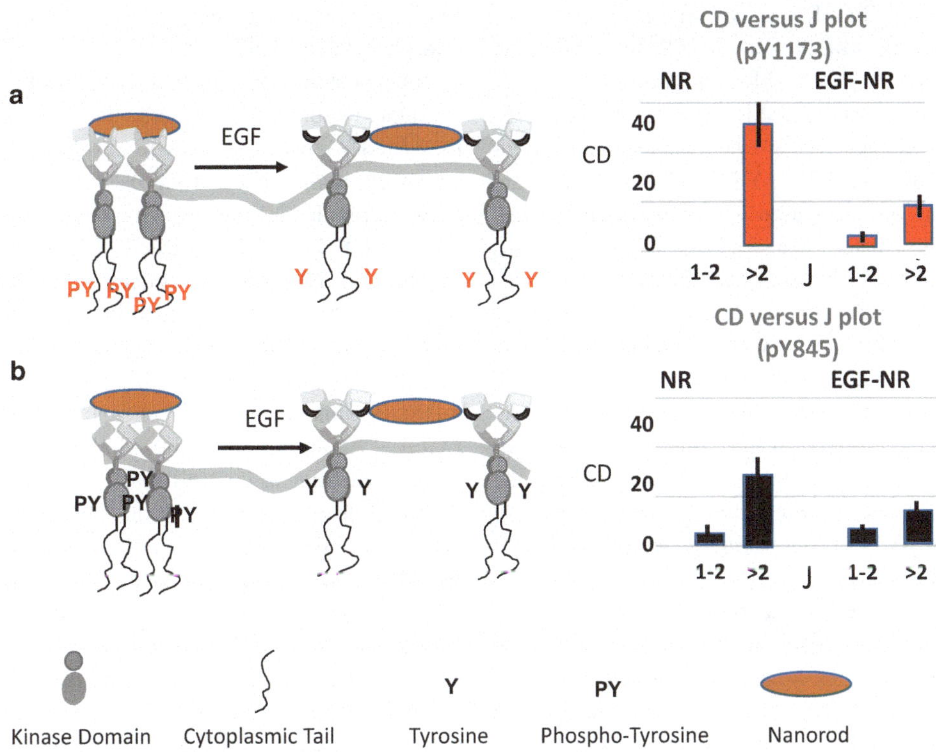

Fig. 4.3 NR- and EGF-NR-induced activation of EGFR on HeLa cells. (**a**) pbICS on mAb-labelled pY1173 detects activated EGFR oligomers in the presence of NRs. EGF-NR decreases higher-order oligomers containing multiple pY1173. (**b**) pbICS on mAb-labelled pY845 detects partially activated EGFR oligomers in the presence of NRs. EGF-NR decreases the activation of the preformed oligomers. Red bars denote total density of detected pY1173 (in units of number of mAb/square micron), and black bars denote total density of pY845 (in units of number of mAb/square micron). 1–2 refers to EGFR oligomers containing 1 or 2 anti-phosphotyrosine mAbs per oligomer, and > 2 refers to EGFR oligomers containing greater than 2 anti-phosphotyrosine mAbs per oligomer

square micron (NR) versus 46 antibodies per square micron (24 nM EGF-T)) but greater proportion of activated receptors in higher-order oligomers (Tables 4.1, 4.2 and Fig. 4.3).

NR-treated EGFR was slightly elevated for pY845 phosphorylation compared to EGF-treated EGFR (24 c/f 19 pY845 per square micron) which was reflected in a greater proportion of phosphorylated higher-order oligomers (Table 4.2, $J = 8.3 \pm 1$ mAb/cluster; Fig. 4.3, 20 mAbs/square micron in oligomers).

In order to assess the influence of EGF spatial organization on EGFR phosphorylation, we activated EGFR with 100-nm-long gold NRs conjugated with EGF. The cluster size distribution of EGFR pY845 and EGFR pY1173 in HeLa cells treated with EGF-NRs is depicted in Fig. 4.3. EGF-NR decreased the average density of detected pY1173 significantly, as compared to the NR control (Table 4.1 from 44 mAb/square micron to 14 mAb/square micron). In the context of average cluster densities and cluster sizes, NR-EGF decreased the density of pY1173 clusters from 9 ± 3 clusters/square micron (NR control) to 2 ± 0.3 clusters/square micron (EGF-NR) significantly (t-test, $N = 20$, $p = 0.0257$). However, the change in average phospho-cluster size was not statistically significant (mean1 = 4.3, SEM1 = 1, mean2 = 5.8, SEM2 = 1, $N = 20$, $p = 0.2955$). Therefore, the large decrease in overall phosphorylation at Y1173 through NR-EGF can be attributed to a change in the density of activated EGFR clusters. We propose that the EGF bound to the NR restricts the EGFR spatially so that access of the cytoplasmic tail on one EGFR to a proximal kinase on another EGFR is minimized.

pY845 detected EGFR clusters remained at the same average density when treated with NR or NR-EGF (both 3 ± 1 pY845 mAb clusters/square micron) but decreased significantly in average phospho-cluster size for the NR-EGF-treated cells (8.3 ± 1 pY845mAb/cluster (NR treated) to 5.2 ± 1 pY845 mAb/cluster (NR-EGF treated) ($p = 0.0346$). The change in overall phosphorylation in Y845 with application of NR-EGF appears to be reflected in a decrease in activation within the clusters. In this case, the geometry of EGFR bound to the nanoparticles also reduces the phosphorylation efficiency of EGFR (as in the case of pY1173) but not the detected number of phospho-clusters.

Taken together the results of the NR-EGF treatment cause a reduction in phosphorylation of residues in the kinase domain and in the tail of the EGFR. We attribute these reductions in phosphorylation to the spacing of EGF on the nanoparticle surface which allows the EGFR to bind locally as a monomer or dimer and prevents trans-phosphorylation between dimers needed for full phosphorylation.

4.4.3 ERK Activation

Phosphorylation of EGFR triggers the activation of different downstream signalling pathways. The MAPK pathway, which includes phosphorylation of Sos, Ras, Raf, Mek and Erk proteins, is critical for the regulation of cell proliferation and survival (Oyarizin et al. 2014). To investigate the link between EGFR Tyr phosphorylation and activation of MAPK pathway, we measured activation (phosphorylation) of nuclear Erk1/2 (ppErk1/2) using an immunofluorescence approach applied previously (Oyarizin et al. 2014).

Figure 4.4a shows a typical image of HeLa cells labelled for ppErk1/2 (green) and propidium iodide (PI, red) after stimulation with 24 nM of EGF-T confirming the nuclear localization of the phosphorylated ERK.

As expected (Oyarizin et al. 2014), HeLa cells stimulated with EGF-T triggered the activation of the MAPK pathway which resulted in an enhanced fluorescence from the anti-phospho-ERK antibody in the nucleus (Fig. 4.4b, c, 24 nM EGF-T, nuclear p-ERK 75% increase).

The EGF-NRs revealed interesting behaviour (Fig. 4.4c). EGF-NRs decreased the activation of ERK relative to NR control (nuclear fluorescence of ppErk1/2 decreased by about 20%) which might be linked to the significant decrease in EGFR cytoplasmic tail phosphorylation (total pY1173 (relative to control)).

We looked for correlations between the parameters defining the phosphorylation of EGFR and the phosphorylation of ERK. Amongst all the parameters, we found a positive linear correlation ($R^2 = 0.92$) between the average density of activated higher-order phospho-clusters (pY1173) (Figs. 4.2a and 4.3a) from pbICS measurements and nuclear p-ERK fluorescence for control, NR, NR-EGF and EGF samples. The relevant values for the density of HO phospho-clusters (mAbs/square micron) were in order of increasing values 0, 12, 32 and 40, and the corresponding p-ERK (immunofluorescence, a.u.) values were 425, 650, 750 and 830 for control and NR-EGF-, EGF- and NR-treated cells.

Taken together, these data supported the hypothesis that the nanotechnology-driven perturbation of the spatial organization of EGFRs at the cell membrane (with gold EGF-NR) can influence the phosphorylation pattern of the EGFR and downstream signalling activation. To our knowledge, this is the first experimental validation of this phenomenon for the EGFR.

4.5 Conclusion

We employed a powerful new tool, pbICS, to quantitatively determine densities and sizes of labelled activated receptor complexes on the sur-

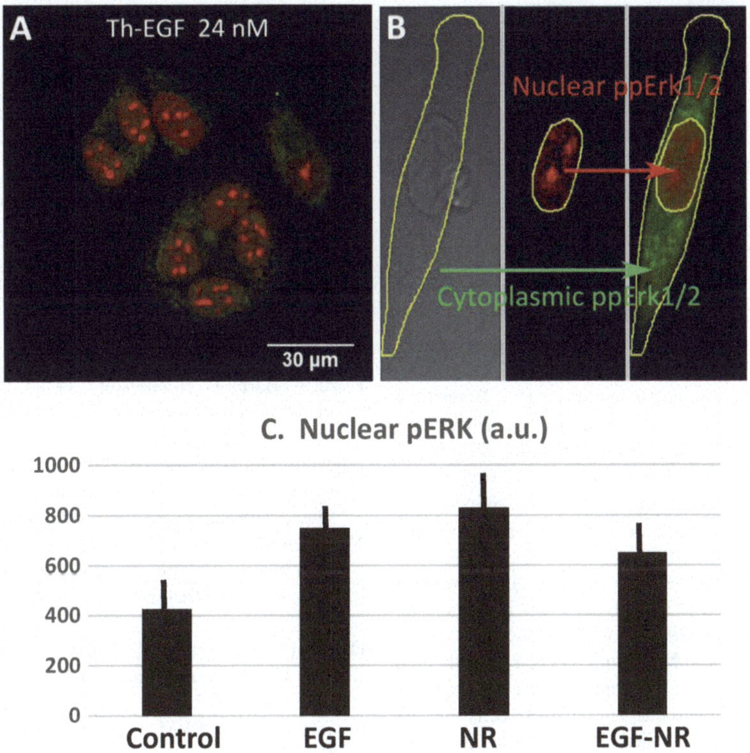

Fig. 4.4 Activation of MAPK with EGF, NR and EGF-NR (**a**) Typical image of HeLa cells after stimulation with 24 nM of EGF-T for 15 min at 37 °C. Cells were fixed and labelled for ppErk1/2 (in green) and PI (in red) prior to confocal imaging. Scale bar is 30 μm. (**b**) Example of analysis to distinguish between cytoplasmic and nuclear activation of ppErk1/2. Cell borders were contoured in DIC imaging mode, while nuclear regions were highlighted by PI staining. (**c**) Quantification of nuclear activation of ppErk1/2 in HeLa cells after serum starvation (control), EGF-T treatment, NR treatment and EGF-NR treatment for 15 min at 37 °C. Length of bars represent fluorescence signal

face of cervical cancer cells. Our studies add to a body of work linking higher-order EGFR oligomerization with phosphorylation. First, full phosphorylation of pY1173 and partial phosphorylation of pY845 correlated with higher-order EGFR oligomers in cervical cancer cells. EGF or NR could induce higher-order EGFR oligomerization which increased the phosphorylation of pY1173, while pY845 was associated with preformed higher-order oligomers in the cancer cell environment and, therefore, less affected by EGF or NR treatments. These observations are consistent with the work of Martin-Fernandez and Kuriyan laboratories who both concluded that phosphorylation of tyrosines proximal to the EGFR kinase required higher-order oligomers and tyrosines proximal to the C-terminus was enhanced in higher-order oligomers (relative to dimers) (Needham et al. 2016; Huang et al. 2016). EGF-NR produced distinct changes to the distribution of activated EGFR, reducing the density of pY1173 clusters while at the same time reducing the cluster size of pY845 clusters. These results suggest that the linkage between the cell receptor surface distribution and phosphorylation site is complex and warrants further investigation.

The findings reported here represent a step towards understanding how receptor clustering affects the EGFR signalling cascades and provide proof of concept for new drug-targeting strategies that prevent or disrupt receptor higher-order oligomerization and activation. Given that EGFR is associated with proliferation and tumorigenicity in cancer cells (Abulrob et al. 2010), the ability to block phosphorylation of this receptor using nanotechnology via interference with higher-order oligomers may find potential application in future clinical settings.

Acknowledgements AHAC and JWMC gratefully acknowledge the Australian Research Council for funding this research (Grant Number: DP130101475).

Disclosure The author reports no conflicts of interest in this work.

References

Abulrob A, Lu Z, Baumann E et al (2010) Nanoscale imaging of epidermal growth factor receptor clustering: effects of inhibitors. J Biol Chem 285:3145–3156

Berkers JA, van Bergen en Henegouwen PM, Boonstra J (1991) Three classes of epidermal growth factor receptors on HeLa cells. J Biol Chem 266:922–927

Ciccotosto GD, Kozer N, Chow TTY et al (2013) Aggregation distributions on cells determined by photobleaching image correlation spectroscopy. Biophys J 104:1056–1064

Clayton AHA, Orchard SG, Nice EC et al (2008) Predominance of activated EGFR higher-order oligomers on the cell surface. Growth Factors 26:316–324

Clayton AHA, Tavarnesi ML, Johns TG (2007) Unligated epidermal growth factor receptor forms higher order oligomers within microclusters on A431 cells that are sensitive to tyrosine kinase inhibitor binding. Biochemistry 46:4589–4597

Clayton AHA, Walker F, Orchard SG et al (2005) Ligand-induced dimer-tetramer transition during the activation of the cell surface epidermal growth factor receptor – a multidimensional microscopy analysis. J Biol Chem 280:30392–30399

Crow MJ, Seekell K, Ostrander JH et al (2011) Monitoring of receptor dimerization using plasmonic coupling of gold nanoparticles. ACS Nano 11:8532–8540

Delcassian D, Depoil D, Rudnicka D et al (2013) Nanoscale ligand spacing influences receptor triggering in T cells and NK cells. Nano Lett 13:5608–5614

Dreaden EC, Alkilany AM, Huang X et al (2012) The golden age: gold nanoparticles for biomedicine. Chem Soc Rev 41:2740–2779

Flynn JF, Wong C, Wu J (2009) Anti-EGFR therapy: mechanism and advances in clinical efficacy in breast cancer. J Oncol . https://doi.org/10.1155/2009/526963

Huang Y, Bharill S, Karandur D et al (2016) Molecular basis for multimerization in the activation of the epidermal growth factor receptor. elife 5:e14107. https://doi.org/10.7554/eLife.14107

Ichinose J, Muratab M, Yanagida T et al (2004) EGF signalling amplification induced by dynamic clustering of EGFR. Biochem Biophys Res Comm 324:1143–1149

Kolch W, Pitt A (2010) Functional proteomics to dissect tyrosine kinase signalling pathways in cancer. Nat Rev Cancer 10:618–629

Kovacs E, Das R, Wang Q et al (2015) Analysis of the role of the C-terminal tail in the regulation of the epidermal growth factor receptor. J Mol Cell Biol 35:3083–3102

Kozer N, Barua D, Orchard S et al (2013) Exploring higher-order EGFR oligomerisation and phosphorylation – a combined experimental and theoretical approach. Mol BioSyst 9:1849–1863

Kozer N, Barua D, Henderson C et al (2014) Recruitment of the adaptor protein Grb2 to EGFR tetramers. Biochemistry 53:2594–2604

Lajevardipour A, Chon JWM, Clayton AHA et al (2015) Complex aggregation distributions by photobleaching image correlation spectroscopy. AIMS Biophys 2:1–7

Needham SR, Roberts SK, Arkhipov A et al (2016) EGFR oligomerization organizes kinase-active dimers into competent signalling platforms. Nat Commun 7:13307. https://doi.org/10.1038/ncomms13307

Oyarzún DA, Bramhall JL, Lopez-Caamal F et al (2014) The EGFR demonstrates linear signal transmission. Integr Biol 6:736–742

Paviolo C, Chon JWM, Clayton AHA (2015) Inhibiting EGFR clustering and cell proliferation with gold nanoparticles. Small 11:1638–1643

Peckys DB, Baudoin J, Eder M et al (2013) Epidermal growth factor receptor subunit locations determined in hydrated cells with environmental scanning electron microscopy. Sci Rep 3:2626

Rauch J, Kolch W, Mahmoudi M (2012) Cell type-specific activation of AKT and ERK signaling pathways by small negatively-charged magnetic nanoparticles. Sci Rep 2:1–9

Shaw A, Lundin V, Petrova E et al (2014) Spatial control of membrane receptor function using ligand nanocalipers. Nat Meth 11:841–846

Tice DA, Biscardi JS, Nickles AL et al (1999) Mechanism of biological synergy between cellular Src and epidermal growth factor receptor. PNAS 96:1415–1420

Wang Y, Gao J, Guo X et al (2014) Regulation of EGFR nanocluster formation by ionic protein-lipid interaction. Cell Res 24:959–976

Wilson KJ, Gilmore JL, Foley J et al (2009) Functional selectivity of EGF family peptide growth factors: implications for cancer. Pharmacol Ther 122:1–8

Factors Influencing the Chaperone-Like Activity of Major Proteins of Mammalian Seminal Plasma, Equine HSP-1/2 and Bovine PDC-109: Effect of Membrane Binding, pH and Ionic Strength

5

Cheppali Sudheer Kumar, Bhanu Pratap Singh, Sk. Alim, and Musti J. Swamy

Abstract

HSP-1/2 and PDC 109 belong to a family of fibronectin type II proteins, present in high concentrations in bovine and equine seminal plasma, respectively. These proteins act as extracellular small heat shock proteins and protect target/client proteins against various kinds of stress. They also exhibit characteristic binding to choline phospholipids present on the sperm plasma membrane and cause efflux of choline phospholipids and cholesterol, resulting in sperm capacitation. The current study demonstrates that hypersaline conditions decrease the chaperone-like activity (CLA) of HSP-1/2. On the other hand, lipoprotein aggregates formed by the binding of choline phospholipids to this protein exhibit higher CLA than HSP-1/2 alone *in vitro*; the increased CLA can be correlated to the increased surface hydrophobicity of the lipoprotein aggregates. Presence of cholesterol in the membrane was found to decrease such enhancement in the CLA. We have also observed that salinity of the medium affects the chaperone activity by altering the polydisperse nature of the HSP-1/2. Together these results indicate that hydrophobicity and polydispersity are important for the chaperone-like activity of HSP-1/2 and factors that can alter these properties of HSP-1/2 can modulate its CLA. Further, studies on PDC-109 show that the chaperone-like and membrane-destabilizing activities of this protein are differentially affected by change in pH.

Keywords

Seminal plasma protein · Molecular chaperone · Fluorescence spectroscopy · Circular dichroism · Aggregation assay · pH switch · Capacitation · Membrane destabilization · Hydrophobicity

5.1 Introduction

Peripheral proteins are a diverse class of proteins which interact with the membrane surface or exhibit partial penetration, but do not span its entire thickness. Their interaction with membranes may be mediated by electrostatic

C. S. Kumar
State Key Laboratory of Biomembrane and Membrane biotechnology, School of Lifesciences, Tsinghua University, Beijing, China

B. P. Singh · S. Alim · M. J. Swamy (✉)
School of Chemistry, University of Hyderabad, Hyderabad, India
e-mail: mjswamy@uohyd.ac.in;
http://chemistry.uohyd.ac.in/~mjs/

K. Chattopadhyay, S. C. Basu (eds.), *Biochemical and Biophysical Roles of Cell Surface Molecules*, Advances in Experimental Medicine and Biology 1112, https://doi.org/10.1007/978-981-13-3065-0_5

and/or hydrophobic interactions, anchoring molecules or specific lipid-protein interactions (Marsh et al. 2002; Whited and Johs 2015). One such specific interaction that regulates sperm maturation is the binding of seminal fibronectin type II (FnII) proteins to the choline phospholipids present on the sperm plasma membrane. The seminal FnII proteins are secreted from the accessory sex glands into seminal plasma, which is the semi-liquid portion of the semen. Sperm cells as such cannot interact with the egg, but during their stay in the female reproductive tract, they undergo capacitation, a maturation process in which morphological and biochemical changes occur in the sperm plasma membrane enabling sperm to interact with egg. During ejaculation, seminal FnII proteins attach to sperm surface and play a major role in sperm capacitation. They bind to the sperm surface through choline phospholipids and remove cholesterol (Chol) and phospholipids from the plasma membrane, a process termed *cholesterol efflux*. This efflux increases the fluidity of the plasma membrane, which in turn facilitates the acrosome reaction (Manjunath and Therein 2002; Plante et al. 2016).

PDC-109 and HSP-1/2 are major FnII proteins present at high concentrations in bovine and equine seminal plasma, respectively. PDC-109 is the best characterized protein of this family. The interaction between PDC-109 and choline phospholipids and various factors affecting it have been extensively characterized using various biophysical and biochemical methods, as this interaction is crucial for sperm capacitation (Swamy 2004; Desnoyers and Manjunath 1992; Gasset et al. 2000; Ramakrishnan et al. 2001; Greube et al. 2001; Wah et al. 2002; Swamy et al. 2002; Thomas et al. 2003; Anbazhagan and Swamy 2005; Anbazhagan et al. 2008, 2011; Damai et al. 2015). The major protein of horse seminal plasma, HSP-1/2, is the equine analogue of PDC-109. It is a mixture of two proteins, HSP-1 and HSP-2, which have the same amino acid sequence but differ in N-terminal flanking region and degree of glycosylation (Calvete et al. 1995). Studies on the interaction of HSP-1/2 with choline phospholipids have shown that the binding involves tryptophan residues and that the protein partially penetrates into the membrane hydrophobic core, which disrupts the membrane integrity, resulting in the formation of lipoprotein aggregates/particles (Greube et al. 2004; Kumar et al. 2016; Kumar and Swamy 2016a).

It is now well established that the major FnII proteins of bovine and equine seminal plasma exhibit chaperone-like activity (CLA). Studies carried out on the CLA of PDC-109 and HSP-1/2 have shown that they can protect various proteins against thermal and chemical stress (Sankhala and Swamy 2010; Sankhala et al. 2011, 2012). HSP-1/2 was also shown to protect target proteins and small molecules against oxidative stress (Kumar and Swamy 2016b). The chaperone activity of HSP-1/2 was found to be inversely correlated to its membrane-destabilizing activity, with the two activities being regulated by a *pH switch* (Kumar and Swamy 2016a). Various external factors such as surfactants and L-carnitine were found to modulate the CLA of HSP-1/2 (Kumar and Swamy 2017a, b). Modulation of the CLA of HSP-1/2 by phosphorylcholine (PrC) and L-carnitine indicates that molecules containing choline/choline-mimicking moiety can alter the CLA of HSP-1/2. In the present study, we have investigated the effect of choline phospholipids, which are the physiological ligands of HSP-1/2, as well as the effect of high salt in the medium (hypertonic saline) on the CLA of HSP-1/2. We have also investigated the effect of varying the pH on the membranolytic and chaperone-like activities of PDC-109, which revealed an inverse relation between these two activities regulated by a pH switch, similar to that observed with HSP-1/2 (Kumar and Swamy 2016a).

5.2 Materials and Methods

5.2.1 Materials

Carbonic anhydrase, phosphorylcholine chloride calcium salt, heparin-agarose type I beads, 8-anilinonaphthalene sulphonic acid (ANS), 4,4′-Dianilino-1,1′-binaphthyl-5,5′-disulphonic acid (bis-ANS) and DEAE Sephadex A-25 were obtained from Sigma (St. Louis, MO, USA). p-Aminophenyl phosphorylcholine-agarose column was obtained from Pierce Chemical Co. (Oakville, Ontario, Canada). Dimyristoyl phosphatidylcholine (DMPC), dimyristoyl phosphatidylglycerol (DMPG) and cholesterol were obtained from Avanti Polar Lipids (Alabaster, AL, USA). Lactate dehydrogenase (LDH), alcohol dehydrogenase (ADH), enolase (ENL), tris base and other chemicals were purchased from local suppliers and were of the highest purity available.

5.2.2 Purification of HSP-1/2 and PDC-109

HSP-1/2 was purified from seminal plasma collected from healthy horses by affinity chromatography on heparin-agarose and *p*-aminophenyl phosphorylcholine followed by reverse phase HPLC as described earlier (Calvete et al. 1995; Sankhala et al. 2012). The purified protein was dialyzed against 50 mM tris buffer, pH 7.4, containing 0.15 M NaCl and 5 mM EDTA (TBS). PDC-109 was purified from bovine seminal plasma by gel filtration on Sephadex G-50, followed by affinity chromatography on DEAE Sephadex A-25, essentially as described earlier (Calvete et al. 1996; Ramakrishnan et al. 2001).

5.2.3 Aggregation Assay to Investigate the Chaperone Activity of HSP-1/2

Chaperone activity of HSP-1/2 was assayed by monitoring its ability to prevent thermal aggregation employing carbonic anhydrase (CA) as the substrate protein. CA was incubated at 48 °C, and its aggregation was monitored by recording light scattering at 360 nm as a function of time in a Cary 100 UV/visible spectrophotometer. A fixed concentration of HSP-1/2 was pre-incubated with CA in TBS for 5 min at room temperature, and then aggregation assays were performed as described in previous studies with other client proteins (Sankhala et al. 2012). Aggregation profile for the native enzyme was taken as 100%, and percent aggregation of other samples was calculated with respect to native enzyme. Percent protection was obtained by the expression: percent protection = 100 – percent aggregation.

5.2.4 Effect of Salt on the CLA of HSP-1/2

Effect of high ionic strength medium on the CLA of HSP-1/2 was investigated by aggregation assays performed at 48 °C as described above, using CA and alcohol dehydrogenase (ADH) as target proteins. HSP-1/2, dialyzed extensively against 10 mM Mops buffer containing 50 mM EDTA and 0.5 M NaCl (pH 7.4), was used in these studies. The target enzymes were also suspended in the same buffer. HSP-1/2 in 10 mM Mops buffer (pH 7.4) without salt and EDTA was used as the control. Relative chaperone activity was calculated in terms of the protection offered by HSP-1/2 with respect to the target enzyme's aggregation alone.

5.2.5 Preparation of Liposomes

Lipids taken in a glass tube were dissolved in either dichloromethane or dichloromethane-methanol mixture and were dried under a gentle stream of nitrogen gas. After removing the remaining traces of solvent by vacuum desiccation for 3–4 h, the lipid film was hydrated with TBS to give the desired lipid stock concentration. Small unilamellar vesicles (SUVs) were prepared by subjecting the hydrated lipid suspension to 3–4 freeze-thaw cycles followed by sonication for 30 min in a bath sonicator above the gel-liquid crystalline phase transition temperature.

5.2.6 Investigating the Effect of Phospholipids on the CLA of HSP-1/2

Chaperone activity of HSP-1/2 was measured by monitoring protection offered by HSP-1/2 to substrate proteins (LDH and enolase) in thermal aggregation assays (Sankhala et al. 2012). To investigate the effect of lipid binding on the CLA of HSP-1/2, 0–6 μM of individual lipids (DMPC or DMPG) was mixed with a fixed concentration of HSP-1/2 and incubated for 5–10 min. The target protein (LDH/enolase) was incubated with this mixture for 5 min, and the aggregation assay was performed by monitoring the turbidity resulting from the formation of aggregates, as described above. In order to investigate the effect of cholesterol incorporation in DMPC vesicles on the CLA of HSP-1/2, small unilamellar vesicles of DMPC/cholesterol (3:1, mol/mol) were prepared as described above, and the chaperone assay was performed by pre-incubating HSP-1/2 with the DMPC/cholesterol mixture followed by the aggregation assay using LDH/enolase as the target protein. Concentrations of HSP-1/2, LDH, enolase and the lipids used are given in the appropriate figure legends.

5.2.7 Fluorescence Spectroscopy

Fluorescence spectroscopic studies were carried out on a Spex Fluoromax-4 fluorescence spectrometer at room temperature with excitation and emission band-pass filters set at 2 and 3 nm, respectively. Studies on the binding of ANS to HSP-1/2 were performed by adding small aliquots of the probe from a 1 mM stock solution to ~1.1–1.6 μM HSP-1/2, taken in a 1 cm × 1 cm path length rectangular quartz cuvette. Samples were excited at 350 nm, and emission spectra were recorded from 450 to 600 nm. Fluorescence intensities were corrected for dilution effects and for contribution from the free ligand determined in parallel titrations of ANS with buffer alone. The results presented are average values from three independent experiments.

Interaction of bis-ANS with HSP-1/2, phospholipids and their mixture was investigated by fluorescence spectroscopy as described above. Samples containing HSP-1/2, phospholipids and their mixtures as described in the legend to Fig. 5.3 were prepared. The concentrations used were HSP-1/2 (~3 μM), bis-ANS (5 μM) and phospholipids (~10 μM). Samples were excited at 385 nm, and emission spectra were recorded between 450 and 650 nm. Slit widths of 2 and 3 nm were used on the excitation and emission monochromators, respectively. All results reported are average values from two independent experiments (with standard deviations <5%).

5.2.8 Circular Dichroism Spectroscopy

CD spectra of HSP-1/2 alone and in the presence of different lipids (DMPC, DMPG and DMPC: cholesterol (4:1, mol/mol)) were recorded on an Aviv model 420 CD spectrometer (Aviv Biomedical, Lakewood, NJ, USA). All measurements were performed using a 2 mm path length rectangular quartz cell. Concentration of HSP-1/2 was 0.5 mg/ml (~33 μM), and that of different lipids was 0.1 mM. All samples were dispersed in TBS.

5.2.9 Effect of pH on the CLA of PDC-109

To investigate the effect of pH on the CLA of PDC-109, aggregation assays were carried out at different pH essentially as described earlier for studies with HSP-1/2 (Kumar and Swamy 2016a). At each pH, the protection offered by PDC-109 to the target protein (ADH) alone was calculated as described in the above sections.

5.2.10 Erythrocyte Lysis Assay

Disruption of erythrocyte membranes by PDC-109 was investigated by absorption spectroscopy as described earlier for HSP-1/2 (Kumar and Swamy 2016a). Different amounts of PDC-109 and 100 μL of 4% erythrocyte suspension in TBS were mixed in a test tube, and the volume was adjusted to 1.0 mL with TBS. The sample was then incubated for 1 h and centrifuged at 1500 × g for 5 min. The supernatant was collected, and its absorbance at 415 nm (corresponding to the haem moiety) was measured using an Agilent Cary 100 UV-Vis spectrophotometer. For studying the pH dependence, 200 μL of 4% erythrocyte suspension in TBS was taken, and 800 μL of buffer of desired pH was added and centrifuged at 1500 × g for 5 min. The clear supernatant was decanted, and total volume was made up to 1.0 mL with buffer of desired pH. This was repeated twice to ensure complete exchange of the buffer, and then the assay was carried out as described above. Results from a minimum of three independent experiments are presented together with standard deviations.

5.3 Results and Discussion

5.3.1 HSP-1/2 Inhibits Thermal Aggregation of CA

Aggregation assays aimed at investigating the effect of HSP-1/2 on the thermal aggregation of CA are shown in Fig. 5.1. Similar to many other proteins, CA aggregates and precipitates when incubated at high temperatures resulting in an increase in the turbidity of the sample as shown in curve 1. In the presence of HSP-1/2, this aggregation decreases in a concentration-dependent manner. In the presence of 0.02 mg/mL HSP-1/2, ~72% aggregation was observed compared to native enzyme alone. In the presence of 0.05 mg/ml and 0.075 mg/ml of HSP-1/2, the aggregation decreased to 58% and 17%, respectively. These results further support our previous reports that HSP-1/2 exhibits CLA by inhibiting the thermal aggregation of a variety of proteins (Sankhala et al. 2012; Kumar and Swamy 2016a).

5.3.2 Effect of High Salinity on CLA of HSP-1/2

Results of turbidimetric assays aimed at investigating the effect of high salinity on the CLA of HSP-1/2 using CA and ADH as target proteins are shown in Fig. 5.2. Incubation of CA at 48 °C resulted in an increase in the turbidity of the solution, and pre-incubation with HSP-1/2- resulted in a significant decrease in the turbidity, with only 25% aggregation being seen, as compared to that observed with CA alone. When the assay was performed with CA in the same buffer containing 50 mM EDTA and 0.5 M NaCl, significant increase in the turbidity is observed compared to that observed with CA in buffer without EDTA and NaCl. Decrease in the aggregation was rather small (~15%) when CA was pre-incubated with HSP-1/2 in buffer containing 0.5 M NaCl and 50 mM EDTA as compared to that observed (~75%) when the assay was done in buffer with no added salts. This could be because of two reasons. One reason for this could be that the increase in the turbidity of the target protein indicates more aggregates of CA are formed in the presence of high salt that may exceed the turnover of HSP-1/2. Hence HSP-1/2 cannot interact with all the aggregates and decrease the sample turbidity resulting in decrease in CLA. Another reason is related to the structural changes associated with the HSP-1/2 in the hypersaline conditions. Previous reports have shown that, in the presence of high salt concentration, FnII proteins (PDC-109 and HSP-1/2) lose their polydisperse nature and become monomeric/dimeric in nature (Gasset et al. 1997; Sankhala et al. 2011; Calvete et al. 1997). As polydispersity is shown to be important for the CLA of HSP-1/2, this decrease in polydispersity can be correlated to the decrease in the CLA of HSP-1/2 (Sankhala et al. 2012). To investigate this further, we have carried out a similar assay with ADH as target protein. ADH alone has shown somewhat similar aggregation in the absence and in the presence of salt (as compared to CA, which showed large differences in the aggregation levels in the presence and absence of salt). Pre-incubation of ADH with HSP-1/2 in buffer containing high ionic strength resulted in only 17% decrease in the aggregation (activity), whereas incubation of ADH with HSP-1/2 in buffer containing no added salt resulted in ~81% decrease in the aggregation. These results indicate that decrease in the CLA of HSP-1/2 in high ionic strength buffers is most likely due to structural changes associated

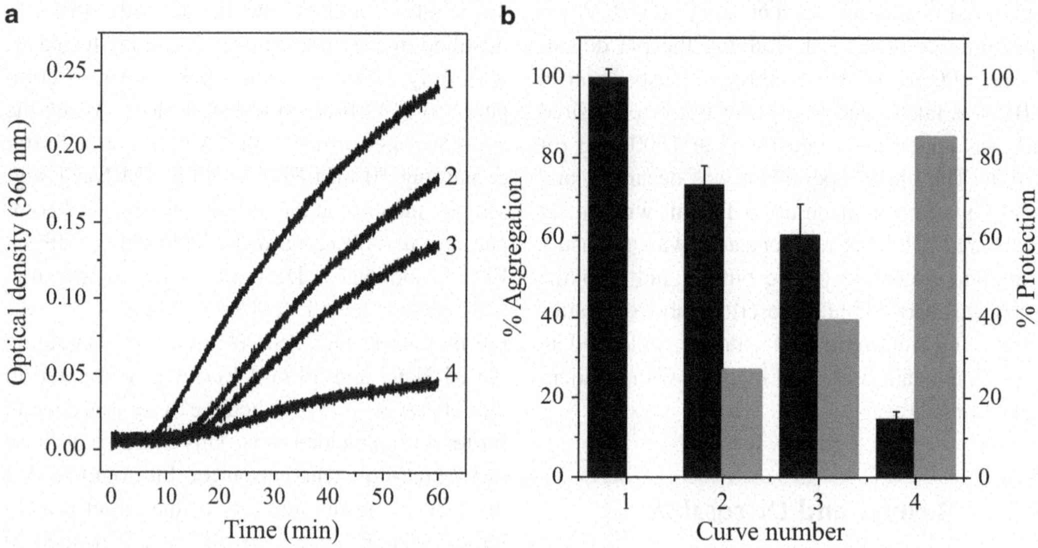

Fig. 5.1 Chaperone-like activity of HSP-1/2. (**a**) Aggregation profiles shown correspond to (1) CA alone, (2) CA+ 0.02 mg/mL HSP-1/2, (3) CA+ 0.05 mg/mL HSP-1/2 and (4) CA+ 0.075 mg/mL HSP-1/2. Concentration of CA was 0.1 mg/mL in all samples. (**b**) Bar diagram representation of results shown in panel (**a**). Black bars show percent aggregation, and grey bars represent percent protection

with HSP-1/2 and not because of increase in turbidity of the target protein. The present observation that high ionic strength decreases the CLA of HSP-1/2, taken together with previous reports that binding of PrC also decreases the chaperone activity of this protein (Sankhala et al. 2012), indicates that polydispersity is an important factor for the CLA of HSP-1/2.

5.3.3 Effect of Phospholipid Binding on the CLA of HSP-1/2

The effect of phospholipid binding on the CLA of HSP-1/2 was assessed by turbidimetry, and the results obtained are presented in Fig. 5.3. When incubated at 48 °C, LDH aggregates resulting in an increase in the turbidity of the sample as shown in Fig. 5.3a, curve 1. Presence of HSP-1/2 decreases this aggregation. At a LDH-to-HSP-1/2 ratio of 3:1, the aggregation decreased to 55% as compared to that of LDH alone (curve 2). When HSP-1/2 was pre-incubated with 2 μM DMPC, the aggregation decreased to 32.5%, whereas in the presence of 6 μM DMPC, the aggregation decreased further to ~15% (curves 3 and 4, respectively). HSP-1/2 incubated with the same concentrations of DMPC did not show any aggregation (data not shown). These results indicate that presence of DMPC increases the CLA of HSP-1/2. To investigate if this effect is specific to the choline phospholipids, a similar assay was carried out with HSP-1/2 pre-incubated with DMPG vesicles. The results are shown in Fig. 5.3c, d. From the figure, it can be clearly seen that addition of DMPG did not have any significant effect on the CLA of HSP-1/2; rather a slight increase in the aggregation was observed, indicating a slight negative modulation of the CLA of HSP-1/2 by DMPG. Neither DMPC nor DMPG has any significant effect on the aggregation profile of LDH alone.

Similar results were obtained when enolase (ENL) was used as the target protein (Fig. 5.3e, f). HSP-1/2 decreases the heat-induced aggregation of ENL to 48% compared to native ENL. The aggregation further decreased to ~15% in the presence of HSP-1/2 pre-incubated with 4 μM DMPC. No such increase in the protective activity of HSP-1/2 upon interaction with DMPG was observed. These results indicate that interaction of DMPC modulates the CLA of HSP-1/2. This

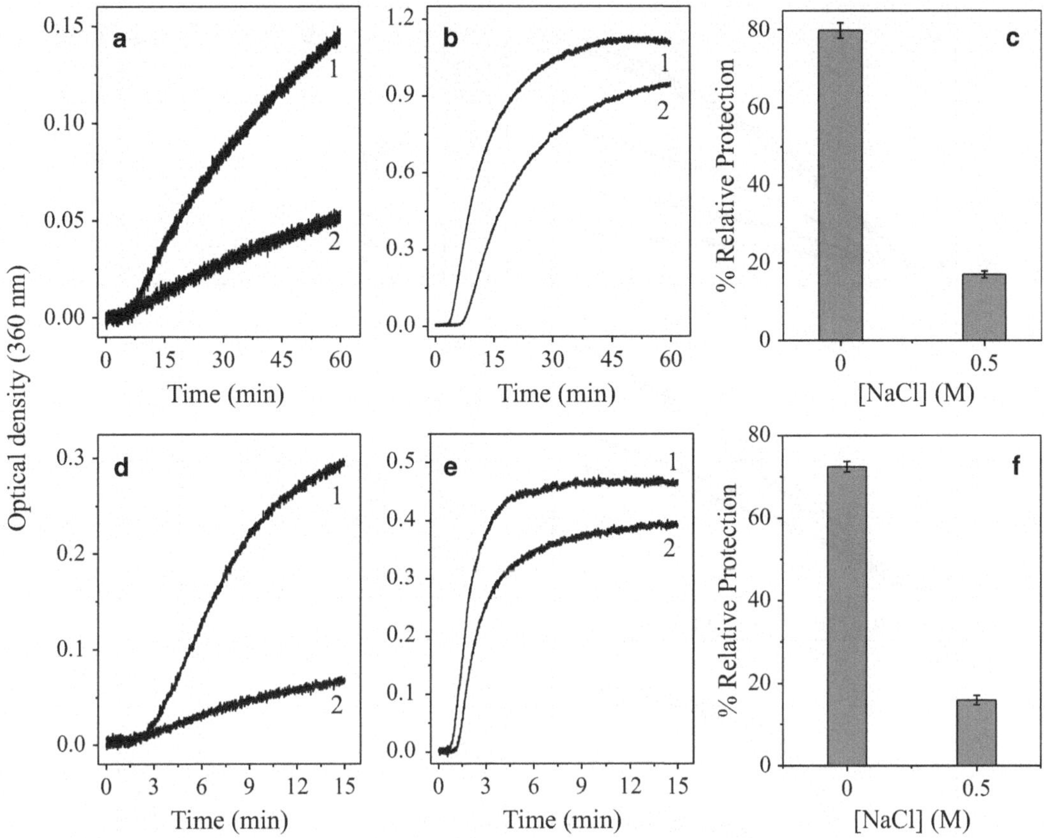

Fig. 5.2 Effect of high ionic strength on the CLA of HSP-1/2. Aggregation profiles of CA (curve 1) and CA+ 0.05 mg/ml HSP-1/2 (curve 2) without salt (**a**) and in presence of 50 mM EDTA and 0.5 M NaCl (**b**). Relative percent protection offered by HSP-1/2 towards CA (**c**). Aggregation profiles of ADH (curve 1) and ADH + 0.015 mg/ml HSP-1/2 (curve 2) without salt (**d**) and in the presence of 50 mM EDTA and 0.5 M NaCl (**e**). Relative percent protection offered by HSP-1/2 towards ADH (**f**). Concentrations of CA and ADH were 0.1 mg/mL and 0.05 mg/mL, respectively

could be because of two reasons. One is the binding of DMPC induces structural changes in HSP-1/2, which may in turn lead to changes in its chaperone activity. Results of CD spectroscopic studies show that binding of phospholipids, especially DMPC, induces changes in the tertiary structure of HSP-1/2 (Fig. 5.4a). These structural changes could be responsible for the observed change in the CLA of HSP-1/2. The second reason could be the formation of lipoprotein aggregates of HSP-1/2 and DMPC. From our earlier studies, it is clear that HSP-1/2 can destabilize and solubilize the membrane and this binding leads to the formation of lipoprotein particles/complexes (Kumar and Swamy 2016a). This formation of lipoprotein complexes can alter the changes in the hydrophobicity of the overall system, as discussed in the following section.

5.3.4 Surface Hydrophobicity of HSP-1/2

Surface hydrophobicity is an important factor that is directly correlated to the chaperone activity of many extracellular chaperones (Bakthisaran et al. 2015; Sheluho and Ackerman 2001). Chaperones interact with their client proteins mostly through hydrophobic patches present on their surface and the hydrophobic patches on target protein, which get exposed when subjected to stress (Moparthi et al. 2010; Lee et al. 1997; Voellmy and

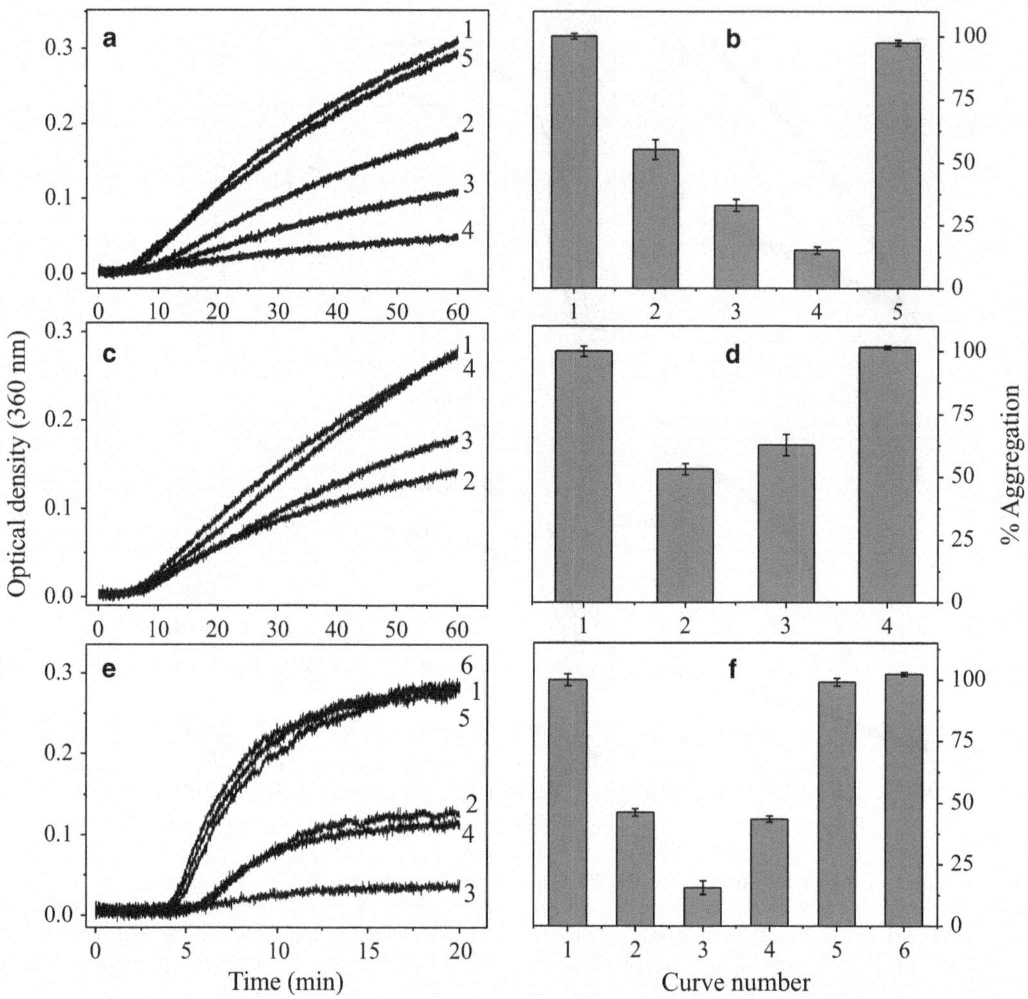

Fig. 5.3 Effect of phospholipid binding on CLA of HSP-1/2. (**a**) Aggregation profiles correspond to (1) LDH, (2) + HSP-1/2, (3) + HSP-1/2 + 2 μM DMPC, (4) + HSP-1/2+ 6 μM DMPC and (5) + 6 μM DMPC. (**b**) Bar diagram showing percent aggregation corresponding to curves in panel (**a**). (**c**) Aggregation profiles correspond to (1) LDH, (2) + HSP-1/2, (3) + HSP-1/2 + 4 μM DMPG and (4) + 4 μM DMPG. (**d**) Bar diagram showing percent aggregation corresponding to curves in panel (**c**). (**e**) Aggregation profiles correspond to (1) ENL, (2) + HSP-1/2, (3) + HSP-1/2 + 4 μM DMPC, (4) + HSP-1/2 + 4 μM DMPG, (5) +4 μM DMPC and (6) + 4 μM DMPG. (**f**) Bar diagram showing percent aggregation corresponding to curves in panel (**e**). The concentrations of LDH, ENL and HSP-1/2 used in the assays are 0.075, 0.1 and 0.025 mg/ml, respectively

Boellmann 2007). Hence it is very important to characterize the surface hydrophobicity of HSP-1/2 in order to understand the interface of chaperone-client complexes. ANS and its dimer bis-ANS are fluorescence probes which specifically bind to the hydrophobic patches present on proteins and show a characteristic increase in their quantum yield upon binding. As hydrophobicity is found to be an important factor and could be correlated with the chaperone activity of HSP-1/2 (Kumar and Swamy 2016a), we have characterized the surface hydrophobicity of HSP-1/2 by monitoring the interaction of ANS and bis-ANS with HSP-1/2.

Fluorescence spectra and binding curve corresponding to the titrations of HSP-1/2 with ANS are shown in Fig. 5.4b, c. From these figures, it is clear that addition of ANS to

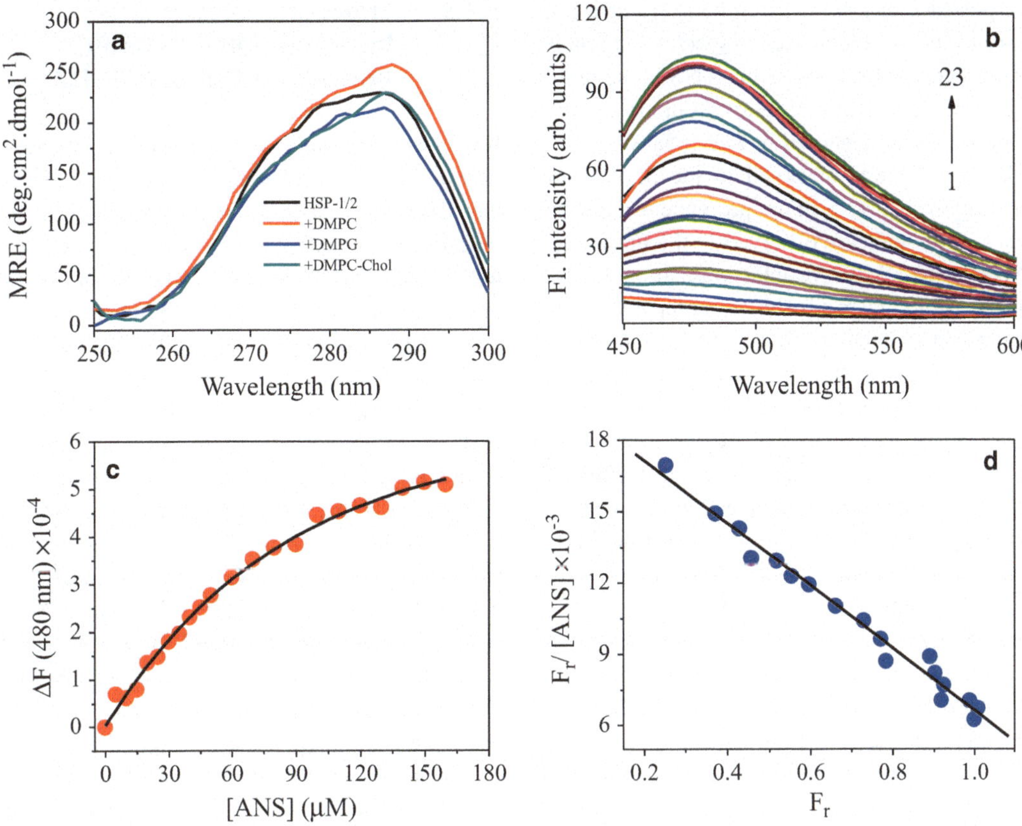

Fig. 5.4 Phospholipid interaction and ANS binding to HSP-1/2. (**a**) Near-UV CD spectra of HSP-1/2 under native conditions and in the presence of DMPC, DMPG and DMPC/Chol vesicles. Concentration of HSP-1/2 was 33 μM, and that of the total lipid in each sample was 0.1 mM. (**b**) Fluorescence titration of HSP-1/2 with ANS. Spectrum 1 (lowest intensity) corresponds to HSP-1/2 alone (1.1 μM), and spectra 2–22 correspond to those obtained in the presence of increasing concentrations of ANS. (**c**) Binding curve obtained by plotting change in fluorescence intensity (ΔF) vs ANS concentration. (**d**) Linear Scatchard plot for the binding data (plot of F_r/[ANS] versus F_r), where F_r is the ratio of corrected fluorescence intensity and maximum fluorescence intensity for the binding data. Slope of this plot gives association constant for the interaction, and the X-intercept gives the stoichiometry of binding

HSP-1/2 results in an increase in the fluorescence intensity and leads to saturation at higher concentrations. In these experiments, the concentration of HSP-1/2 was kept very low such that $[ANS]_{bound} << [ANS]_{total}$ and $[ANS]_{total}$ can be considered as the free [ANS] for analysis and the data were analysed by the Scatchard method as described earlier for PDC-109 (Sankhala et al. 2011). From the slope of the Scatchard plot given in Fig. 5.4d, the binding constant for the association of ANS to HSP-1/2 was estimated to be 1.06 (±0.2) × 10^4 M^{-1}. From the intercept, a number of binding pockets were found to be 1.16 indicating equimolar stoichiometry for the ANS-HSP-1/2 association (Fig. 5.4d), which is very similar to the interaction of ANS with other small heat shock proteins such as PDC-109 and α-crystallin (Sankhala et al. 2011; Liang and Li 1991). These results indicate that HSP-1/2-contains hydrophobic patches on its surface that accommodate one ANS molecule per HSP-1/2 protomer.

In further experiments we investigated changes in surface hydrophobicity of HSP-1/2 in the presence of DMPC using bis-ANS as a fluorescence probe. Fluorescence spectra and relative fluorescence intensities of bis-ANS alone and in the presence of HSP-1/2, phospholipids and their combinations are shown in Fig. 5.5a, b. A fourfold increase was observed in the fluorescence intensity of bis-ANS in the presence of HSP-1/2, whereas the increase observed in the presence of DMPG and DMPC vesicles was about 1.5-fold and 2.5-fold, respectively, as compared to the fluorescence intensity of bis-ANS in buffer. A slight increase in the fluorescence enhancement was observed for HSP-1/2-DMPG mixture (5.5 times) compared to that of HSP-1/2 alone (4.5 times). On the other hand, in case of HSP-1/2-DMPC complex, the increase in fluorescence intensity was observed to be ~1.6-fold higher than that observed with HSP-1/2 alone and > seven-fold higher as compared to buffer alone.

The larger increase in the fluorescence intensity of the fluorescent probe indicates greater exposure of the hydrophobic surfaces. It is well known that HSP-1/2 intercalates into lipid membranes, which results in their destabilization and subsequent formation of the lipoprotein aggregates (Kumar and Swamy 2016a). From the above results, the lipoprotein aggregates formed appear to have higher hydrophobicity compared to HSP-1/2 alone, and this increase in hydrophobicity could facilitate the interaction of HSP-1/2 with client proteins, resulting in a suppression of their aggregation. Interaction of HSP-1/2 with DMPG is weaker as compared to that with DMPC, due to which HSP-1/2 may not intercalate into the hydrophobic core of DMPG membranes to the same extent as with DMPC membranes. Hence the exposure of hydrophobic core of DMPG may not be the same as that of DMPC, which in turn results in smaller changes in the fluorescence intensity for the HSP-1/2-DMPG complex compared to that of HSP-1/2-DMPC complex, which could be correlated with the marginal effect of DMPG binding on the CLA of HSP-1/2.

5.3.5 Effect of Cholesterol Incorporation into DMPC Vesicles on CLA of HSP-1/2

It is clear from the above experiments that addition of DMPC vesicles increases (or potentiates) the CLA of HSP-1/2. In further studies, we investigated the effect of incorporation of cholesterol – an important constituent of the sperm plasma membrane besides choline phospholipids – on the CLA of HSP-1/2. Results obtained from studying this aspect using aggregation assays are shown in Fig. 5.6. Incubation of LDH with HSP-1/2 decreases its aggregation to 62% as compared to native LDH, which is considered as 100% (Fig. 5.6a, b). Pre-incubation of HSP-1/2 with DMPC vesicles reduced this aggregation to ~31%, supporting the above results that DMPC potentiates the CLA of HSP-1/2. When DMPC vesicles containing cholesterol in a ratio of 3:1 were added, 45% aggregation was observed. Similar results were obtained when enolase (ENL) was used as the target protein (Fig. 5.6c, d).

The above results indicate that the presence of cholesterol has a negative effect compared to DMPC alone on the CLA of HSP-1/2; however, in the presence of DMPC/Chol mixture, HSP-1/2 still showed higher CLA than HSP-1/2 alone. These results further support our model that the lipoprotein complex formation is the reason for the enhanced chaperone activity of HSP-1/2 in the presence of DMPC. As cholesterol incorporation imparts stability and rigidity to the DMPC membrane, the membrane-destabilizing/membrane-solubilizing ability of HSP-1/2 is reduced leading to the formation of fewer lipoprotein aggregates (Kumar and Swamy 2016a). This was further supported by the decreased hydrophobicity of HSP-1/2-DMPC/Chol complex compared to that of HSP-1/2-DMPC (Fig. 5.5b). Hence the addition of DMPC/Chol vesicles did not lead to the increase in the CLA of HSP-1/2 to the same extent as DMPC vesicles alone. However, the possibility that interaction of

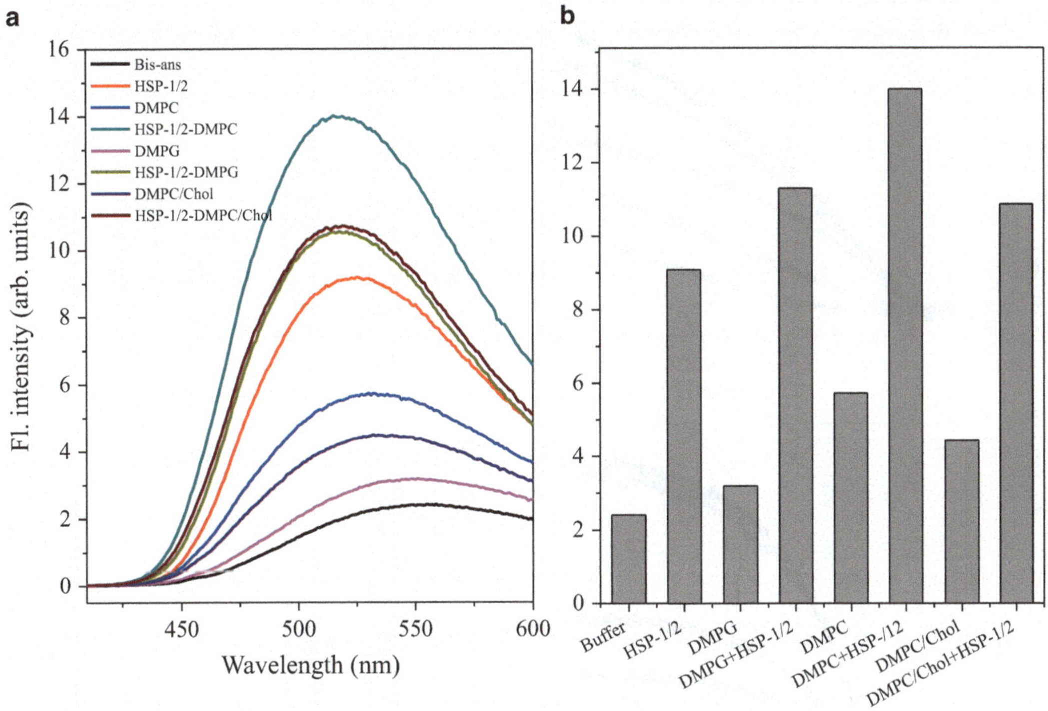

Fig. 5.5 Changes in the hydrophobicity of HSP-1/2 upon binding to phospholipids. (**a**) Fluorescence spectra correspond to bis-ANS interaction with HSP-1/2, phospholipids and their combinations. (**b**) The relative fluorescence intensity of various samples in panel (**a**). The concentrations of bis-ANS, HSP-1/2 and individual lipids were 10 μM, 25 μg/ml and 5 μM, respectively

cholesterol with CRAC (cholesterol recognition amino acid consensus) motifs on the protein may be responsible for its observed effect on the CLA of HSP-1/2 (Scolari et al. 2010) cannot be ruled out.

5.3.6 Effect of pH on the Chaperone-Like and Membranolytic Activities of PDC-109

It is known that environmental factors such as salinity, temperature and other small molecules affect the activity of chaperones (Bakthisaran et al. 2015). pH is also an important factor that was found to modulate the activity of various small heat shock proteins including HSP-1/2 (Kumar and Swamy 2016a). In order to investigate whether the dependence of CLA on pH of the medium is a common mechanism under operation in the FnII proteins of mammalian seminal plasma, we investigated the dual activities of PDC-109 as a function of pH. Effect of the membrane-destabilizing activity of PDC-109 was investigated by using human erythrocytes as model cell membranes as they share similar composition and stoichiometry of various lipids to that of sperm plasma membrane. It is already known that PDC-109 destabilizes the erythrocyte membrane in a time- and concentration-dependent manner (Damai et al. 2010). As shown in Fig. 5.7, erythrocyte lytic activity of PDC-109 was found to be high at the mildly acidic pH of 6.0, which gradually decreased with increase in the pH, and very low erythrocyte lysis was observed at the mildly basic pH of 8.0.

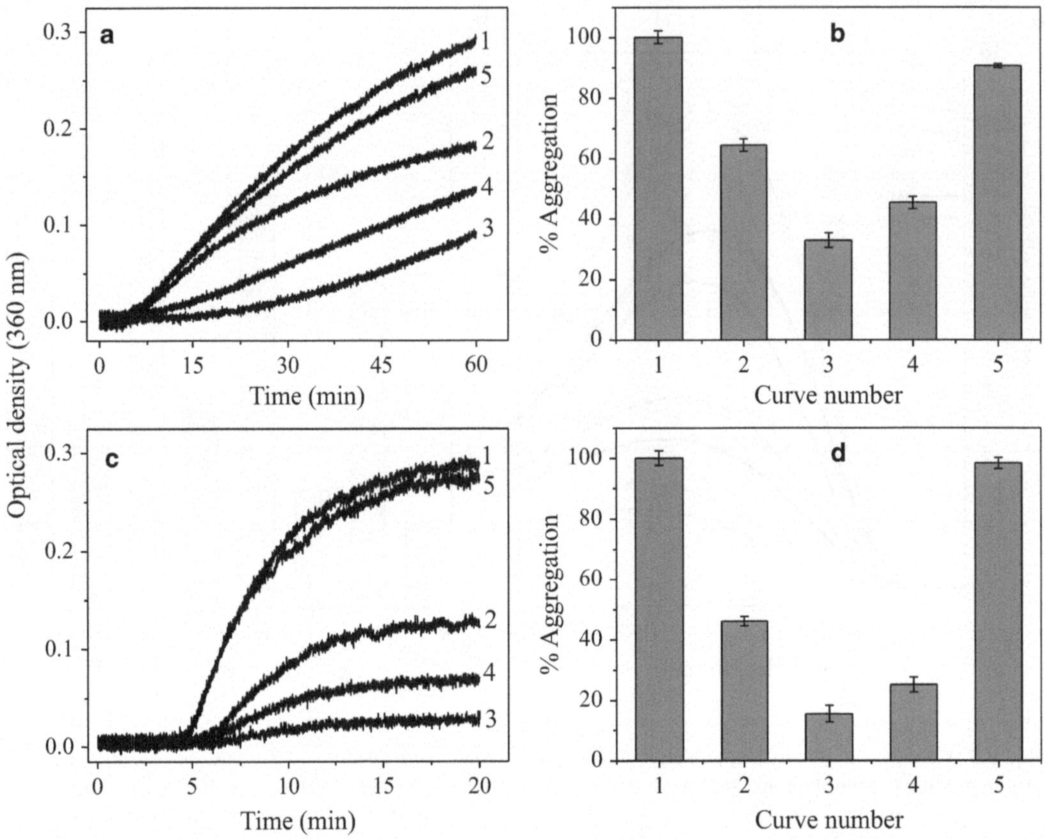

Fig. 5.6 Effect of binding of phospholipid vesicles containing cholesterol on the CLA of HSP-1/2. (**a**) Aggregation profiles correspond to (1) LDH, (2) + HSP-1/2, (3) + HSP-1/2 + 2 μM DMPC, (4) + HSP-1/2 + 2 μM DMPC/Chol and (5) + 2 μM DMPC/Chol. (**b**) Bar diagram showing percent aggregation corresponding to curves in panel **A**. The concentrations of LDH and HSP-1/2 used in this assay are 0.075 mg/ml and 0.015 mg/ml, respectively. (**c**) Aggregation profiles correspond to (1) ENL, (2) + HSP-1/2, (3) + HSP-1/2 + 4 μM DMPC, (4) + HSP-1/2 + 4 μM DMPC/Chol and (5) + 4 μM DMPC/Chol. The concentrations of ENL and HSP-1/2 used in this assay are 0.1 mg/ml and 0.02 mg/ml, respectively

This observation is very similar to the pH dependence of the erythrolytic activity of HSP-1/2, which showed a significant decrease when the pH was increased from pH 6.0 to 8.5 (Kumar and Swamy 2016a).

Further, effect of pH on the chaperone-like activity of PDC-109 (Sankhala and Swamy 2010) was investigated using ADH as a target protein (Fig. 5.8a, b). From this figure, it can be clearly seen that the CLA of PDC-109 is low at mildly acidic pH (pH 6.5 and 6.0) and increases with increase in the pH with the highest activity been seen at pH 8.0 (Fig. 5.8c). Taken together these results clearly demonstrate that both activities of PDC-109, i.e. membranolytic and chaperone-like activities, are inversely related to each other and both activities are regulated by a pH switch (Fig. 5.8d). This is an important observation as pH is known to play a key role in various reproductive processes such as intracellular Ca^{2+} movement, protein tyrosine phosphorylation and hypermotility of spermatozoa (Kirichok and Lishko 2011; McPartlin et al. 2009; Leemans et al. 2014). As PDC-109 is present in the seminal plasma in high concentrations as polydisperse oligomers and also in the membrane-bound form (attached peripherally to the sperm plasma membrane), any change in the pH of the surrounding medium may affect its activities. Since the pH of mammalian seminal

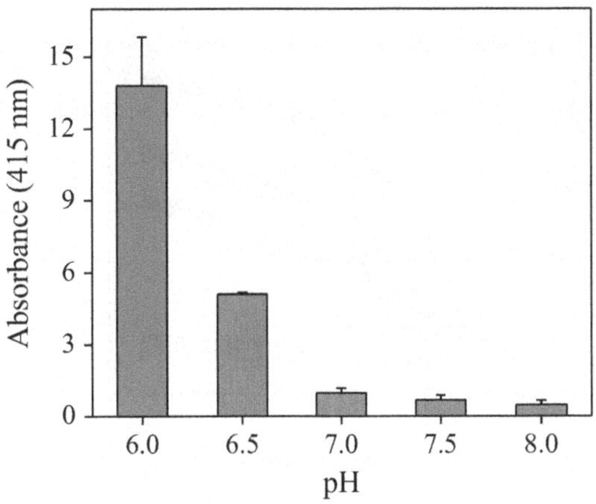

Fig. 5.7 Effect of changing pH on the erythrocyte lysis by PDC-109. Erythrocyte lysis was monitored by measuring the absorbance of the supernatant at 450 nm (corresponding to the haemoglobin released upon lysis) after centrifuging the sample to pellet the intact erythrocytes and cell membranes of lysed erythrocytes. See text for details

plasma is mildly basic in nature, the reduced cell membrane destabilization by PDC-109 at mildly basic conditions is physiologically relevant as it prevents premature destabilization of sperm plasma membrane before reaching the egg. The oviduct-bound sperm shows a time-dependent rise in the pH (Leemans et al. 2014), which may weaken the interaction of PDC-109 with phospholipids and minimize the possibility of premature capacitation. On the other hand, since the spermatozoa encounter a hostile environment (acidic pH) during their passage thorough the female reproductive tract, increased CLA of PDC-109 at low pH is expected to protect the proteins of the seminal plasma and sperm membrane, thus ensuring maintenance of the spermatozoa in a healthy and reproductively active state. In a previous study, we reported that the membranolytic and chaperone-like activities of the major equine seminal plasma protein, HSP-1/2, are also regulated in a similar way, with the CLA being high at mildly basic pH but low at mildly acidic pH, whereas the membranolytic activity was high at mildly acidic pH but low at mildly basic pH, that is, the two activities are inversely related and that a pH switch regulates them (Kumar and Swamy 2016a). The present results demonstrate that a similar mechanism operates in the bovine system, suggesting a common mode of mechanistic regulation of the dual functionality of seminal FnII proteins across species.

In summary, the present study shows that the CLA of HSP-1/2 can be altered/modified by lipids and salinity of the medium. High salt medium results in a decrease of the polydisperse nature of HSP-1/2 and decreases its CLA, whereas binding of choline phospholipids, the physiological ligands of HSP-1/2, results in the formation of lipoprotein complexes which exhibit higher chaperone activity compared to that of HSP-1/2 alone. This increase in CLA of HSP-1/2 can be correlated with increased hydrophobicity of the system in the presence of choline phospholipids. Presence of cholesterol in the membrane decreases the enhancement of the CLA, which could be attributed to the increase in the membrane rigidity and subsequent decrease in the ability of HSP-1/2 to form the lipoprotein complexes or direct interaction with CRAC domain of HSP-1/2 or a combination of both. This is an important result as the *cholesterol efflux* caused by HSP-1/2 results in the formation of lipoprotein aggregates *in vivo*. We propose that such aggregates formed by the interaction of seminal FnII proteins are functionally significant in that they protect other proteins of the seminal plasma against various kinds of stress. The present results together with those obtained earlier with PDC-109 suggest that membrane interaction plays an important role in modulating the activity of membrane-associated chaperones such as Hsp12, spectrin and PDC-109 (Bhattacharyya et al. 2004; Welker et al. 2010; Sankhala et al. 2011). Studies on the pH dependence of the activities of PDC-109 indicate that a pH switch regulates the membrane-destabilizing and chaperone-like activities, with the two activities being inversely correlated to each other. Our results demonstrate that changing physiological conditions such as ionic strength and pH and membrane binding modulate the CLA of seminal FnII proteins and impart new insights into our understanding of the regulation of chaperone

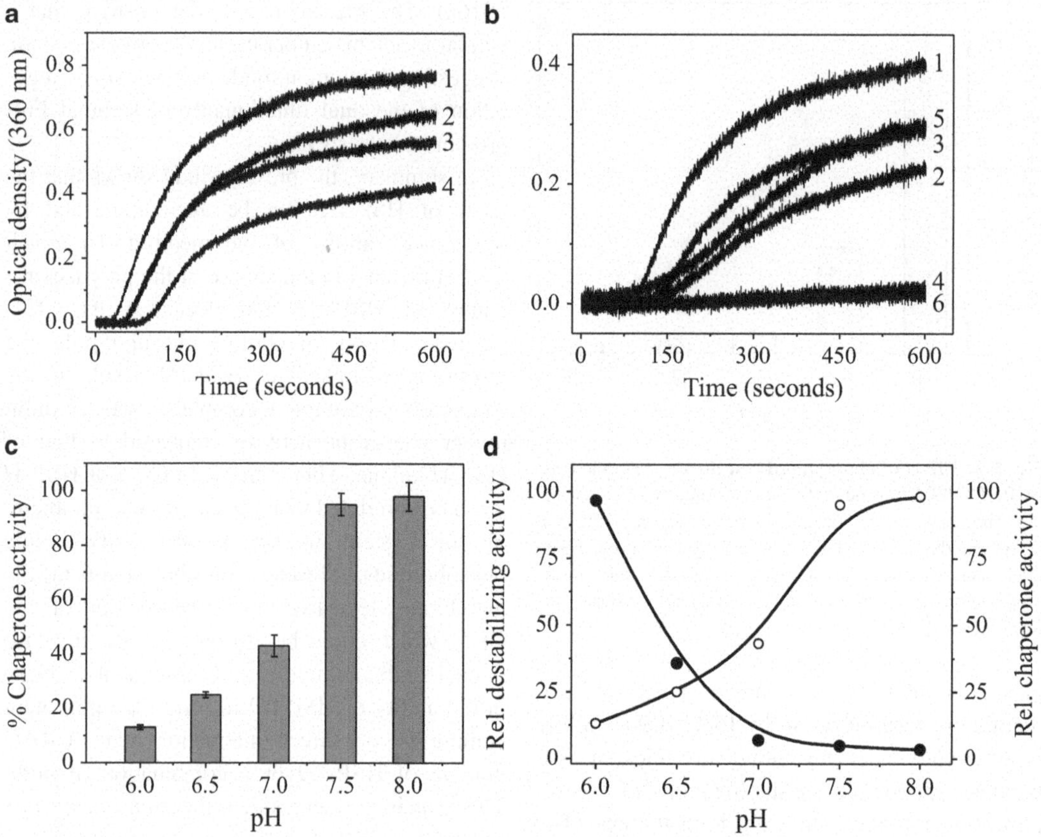

Fig. 5.8 Effect of pH on the CLA and membranolytic activity of PDC-109. (a) Aggregation assay at pH 6.0 (curves 1and 2) and 6.5 (curves 3and 4). Curves 1 and 3: ADH alone. Curves 2 and 4: ADH + PDC-109. (**b**) Aggregation assay at pH 7.0 (curves 1 and 2), 7.5 (curves 3 and 4) and 8.0 (curves 5and 6). Curves 1, 3 and 5: ADH alone. Curves 2, 4 and 6: ADH + PDC-109. Concentrations of ADH and PDC-109 used in (**a**) and (**b**) were 0.05 mg/mL and 0.04 mg/mL, respectively. (**c**) Bar diagram showing % chaperone activity (protection) offered by PDC-109 at different pH. (**d**) pH dependence of erythrocyte membrane-destabilizing activity (open circles) and chaperone-like activity (closed circles)

activity of these proteins by subtle changes in the surrounding environment.

Acknowledgements This work was supported by a research grant from the Department of Science and Technology to MJS. CSK and BPS were supported by Senior Research Fellowships from the CSIR (India) and UGC (India), respectively. We thank Dr. G. Vinu (Lam Farm, Guntur, Andhra Pradesh) for providing samples of bovine semen and Dr. Sanjay K. Ravi (ICAR-National Research Centre on Equines, Bikaner, India) for providing samples of horse semen. Financial and infrastructural support from the University Grants Commission, New Delhi (through the UPE-II and CAS programmes), and the Department of Science and Technology, New Delhi (through the PURSE and FIST programmes), are gratefully acknowledged.

References

Anbazhagan V, Swamy MJ (2005) Thermodynamics of phosphorylcholine and lysophosphatidylcholine binding to the major protein of bovine seminal plasma, PDC-109. FEBS Lett 579:2933–2938

Anbazhagan V, Damai RS, Paul A, Swamy MJ (2008) Interaction of the major protein from bovine seminal plasma, PDC-109 with phospholipid membranes and soluble ligands investigated by fluorescence approaches. Biochim Biophys Acta 1784:891–899

Anbazhagan V, Sankhala RS, Singh BP, Swamy MJ (2011) Isothermal titration calorimetric studies on the interaction of the major bovine seminal plasma protein, PDC-109 with phospholipid membranes. PLoS One 6: e25993

Bakthisaran R, Tangirala R, Rao Ch M (2015) Small heat shock proteins: role in cellular functions and physiology. Biochim Biophys Acta 1854:291–319

Bhattacharyya M, Ray S, Bhattacharya S, Chakrabarthi A (2004) Chaperone activity and prodan binding at the self-association domain of erythroid spectrin. J Biol Chem 53:55080–55088

Calvete JJ, Mann KH, Schafer W, Sanz L, Reinert M, Nessau S, Töpfer-Petersen E (1995) Amino acid sequence of HSP-1, a major protein of stallion seminal plasma: effect of glycosylation on its heparin- and gelatin-binding capabilities. Biochem J 310:615–622

Calvete JJ, Paloma FV, Sanz L, Romero A (1996) A procedure for the large-scale isolation of major bovine seminal plasma proteins. Protein Expr Purif 8:48–56

Calvete JJ, Raida M, Gentzel M, Urbanke C, Sanz L, Töpfer-Petersen E (1997) Isolation and characterization of heparin- and phosphorylcholine-binding proteins of boar and stallion seminal plasma. Primary structure of porcine pB1. FEBS Lett 407:201–206

Damai RS, Sankhala RS, Anbazhagan V, Swamy MJ (2010) ^{31}P-NMR and AFM studies on the cell and model membranes by the major bovine seminal plasma protein, PDC-109. IUBMB Life 62:841–851

Damai RS, Tarafdar PK, Singh BP, Reddy ST, Swamy MJ (2015) Biophysical characterization of the interaction of *O*-acylcholines with the major bovine seminal plasma protein, PDC-109. Adv Exp Med Biol 842:279–292

Desnoyers L, Manjunath P (1992) Major proteins of bovine seminal plasma exhibit novel interaction with phospholipids. J Biol Chem 267:10149–10155

Gasset M, Saiz JL, Sanz L, Gentzel M, Töpfer-Petersen E, Calvete JJ (1997) Conformational features and thermal stability of bovine seminal plasma protein PDC-109 oligomers and phosphoryl choline bound complexes. Eur J Biochem 250:735–744

Gasset M, Magdaleno L, Calvete JJ (2000) Biophysical study of the perturbation of model membrane structure caused by seminal plasma protein PDC-109. Arch Biochem Biophys 374:241–247

Greube A, Müller K, Töpfer-Petersen E, Herrmann A, Müller P (2001) Influence of the bovine seminal plasma protein PDC-109 on the physical state of membrane. Biochemistry 40:8326–8334

Greube A, Müller K, Töpfer-Petersen E, Herrmann A, Müller P (2004) Interaction of Fn type II proteins with membranes: the stallion seminal plasma protein SP-1/2. Biochemistry 43:464–472

Kirichok Y, Lishko PV (2011) Rediscovering sperm ion channels with patch-clamp technique. Mol Hum Reprod 17:478–499

Kumar CS, Swamy MJ (2016a) A pH switch regulates the inverse relationship between membranolytic and chaperone-like activities of HSP-1/2, a major protein of horse seminal plasma. Biochemistry 55:3650–3657

Kumar CS, Swamy MJ (2016b) HSP-1/2 a major horse seminal plasma protein, acts as a chaperone against oxidative stress. Biochem Biophys Res Commun 473:1058–1063

Kumar CS, Swamy MJ (2017a) Differential modulation of chaperone-like activity of HSP-1/2, a major protein of horse seminal plasma by anionic and cationic surfactants. Int J Biol Macromol 96:524–531

Kumar CS, Swamy MJ (2017b) Modulation of chaperone-like and membranolytic activities of major horse seminal plasma protein, HSP-1/2 by L-carnitine. J Biosci 42:469–479

Kumar CS, Sivaramakrishna D, Ravi SK, Swamy MJ (2016) Fluorescence investigations on choline phospholipid binding and chemical unfolding of HSP-1/2, a major protein of horse seminal plasma. J Photochem Photobiol B Biol 158:89–98

Lee GJ, Roseman AM, Saibil HR, Vierling E (1997) A small heat shock protein stably binds heat-denatured model substrates and can maintain a substrate in a folding-competent state. EMBO J 16:659–671

Leemans B, Gadella BM, Sostaric E, Neils H, Stout TA, Hoogewijis M, Van Soom A (2014) Oviduct binding and elevated pH induce protein tyrosine phosphorylation in stallion spermatozoa. Biol Reprod 91(13):1–12

Liang JN, Li XY (1991) Interaction and aggregation of lens crystallins. Exp Eye Res 53:61–66

Manjunath P, Thérein I (2002) Role of seminal plasma phospholipid-binding proteins in sperm membrane lipid modification that occurs during capacitation. J Reprod Immunol 53:109–119

Marsh D, Horváth LI, Swamy MJ, Mantripragada S, Kleinschmidt JH (2002) Interaction of membrane-spanning proteins with peripheral and lipid-anchored membrane proteins: perspectives from protein-lipid interactions. Mol Membr Biol 19:247–255

McPartlin LA, Suarez SS, Czaya CA, Hinrichs K, Bedford-Guaus SJ (2009) Hyper activation of stallion sperm is required for the successful in vitro fertilization of equine oocytes. Biol Reprod 81:199–206

Moparthi SB, Fristedt R, Mishra R, Almstedt K, Karlsson M, Hammarstrom P, Carlsson U (2010) Chaperone activity of Cyp18 through hydrophobic condensation that enables rescue of transient misfolded molten globule intermediates. Biochemistry 49:1137–1145

Plante G, Prud'homme B, Fan J, Lafleur M, Manjunath P (2016) Evolution and function of mammalian binder of sperm proteins. Cell Tissue Res 363:105–127

Ramakrishnan M, Anbazhagan V, Pratap TV, Marsh D, Swamy MJ (2001) Membrane insertion and lipid–protein interactions of bovine seminal plasma protein, PDC-109 investigated by spin label electron spin resonance spectroscopy. Biophys J 81:2215–2225

Sankhala RS, Swamy MJ (2010) The major protein of bovine seminal plasma, PDC-109 is a molecular chaperone. Biochemistry 49:3908–3918

Sankhala RS, Damai RS, Swamy MJ (2011) Correlation of membrane binding and hydrophobicity to the chaperone-like activity of PDC-109, the major protein of bovine seminal plasma. PLoS One 6:e17330

Sankhala RS, Kumar CS, Singh BP, Arangasamy A, Swamy MJ (2012) HSP-1/2, a major protein of equine seminal plasma, exhibits chaperone-like activity. Biochem Biophys Res Commun 427:18–23

Scolari S, Müller K, Bittman R, Hermann A, Müller P (2010) Interaction of mammalian seminal plasma protein, PDC-109 with cholesterol: implications for a putative CRAC domain. Biochemistry 49:9027–9031

Sheluho D, Ackerman SH (2001) An accessible hydrophobic surface is a key element of molecular chaperone action of Atp11p. J Biol Chem 276(43):39945–39949

Swamy MJ (2004) Interaction of bovine seminal plasma proteins with model membranes and sperm plasma membranes. Curr Sci 87:203–211

Swamy MJ, Marsh D, Anbazhagan V, Ramakrishnan M (2002) Effect of cholesterol on the interaction of seminal plasma protein, PDC-109 with phosphatidylcholine membranes. FEBS Lett 528:230–234

Thomas CJ, Anbazhagan V, Ramakrishnan M, Sultan N, Surolia I, Swamy MJ (2003) Mechanism of membrane binding by the bovine seminal plasma protein, PDC-109. A surface plasmon resonance study. Biophys J 84:3037–3044

Voellmy R, Boellmann F (2007) Chaperone regulation of heat shock protein response. Adv Exp Med Biol 594:88–99

Wah DA, Tornero CF, Sanz L, Romero A, Calvete JJ (2002) Sperm coating mechanism from the 1.8 Å crystal structure of PDC-109-phosphorylcholine complex. Structure 10:505–514

Welker S, Rudolph B, Frenzel E, Hagn F, Liebisch G, Schmitz G, Scheuring J, Kerth A, Blume A, Weinkauf S, Haslback M (2010) Hsp12 is an intrinsically unstructured stress protein that folds upon membrane association and modulates membrane function. Mol Cell 39:507–552

Whited AM, Johs A (2015) The interactions of peripheral membrane proteins with biological membranes. Chem Phys Lipids 192:51–59

6 Exploring the Mechanism of Viral Peptide-Induced Membrane Fusion

Gourab Prasad Pattnaik, Geetanjali Meher, and Hirak Chakraborty

Abstract

Membrane fusion is essential in several cellular processes in the existence of eukaryotic cells such as cellular trafficking, compartmentalization, intercellular communication, sexual reproduction, cell division, and endo- and exocytosis. Membrane fusion proceeds in model membranes as well as biological membranes through the rearrangement of lipids. The stalk hypothesis provides a picture of the general nature of lipid rearrangement based on mechanical properties and phase behavior of water-lipid mesomorphic systems. In spite of extensive research on exploring the mechanism of membrane fusion, a clear molecular understanding of intermediate and pore formation is lacking. In addition, the mechanism by which proteins and peptides reduce the activation energy for stalk and pore formation is not yet clear though there are several propositions on how they catalyze membrane fusion. In this review, we have discussed about various putative functions of fusion peptides by which they reduce activation barrier and thus promote membrane fusion. A careful analysis of the discussed effects of fusion peptides on membranes might open up new possibilities for better understanding of the membrane fusion mechanism.

G. P. Pattnaik · G. Meher · H. Chakraborty (✉)
School of Chemistry, Sambalpur University, Burla, Odisha, India
e-mail: hirak@suniv.ac.in

Keywords

Membrane fusion · Bending energy · Void space · Membrane curvature · Depth-dependent membrane ordering · Fusion peptide

6.1 Introduction

Membrane fusion, the merging of two closely apposed lipid bilayers into a single bilayer, is essential for the life of eukaryotic cells. Several events such as cellular trafficking, sexual reproduction, compartmentalization, intercellular communication, endo- and exocytosis, and cell division are dependent on this basic process (Primakoff and Myles 2002; Risselada et al. 2011; Sollner and Rothman 1994; Stein et al. 2004; Verkleij and Post 2000). Additionally, enveloped viruses invade the host cell through membrane fusion (Hughson 1997). Extensive studies on understanding the fusion process led to the agreement that fusion process proceeds in both model membranes and protein-laden biological membranes through rearrangement of lipids (Chernomordik and Kozlov 2008; Lee and Lentz 1998). The stalk hypothesis provides a picture of the general nature of lipid rearrangement based on mechanical properties and phase behavior of water-lipid mesomorphic systems (Kozlov et al. 1989; Malinin and Lentz 2004). According to the stalk hypothesis, the outer

K. Chattopadhyay, S. C. Basu (eds.), *Biochemical and Biophysical Roles of Cell Surface Molecules*, Advances in Experimental Medicine and Biology 1112, https://doi.org/10.1007/978-981-13-3065-0_6

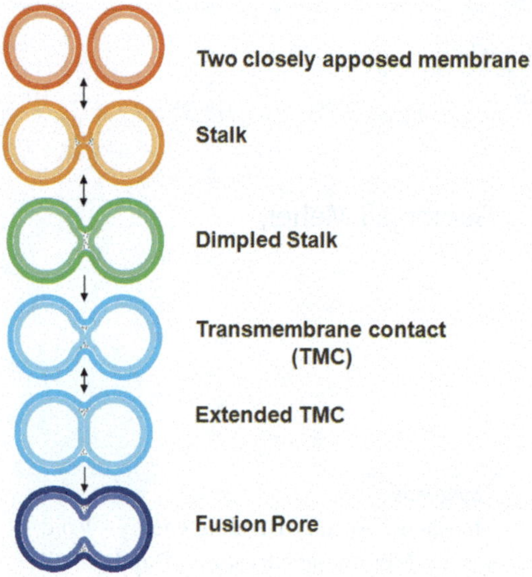

Fig. 6.1 Schematic representation of stepwise fusion process

leaflets of each bilayer merge to form the semi-stable stalk intermediate (Yang and Huang 2003), where the lipid arrangements are more similar to the non-lamellar phase than the lamellar lipid phases. The stalk evolves via a transmembrane contact (TMC) to a semi-stable intermediate (extended transmembrane contact, ETMC) followed by the opening of a fusion pore (Lee and Lentz 1997; Siegel 1999). A schematic representation of stepwise fusion process has been shown in Fig. 6.1.

Enveloped viruses exploit membrane fusion to enter the host cell. Some viruses fuse with the plasma membrane at the cell surface (Dimitrov et al. 1991; Freed et al. 1990), whereas others gain entry by receptor-mediated endocytosis followed by fusion with the endosome (Earp et al. 2005; Skehel and Wiley 2000). The fusion process is triggered by either receptor-binding (as in the entry of HIV) or low pH that prevails in the endosome (as in the case of influenza). A special class of proteins, namely, fusion proteins, present in the viral envelope, play a crucial role in the fusion of virus with the host cell. Furthermore, a stretch of 20–25 amino acids at the N-terminus of the fusion protein plays a key role in promoting membrane fusion (Chakraborty et al. 2013, 2017; Gething et al. 1986). This stretch of amino acids is termed as "fusion peptide," and mutations in this region have been shown to block fusion-mediated viral infection (Danieli et al. 1996; Epand 2003; Li et al. 2005).

Despite extensive research on membrane fusion, a decent molecular understanding of the two key steps in the process, initial intermediate formation and final pore formation, is lacking. Moreover, the mechanism of reduction of activation energy for stalk and fusion pore formation, by the membrane proteins and peptides, remains unclear. The major challenge in this issue is posed by the transient and unfavorable fluctuation in lipid/water arrangement, which is difficult to assess and characterize experimentally. In mechanistic terms, such unstable transient molecular arrangements are viewed as activated intermediates that provide free energy barrier for a reaction. It is essential to understand these transient molecular arrangements (i.e., activation barriers) if we are to identify the role of fusion proteins in reducing these free energy barriers to catalyze in vivo fusion process (Lentz et al. 2000). Several putative functions of fusion peptides, in terms of their effect on the physical properties and dynamics of the membrane, have been proposed. In this report, we have discussed the various putative functions of fusion peptides in promoting membrane fusion.

6.2 Fusion Peptide Reduces Bending Energy

To elucidate the effect of fusion peptide on bending energy, Tristram-Nagle et al. measured the bending modulus of pure lipid bilayer both in absence and presence of gp41 fusion peptide utilizing diffuse x-ray scattering (Tristram-Nagle and Nagle 2007). The diffuse x-ray scattering avoids the complication of quantitative structure determination of fully hydrated lipid bilayer, which is essential to mimic the physiological environment (Kucerka et al. 2005a, b, 2006; Liu

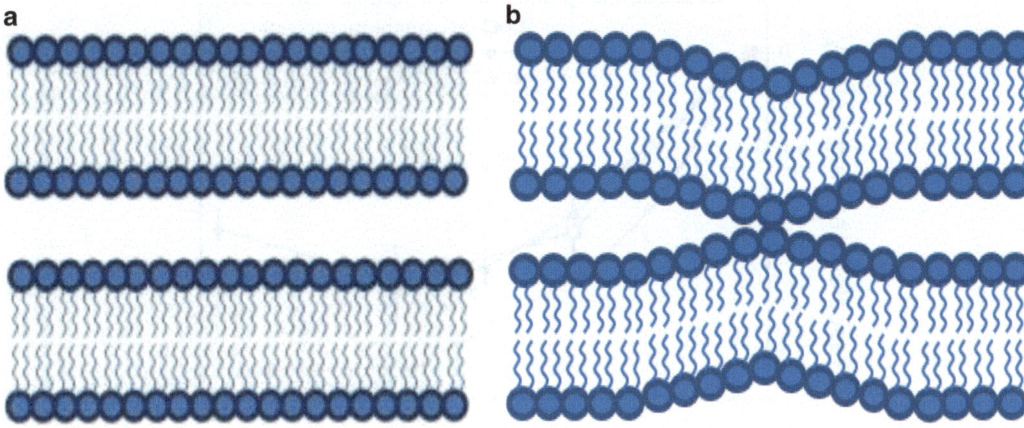

Fig. 6.2 Schematic representation of (**a**) two closely apposed bilayers and (**b**) bending of two closely apposed bilayers, a necessary step for inducing membrane fusion

and Nagle 2004). Diffuse x-ray scattering provides information about membrane-bending modulus K_C, which is a measure of the amount of energy required to bend the membrane $\left(E = \frac{1}{2}K_C C^2\right.$, where "$C$" is the curvature of the membrane). Figure 6.2 shows the bending of two bilayers, which is a key step in the intermediate formation; the ease of bending is expected to reduce the activation barrier for the intermediate formation. It has been shown that gp41 fusion peptide (a 23-amino acid stretch at the N-terminus of gp41 protein) reduces bending modulus by a factor of 13 for the thicker and stiffer 1,2-*sn*-dierucoylphosphatidylcholine bilayer and by a factor of 3 for 1,2-*sn*-dioleoylphosphatidylcholine bilayer (Tristram-Nagle and Nagle 2007). In addition, the bending modulus reduces exponentially with increasing peptide concentration.

The bending modulus has further been studied using fluctuation analysis and aspiration with micropipettes (Shchelokovskyy et al. 2011). The effect of gp41 fusion peptide on the bending modulus has been measured in giant unilamellar vesicles, which is a better model for cell membrane. Results from both the methods showed that membrane stiffness gradually reduces with increasing concentration of gp41 fusion peptide. Figure 6.3 shows the effect of gp41 fusion peptide on the bending modulus, measured using different techniques.

Small-angle x-ray diffraction data further ascertains the ability of fusion peptide to reduce bending energy, which eventually favors the formation of highly curved intermediates. The small-angle x-ray diffraction measurement of dipalmitoleoylphosphatidylethanolamine (DPoPE) bilayer showed the formation of cubic phase (Q_{II}) in the presence of 20-amino acid N-terminal stretch of hemagglutinin fusion peptide (Tenchov et al. 2013). In addition, the lattice parameter of the cubic phase changes with the concentration of hemagglutinin fusion peptide. Interestingly, the fusion-inefficient mutant hemagglutinin peptides have insignificant effect on the cubic phase formation.

6.3 Fusion Peptide Fills the Void Space in the Intermediates

The void space between the bilayer leaflets, formed in the hemifusion intermediate, makes hydrophobic interstices energetically unfavorable (Siegel 1999). Hydrocarbons such as hexadecane and tetradecane, which can occupy these voids, are known to facilitate vesicular fusion (Basanez et al. 1998; Chakraborty et al. 2013; Chanturiya et al. 1999; Chen and Rand 1998; Rand et al. 1990; Walter et al. 1994) by reducing the energy of the hydrophobic interstices. The same

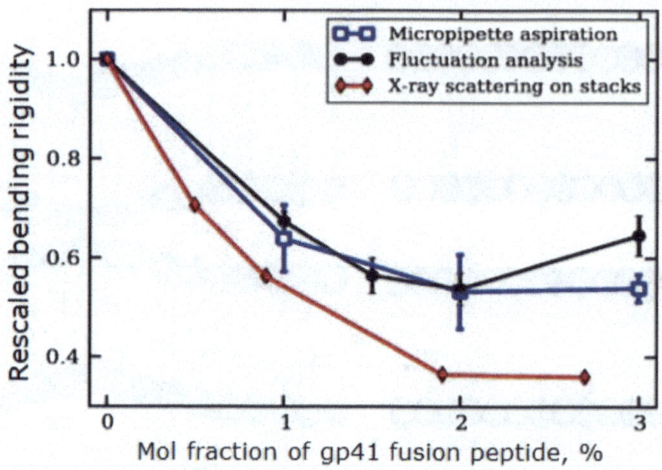

Fig. 6.3 Relative change of bending rigidity as a function of the concentration of gp41 fusion peptide in DOPC bilayers. Different datasets correspond to the results obtained with different methods, fluctuation analysis (filled circles), micropipette aspiration (open squares), and diffuse x-ray scattering from Tristram-Nagle and Nagle 2007 (filled diamonds). (Adapted and modified from Shchelokovskyy et al. 2011 with permission from Attribution-NonCommercial-ShareAlike 3.0 Unported (CC BY-NC-SA 3.0) of creative commons)

mechanism is followed by the fusion peptides in promoting hemifusion intermediate.

The ability of fusion peptide to fill the void volume has been estimated by measuring fluorescence lifetime of C_6-NBD-PC. The external addition of C_6-NBD-PC into preformed vesicles allows it to partition into the membrane, the fraction partitioned being proportional to the void space available in the bilayer (Haque et al. 2005). The same group also demonstrated that both gp41 and hemagglutinin fusion peptides reduce the membrane partitioning of C_6-NBD-PC and rather promote lipid mixing in polyethylene glycol-mediated fusion of small unilamellar vesicles (Haque et al. 2005). This leads to the assumption that both the fusion peptides fill the void space available in the bilayer leaflet that impedes the partitioning of C_6-NBD-PC in the membrane. The stability of curved intermediate promotes lipid mixing in the membrane fusion reaction. Figure 6.4 shows a schematic representation of how the fusion peptide can occupy the void space in nonlinear intermediates.

The transmembrane domain of vesicular stomatitis virus fusion protein (TMD-VSV) is also known to fill the void space of hemifusion intermediate to promote membrane fusion (Sengupta et al. 2014). This claim has been validated further by the fact that TMD-VSV competes with hexadecane in promoting the rate of intermediate formation. The stabilization of the intermediate structure should be optimized such that it facilitates hemifusion intermediate formation, but does not inhibit the downstream steps of the process. It is important to mention that G1S mutant of hemagglutinin fusion peptides promotes hemifusion intermediate formation, but restricts complete pore opening due to excessive stabilization of the intermediate (Li et al. 2005).

6.4 Fusion Peptide Alters the Curvature of the Membrane

Although the elastic property of lipid bilayers does not allow formation of spontaneous curvature, there are specific lipids, proteins, and cytoskeletal components that play important role in stabilizing the membrane curvature. Depending on the lipid geometry, i.e., headgroup-to-acyl

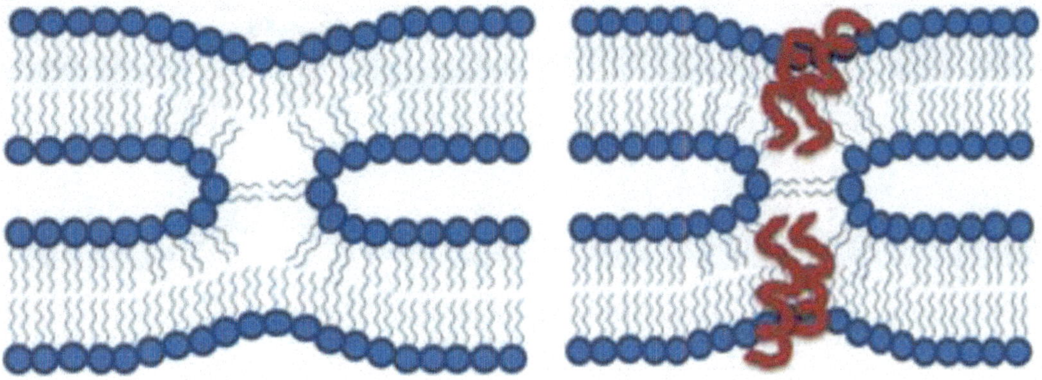

Fig. 6.4 Schematic representation of (**a**) formation of void volume in the fusion intermediate and (**b**) filling of void volume by peptide, partitioning into the membrane, hence stabilizing the fusion intermediate

chain length ratio, membrane curvature gets established (McMahon and Boucrot 2015). Lipids having smaller headgroup-to-acyl chain ratio, such as phosphatidylethanolamine (PE), form a conical shape and allow closer proximity of headgroups to stabilize a negative curvature of the membrane. On the other hand, lipids having higher headgroup-to-acyl chain ratio, such as phosphatidylinositol (PI), form inverted cones where acyl chains could be more compact compared to the headgroups, leading to positive curvature of the membrane (Chernomordik and Kozlov 2003; Di Paolo and De Camilli 2006; Zimmerberg and Kozlov 2006). This lipid-dependent membrane curvature is, however, not very relevant in the context of cellular membranes, because of the heterogeneous lipid composition and natural elasticity of the cell membrane. It has been shown that coat proteins such as clathrin, COPI, and COPII are key factors for stabilizing the membrane during vesicle budding (Jensen and Schekman 2011; Kirchhausen 2000; McMahon and Boucrot 2011). Membrane binding of crescent-shaped BAR (Bin/Amphiphysin/Rvs) domains and their oligomerization induce membrane curvature. BAR domains favor the formation of membrane tubules but inhibit membrane scission, which requires extreme curvature (Boucrot et al. 2012; Peter et al. 2004).

According to the stalk model, two membranes must join to form a stalk before proceeding to a transmembrane contact and eventually form a fusion pore (Chernomordik and Kozlov 2005; Kozlov et al. 1989; Markin and Albanesi 2002). The hourglass-shaped, high-energy fusion-stalk intermediate has a greater degree of concave (negative) curvature along one axis and that of convex (positive) curvature along the other. The inverted-wedge-shaped dioleoylphosphatidylethanolamine (DOPE) stabilizes the negative curvature and promotes vesicular fusion (Leikin et al. 1996; Zimmerberg and Kozlov 2006), whereas wedge-shaped lysophosphatidylcholine (LPC) promotes the positive curvature, thereby inhibiting membrane fusion (Chernomordik 1996; Epand 1985). Detailed crystallographic and differential scanning calorimetric experiments have demonstrated that hemagglutinin fusion peptide promotes negative curvature and hence the formation of hexagonal and cubic phases of lipids (Epand and Epand 1994; Siegel and Epand 2000; Tenchov et al. 2013). The hemagglutinin fusion peptide has been shown to order the headgroup and condense the interfacial region (Ge and Freed 2009). NMR study has further shown that, wild-type hemagglutinin fusion peptide and its F3G mutant, both assume helical hairpin structure in order to bind to the membrane with equal efficiency. On the other hand, the F3G mutant exhibits a reduced propensity to stabilize the negative curvature compared to the wild-type peptide, resulting in attenuated fusion activity in the former relative to the latter (Smrt et al. 2015).

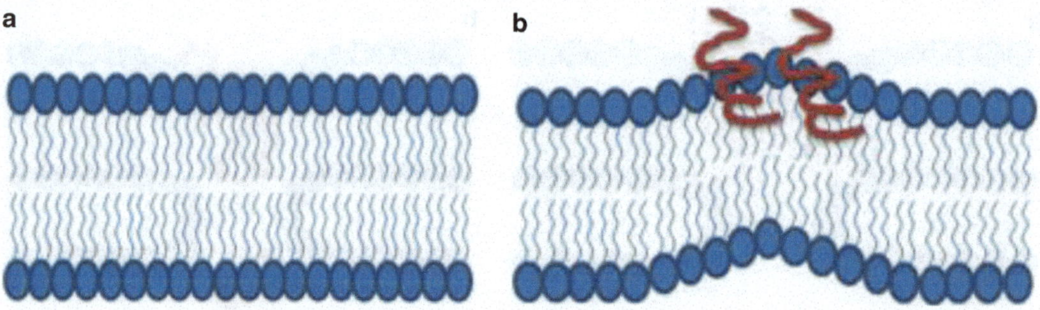

Fig. 6.5 Schematic representation of (**a**) flat lipid bilayer and (**b**) membrane curvature induced by peptide. The curvature formation is essential to initiate membrane fusion and its completion

Further, the extent of curvature formed at the region of insertion depends on the depth of insertion; shallower insertion leads to maximum change in curvature (Campelo et al. 2008; Kozlov et al. 2014).

Progress of the stalk toward transmembrane contact and fusion pore requires the generation of positive curvature. The ability of a peptide to induce negative curvature at the initial stage and positive curvature thereafter makes it a suitable candidate as fusion peptide. The hemagglutinin fusion peptide possesses this unique ability to alter the membrane curvature to accomplish membrane fusion (Chakraborty et al. 2013). The fusion domain of the feline leukemia virus, simian immunodeficiency virus (SIV), reduces the transition temperature of dipalmitoleoylphosphatidylethanolamine from lamellar to hexagonal phase, which is characterized with higher negative curvature (Davies et al. 1998; Epand et al. 1994). Figure 6.5 shows a schematic representation of how the fusion peptide can promote membrane curvature and induce formation of nonlinear intermediates.

6.5 Fusion Peptide Induces Depth-Dependent Membrane Ordering

The electron spin resonance recordings, using spin-labeled lipid probes, support the hypothesis that wild-type hemagglutinin fusion peptide orders the headgroup region and promotes negative intrinsic curvature (Ge and Freed 2009; Lai and Freed 2014; Lai et al. 2006). However, the peptide-induced ordering of headgroup region could not be directly correlated with the fusogenic ability of the peptide according to the fluorescence studies using fluorescent probes like DPH, TMA-DPH, and C_6-NBD-PC (Chakraborty et al. 2013; Haque et al. 2005, 2011). This contradiction indicates that while the peptide-induced ordering is important, the exact location of bilayer ordering that helps in membrane fusion also needs further attention.

Depth-dependent membrane ordering, induced by fusion peptides, has been estimated using n-(9-anthroyloxy)-stearic acid (n-AS) probes, in which the anthroyloxy group is attached at various positions of the alkyl chain of stearic acid (Chakraborty et al. 2017). These probes are known to be located at a varying depths in the bilayer, depending on the location of the attached anthroyloxy group on the fatty acid chain (Abrams et al. 1992; Abrams and London 1993; Chakraborty et al. 2015; Haldar et al. 2012). The wild-type hemagglutinin fusion peptide was found to order the bilayer between 10 and 14 Å from the center of the bilayer; this ordering seems to be crucial to manifest the fusogenic ability of the peptide. Two other mutants of hemagglutinin fusion peptide, G1S and W14A, do not order this region and hence fail to promote fusion pore formation, despite the stalk formation (Fig. 6.6). G1V promotes the same region as wild type, but the extent of interfacial ordering is relatively low. As it is evident from Fig. 6.1, the elastic stability of the bilayer is essential to allow the expansion of the transmembrane contact to open the fusion

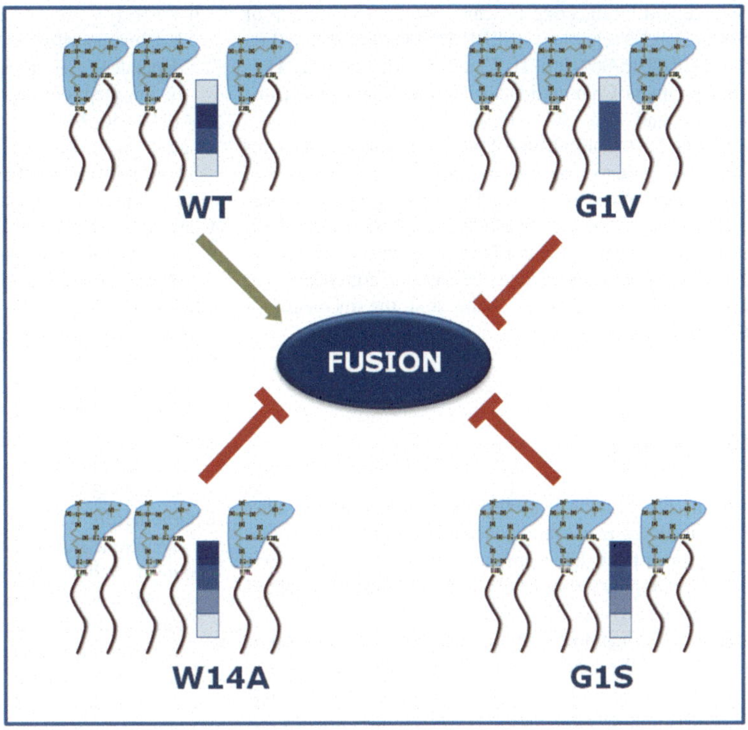

Fig. 6.6 Schematic representation of depth-dependent membrane ordering by wild-type and mutant hemagglutinin fusion peptides. Darker color indicates stronger peptide-induced effect at that location. Wild-type peptide significantly orders the interfacial region, whereas another fusion-inefficient mutant fails to do that. (Adapted with permission from Chakraborty et al. 2017. Copyright (2017) American Chemical Society)

pore. Wild-type fusion peptide increases the mechanical stability of the interfacial region by increasing its order, hence enabling expansion of the transmembrane contact (Chakraborty et al. 2013; Malinin and Lentz 2004). Taken together, it is important to order the bilayer at a particular region to enhance the mechanical stability of the transmembrane contact, enabling it to expand and open the fusion pore.

6.6 Conclusion and Future Perspectives

Although we have discussed only representative examples regarding the possible role of fusion peptides in promoting membrane fusion, there exist many other pathways that could potentially provide the favorable energy to promote membrane fusion. With the advent of x-ray scattering, calorimetric, and spectroscopic methodologies, the understanding of peptide-lipid interaction in the context of membrane fusion is gaining momentum. The limitation of these measurements, however, is that the effect of fusion peptides on bilayers is being recorded under equilibrium conditions, rather than during the actual course of fusion; it would be more relevant to monitor the membrane properties during the fusion process. Moreover, kinetic measurements related to the generation of different intermediate states could provide more detailed information about the possible role of fusion peptides in promoting membrane fusion. Recently, efforts have been made to understand the structure of the transition states during the fusion pathway utilizing activation free energy, enthalpy, and entropy calculations from temperature-dependent kinetic measurements (Chakraborty et al. 2012, 2014, 2013; Sengupta et al. 2014). Careful analysis of these results might open up new possibilities for better understanding of the membrane fusion mechanism.

Acknowledgements This work was supported by research grants from the University Grants Commission, New Delhi (File No. F.4-5(138-FRP)/2014(BSR)), and

Science and Engineering Research Board, Department of Science and Technology (SERB-DST), New Delhi (File No. ECR/2015/000195). H. C. and G. M. thank the University Grants Commission for UGC-Assistant Professor position and UGC-BSR Research Fellowship, respectively. G. P. P. thanks SERB-DST for his project assistantship. We thank the Department of Science and Technology, New Delhi, and UGC for providing instrument facility to the School of Chemistry, Sambalpur University, under the FIST and DRS programs, respectively. We gratefully acknowledge the critical comments and discussions by Dr. S. N. Sahu and the members of Chakraborty group.

References

Abrams FS, London E (1993) Extension of the parallax analysis of membrane penetration depth to the polar region of model membranes: use of fluorescence quenching by a spin-label attached to the phospholipid polar headgroup. Biochemistry 32:10826–10831

Abrams FS, Chattopadhyay A, London E (1992) Determination of the location of fluorescent probes attached to fatty acids using parallax analysis of fluorescence quenching: effect of carboxyl ionization state and environment on depth. Biochemistry 31:5322–5327

Basanez G, Goni FM, Alonso A (1998) Effect of single chain lipids on phospholipase C-promoted vesicle fusion. A test for the stalk hypothesis of membrane fusion. Biochemistry 37:3901–3908

Boucrot E, Pick A, Camdere G, Liska N, Evergren E, McMahon HT, Kozlov MM (2012) Membrane fission is promoted by insertion of amphipathic helices and is restricted by crescent BAR domains. Cell 149:124–136

Campelo F, McMahon HT, Kozlov MM (2008) The hydrophobic insertion mechanism of membrane curvature generation by proteins. Biophys J 95:2325–2339

Chakraborty H, Tarafdar PK, Bruno MJ, Sengupta T, Lentz BR (2012) Activation thermodynamics of poly (ethylene glycol)-mediated model membrane fusion support mechanistic models of stalk and pore formation. Biophys J 102:2751–2760

Chakraborty H, Tarafdar PK, Klapper DG, Lentz BR (2013) Wild-type and mutant hemagglutinin fusion peptides alter bilayer structure as well as kinetics and activation thermodynamics of stalk and pore formation differently: mechanistic implications. Biophys J 105:2495–2506

Chakraborty H, Sengupta T, Lentz BR (2014) pH alters PEG-mediated fusion of phosphatidylethanolamine-containing vesicles. Biophys J 107:1327–1338

Chakraborty H, Haldar S, Chong PL, Kombrabail M, Krishnamoorthy G, Chattopadhyay A (2015) Depth-dependent organization and dynamics of archaeal and eukaryotic membranes: development of membrane anisotropy gradient with natural evolution. Langmuir 31:11591–11597

Chakraborty H, Lentz BR, Kombrabail M, Krishnamoorthy G, Chattopadhyay A (2017) Depth-dependent membrane ordering by hemagglutinin fusion peptide promotes fusion. J Phys Chem B 121:1640–1648

Chanturiya A, Leikina E, Zimmerberg J, Chernomordik LV (1999) Short-chain alcohols promote an early stage of membrane hemifusion. Biophys J 77:2035–2045

Chen Z, Rand RP (1998) Comparative study of the effects of several n-alkanes on phospholipid hexagonal phases. Biophys J 74:944–952

Chernomordik L (1996) Non-bilayer lipids and biological fusion intermediates. Chem Phys Lipids 81:203–213

Chernomordik LV, Kozlov MM (2003) Protein-lipid interplay in fusion and fission of biological membranes. Annu Rev Biochem 72:175–207

Chernomordik LV, Kozlov MM (2005) Membrane hemifusion: crossing a chasm in two leaps. Cell 123:375–382

Chernomordik LV, Kozlov MM (2008) Mechanics of membrane fusion. Nat Struct Mol Biol 15:675–683

Danieli T, Pelletier SL, Henis YI, White JM (1996) Membrane fusion mediated by the influenza virus hemagglutinin requires the concerted action of at least three hemagglutinin trimers. J Cell Biol 133:559–569

Davies SM, Epand RF, Bradshaw JP, Epand RM (1998) Modulation of lipid polymorphism by the feline leukemia virus fusion peptide: implications for the fusion mechanism. Biochemistry 37:5720–5729

Di Paolo G, De Camilli P (2006) Phosphoinositides in cell regulation and membrane dynamics. Nature 443:651–657

Dimitrov DS, Golding H, Blumenthal R (1991) Initial stages of HIV-1 envelope glycoprotein-mediated cell fusion monitored by a new assay based on redistribution of fluorescent dyes. AIDS Res Hum Retrovir 7:799–805

Earp LJ, Delos SE, Park HE, White JM (2005) The many mechanisms of viral membrane fusion proteins. Curr Top Microbiol Immunol 285:25–66

Epand RM (1985) Diacylglycerols, lysolecithin, or hydrocarbons markedly alter the bilayer to hexagonal phase transition temperature of phosphatidylethanolamines. Biochemistry 24:7092–7095

Epand RM (2003) Fusion peptides and the mechanism of viral fusion. Biochim Biophys Acta 1614:116–121

Epand RM, Epand RF (1994) Relationship between the infectivity of influenza virus and the ability of its fusion peptide to perturb bilayers. Biochem Biophys Res Commun 202:1420–1425

Epand RF, Martin I, Ruysschaert JM, Epand RM (1994) Membrane orientation of the SIV fusion peptide determines its effect on bilayer stability and ability to promote membrane fusion. Biochem Biophys Res Commun 205:1938–1943

Freed EO, Myers DJ, Risser R (1990) Characterization of the fusion domain of the human immunodeficiency virus type 1 envelope glycoprotein gp41. Proc Natl Acad Sci U S A 87:4650–4654

Ge M, Freed JH (2009) Fusion peptide from influenza hemagglutinin increases membrane surface order: an electron-spin resonance study. Biophys J 96:4925–4934

Gething MJ, Doms RW, York D, White J (1986) Studies on the mechanism of membrane fusion: site-specific mutagenesis of the hemagglutinin of influenza virus. J Cell Biol 102:11–23

Haldar S, Kombrabail M, Krishnamoorthy G, Chattopadhyay A (2012) Depth-dependent heterogeneity in membranes by fluorescence lifetime distribution analysis. J Phys Chem Lett 3:2676–2681

Haque ME, Koppaka V, Axelsen PH, Lentz BR (2005) Properties and structures of the influenza and HIV fusion peptides on lipid membranes: implications for a role in fusion. Biophys J 89:3183–3194

Haque ME, Chakraborty H, Koklic T, Komatsu H, Axelsen PH, Lentz BR (2011) Hemagglutinin fusion peptide mutants in model membranes: structural properties, membrane physical properties, and PEG-mediated fusion. Biophys J 101:1095–1104

Hughson FM (1997) Enveloped viruses: a common mode of membrane fusion? Curr Biol 7:R565–R569

Jensen D, Schekman R (2011) COPII-mediated vesicle formation at a glance. J Cell Sci 124:1–4

Kirchhausen T (2000) Three ways to make a vesicle. Nat Rev Mol Cell Biol 1:187–198

Kozlov MM, Leikin SL, Chernomordik LV, Markin VS, Chizmadzhev YA (1989) Stalk mechanism of vesicle fusion. Intermixing of aqueous contents. Eur Biophys J 17:121–129

Kozlov MM, Campelo F, Liska N, Chernomordik LV, Marrink SJ, McMahon HT (2014) Mechanisms shaping cell membranes. Curr Opin Cell Biol 29:53–60

Kucerka N, Liu Y, Chu N, Petrache HI, Tristram-Nagle S, Nagle JF (2005a) Structure of fully hydrated fluid phase DMPC and DLPC lipid bilayers using X-ray scattering from oriented multilamellar arrays and from unilamellar vesicles. Biophys J 88:2626–2637

Kucerka N, Tristram-Nagle S, Nagle JF (2005b) Structure of fully hydrated fluid phase lipid bilayers with monounsaturated chains. J Membr Biol 208:193–202

Kucerka N, Tristram-Nagle S, Nagle JF (2006) Closer look at structure of fully hydrated fluid phase DPPC bilayers. Biophys J 90:L83–L85

Lai AL, Freed JH (2014) HIV gp41 fusion peptide increases membrane ordering in a cholesterol-dependent fashion. Biophys J 106:172–181

Lai AL, Park H, White JM, Tamm LK (2006) Fusion peptide of influenza hemagglutinin requires a fixed angle boomerang structure for activity. J Biol Chem 281:5760–5770

Lee J, Lentz BR (1997) Evolution of lipidic structures during model membrane fusion and the relation of this process to cell membrane fusion. Biochemistry 36:6251–6259

Lee J, Lentz BR (1998) Secretory and viral fusion may share mechanistic events with fusion between curved lipid bilayers. Proc Natl Acad Sci U S A 95:9274–9279

Leikin S, Kozlov MM, Fuller NL, Rand RP (1996) Measured effects of diacylglycerol on structural and elastic properties of phospholipid membranes. Biophys J 71:2623–2632

Lentz BR, Malinin V, Haque ME, Evans K (2000) Protein machines and lipid assemblies: current views of cell membrane fusion. Curr Opin Struct Biol 10:607–615

Li Y, Han X, Lai AL, Bushweller JH, Cafiso DS, Tamm LK (2005) Membrane structures of the hemifusion-inducing fusion peptide mutant G1S and the fusion-blocking mutant G1V of influenza virus hemagglutinin suggest a mechanism for pore opening in membrane fusion. J Virol 79:12065–12076

Liu Y, Nagle JF (2004) Diffuse scattering provides material parameters and electron density profiles of biomembranes. Phys Rev E Stat Nonlinear Soft Matter Phys 69:040901

Malinin VS, Lentz BR (2004) Energetics of vesicle fusion intermediates: comparison of calculations with observed effects of osmotic and curvature stresses. Biophys J 86:2951–2964

Markin VS, Albanesi JP (2002) Membrane fusion: stalk model revisited. Biophys J 82:693–712

McMahon HT, Boucrot E (2011) Molecular mechanism and physiological functions of clathrin-mediated endocytosis. Nat Rev Mol Cell Biol 12:517–533

McMahon HT, Boucrot E (2015) Membrane curvature at a glance. J Cell Sci 128:1065–1070

Peter BJ, Kent HM, Mills IG, Vallis Y, Butler PJ, Evans PR, McMahon HT (2004) BAR domains as sensors of membrane curvature: the amphiphysin BAR structure. Science 303:495–499

Primakoff P, Myles DG (2002) Penetration, adhesion, and fusion in mammalian sperm-egg interaction. Science 296:2183–2185

Rand RP, Fuller NL, Gruner SM, Parsegian VA (1990) Membrane curvature, lipid segregation, and structural transitions for phospholipids under dual-solvent stress. Biochemistry 29:76–87

Risselada HJ, Kutzner C, Grubmuller H (2011) Caught in the act: visualization of SNARE-mediated fusion events in molecular detail. Chembiochem 12:1049–1055

Sengupta T, Chakraborty H, Lentz BR (2014) The transmembrane domain peptide of vesicular stomatitis virus promotes both intermediate and pore formation during PEG-mediated vesicle fusion. Biophys J 107:1318–1326

Shchelokovskyy P, Tristram-Nagle S, Dimova R (2011) Effect of the HIV-1 fusion peptide on the mechanical properties and leaflet coupling of lipid bilayers. New J Phys 13:25004

Siegel DP (1999) The modified stalk mechanism of lamellar/inverted phase transitions and its implications for membrane fusion. Biophys J 76:291–313

Siegel DP, Epand RM (2000) Effect of influenza hemagglutinin fusion peptide on lamellar/inverted phase transitions in dipalmitoleoylphosphatidylethanolamine:

implications for membrane fusion mechanisms. Biochim Biophys Acta 1468:87–98
Skehel JJ, Wiley DC (2000) Receptor binding and membrane fusion in virus entry: the influenza hemagglutinin. Annu Rev Biochem 69:531–569
Smrt ST, Draney AW, Lorieau JL (2015) The influenza hemagglutinin fusion domain is an amphipathic helical hairpin that functions by inducing membrane curvature. J Biol Chem 290:228–238
Sollner T, Rothman JE (1994) Neurotransmission: harnessing fusion machinery at the synapse. Trends Neurosci 17:344–348
Stein KK, Primakoff P, Myles D (2004) Sperm-egg fusion: events at the plasma membrane. J Cell Sci 117:6269–6274
Tenchov BG, MacDonald RC, Lentz BR (2013) Fusion peptides promote formation of bilayer cubic phases in lipid dispersions. An x-ray diffraction study. Biophys J 104:1029–1037
Tristram-Nagle S, Nagle JF (2007) HIV-1 fusion peptide decreases bending energy and promotes curved fusion intermediates. Biophys J 93:2048–2055
Verkleij AJ, Post JA (2000) Membrane phospholipid asymmetry and signal transduction. J Membr Biol 178:1–10
Walter A, Yeagle PL, Siegel DP (1994) Diacylglycerol and hexadecane increase divalent cation-induced lipid mixing rates between phosphatidylserine large unilamellar vesicles. Biophys J 66:366–376
Yang L, Huang HW (2003) A rhombohedral phase of lipid containing a membrane fusion intermediate structure. Biophys J 84:1808–1817
Zimmerberg J, Kozlov MM (2006) How proteins produce cellular membrane curvature. Nat Rev Mol Cell Biol 7:9–19

Amyloids Are Novel Cell-Adhesive Matrices

7

Reeba S. Jacob, Subhadeep Das, Namrata Singh, Komal Patel, Debalina Datta, Shamik Sen, and Samir K. Maji

Abstract

Amyloids are highly ordered peptide/protein aggregates traditionally associated with multiple human diseases including neurodegenerative disorders. However, recent studies suggest that amyloids can also perform several biological functions in organisms varying from bacteria to mammals. In many lower organisms, amyloid fibrils function as adhesives due to their unique surface topography. Recently, amyloid fibrils have been shown to support attachment and spreading of mammalian cells by interacting with the cell membrane and by cell adhesion machinery activation. Moreover, similar to cellular responses on natural extracellular matrices (ECMs), mammalian cells on amyloid surfaces also use integrin machinery for spreading, migration, and differentiation. This has led to the development of biocompatible and implantable amyloid-based hydrogels that could induce lineage-specific differentiation of stem cells. In this chapter, based on adhesion of both lower organisms and mammalian cells on amyloid nanofibrils, we posit that amyloids could have functioned as a primitive extracellular matrix in primordial earth.

Keywords

Amyloids · Protein aggregates · Cell adhesion · Extracellular matrix · Tissue engineering

7.1 Amyloids: An Introduction

Amyloids are fibrillar peptide/protein aggregates with β-sheet-rich secondary structure (Chiti and Dobson 2006). These highly organized structures are of particular interest because of their association with many biological functions in diverse organisms including mammals, though originally amyloids were considered as pathogenic entities in many neurodegenerative disorders (Chiti and Dobson 2006). Amyloid fibrils possess distinct characteristics in terms of biophysical as well as histological properties due to their unique secondary structure. The β-sheets present in amyloid fibrils are arranged such that individual β-strands are orthogonal, whereas the β-sheets are parallel to the fibril axis (Sunde and Blake 1997; Sunde et al. 1997). This arrangement termed as cross-β-sheet motif makes amyloids highly ordered and enables dense peptide packing making these aggregates inherently inert and resistant to harsh environmental conditions such as extreme pH, temperature, as well as proteases (Meersman and Dobson 2006; Mesquida et al. 2007; Zurdo et al. 2001).

R. S. Jacob · S. Das · N. Singh · K. Patel · D. Datta · S. Sen
S. K. Maji (✉)
Department of Biosciences and Bioengineering, Indian Institute of Technology Bombay, Mumbai, Maharashtra, India
e-mail: reeba.s.jacob@iitb.ac.in; samirmaji@iitb.ac.in

K. Chattopadhyay, S. C. Basu (eds.), *Biochemical and Biophysical Roles of Cell Surface Molecules*, Advances in Experimental Medicine and Biology 1112, https://doi.org/10.1007/978-981-13-3065-0_7

The cross-β-sheet structure allows amyloids to bind to dyes such as Congo red (CR) (Westermark et al. 1999) and thioflavin T (LeVine 1993). Amyloids, upon CR binding, produce a yellow-green birefringence under cross-polarized light (Westermark et al. 1999). In amyloid disorders, these fibrillar aggregates get deposited in patient's tissues, which clinicians stain with CR for disease diagnosis (Eisenberg and Jucker 2012). Most disease-associated amyloids such as amyloid-beta (Aβ) in Alzheimer's disease (AD), prions (PrP) in spongiform encephalopathy, and islet amyloid polypeptide (IAPP) in type 2 diabetes were identified using this method (Chiti and Dobson 2006). However, the biophysical study of the amyloids revealed a more structural understanding of these aggregates. In circular dichroism (CD) spectroscopy, amyloids exhibit a minimum at ~218 nm (Jha et al. 2013; Anoop et al. 2014), and in Fourier transform infrared (FTIR) spectroscopy, these aggregates show peaks around 1630 cm^{-1} and 1620 cm^{-1}, respectively, which are characteristic peaks of β-sheet structures (Hiramatsu and Kitagawa 2005; Ghosh et al. 2014; Fradinger et al. 2005). The X-ray diffraction of the amyloids yields a characteristic diffraction pattern with a meridional reflection at 4.7 Å and an equatorial reflection at around 8–11 Å representing a cross-β-sheet diffraction pattern (Sunde and Blake 1997; Sunde et al. 1997). Electron microscopy studies of amyloids showed that these protein aggregates are fibrillar in appearance, ~6–12 nm in diameter, unbranched, and a few micrometers in length (Fig. 7.1) (Sunde and Blake 1997; Sunde et al. 1997). Mostly, the fibrils are composed of protofilaments that are either laterally associated or helically twisted with each other. A combination of biophysical and histological tools of amyloid detection has identified many proteins that can form amyloids irrespective of their association with native functions or diseases.

Amyloid formation is considered as a nucleation-dependent polymerization wherein protein monomers either natively folded or unfolded participate in the self-assembly process to form stable amyloid fibrils (Harper and Lansbury 1997; Jarrett and Lansbury 1993; Ferrone 1999). This aggregation kinetics is often studied using ThT binding and/or light scattering experiments (Serio et al. 2000; Naiki et al. 1997; Uversky et al. 2002), which yield a sigmoidal growth curve showing three distinct phases: lag phase, where monomeric protein associates slowly to form aggregation-competent nuclei; elongation phase, where aggregation-competent nuclei grow to form fibrils; and at the end, sta-

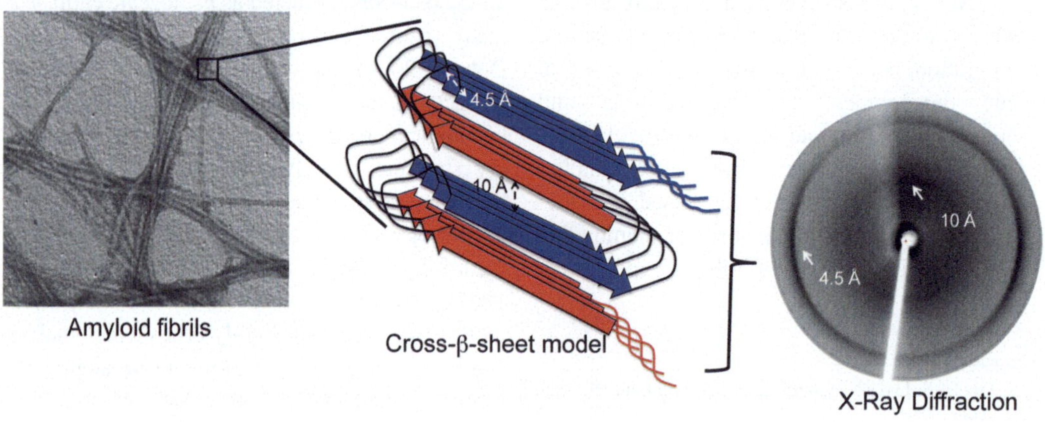

Fig. 7.1 Amyloid structure. Amyloid fibrils showing typical diffraction pattern of a cross-β-sheet motif with characteristic meridional reflection at 4.7 Å and an equatorial reflection at around 8–11 Å

tionary phase, where matured fibrils remain in equilibrium with monomeric protein (Harper and Lansbury 1997; Jarrett and Lansbury 1993; Ferrone 1999). During protein aggregation, proteins can exist as an ensemble of conformers, in which some of them associate to form oligomers that could be toxic to cells (Haass and Selkoe 2007; Kayed et al. 2003; Bucciantini et al. 2002; Walsh et al. 2002). In this regard, an amyloid formation may play a pacifying role wherein the toxic oligomers are scavenged and converted to stable amyloid fibrils (Kopito 2000; Carrotta et al. 2005; Kirkitadze et al. 2002). Recent studies in amyloid biology also support this hypothesis and suggest that protein/peptide oligomers rather than mature fibrils are more cytotoxic and implicated in disease (Haass and Selkoe 2007; Kayed et al. 2003; Bucciantini et al. 2002; Walsh et al. 2002).

Though historically amyloids are associated with more than two dozen of diseases in humans, recent investigations have revealed amyloids can perform native biological functions in several hosts spanning from bacteria to mammals (Chiti and Dobson 2006; Fowler et al. 2007). In these organisms, amyloids are involved in various functional roles rather than causing disease. For example, in several bacteria, including *E.coli*, amyloids are involved in biofilm formation, which helps in their colonization and attachment to surfaces (Chapman et al. 2002). In fungi, in addition to surface adhesion, amyloids also provide support for hyphae/spore formation (Hammer et al. 2008; Gebbink et al. 2005), and fungal prions offer a selective advantage for growth during stress conditions for its host (Osherovich and Weissman 2002; True and Lindquist 2000; Uptain and Lindquist 2002; Chien et al. 2004). In higher organisms such as mammals, Kelly and coworkers discovered the occurrence of Pmel amyloid fibrils in melanosomes (Fowler et al. 2006). In this study, the authors suggested that the amyloid Pmel17 reduces toxicity by accelerating the polymerization of melanin precursor into melanin (Fowler et al. 2006). In addition, it was proposed that hormones stored in pituitary gland could form amyloidogenic structure under in vitro and in vivo conditions (Maji et al. 2009a). The unique structural property of cross-β-sheet could aid in packaging and storage of peptide/protein into hormone-specific granules. Hormones in amyloidogenic form are also capable of releasing functionally active monomers. Thus, the robust organization of amyloid could be important for hormone storage in secretory granules as well as for regulated secretion of hormones (Maji et al. 2009a; Anoop et al. 2014; Jacob et al. 2016a).

7.2 Amyloids as Functional Biomaterials

Extensive biochemical and structural studies on amyloid formation have led to the understanding that amyloids are not rare phenomena associated with diseases but a well-defined structure of proteins and polypeptides. Amyloids are now better referred to as a "state" of proteins, which has a cross-β fiber diffraction pattern (Eisenberg and Jucker 2012; Knowles et al. 2014). In principle, (m)any peptide/proteins could self-assemble to form amyloid structures, under certain conditions (Chiti and Dobson 2006; Knowles et al. 2014). Due to their highly organized nature, increased stability, and mechanical stiffness comparable to silk and steel, amyloids can be used for bio- and nanotechnological applications (Sunde and Blake 1997; Smith et al. 2006). In addition to tuning physicochemical properties of amyloid fibrils by modifying amino acid sequences (Maji et al. 2008), specific functional groups can also be added to their amino acid side chains for desired purposes (Gras 2009; Gras et al. 2008; Scheibel et al. 2003). Recent advances in computational algorithms, which can also be employed to predict different properties such as hydrophobicity, secondary structure, and aggregation propensity of proteins from their sequence information, can be suitably utilized to design and synthesize amyloid-based materials with specific properties (Mankar et al. 2011). By virtue of amyloid's characteristic traits, efforts have been made to design functional biomaterials such as biosensors, tissue culture scaffolds, nanowires/optoelectronics, and drug delivery depots (Cherny and Gazit 2008; Knowles et al. 2010; Li and Mezzenga 2013; Gras 2009; Mankar et al. 2011).

The highly organized assembly or fibrillar structure of amyloids makes them a natural nanomaterial suited for a variety of functions. A number of applications including generation of nanowires, light-emitting diodes (LEDs), and biosensors have been reported using amyloid nanofibrils (Reches and Gazit 2003; Tanaka et al. 2008; Li et al. 2012). For example, Reches and Gazit have utilized dipeptide, Phe-Phe as a template for metal nanowire synthesis (Reches and Gazit 2003, 2006; Gazit 2007). This template region is derived from the middle segment of Alzheimer's disease-related amyloid β-peptide (Reches and Gazit 2003). Likewise, Han et al. constructed graphene hollow shell nanowires. These nanotubes can be easily synthesized and are biodegradable in nature (Han et al. 2010). In another study, yeast prion protein Sup35 was genetically modified to have accessible cysteine residues post-fibrillation such that they can be conjugated with gold or silver colloids/nanoparticle to develop fine nanowires (Scheibel et al. 2003). Although nanowire development has also been reported from other biomolecules such as DNA, proteins, and carbon nano-materials, due to complex manufacturing procedures and difficulty in controlled functionalization, such techniques pose several challenges (Li and Mezzenga 2013). Nanowires developed from amyloids have gained attraction due to facile methods of harnessing their self-assembly behavior (Cherny and Gazit 2008). Amyloids in conjugation with various polymers can also enhance their electrical/optical properties. In one study, the addition of insulin amyloid fibrils increased the external quantum efficiency of a luminescent polymer tenfold (Tanaka et al. 2008). Amyloid fibril usage has also been reported in photovoltaic cells, where the β-lactoglobulin amyloid-TiO_2 hybrid material was used in heterojunction photovoltaic devices (Bolisetty et al. 2012). Integrating polymers and amyloid fibrils leads to enhanced performance and sensitivity of biosensors. For example, Mezzenga and coworkers used β-lactoglobulin amyloid fibrils, which were adsorbed on graphene sheets to develop enzyme-based biosensors (Li et al. 2012). Moreover, Sasso et al. functionalized whey protein nanofibrils (WPNFs) by attaching it with quantum dots and gold nanoparticles for designing thiolated glucose oxidase (GOx)-functionalized WPNFs to develop electrochemical enzymatic biosensor with increased sensitivity (Sasso et al. 2014), suggesting that such approaches can be utilized for developing multienzyme-based biosensors (Sasso et al. 2014).

Protein- and peptide-based drugs are often rendered inactive after administration due to protease activity in our body (Langer 1998). Thus, improvement of the half-life of such drugs is an active area of research. Also, sustained or controlled delivery of the drug over a period of time would make it more effective. Infusion pumps, liposomes, polymer depots, and biodegradable hydrogels are currently used to serve the purpose (Langer 1990, 1998, 2001). However, recent studies have suggested that self-assembled proteins/peptides can be used for sustained and effective drug delivery. For example, insulin and TGF-β_3 microcrystals were able to serve as protected reservoirs of the drugs as well as the controlled release of active drugs (Brader et al. 2002; Jen et al. 2002). On similar trend, amyloid conformation can also serve either as a drug delivery vehicle or a drug reservoir/depot itself. It was recently observed that analogs of gonadotrophin-releasing hormone (GnRH) were able to release active monomers from its amyloid form (Maji et al. 2008). These studies suggest that either encapsulation of drug within amyloid fibrils or amyloid fibrils developed from peptide drugs could be used as a vehicle for drug delivery application. In addition to these, several studies have also demonstrated the potential of amyloid and amyloid-derived scaffolds for cell culture and tissue engineering, which will be further detailed in the following sections.

7.3 Native Extracellular Matrices: Composition and Properties

The extracellular matrix (ECM) in higher organisms is a meshwork of proteins (~200) and polysaccharides forming an elaborate network that not only serves as a scaffold integrating multiple cell types to form organs but also plays key roles in regulating multiple aspects of cellular functions including survival, growth, migration, and differentiation via sequestration of growth factors. The ECM is a dynamic environment whose composition and organization are tissue-specific and determined by its interaction with various cell types within a given tissue. ECM integrity is maintained by multiple structural proteins including collagen, fibronectin, laminin, and elastin, with collagen being the most abundant. Though 28 different types of collagen have been identified to date (Myllyharju and Kivirikko 2004; Ricard-Blum 2011), fibrillar collagen I is the principal component of interstitial matrices accounting for ~90% of the protein content of connective tissues (Wolf et al. 2009) and is upregulated in various cancers (Paszek et al. 2005). In comparison, the brain is enriched with laminin, hyaluronic acid, and proteoglycans (Reinhard et al. 2016). Collagen IV and laminin represent two of the main ingredients of sheet-like basement membrane (Kalluri 2003), which segregate tissues and provide structural support.

The physical properties of a given tissue are largely dictated by the composition and organization of the ECM. For example, collagen I levels have been shown to exhibit a power-law dependence on tissue stiffness, with soft brain tissue possessing low levels of collagen and stiff tissues possessing increased levels (Swift et al. 2013). Over the last decade, a plethora of research has demonstrated the profound influence of ECM stiffness in regulating various cellular processes including cell spreading (Engler et al. 2004a; Yeung et al. 2005), cell migration (Pelham and Wang 1997), proliferation (Klein et al. 2009; Ulrich et al. 2009), and differentiation (Engler et al. 2004b; Engler et al. 2006; Gilbert et al. 2010). These findings have led to development of hydrogel platforms as suitable mimics of natural ECMs for tissue engineering platforms, where the stiffness is regulated using synthetic (e.g., PEG (Mahoney and Anseth 2006), PDMS (Trappmann et al. 2012), or natural-derived (e.g., alginate) (Rowley et al. 1999)) polymers and linking with cell-adhesive peptides (e.g., RGD (Dennes et al. 2007), IKVAV (Santiago et al. 2006)). An additional feature of collagen and fibrin gels is their nonlinear elasticity (van Oosten et al. 2016), i.e., stiffening under force, induced by alignment of matrix fibers. Such stiffening not only protects tissues from damage but enhances long-range interaction between cells in the matrix (Winer et al. 2009).

One aspect of ECM networks that have not been adequately addressed is their viscoelasticity (Chaudhuri 2017). It has long been known that natural ECMs are not such elastic but viscoelastic, i.e., their deformation in response to stresses is time-dependent. Such time-dependent behavior may arise from force-dependent dissociation and rebinding of cross-linking agents leading to ECM reorganization and/or force-driven protein unfolding. Though studies on how tissue viscoelasticity impacts cell behavior are much less, recent reports have demonstrated the role of viscoelasticity in driving cell spreading on soft alginate gels (Chaudhuri et al. 2015) and dextran gels (Baker et al. 2015), possibly through increased integrin clustering. Under 3D culture conditions, viscoelasticity has been shown to mediate cell spreading and enable embryonic stem cell-derived motor neurons to extend neurites in stress-relaxing PEG hydrogels (McKinnon et al. 2014), as well as drive optimal osteogenic differentiation of mesenchymal stem cells in ionically cross-linked alginate hydrogels (Chaudhuri et al. 2016).

Another key physical feature of the ECM is topography, i.e., the orientation of matrix fibers. Numerous studies have shown that cells in fibrillar matrices elongate, polarize, and migrate in a persistent direction using matrix fibers as contact guidance cues (Simitzi et al. 2017; Vargas et al. 2017). Fibroblasts are capable of responding to grooved patterns with features comparable to single collagen fibrils (~30–100 nm) (Kim et al. 2009). These studies suggest that physical cross talk with ECM features can activate signaling independent of chemical factors in driving directional migration. Indeed, studies with various types of stem cells have shown that topography can not only drive directed migration but also regulate stem cell fate by inducing changes in cell morphology (Leung et al. 2014; Wang et al. 2012). In particular, techniques such as microfabrication and electrospinning have gained rising popularity for fabricating substrates/scaffolds encoding cues across multiple length scales. These technologies have enabled studies that address fundamental science as well as led to tissue engineering applications.

7.4 Amyloids Closely Mimic Several Features of the ECM

Similar to ECM, amyloids are fibrillar in nature. The unique surface topography of amyloid fibrils can be used for developing cell-adhesive materials. In this regard, it has been reported that amyloid fibrils alone or in association with functional moieties or coated with fibronectin- and laminin-like ECM proteins provide suitable scaffolds for cell adhesion and growth [43]. For instance, Gras and coworkers constructed functionalized fibrils derived by combining transthyretin protein sequence with the tripeptide Arg-Gly-Asp (RGD), an integrin recognition motif found in fibronectin for enhancing cell-adhesive ability of amyloid fibrils (Gras et al. 2008). Moreover, proteins from demineralized enamel matrices were also reported to form cross-β-sheet structure, which could be used in hard tissue engineering (Glimcher et al. 1961, 1965).

Further, to decouple nano-topography from the surface chemistry of amyloid fibrils, Reynolds et al. designed amyloid fibrils layered with plasma polymer (Reynolds et al. 2013). The surface created by coating plasma layer on amyloid fibrils retains the amyloid topography but masks the surface chemistry of the fibrils. The authors reported more increase in cell adhesion on amyloid fibril/conjugated polymer surface than on polymer alone, suggesting that the topography of amyloid fibrils favors cell adhesion (Reynolds et al. 2013). Moreover, in a subsequent study, the authors showed that the denser the fibril coverage, the more cells adhered to amyloid fibril-coated surfaces (Reynolds et al. 2014). Alternatively, another reason for enhanced cell adhesion could be due to the increased deposition of serum proteins on the amyloid fibril-coated surfaces. The study also showed that cell adhesion property was enhanced by increasing lysozyme fibril concentration (Reynolds et al. 2014). Specifically, in this study, the authors showed that when the fibril-coated surface had 2–40% coverage with lysozyme fibril network, the attached cells showed less spreading area. However, when the fibril surface coverage was more than 80%, the spreading area of the cells increased substantially. Moreover, this study also noted that fibril coverage more than the thickness of the individual fibrils dictated cell adhesion (Reynolds et al. 2014). Overall, these studies suggest that nanotopography of amyloids enables cell adhesion and cells on amyloid surfaces show density-dependent adhesion responses similar to ECM proteins.

However, what makes the amyloid surface adhesive for cells? Is it because of sequence similarity of some of these proteins to cell recognition motifs? Is the surface property of amyloid itself enough to initiate the cell adhesion? To answer these questions, recently we studied cell adhesion on amyloids using more than 20 proteins/peptides exhibiting different primary structures and sequences (Jacob et al. 2016b). Many of which are already reported to form amyloid in secretory granule-relevant conditions and are nontoxic (Maji et al. 2009a). The study demonstrated that mammalian cells were capable of adhering and spreading on amyloid fibrils and the adhesion is not specific for single-cell type but for many including stem cells and non-adherent cells like RBC (Jacob et al. 2016b). This could also suggest that membrane-fibril interaction could be the starting step for cell adhesion on amyloids. Furthermore, in this study, we showed that the conversion of cell-repulsive protein BSA, to its amyloid form, resulted in making the protein cell adhesive, suggesting cell adhesion to be a generic property of amyloids (Jacob et al. 2016b).

Cells adhere to the underlying ECM through integrin-mediated junctions called focal adhesions, which act as contact points between the cell cytoskeleton and the ECM (Choi et al. 2011). To understand cell adhesion mechanism on amyloid fibrils, we investigated cytoskeletal arrangement along with focal adhesion machinery of amyloid-bound cells. The study demonstrated that cells adhering to amyloid surfaces were able to make integrin-based adhesions and activated focal adhesion kinase (FAK), a downstream integrin-signaling molecule. Moreover, cells adhering on amyloid surfaces had a greater number and larger focal adhesions compared to cells cultured on collagen I. Further, integrin blocking by antibody or peptides caused a reduction in cell spreading on fibril surface, suggesting that integrin-mediated cell adhesion is key for cell spreading mechanism on amyloid surfaces (Jacob et al. 2016b). Additionally, seeding of NIH/3T3 fibroblast cells on amyloid fibril coatings of different densities, a biphasic spreading response of the cells was observed, which was comparable to ECM density-dependent cell spreading. Moreover, these cells were also motile on amyloid surfaces (Jacob et al. 2016b). Taken together, it can be suggested that behavior of cells on amyloid nanofibrils and ECM could be similar.

The charge of a surface is also known to regulate cell adhesion. Poly-L-lysine (PLL) is a synthetic substrate routinely used on surfaces to promote cell adhesion. However, cell adhesion on PLL-coated surfaces is not integrin mediated. In our previous study, differently charged and uncharged polymers and their amyloid counterparts were used to demarcate the role of charge and cross-β-sheet motif on cell adhesion (Jacob et al. 2016b). The data showed that cells adhered to charged polypeptide surfaces, but the adhesion was greater on their amyloid counterpart compared to monomeric polymers. Moreover, the amyloid surfaces could also recruit integrin-mediated focal adhesion, whereas, in the monomeric counterpart of the same polymer, integrin expression was much reduced. This study suggests that adhesion could be mediated by a combination of cell membrane-fibril interactions as well as integrin-mediated focal adhesion complex formation (Fig. 7.2) and amyloids similar to ECM proteins induce integrin-mediated cell signaling pathways.

In addition, Jacob et al. (2016b) also showed that the cell adhesivity of amyloids is not only restricted to nontoxic amyloids but also extends to disease-associated amyloids such as fibrils of Parkinson's disease-associated α-synuclein (α-Syn) and Alzheimer's disease-associated $A\beta_{42}$ fibrils (Goedert 2001; Hardy and Selkoe 2002). However, cell adhesion was less on fibrils of $A\beta_{(25-35)}$, a neurotoxic sequence in Aβ peptide (Pike et al. 1995). This difference in cell adhesion on fibrils formed from related peptides could be due to more exposure of toxic epitope in $A\beta_{(25-35)}$ fibrils, whereas it might be comparatively masked in fibrils formed by full-length $A\beta_{42}$ peptide. In addition to nano-topography, the surface roughness of the coated amyloid fibrils can play a vital role in death/apoptosis of adherent cells (Gras et al. 2008). Reynold et al. also found similar observations, where they showed that the cells adhered on rough surfaces are less viable (Reynolds et al. 2015). Together, this suggests

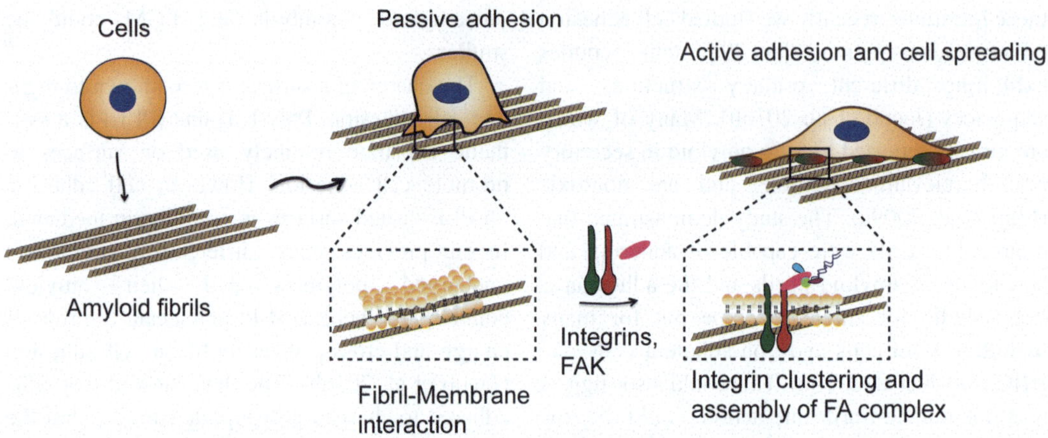

Fig. 7.2 Mammalian cell adhesion on amyloid fibrils. Adhesion of cells on amyloid fibrils is proposed to be initiated by lipid-fibril interaction-mediated passive adhesion followed by integrin clustering and activation of downstream signaling cascades leading to the formation of focal adhesions and cell spreading

that though amyloid surface favors cell adhesion, the surface chemistry of the fibrils determines the viability of the adhered cells (Reynolds et al. 2015). Moreover, Jacob et al. demonstrated that the surface topography of amyloids positively influences the differentiation of SH-SY5Y cells to neurons (Jacob et al. 2016c). Thus, other than fibrillar nature, a feature shared by many ECM proteins and amyloid fibrils, these higher-order protein aggregates are akin to ECM in many ways. This raises an exciting possibility that during evolution amyloids might have served as primitive cell-adhesive substrates for templating cell/tissue organization along with other cellular functions.

7.5 Amyloid-Based Scaffolds for Tissue Engineering

The fabrication of synthetic substrates with similar physiochemical properties of ECM, in order to mimic the biological microenvironment for cell adhesion, proliferation, and differentiation, is one of the important aspects of tissue engineering (Langer and Tirrell 2004; Patterson et al. 2010). In this regard, chemically and biologically derived polymer scaffolds have promising significance (Langer 2000; Hubbell 1995). However, some of them show restricted in vivo applications due to their toxicity, immunogenicity, or biological contamination from polymers derived from tissues (Langer and Tirrell 2004). In this line of thought, synthetic peptide–/protein-derived hydrogels are of importance because they negate the possibility of biological contamination and their scaffolds can mimic ECM-like environment (Patterson et al. 2010; Zhang 2002; Zhang et al. 2002). In recent years, extensive research has been carried out for designing hydrogels with various utilities especially in the area of tissue engineering and drug delivery (Jacob et al. 2015; Das et al. 2016; Benoit et al. 2008; Annabi et al. 2014). Hydrogels are routinely used as carriers for encapsulating cells and drugs/growth factors to target sites for cell regeneration at the injury site. Their characteristic microstructure influences the signaling pathways, hence modulating cellular fate (Peppas et al. 2006; Smith 2010; Lee et al. 2000).

Peptide hydrogels are a special class of hydrogel based on self-assembly of the peptide through non-covalent interactions, which form higher-order networks and subsequently water entrapment to form gels (Meital Reches 2006). Peptide-based hydrogels are also reversible in nature as they employ non-covalent intermolecular forces such as hydrophobic interaction,

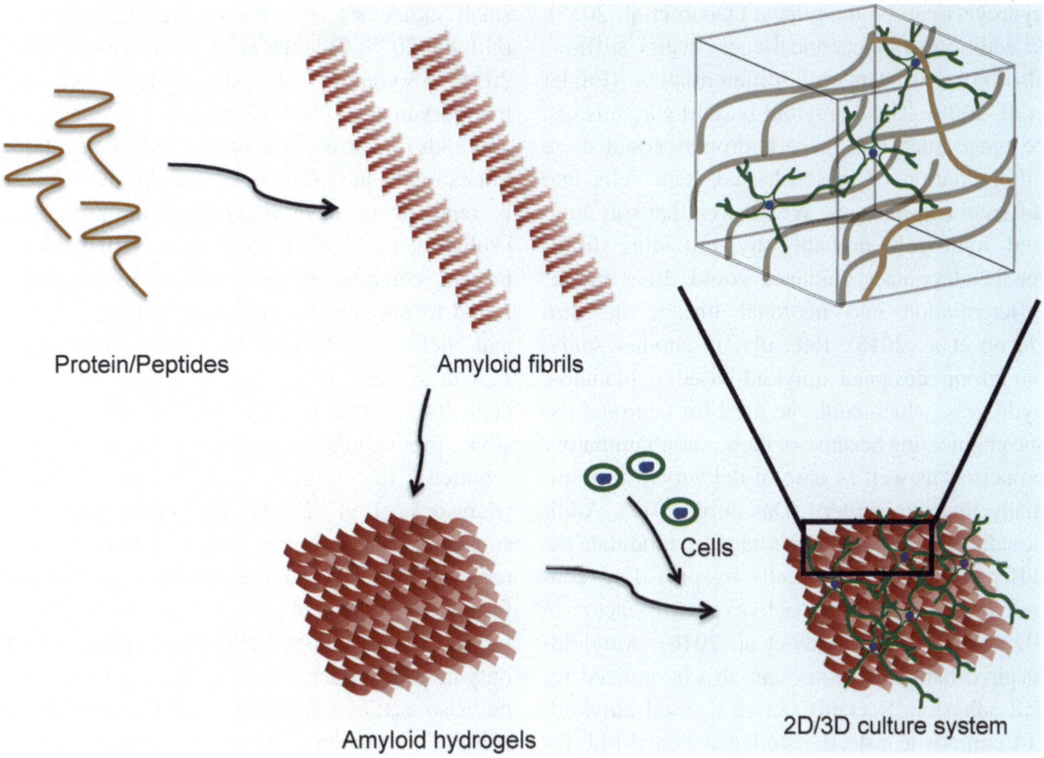

Fig. 7.3 Amyloid-based scaffolds for tissue engineering. Proteins/peptides can self-assemble to form amyloid fibrils. Such amyloid fibrils can bundle and form higher-order aggregates that can lead to the complex network formation to form gels and films that can be used for preparing hydrogels and scaffolds for directed stem cell differentiation

hydrogen bonds, and ionic interactions. Hence, peptide hydrogels are easily biodegradable and have less potential for inflammation compared to polymer-based hydrogels. Moreover, the properties of hydrogel could be easily tuned through side-chain modification to the desired functionality (Gras 2009; Mankar et al. 2011; Nisbet and Williams 2012). There is much similarity between amyloid fibrils and natural ECMs such as their nano-fibrillar morphology, integrin engagement and signaling, and density-dependent responses (Jacob et al. 2016b; Reynolds et al. 2014, 2015; Gras et al. 2008). These suggest that hydrogel based on amyloid fibrils could be promising candidates for developing biomaterials for tissue engineering applications (Mankar et al. 2011).

In this regard, Yan et al. have shown that amyloid hydrogel can aid in adherence and proliferation of fibroblasts in the absence of any cell recognition motif or coating of ECM proteins (Yan et al. 2008). Moreover, recent studies by our group have shown that biocompatible and self-healing amyloid hydrogels can be designed from peptides derived from Aβ_{42} C-terminus (Jacob et al. 2015). The study also demonstrated that these amyloid hydrogels could support adhesion and spread of diverse cell types including human mesenchymal stem cells. Since these nanofibril hydrogels are thixotropic, it can be easily formulated to a 3D cell culture system that can entrap growth media and support the growth of cells within the hydrogel (Fig. 7.3). Further, alteration in the concentration of peptides and salt and the stiffness of amyloid-based hydrogels can be modulated (Jacob et al. 2015). In addition to biochemical cues, matrix stiffness also governs stem cell differentiation (Engler et al. 2006). Thus, amyloid-based hydrogels can be tuned such that these hydrogels could drive differentiation of mesenchymal stem cells into different cell lineages. We showed that soft amyloid hydrogels probably by providing fibril-mediated contact guidance could drive hMSC differentiation into neuronal lineage in vitro (Jacob et al. 2015). Recently, in another study, our group designed amyloid-based implantable hydrogels, which could be used for neuronal tissue engineering because of their noninflammatory properties as well as ease of delivery with minimally invasive surgery (Das et al. 2017). Additionally, amyloid hydrogels can also modulate the differentiation of stem cells by providing controlled exposure of bioactive growth factor in 3D culture system (Das et al. 2016). Amyloid-inspired nanocomposites can also be utilized for cell adhesion. Recently, Li et al. used amyloid-HA composite instead of collagen as scaffolds for regeneration of bone tissue (Li et al. 2014). Their work demonstrated that preosteoblasts can adhere and proliferate on the designed amyloid-HA nanocomposites (Li et al. 2014). These studies infer that amyloid-based hydrogels do not require cell-adhesive moiety and hence find its application in tissue engineering.

7.6 Amyloids as Ancient ECM

The ability to self-replicate (Cohen and Prusiner 1998; Jarrett and Lansbury 1993), and the function of amyloids as biochemical catalysts (Rufo et al. 2014), has entitled amyloids to be one of the most ancient protein folds (Greenwald and Riek 2012; Carny and Gazit 2005). The sequence-dependent distribution of polar and nonpolar residues depending on the primary sequences gives amyloid fibrils distinct surface properties (Nelson and Eisenberg 2006; Wasmer et al. 2008; Maji et al. 2009b). This property may enable amyloids to bind to a varied range of small molecules and macromolecules/polymers (Nilsson 2009; Calamai et al. 2006; Ghosh et al. 2014; Solomon et al. 2011). Various microorganisms use the adhesive property of amyloids for surface attachment and colonization. For example, in sea barnacle, the adhesive cement is reported to have amyloidogenic properties (Sullan et al. 2009; Barlow et al. 2010). Many protein components of bacterial biofilms are found to polymerize into amyloid fibrils (Otzen and Nielsen 2008), with curli amyloid made by *E. coli* serving as a prime example (Chapman et al. 2002). In yeast, adhesins, which are responsible for cellular aggregation of yeast, are reported to possess amyloid-like nature (Ramsook et al. 2010). The extra polymeric substances of certain green algae were previously reported to be amyloidogenic, inducing colony formation in these algae (Mostaert et al. 2006). In the abovementioned examples, amyloids not only help in surface adhesion of these organisms but also act as extracellular matrix helping in colony formation, thereby ensuring better survival.

Over the last few decades, researchers are involved in understanding the prebiotic atmosphere and the origin of life (Sutherland 2017). Recent advancements led to the finding that protein-like molecules with cross-β-sheet structure(s) could be the first biomolecules capable of self-propagating and information-processing, which evolved under extreme conditions of the primitive earth environment (Maury 2009; Greenwald and Riek 2010, 2012; Greenwald et al. 2016). For example, Miller and coworkers demonstrated the synthesis of amino acids under the primordial conditions (Miller 1953). The peptides, which can form amyloid, can escape harsh environments such as hydrolysis, chemical modification, volcanic gases, and adverse environments (Leman et al. 2004; Schwendinger and Rode 1992). Likewise, Greenwald and coworkers have shown that alanine and valine, the most abundant amino acids in the prebiotic environment, can form amyloid fibers consisting of cross-β structure (Greenwald and Riek 2012;

Greenwald et al. 2016). This indicates that primitive life forms that were prevalent during harsh conditions of the primordial earth would have selected protein/peptide structures or conformations that could be stable and sustainable in extreme conditions. The structural stability of β-sheet fold renders it the resistance against radiation and high temperature (Maury 2015). Thus, there are several arguments, which favor the idea of a possible key role of protein aggregation during the early evolution and might have served as the mother of any protein fold (Maury 2009; Chernoff 2001; True and Lindquist 2000). In addition, it is believed that this stable β-sheet structure carried out both the functions of replication and information transmission in a prion-like mode. All these suggestions are indicative of amyloids being an entity of biomolecular evolution.

One of the pivotal transitions in primitive earth was the evolution of multicellular eukaryotic organisms from single-cell ancestors (Michod 2007). Studies indicate that the ECM played a key role in orchestrating this transition (Ozbek et al. 2010). The major ancestral ECM proteins that diversified and evolved into present-day ECM proteins include collagen, laminin, perlecan, and fibrillin. It has been shown that these proteins are conserved from sponges to metazoans (Ozbek et al. 2010). The relatively harsh conditions of the primitive earth, the stability of the amyloid state of proteins, and the gradual evolution of the extracellular matrix all point towards a possibility that ancestral ECM proteins were amyloidogenic in nature and existed in the amyloid state in the primitive era. This assumption led us to investigate the presence of any amyloidogenic region in the primary sequence of ECM related proteins in primitive organisms such as *Chlamydomonas*, which could probably be carried over in higher organisms with evolution. To evaluate the presence of amyloidogenic region, we used "WALTZ," a web-based tool developed by Sebastian Maurer-Stroh et al., for prediction of the amyloid-forming tendency of proteins, based on amino acid sequence (Maurer-Stroh et al. 2010). We used this algorithm for predicting the amyloid-forming propensity of ECM-related proteins, specifically those that have a role in cell adhesion. To do so, protein sequences of ECM proteins such as collagen, fibrillin, laminin, and perlecan protein were taken from UniProt database and were compared to its homologous sequences in higher organisms. For example, primary sequences of perlecan and laminin proteins present in *Caenorhabditis elegans* as well as *Homo sapiens* were compared, and both were found to exhibit extensive stretches of amyloid-prone regions (Fig. 7.4a, b). Since, for unicellular organisms, these proteins were not well annotated in the database, we decided to use *C. elegans* as an example of a lower organism for comparison. Similarly, fibrillin, another ECM protein present in *Chlamydomonas reinhardtii* and its homolog in humans (fibrillin 1), also showed regions of amyloid-forming propensities (Fig. 7.4c). Moreover, comparison of collagen sequences from several organisms including *Chlamydomonas reinhardtii, Volvox carteri, Caenorhabditis elegans*, and *Homo sapiens* also revealed the presence of amyloid-prone regions (Fig. 7.4d). However, the frequency of occurrence of the amyloidogenic region varied across the organisms. Our preliminary study on ECM protein sequence showed that amyloid-prone regions are not only present in ECM proteins of lower organisms but are also prevalent in the higher homologs. Our bioinformatics analysis is in support of the possibility that, during early life, proteins related to the ECM could exist in an amyloid state and the modern homologs of such proteins retain amyloid-forming capability. For example, CD2, a cell adhesion molecule of the rat, was shown to form amyloid fibrils (Carroll et al. 2006). However, further studies of this hypothesis, by extrapolating with predictions based on structural information, for example, the calculation of Rosetta energy (Goldschmidt et al. 2010), simulations with short primitive peptide sequences, or correlation studies with proteins from a particular phylogenetic tree, are required to confirm this hypothesis.

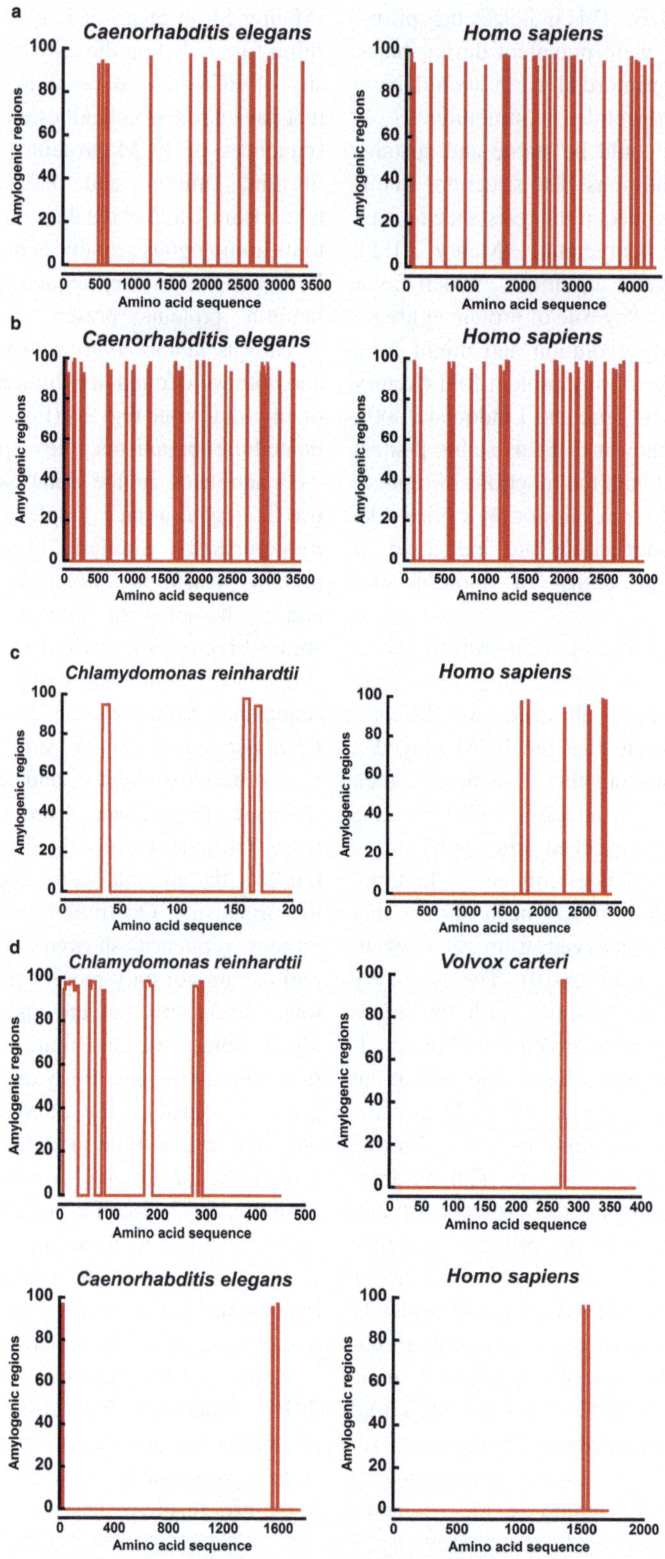

Fig. 7.4 The amyloid propensity of ECM proteins. WALTZ algorithm was used to analyze the aggregation-prone regions of various ECM proteins of the lower organism and its homologs in humans. (**a**) Basement membrane proteoglycan protein of *Caenorhabditis elegans* (Q06561) and basement membrane-specific

7.7 Conclusion

Amyloids are ancient protein fold that is not only associated with diseases but can also perform functions in host organisms. Amyloids, due to its characteristic cross-β-sheet structure and distinctive surface properties, are utilized by various organisms for biological functions such as adhesion, colonization, and storage. In this regard, our preliminary sequence analysis study on various ECM proteins of lower and higher organisms suggests that amyloidogenic regions are a common feature in many ECM proteins. In addition, mammalian cells were also shown to adhere onto amyloid fibrils due to the unique stickiness of the amyloid fibrils and ECM-mimetic nano-topography. Moreover, the superior material properties of the amyloids are also currently being exploited for different nano-biotechnological applications. In line with these findings, proteins/peptides could be used as building blocks to construct amyloid-based substrates that are cell adhesive and could lead to the development of a novel class of scaffolds that have applications in various fields of tissue engineering and biomaterial design.

Acknowledgments The authors wish to acknowledge DBT (BT/PR9797/NNT/28/774/2014), Government of India, and Wadhwani Research Center for Bioengineering (WRCB) for their financial support.

References

Annabi N, Tamayol A, Uquillas JA, Akbari M, Bertassoni LE, Cha C, Camci-Unal G, Dokmeci MR, Peppas NA, Khademhosseini A (2014) 25th anniversary article: rational design and applications of hydrogels in regenerative medicine. Adv Mater 26(1):85–123

Anoop A, Ranganathan S, Das Dhaked B, Jha NN, Pratihar S, Ghosh S, Sahay S, Kumar S, Das S, Kombrabail M, Agarwal K, Jacob RS, Singru P, Bhaumik P, Padinhateeri R, Kumar A, Maji SK (2014) Elucidating the role of disulfide bond on amyloid formation and fibril reversibility of somatostatin-14: relevance to its storage and secretion. J Biol Chem 289(24):16884–16903. https://doi.org/10.1074/jbc.M114.548354

Baker BM, Trappmann B, Wang WY, Sakar MS, Kim IL, Shenoy VB, Burdick JA, Chen CS (2015) Cell-mediated fibre recruitment drives extracellular matrix mechanosensing in engineered fibrillar microenvironments. Nat Mater 14(12):1262–1268. https://doi.org/10.1038/nmat4444

Barlow DE, Dickinson GH, Orihuela B, Kulp JL, Rittschof D, Wahl KJ (2010) Characterization of the adhesive plaque of the barnacle Balanus amphitrite: Amyloid-Like Nanofibrils are a major component. Langmuir 26(9):6549–6556. https://doi.org/10.1021/la9041309

Benoit DS, Schwartz MP, Durney AR, Anseth KS (2008) Small functional groups for controlled differentiation of hydrogel-encapsulated human mesenchymal stem cells. Nat Mater 7(10):816–823. https://doi.org/10.1038/nmat2269

Bolisetty S, Adamcik J, Heier J, Mezzenga R (2012) Amyloid directed synthesis of titanium dioxide nanowires and their applications in hybrid photovoltaic devices. Adv Funct Mat 22(16):3424–3428

Brader ML, Sukumar M, Pekar AH, McClellan DS, Chance RE, Flora DB, Cox AL, Irwin L, Myers SR (2002) Hybrid insulin cocrystals for controlled release delivery. Nat Biotechnol 20(8):800–804

Bucciantini M, Giannoni E, Chiti F, Baroni F, Formigli L, Zurdo J, Taddei N, Ramponi G, Dobson CM, Stefani M (2002) Inherent toxicity of aggregates implies a common mechanism for protein misfolding diseases. Nature 416(6880):507–511

Calamai M, Kumita JR, Mifsud J, Parrini C, Ramazzotti M, Ramponi G, Taddei N, Chiti F, Dobson CM (2006) Nature and significance of the interactions between amyloid fibrils and biological polyelectrolytes. Biochemistry 45(42):12806–12815

Carny O, Gazit E (2005) A model for the role of short self-assembled peptides in the very early stages of the origin of life. FASEB J 19(9):1051–1055 doi:19/9/1051 [pii]111096/fj.04-3256hyp

Fig. 7.4 (continued) heparan sulfate proteoglycan core protein of *Homo sapiens* (P98160). (**b**) Laminin alpha protein of *Caenorhabditis elegans* (G5EEV6) and laminin subunit alpha-1 protein of *Homo sapiens* (P25391). (**c**) Fibrillin protein of *Chlamydomonas reinhardtii* (A8IRS0) and fibrillin-1 protein of *Homo sapiens* (P35555). (**d**) Collagen-like protein of *Chlamydomonas reinhardtii* (A8HND8), collagen-related protein of *Volvox carteri f. nagariensis* (D8TNW6), collagen alpha-2-(IV) chain protein of *Caenorhabditis elegans* (P17140), and *Homo sapiens* (P08572). The study shows that amyloid-prone regions are not only present in the ECM proteins of lower organisms but also prevalent in the higher homologs. *Corresponding UniProtKB accession numbers are mentioned in paratheses "( )" following the respective protein details*

Carroll A, Yang W, Ye Y, Simmons R, Yang JJ (2006) Amyloid fibril formation by a domain of rat cell adhesion molecule. Cell Biochem Biophys 44(2):241–249. https://doi.org/10.1385/cbb:44:2:241

Carrotta R, Manno M, Bulone D, Martorana V, San Biagio PL (2005) Protofibril formation of amyloid beta-protein at low pH via a non-cooperative elongation mechanism. J Biol Chem 280(34) (0021–9258 (Print)):30001–30008

Chapman MR, Robinson LS, Pinkner JS, Roth R, Heuser J, Hammar M, Normark S, Hultgren SJ (2002) Role of Escherichia coli curli operons in directing amyloid fiber formation. Science 295 (5556):851–855. https://doi.org/10.1126/science.1067484 295/5556/851 [pii]

Chaudhuri O (2017) Viscoelastic hydrogels for 3D cell culture. Biomater Sci 5(8):1480–1490. https://doi.org/10.1039/c7bm00261k

Chaudhuri O, Gu L, Darnell M, Klumpers D, Bencherif SA, Weaver JC, Huebsch N, Mooney DJ (2015) Substrate stress relaxation regulates cell spreading. Nat Commun 6:6364. https://doi.org/10.1038/ncomms7365

Chaudhuri O, Gu L, Klumpers D, Darnell M, Bencherif SA, Weaver JC, Huebsch N, Lee HP, Lippens E, Duda GN, Mooney DJ (2016) Hydrogels with tunable stress relaxation regulate stem cell fate and activity. Nat Mater 15(3):326–334. https://doi.org/10.1038/nmat4489

Chernoff YO (2001) Mutation processes at the protein level: is Lamarck back? Mutat Res 488(1):39–64

Cherny I, Gazit E (2008) Amyloids: not only pathological agents but also ordered nanomaterials. Angew Chem Int Ed Engl 47(22):4062–4069. https://doi.org/10.1002/anie.200703133

Chien P, Weissman JS, DePace AH (2004) Emerging principles of conformation-based prion inheritance. Annu Rev Biochem 73:617–656. https://doi.org/10.1146/annurev.biochem.72.121801.161837

Chiti F, Dobson CM (2006) Protein misfolding, functional amyloid, and human disease. Annu Rev Biochem 75:333–366

Choi Y, Chung H, Jung H, Couchman JR, Oh ES (2011) Syndecans as cell surface receptors: unique structure equates with functional diversity. Matrix Biol 30 (2):93–99

Cohen FE, Prusiner SB (1998) Pathologic conformations of prion proteins. Annu Rev Biochem 67:793–819

Das S, Kumar R, Jha NN, Maji SK (2017) Controlled exposure of bioactive growth factor in 3D amyloid hydrogel for stem cells differentiation. Adv Healthc Mater 6(18). https://doi.org/10.1002/adhm.201700368

Das S, Zhou K, Ghosh D, Jha NN, Singh PK, Jacob RS, Bernard CC, Finkelstein DI, Forsythe JS, Maji SK (2016) Implantable amyloid hydrogels for promoting stem cell differentiation to neurons. Npg Asia Materials 8:e304. https://doi.org/10.1038/am.2016.116

Dennes TJ, Hunt GC, Schwarzbauer JE, Schwartz J (2007) High-yield activation of scaffold polymer surfaces to attach cell adhesion molecules. J Am Chem Soc (0002-7863 (Print)). doi:D - NLM: NIHMS61762

Eisenberg D, Jucker M (2012) The amyloid state of proteins in human diseases. Cell 148(6):1188–1203

Engler A, Bacakova L, Newman C, Hategan A, Griffin M, Discher D (2004a) Substrate compliance versus ligand density in cell on gel responses. Biophys J 86(1 Pt 1):617–628. https://doi.org/10.1016/s0006-3495(04)74140-5

Engler AJ, Griffin MA, Sen S, Bonnemann CG, Sweeney HL, Discher DE (2004b) Myotubes differentiate optimally on substrates with tissue-like stiffness: pathological implications for soft or stiff microenvironments. J Cell Biol 166(6):877–887. https://doi.org/10.1083/jcb.200405004

Engler AJ, Sen S, Sweeney HL, Discher DE (2006) Matrix elasticity directs stem cell lineage specification. Cell 126(4):677–689. https://doi.org/10.1016/j.cell.2006.06.044 S0092-8674(06)00961-5 [pii]

Ferrone F (1999) Analysis of protein aggregation kinetics. Methods Enzymol 309:256–274

Fowler DM, Koulov AV, Alory-Jost C, Marks MS, Balch WE, Kelly JW (2006) Functional amyloid formation within mammalian tissue. PLoS Biol 4(1). https://doi.org/10.1371/journal.pbio.0040006

Fowler DM, Koulov AV, Balch WE, Kelly JW (2007) Functional amyloid–from bacteria to humans. Trends Biochem Sci 32(5):217–224. https://doi.org/10.1016/j.tibs.2007.03.003

Fradinger EA, Maji SK, Lazo ND, Teplow DB (2005) Studying amyloid beta-protein assembly. In: WXaH X (ed) Amyloid precursor protein. CRC Pres, Boca Ration/London/New York/Washington, DC, pp 83–110

Gazit E (2007) Use of biomolecular templates for the fabrication of metal nanowires. FEBS J 274 (2):317–322 doi:EJB5605 [pii]371111/j.1742-4658.2006.05605.x

Gebbink MF, Claessen D, Bouma B, Dijkhuizen L, Wosten HA (2005) Amyloids–a functional coat for microorganisms. Nat Rev Microbiol 3(4):333–341 doi:nrmicro1127 [pii]391038/nrmicro1127

Ghosh D, Dutta P, Chakraborty C, Singh PK, Anoop A, Jha NN, Jacob RS, Mondal M, Mankar S, Das S, Malik S, Maji SK (2014) Complexation of amyloid fibrils with charged conjugated polymers. Langmuir 30(13):3775–3786. https://doi.org/10.1021/la404739f

Gilbert PM, Havenstrite KL, Magnusson KE, Sacco A, Leonardi NA, Kraft P, Nguyen NK, Thrun S, Lutolf MP, Blau HM (2010) Substrate elasticity regulates skeletal muscle stem cell self-renewal in culture. Science 329(5995):1078–1081. https://doi.org/10.1126/science.1191035

Glimcher MJ, Bonar LC, Daniel EJ (1961) The molecular structure of the protein matrix of bovine dental enamel. J Mol Biol 3:541–546

Glimcher MJ, Levine PT, Bonar LC (1965) Morphological and biochemical considerations in structural studies of the organic matrix of enamel. J Ultrastruct Res 13 (3):281–295

Goedert M (2001) Alpha-synuclein and neurodegenerative diseases. Nat Rev Neurosci 2(7):492–501

Goldschmidt L, Teng PK, Riek R, Eisenberg D (2010) Identifying the amylome, proteins capable of forming amyloid-like fibrils. Proc Natl Acad Sci U S A 107 (8):3487–3492. https://doi.org/10.1073/pnas.0915166107

Gras SL (2009) Surface- and solution-based assembly of amyloid fibrils for biomedical and nanotechnology applications. Adv Chem Eng 35:161–209

Gras SL, Tickler AK, Squires AM, Devlin GL, Horton MA, Dobson CM, MacPhee CE (2008) Functionalised amyloid fibrils for roles in cell adhesion. Biomaterials 29(11):1553–1562

Greenwald J, Friedmann MP, Riek R (2016) Amyloid aggregates arise from amino acid condensations under prebiotic conditions. Angew Chem Int Ed Engl 55(38):11609–11613. https://doi.org/10.1002/anie.201605321

Greenwald J, Riek R (2010) Biology of amyloid: structure, function, and regulation. Structure 18(10):1244–1260. https://doi.org/10.1016/j.str.2010.08.009

Greenwald J, Riek R (2012) On the possible amyloid origin of protein folds. J Mol Biol 421(4–5):417–426. https://doi.org/10.1016/j.jmb.2012.04.015

Haass C, Selkoe DJ (2007) Soluble protein oligomers in neurodegeneration: lessons from the Alzheimer's amyloid beta-peptide. Nat Rev Mol Cell Biol 8(2):101–112

Hammer ND, Wang X, McGuffie BA, Chapman MR (2008) Amyloids: friend or foe? J Alzheimers Dis 13 (4):407–419

Han TH, Lee WJ, Lee DH, Kim JE, Choi EY, Kim SO (2010) Peptide/graphene hybrid assembly into core/shell nanowires. Adv Mater 22(18):2060–2064. https://doi.org/10.1002/adma.200903221

Hardy J, Selkoe DJ (2002) Medicine – The amyloid hypothesis of Alzheimer's disease: progress and problems on the road to therapeutics. Science 297 (5580):353–356

Harper JD, Lansbury PT Jr (1997) Models of amyloid seeding in Alzheimer's disease and scrapie: mechanistic truths and physiological consequences of the time-dependent solubility of amyloid proteins. Annu Rev Biochem 66:385–407

Hiramatsu H, Kitagawa T (2005) FT-IR approaches on amyloid fibril structure. Biochim Biophys Acta 1753 (1):100–107. https://doi.org/10.1016/j.bbapap.2005.07.008

Hubbell JA (1995) Biomaterials in tissue engineering. Biotechnology (N Y) 13(6):565–576

Jacob RS, Das S, Ghosh S, Anoop A, Jha NN, Khan T, Singru P, Kumar A, Maji SK (2016a) Amyloid formation of growth hormone in presence of zinc: relevance to its storage in secretory granules. Sci Rep 6:23370. https://doi.org/10.1038/srep23370

Jacob RS, George E, Singh PK, Salot S, Anoop A, Jha NN, Sen S, Maji SK (2016b) Cell adhesion on amyloid fibrils lacking integrin recognition motif. J Biol Chem 291(10):5278–5298. https://doi.org/10.1074/jbc.M115.678177

Jacob RS, Ghosh D, Singh PK, Basu SK, Jha NN, Das S, Sukul PK, Patil S, Sathaye S, Kumar A, Chowdhury A, Malik S, Sen S, Maji SK (2015) Self healing hydrogels composed of amyloid nano fibrils for cell culture and stem cell differentiation. Biomaterials 54(0):97–105

Jacob RS, Sen S, Maji SK (2016c) Adhesion of Human Mesenchymal Stem Cells and Differentiation of SH-SY5Y Cells on Amyloid Fibrils. Macromol Symp 369(1):35–42

Jarrett JT, Lansbury PT Jr (1993) Seeding "one-dimensional crystallization" of amyloid: a pathogenic mechanism in Alzheimer's disease and scrapie. Cell 73:1055–1058

Jen A, Madorin K, Vosbeck K, Arvinte T, Merkle HP (2002) Transforming growth factor b-3 crystals as reservoirs for slow release of active TGF-b3. J Control Release 78(1–3):25–34

Jha NN, Anoop A, Ranganathan S, Mohite GM, Padinhateeri R, Maji SK (2013) Characterization of amyloid formation by glucagon-like peptides: role of basic residues in heparin-mediated aggregation. Biochemistry 52(49):8800–8810. https://doi.org/10.1021/bi401398k

Kalluri R (2003) Basement membranes: structure, assembly and role in tumour angiogenesis. Nat Rev Cancer 3 (6):422–433. https://doi.org/10.1038/nrc1094

Kayed R, Head E, Thompson JL, McIntire TM, Milton SC, Cotman CW, Glabe CG (2003) Common structure of soluble amyloid oligomers implies common mechanism of pathogenesis. Science 300(5618):486–489

Kim DH, Han K, Gupta K, Kwon KW, Suh KY, Levchenko A (2009) Mechanosensitivity of fibroblast cell shape and movement to anisotropic substratum topography gradients. Biomaterials 30 (29):5433–5444. https://doi.org/10.1016/j.biomaterials.2009.06.042

Kirkitadze MD, Bitan G, Teplow DB (2002) Paradigm shifts in Alzheimer's disease and other neurodegenerative disorders: the emerging role of oligomeric assemblies. J Neurosci Res 69(5):567–577

Klein EA, Yin L, Kothapalli D, Castagnino P, Byfield FJ, Xu T, Levental I, Hawthorne E, Janmey PA, Assoian RK (2009) Cell-cycle control by physiological matrix elasticity and in vivo tissue stiffening. Curr Biol 19 (18):1511–1518. https://doi.org/10.1016/j.cub.2009.07.069

Knowles TP, Oppenheim TW, Buell AK, Chirgadze DY, Welland ME (2010) Nanostructured films from hierarchical self-assembly of amyloidogenic proteins. Nat

Nanotechnol 5(3):204–207 doi:nnano.2010.26 [pii] 711038/nnano.2010.26
Knowles TPJ, Vendruscolo M, Dobson CM (2014) The amyloid state and its association with protein misfolding diseases. Nat Rev Mol Cell Biol 15 (6):384–396. https://doi.org/10.1038/nrm3810
Kopito RR (2000) Aggresomes, inclusion bodies and protein aggregation [Review]. Trends Cell Biol 10 (12):524–530
Langer R (1990) New methods of drug delivery. Science 249(4976):1527–1533
Langer R (1998) Drug delivery and targeting. Nature 392 (6679 Suppl):5–10
Langer R (2000) Biomaterials in drug delivery and tissue engineering: one laboratory's experience. Acc Chem Res 33(2):94–101 doi:ar9800993 [pii]
Langer R (2001) Drug delivery. Drugs on target. Science 293(5527):58–59
Langer R, Tirrell DA (2004) Designing materials for biology and medicine. Nature 428(6982):487–492. https://doi.org/10.1038/nature02388 nature02388 [pii]
Lee KY, Peters MC, Anderson KW, Mooney DJ (2000) Controlled growth factor release from synthetic extracellular matrices. Nature 408(6815):998–1000. https://doi.org/10.1038/35050141
Leman L, Orgel L, Ghadiri MR (2004) Carbonyl sulfide-mediated prebiotic formation of peptides. Science 306 (5694):283–286. https://doi.org/10.1126/science.1102722
Leung VY, Aladin DM, Lv F, Tam V, Sun Y, Lau RY, Hung SC, Ngan AH, Tang B, Lim CT, Wu EX, Luk KD, Lu WW, Masuda K, Chan D, Cheung KM (2014) Mesenchymal stem cells reduce intervertebral disc fibrosis and facilitate repair. Stem Cells 32 (8):2164–2177. https://doi.org/10.1002/stem.1717
LeVine H III (1993) Thioflavine T interaction with synthetic Alzheimer's disease b-amyloid peptides: detection of amyloid aggregation in solution. Protein Sci 2:404–410
Li C, Adamcik J, Mezzenga R (2012) Biodegradable nanocomposites of amyloid fibrils and graphene with shape-memory and enzyme-sensing properties. Nat Nanotechnol 7(7):421–427. https://doi.org/10.1038/nnano.2012.62
Li C, Born AK, Schweizer T, Zenobi-Wong M, Cerruti M, Mezzenga R (2014) Amyloid-hydroxyapatite bone biomimetic composites. Adv Mater 26 (20):3207–3212. https://doi.org/10.1002/adma.201306198
Li C, Mezzenga R (2013) The interplay between carbon nanomaterials and amyloid fibrils in bio-nanotechnology. Nanoscale 5(14):6207–6218. https://doi.org/10.1039/c3nr01644g
Mahoney MJ, Anseth KS (2006) Three-dimensional growth and function of neural tissue in degradable polyethylene glycol hydrogels. Biomaterials 27 (10):2265–2274. https://doi.org/10.1016/j.biomaterials.2005.11.007
Maji SK, Perrin MH, Sawaya MR, Jessberger S, Vadodaria K, Rissman RA, Singru PS, Nilsson KP, Simon R, Schubert D, Eisenberg D, Rivier J, Sawchenko P, Vale W, Riek R (2009a) Functional amyloids as natural storage of peptide hormones in pituitary secretory granules. Science 325 (5938):328–332. https://doi.org/10.1126/science.1173155 1173155 [pii]
Maji SK, Schubert D, Rivier C, Lee S, Rivier JE, Riek R (2008) Amyloid as a depot for the formulation of long-acting drugs. PLoS Biol 6(2):e17
Maji SK, Wang L, Greenwald J, Riek R (2009b) Structure-activity relationship of amyloid fibrils. FEBS Lett 583 (16):2610–2617. https://doi.org/10.1016/j.febslet.2009.07.003 S0014-5793(09)00528-6 [pii]
Mankar S, Anoop A, Sen S, Maji SK (2011) Nanomaterials: amyloids reflect their brighter side. Nano Rev 2(6032)
Maurer-Stroh S, Debulpaep M, Kuemmerer N, de la Paz ML, Martins IC, Reumers J, Morris KL, Copland A, Serpell L, Serrano L, Schymkowitz JWH, Rousseau F (2010) Exploring the sequence determinants of amyloid structure using position-specific scoring matrices. Nat Methods 7:237. https://doi.org/10.1038/nmeth.1432
Maury CP (2009) The emerging concept of functional amyloid. J Intern Med 265(3):329–334
Maury CPJ (2015) Origin of life. Primordial genetics: information transfer in a pre-RNA world based on self-replicating beta-sheet amyloid conformers. J Theor Biol 382(Supplement C):292–297
McKinnon DD, Domaille DW, Brown TE, Kyburz KA, Kiyotake E, Cha JN, Anseth KS (2014) Measuring cellular forces using bis-aliphatic hydrazone crosslinked stress-relaxing hydrogels. Soft Matter 10 (46):9230–9236. https://doi.org/10.1039/c4sm01365d
Meersman F, Dobson CM (2006) Probing the pressure-temperature stability of amyloid fibrils provides new insights into their molecular properties. Biochim Biophys Acta 1764(3):452–460
Meital Reches EG (2006) Molecular self-assembly of peptide nanostructures: mechanism of association and potential uses. Curr Nanosci 2(2):105–111
Mesquida P, Riener CK, MacPhee CE, McKendry RA (2007) Morphology and mechanical stability of amyloid-like peptide fibrils. J Mater Sci Mater Med 18(7):1325–1331. https://doi.org/10.1007/s10856-006-0075-0
Michod RE (2007) Evolution of individuality during the transition from unicellular to multicellular life. Proc Natl Acad Sci U S A 104(Suppl 1):8613–8618. https://doi.org/10.1073/pnas.0701489104
Miller SL (1953) A production of amino acids under possible primitive earth conditions. Science 117 (3046):528–529
Mostaert AS, Higgins MJ, Fukuma T, Rindi F, Jarvis SP (2006) Nanoscale mechanical characterisation of amyloid fibrils discovered in a natural adhesive. J Biol Phys

32(5):393–401. https://doi.org/10.1007/s10867-006-9023-y
Myllyharju J, Kivirikko KI (2004) Collagens, modifying enzymes and their mutations in humans, flies and worms. Trends Genet 20(1):33–43. https://doi.org/10.1016/j.tig.2003.11.004
Naiki H, Hashimoto N, Suzuki S, Kimura H, Nakakuki K, Gejyo F (1997) Establishment of a kinetic model of dialysis-related amyloid fibril extension in vitro. Amyloid-Int J Exp Clin Investig 4(4):223–232
Nelson R, Eisenberg D (2006) Recent atomic models of amyloid fibril structure. Curr Opin Struct Biol 16 (2):260–265
Nilsson KP (2009) Small organic probes as amyloid specific ligands–past and recent molecular scaffolds. FEBS Lett 583(16):2593–2599. https://doi.org/10.1016/j.febslet.2009.04.016
Nisbet DR, Williams RJ (2012) Self-assembled peptides: characterisation and in vivo response. Biointerphases 7 (1–4):2. https://doi.org/10.1007/s13758-011-0002-x
Osherovich LZ, Weissman JS (2002) The utility of prions. Dev Cell 2(2):143–151
Otzen D, Nielsen PH (2008) We find them here, we find them there: functional bacterial amyloid. Cell Mol Life Sci 65(6):910–927. https://doi.org/10.1007/s00018-007-7404-4
Ozbek S, Balasubramanian PG, Chiquet-Ehrismann R, Tucker RP, Adams JC (2010) The evolution of extracellular matrix. Mol Biol Cell 21(24):4300–4305. https://doi.org/10.1091/mbc.E10-03-0251
Paszek MJ, Zahir N, Johnson KR, Lakins JN, Rozenberg GI, Gefen A, Reinhart-King CA, Margulies SS, Dembo M, Boettiger D, Hammer DA, Weaver VM (2005) Tensional homeostasis and the malignant phenotype. Cancer Cell 8(3):241–254. https://doi.org/10.1016/j.ccr.2005.08.010
Patterson J, Martino MM, Hubbell JA (2010) Biomimetic materials in tissue engineering. Mater Today 13 (1–2):14–22
Pelham RJ Jr, Wang Y (1997) Cell locomotion and focal adhesions are regulated by substrate flexibility. Proc Natl Acad Sci U S A 94(25):13661–13665
Peppas NA, Hilt JZ, Khademhosseini A, Langer R (2006) Hydrogels in biology and medicine: from molecular principles to bionanotechnology. Adv Mater 18:1345–1360
Pike CJ, Walencewicz-Wasserman AJ, Kosmoski J, Cribbs DH, Glabe CG, Cotman CW (1995) Structure-activity analyses of b-amyloid peptides: Contributions of the b25-35 region to aggregation and neurotoxicity. J Neurochem 64(1):253–265
Ramsook CB, Tan C, Garcia MC, Fung R, Soybelman G, Henry R, Litewka A, O'eally S, Otoo HN, Khalaf RA, Dranginis AM, Gaur NK, Klotz SA, Rauceo JM, Jue CK, Lipke PN (2010) Yeast cell adhesion molecules have functional amyloid-forming sequences. Eukaryot Cell 9(3):393–404
Reches M, Gazit E (2003) Casting metal nanowires within discrete self-assembled peptide nanotubes. Science 300(5619):625–627
Reches M, Gazit E (2006) Controlled patterning of aligned self-assembled peptide nanotubes. Nat Nanotechnol 1 (3):195–200. https://doi.org/10.1038/nnano.2006.139 nnano.2006.139 [pii]
Reinhard J, Brosicke N, Theocharidis U, Faissner A (2016) The extracellular matrix niche microenvironment of neural and cancer stem cells in the brain. Int J Biochem Cell Biol 81(Pt A):174–183. https://doi.org/10.1016/j.biocel.2016.05.002
Reynolds NP, Charnley M, Bongiovanni MN, Hartley PG, Gras SL (2015) Biomimetic topography and chemistry control cell attachment to amyloid fibrils. Biomacromolecules 16(5):1556–1565. https://doi.org/10.1021/acs.biomac.5b00114
Reynolds NP, Charnley M, Mezzenga R, Hartley PG (2014) Engineered lysozyme amyloid fibril networks support cellular growth and spreading. Biomacromolecules 15(2):599–608. https://doi.org/10.1021/bm401646x
Reynolds NP, Styan KE, Easton CD, Li Y, Waddington L, Lara C, Forsythe JS, Mezzenga R, Hartley PG, Muir BW (2013) Nanotopographic surfaces with defined surface chemistries from amyloid fibril networks can control cell attachment. Biomacromolecules 14 (7):2305–2316
Ricard-Blum S (2011) The collagen family. Cold Spring Harb Perspect Biol 3(1):a004978. https://doi.org/10.1101/cshperspect.a004978
Rowley JA, Madlambayan G, Mooney DJ (1999) Alginate hydrogels as synthetic extracellular matrix materials. Biomaterials 20(1):45–53
Rufo CM, Moroz YS, Moroz OV, Stohr J, Smith TA, Hu X, Degrado WF, Korendovych IV (2014) Short peptides self-assemble to produce catalytic amyloids. Nat Chem 6(4):303–309. https://doi.org/10.1038/nchem.1894
Santiago LY, Nowak RF, Peter Rubin J, Marra KG (2006) Peptide-surface modification of poly(caprolactone) with laminin-derived sequences for adipose-derived stem cell applications. Biomaterials 15(0142–9612 (Print)):2962–2969
Sasso L, Suei S, Domigan L, Healy J, Nock V, Williams MA, Gerrard JA (2014) Versatile multi-functionalization of protein nanofibrils for biosensor applications. Nanoscale 6(3):1629–1634. https://doi.org/10.1039/c3nr05752f
Scheibel T, Parthasarathy R, Sawicki G, Lin XM, Jaeger H, Lindquist SL (2003) Conducting nanowires built by controlled self-assembly of amyloid fibers and selective metal deposition. Proc Natl Acad Sci U S A 100(8):4527–4532
Schwendinger MG, Rode BM (1992) Investigations on the mechanism of the salt-induced peptide formation. Orig Life Evol Biosphere: J Int Soc Study Orig Life 22 (6):349–359

Serio TR, Cashikar AG, Kowal AS, Sawicki GJ, Moslehi JJ, Serpell L, Arnsdorf MF, Lindquist SL (2000) Nucleated conformational conversion and the replication of conformational information by a prion determinant. Science 289(5483):1317–1321

Simitzi C, Ranella A, Stratakis E (2017) Controlling the morphology and outgrowth of nerve and neuroglial cells: the effect of surface topography. Acta Biomater 51:21–52. https://doi.org/10.1016/j.actbio.2017.01.023

Smith DK (2010) Supramolecular gels: building bridges. Nat Chem 2(3):162–163. https://doi.org/10.1038/nchem.566

Smith JF, Knowles TP, Dobson CM, Macphee CE, Welland ME (2006) Characterization of the nanoscale properties of individual amyloid fibrils. Proc Natl Acad Sci U S A 103(43):15806–15811

Solomon JP, Bourgault S, Powers ET, Kelly JW (2011) Heparin binds 8 kDa gelsolin cross-beta-sheet oligomers and accelerates amyloidogenesis by hastening fibril extension. Biochemistry 50(13):2486–2498. https://doi.org/10.1021/bi101905n

Sullan RM, Gunari N, Tanur AE, Chan Y, Dickinson GH, Orihuela B, Rittschof D, Walker GC (2009) Nanoscale structures and mechanics of barnacle cement. Biofouling 25(3):263–275. https://doi.org/10.1080/08927010802688095

Sunde M, Blake C (1997) The structure of amyloid fibrils by electron microscopy and X-ray diffraction. Adv Protein Chem 50:123–159

Sunde M, Serpell LC, Bartlam M, Fraser PE, Pepys MB, Blake CC (1997) Common core structure of amyloid fibrils by synchrotron X-ray diffraction. J Mol Biol 273(3):729–739

Sutherland JD (2017) Opinion: studies on the origin of life — the end of the beginning. Nat Rev Chem 1:0012. https://doi.org/10.1038/s41570-016-0012

Swift J, Ivanovska IL, Buxboim A, Harada T, Dingal PC, Pinter J, Pajerowski JD, Spinler KR, Shin JW, Tewari M, Rehfeldt F, Speicher DW, Discher DE (2013) Nuclear lamin-A scales with tissue stiffness and enhances matrix-directed differentiation. Science 341(6149):1240104. https://doi.org/10.1126/science.1240104

Tanaka H, Herland A, Lindgren LJ, Tsutsui T, Andersson MR, Inganas O (2008) Enhanced current efficiency from bio-organic light-emitting diodes using decorated amyloid fibrils with conjugated polymer. Nano Lett 8(9):2858–2861. https://doi.org/10.1021/nl801510z

Trappmann B, Gautrot JE, Connelly JT, Strange DG, Li Y, Oyen ML, Cohen Stuart MA, Boehm H, Li B, Vogel V, Spatz JP, Watt FM, Huck WT (2012) Extracellular-matrix tethering regulates stem-cell fate. Nat Mater 11(7):642–649. https://doi.org/10.1038/nmat3339

True HL, Lindquist SL (2000) A yeast prion provides a mechanism for genetic variation and phenotypic diversity. Nature 407(6803):477–483

Ulrich TA, de Juan Pardo EM, Kumar S (2009) The mechanical rigidity of the extracellular matrix regulates the structure, motility, and proliferation of glioma cells. Cancer Res 69(10):4167–4174. https://doi.org/10.1158/0008-5472.can-08-4859

Uptain SM, Lindquist S (2002) Prions as protein-based genetic elements. Annu Rev Microbiol 56:703–741

Uversky VN, Li J, Souillac P, Millett IS, Doniach S, Jakes R, Goedert M, Fink AL (2002) Biophysical properties of the synucleins and their propensities to fibrillate – Inhibition of alpha-synuclein assembly by beta- and gamma-synucleins. J Biol Chem 277(14):11970–11978

van Oosten AS, Vahabi M, Licup AJ, Sharma A, Galie PA, MacKintosh FC, Janmey PA (2016) Uncoupling shear and uniaxial elastic moduli of semiflexible biopolymer networks: compression-softening and stretch-stiffening. Sci Rep 6:19270. https://doi.org/10.1038/srep19270

Vargas P, Barbier L, Saez PJ, Piel M (2017) Mechanisms for fast cell migration in complex environments. Curr Opin Cell Biol 48:72–78. https://doi.org/10.1016/j.ceb.2017.04.007

Walsh DM, Klyubin I, Fadeeva JV, Cullen WK, Anwyl R, Wolfe MS, Rowan MJ, Selkoe DJ (2002) Naturally secreted oligomers of amyloid b protein potently inhibit hippocampal long-term potentiation *in vivo*. Nature 416(6880):535–539

Wang Y, Lee WC, Manga KK, Ang PK, Lu J, Liu YP, Lim CT, Loh KP (2012) Fluorinated graphene for promoting neuro-induction of stem cells. Adv Mater 24(31):4285–4290. https://doi.org/10.1002/adma.201200846

Wasmer C, Lange A, Van Melckebeke H, Siemer AB, Riek R, Meier BH (2008) Amyloid fibrils of the HET-s(218-289) prion form a beta solenoid with a triangular hydrophobic core. Science 319(5869):1523–1526

Westermark GT, Johnson KH, Westermark P (1999) Staining methods for identification of amyloid in tissue. Methods Enzymol 309:3–25

Winer JP, Oake S, Janmey PA (2009) Non-linear elasticity of extracellular matrices enables contractile cells to communicate local position and orientation. PLoS One 4(7):e6382. https://doi.org/10.1371/journal.pone.0006382

Wolf K, Alexander S, Schacht V, Coussens LM, von Andrian UH, van Rheenen J, Deryugina E, Friedl P (2009) Collagen-based cell migration models in vitro and in vivo. Semin Cell Dev Biol 20(8):931–941. https://doi.org/10.1016/j.semcdb.2009.08.005

Yan H, Nykanen A, Ruokolainen J, Farrar D, Gough JE, Saiani A, Miller AF (2008) Thermo-reversible protein fibrillar hydrogels as cell scaffolds. Faraday Discuss 139:71–84 discussion 105–128, 419–120

Yeung T, Georges PC, Flanagan LA, Marg B, Ortiz M, Funaki M, Zahir N, Ming W, Weaver V, Janmey PA (2005) Effects of substrate stiffness on cell

morphology, cytoskeletal structure, and adhesion. Cell Motil Cytoskeleton 60(1):24–34. https://doi.org/10.1002/cm.20041
Zhang S (2002) Emerging biological materials through molecular self-assembly. Biotechnol Adv 20:321–339
Zhang S, Marini DM, Hwang W, Santoso S (2002) Design of nanostructured biological materials through self-assembly of peptides and proteins. Curr Opin Chem Biol 6(6):865–871 doi:S1367593102003915 [pii]
Zurdo J, Guijarro JI, Dobson CM (2001) Preparation and characterization of purified amyloid fibrils. J Am Chem Soc 123(33):8141–8142

The Prospects of Cadherin-23 as a Mediator of Homophilic Cell-Cell Adhesion

8

Malay Kumar Sannigrahi, Sai Srinivas, and Sabyasachi Rakshit

Abstract

Cadherins (calcium-dependent adhesion proteins) constitute a family of cell surface proteins that mediate cell-cell adhesion and actively participate in tissue morphogenesis and in mediating tissue integrity. The ecto-domains of cadherins from opposing cell surfaces interact with each other to form the load-bearing trans-dimers and mechanically hold cells together. The "classical" cadherins and desmosomes that form separate groups in cadherin superfamily are mostly explored for their roles in cell-cell adhesion. However, majority of cadherins in cells belong to "non-classical" group which is poorly explored in the context of their cell-binding properties. This review focuses on the role of "nonclassical" cadherin, cadherin-23, in cell-cell adhesion. Overall, this review highlights the need for further investigations on the role of "non-classical" cadherin-23 in cell-cell adhesion.

M. K. Sannigrahi · S. Srinivas
Department of Chemical Sciences, Indian Institute of Science Education and Research (IISER) Mohali, Mohali, Punjab, India

S. Rakshit (✉)
Department of Chemical Sciences, Indian Institute of Science Education and Research (IISER) Mohali, Mohali, Punjab, India

Centre for Protein Science Design and Engineering, Indian Institute of Science Education and Research (IISER) Mohali, Mohali, Punjab, India
e-mail: srakshit@iisermohali.ac.in

Keywords

Nonclassical cadherins · Cadherin-23 · Cell-cell adhesion · Cancer

8.1 Introduction

Cell-cell adhesion is the foremost requirement for the development of multicellular organisms and subsequently tissue morphogenesis. The overall cell-cell adhesion complex is comprised of various cell surface proteins forming different junctions importantly, adherens junction (AJ), desmosomal junction (DJ), tight junction (TJ) and gap junctions (GJ). Among these, AJs and DJs are known to provide the adhesive strength of the junction and prevent tissues from dissociation into component cells in response to external forces. Proteins that predominantly orchestrate the AJs are cadherins. Cadherins interact homophilically with itself from apposing cell surfaces and form AJs. Initially, they were named after the tissue in which they are preferentially expressed (e.g. epithelial [E-cadherin], neural [N-cadherin], placental [P-cadherin], retinal [R-cadherin] and vascular endothelial [VE-cadherin]). However, now they are broadly classified as classical cadherins. DJs are formed by desmosomal cadherins. The hallmark of these cadherin proteins is the extracellular (EC) domains, connected by conserved linker chains in complex with Ca^{2+} ions that define the

K. Chattopadhyay, S. C. Basu (eds.), *Biochemical and Biophysical Roles of Cell Surface Molecules*, Advances in Experimental Medicine and Biology 1112, https://doi.org/10.1007/978-981-13-3065-0_8

bending rigidity of the proteins. Based on this domain organization, more members known as nonclassical or atypical cadherins (clustered protocadherins, Fat cadherins, Dachsous cadherins, cadherin-23, Flamingo and many more) have been registered in the cadherin superfamily. In general, all cadherins feature a cleavable pro-domain (signalling domain, ~20–26 residues long), varying numbers of EC domains (each domain containing ~110 residues), short transmembrane(TM) region and cytoplasmic domains (CD). Classical cadherins and desmosomal cadherins have five EC domains, clustered protocadherins have six EC domains, and nonclassical cadherins have domain numbers varying from 2 to 34 (Fig. 8.1) (Sotomayor et al. 2014).

Unlike homological extracellular domains, the cytosolic domains of cadherins are different within the family and poorly classified. However, their link to cytoskeleton via the number of interacting proteins is unique and classifies specific adhesion junctions. In case of AJs, the cytosolic region of classical cadherins that is conserved within its subfamily binds to the actin filaments via the armadillo family of proteins (α-catenin, β-catenin, p120-catenin). Similarly, desmosomal cadherins also possess a conserved cytosolic domain within its subfamily and bind to intermediate filaments via plakoglobin family of proteins. For nonclassical cadherins, along with the EC domain numbers, the cytosolic domains also differ for individual proteins and often within isoforms. The protein complex linking to cytoskeleton is also not well-understood for nonclassical cadherins. As a consequence, the involvement of nonclassical cadherins in cell-cell adhesion is not clear, even though they are the largest members in cadherin family of

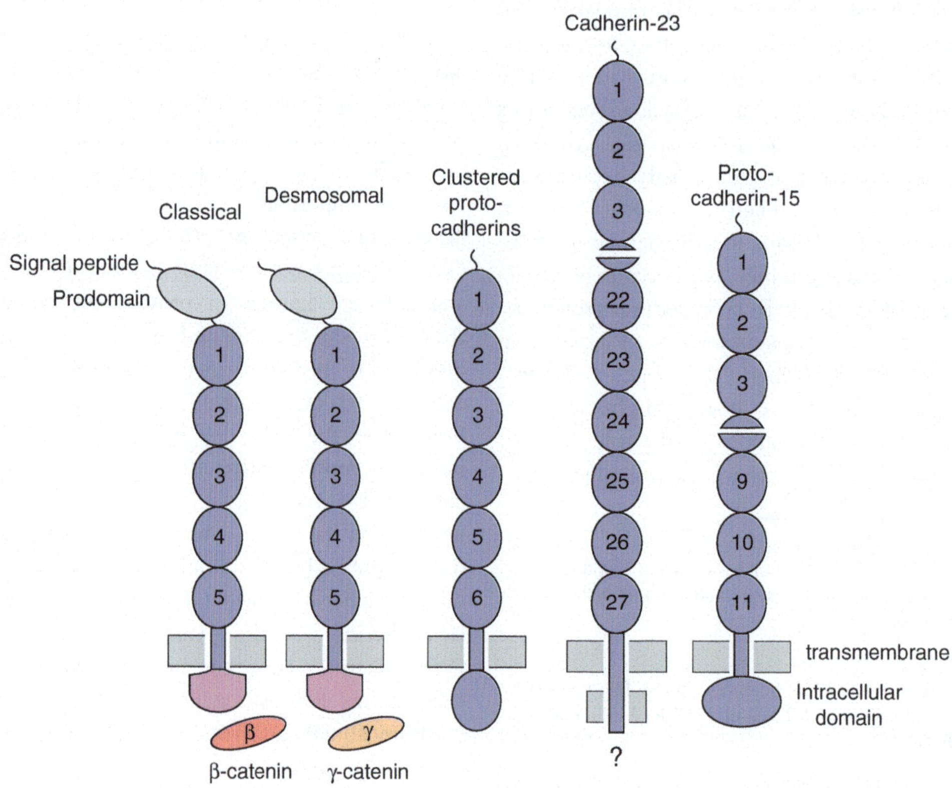

Fig. 8.1 Structural difference in different types of cadherins: Classical cadherins have similar structure, with five extracellular cadherin repeats, a transmembrane domain and an intracellular domain that interact with β-catenin, whereas nonclassical cadherins like Cdh-23 have 27 extracellular domains and distinct cytosolic domains that don't have an β-catenin binding site

proteins. We mainly focus our study on the perspective of a giant cadherin-23 (Cdh-23), as cell-cell adhesion molecule. Cdh-23 is known for its unique nature of engagements in both homophilic interactions and heterophilic interactions, forming homotypic and heterotypic cell-cell junctions, and has been associated with carcinoma, Usher syndrome and many more.

8.2 Cadherin-23

Cdh-23 is one of the tip-link proteins in hair cells and, in partnering with protocadherin-15 (Pcdh-15) in tip-links, conveys the force to mechanosensitive ion channels (Gillespie and Muller 2009) directly as gating springs. During the early development of the stereociliary organization and hair bundle formation in hair cells, Cdh-23 was observed engaged in lateral links however homophilically (Michel et al. 2005). In the retina, Cdh-23 is localized at the synaptic region of the photoreceptive cells and mediates cell-cell adhesion (Lagziel et al. 2005). They were found to be mutated in Usher syndrome, a hereditary deafness and blindness disease in humans, and also in the non-syndromic deafness's DFNB12 and DFNB23 (Sakaguchi et al. 2009).

Crystallographic studies have shown that the first and second extracellular repeats (EC1–2) of Cdh-23 and Pcdh-15 adopt the typical cadherin folds and mediate trans-interactions between the two cadherins (Sotomayor et al. 2012). Electron tomography revealed a unique pattern in the cis-homodimer of Cdh23: a pair of Cdh-23 molecules aligned in the same orientation and intertwined to form a helical complex through interactions between all EC domains except the two terminal ones, EC1 and EC2 (EC1 + 2), which remained exposed outwards to facilitate trans-interactions (Kazmierczak et al. 2007). The amino termini of the two cadherins co-localize on tip-link filaments. They were shown to interact with each other as dimers. Biochemical experiments show that Cdh-23 homodimers interact in trans with Pcdh-15 homodimers to form a filament with structural similarity to tip-links (Kazmierczak et al. 2007). Cdh-23 also lacks the sequence determinants for either S- or X-dimerization. Moreover, the EC1 domain of Cdh-23 has several unique features: a 5 residue long 3_{10}-helix just prior to the A* β-strand, a α-helical loop connecting two β-strands and, most strikingly, an additional Ca^{2+}-binding site towards the N-terminus (Sotomayor et al. 2012).

Apart from neuroepithelial tissues, Cdh-23 is uniformly expressed in organs like the brain, lymph node, kidney, gastrointestinal tract, testis and skin (reported in The Human Proteome Atlas (THPA)) (Fig. 8.2) (Thul et al. 2017; Uhlen et al. 2015, 2017). The Cancer Genome Atlas (TCGA) data showed decreased expression of Cdh-23 in most of the solid cancers including sarcoma, adrenocortical carcinoma, cervical cancer, head and neck squamous cell carcinoma, lung cancer, kidney renal clear cell carcinoma, breast cancer and more (Fig. 8.3). In vitro studies with the MCF-7 breast cancer cell lines have confirmed the involvement of Cdh-23 in homotypic cell-cell adhesion, similar to classical cadherins.

Further, it was observed that Cdh-23 was at the heterotypic cell-cell junctions between MCF-7 and normal breast fibroblasts (NBFs) suggesting its involvement towards tumour metastasis where interaction with connective tissues is the foremost requirements (Apostolopoulou and Ligon 2012). In both homotypic and heterotypic cell-cell junctions, Cdh-23 from opposing cell surfaces was interpreted to be participating in homophilic interactions. The immunoblot of the cell extracts from MCF-7 and NBF cells featured two bands, corresponding to isoform B and isoform C and not the full-length protein (Apostolopoulou and Ligon 2012).

As depicted in Fig. 8.4, isoform B contains the last seven EC domains (20–27), a short TM domain and a distinct cytosolic domain. Isoform C contains only a cytosolic part free from the membrane. It is interesting to note that both these isoforms lacked the outermost domains but still found engaged in cell-cell adhesion junction (Apostolopoulou and Ligon 2012). In contrary, upon using EC1-specific antibodies, authors observed the blockage of Cdh-23-specific interactions at the cell-cell level. The immunoblots are not involved in cell-cell

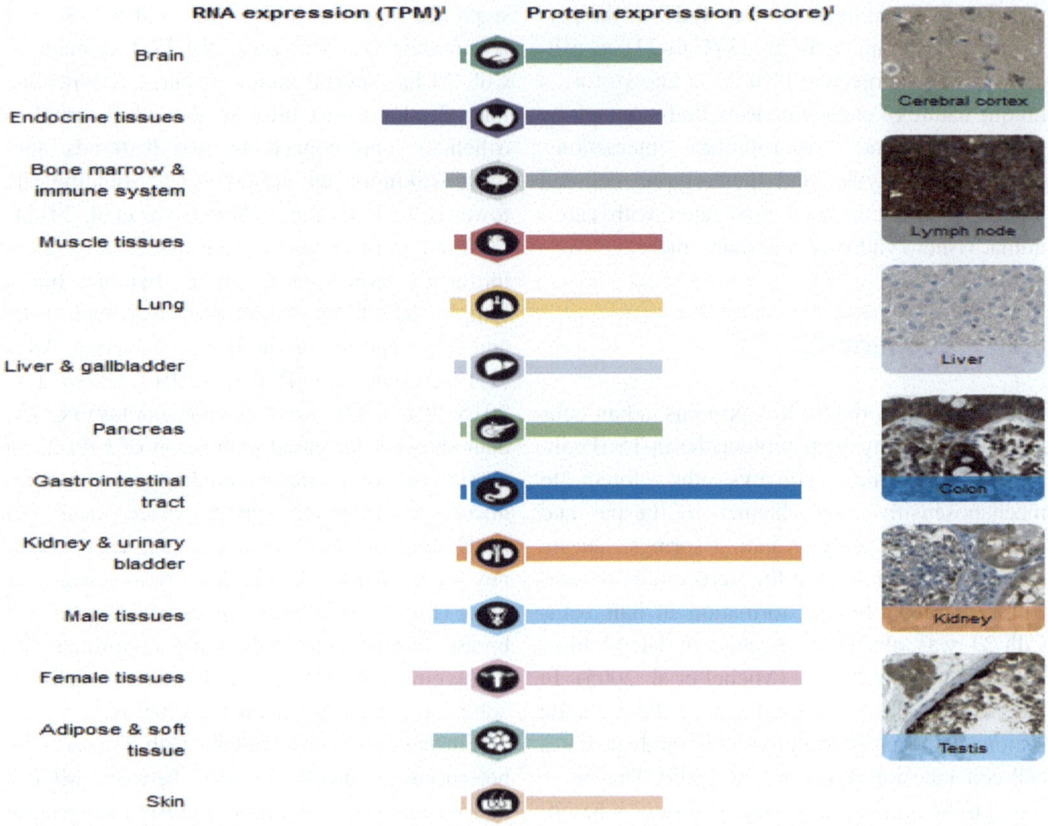

Fig. 8.2 Expression of cadherin-23 in different tissues of humans as observed in The Human Proteome Atlas (THPA)

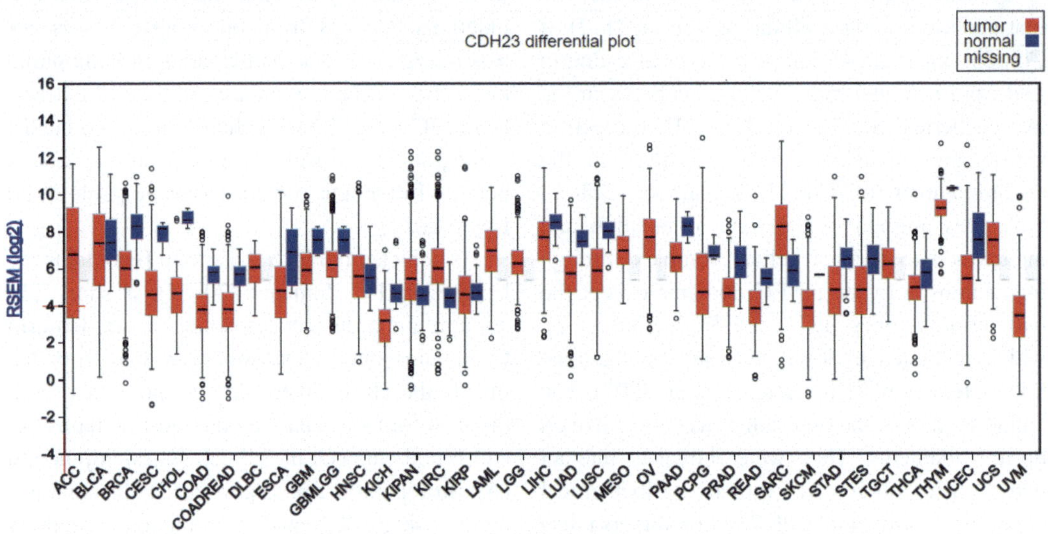

Fig. 8.3 Log2 values of RNA-Seq by Expectation-Maximization (RSEM) of cadherin-23 in different cancer tissue samples and respective control. The data was collected from The Cancer Genome Atlas (TCGA) and analysed by FireBrowse Gene Expression Viewer (**Abbreviations are mentioned in a table)

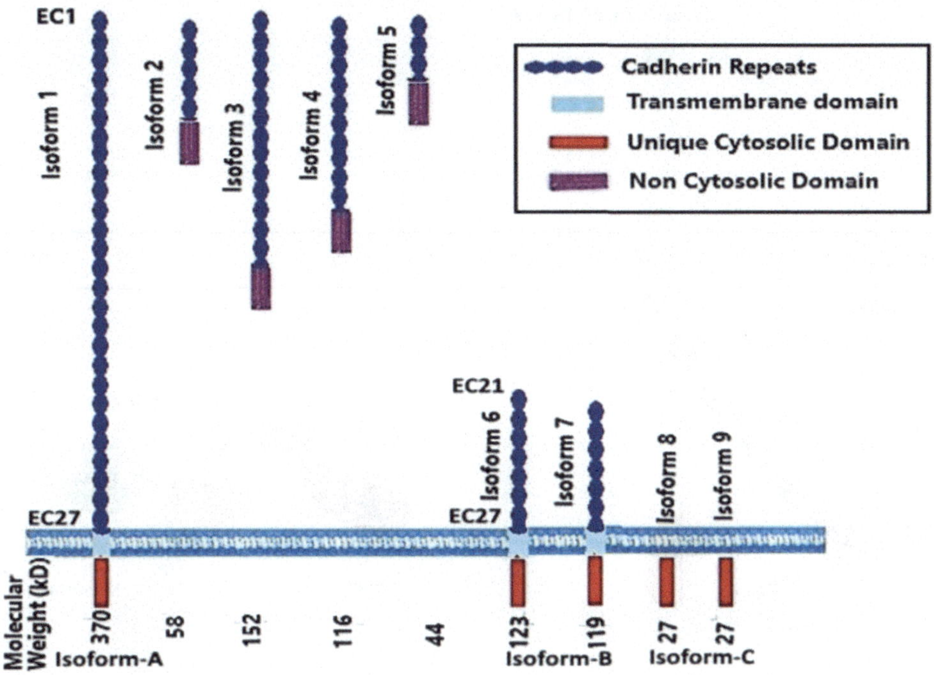

Fig. 8.4 Different isoforms of cadherin-23 as obtained from NCBI (Gene 64,072). Isoform A/1 is the largest isoform (370 Kd) having cytosolic (CD), transmembrane (TM) and 27 extracellular (EC) domains; isoform B/(6 & 7) have less numbers of EC domains than the full-length CD and TM; isoform C/(8 & 9) have only CD; and isoform 2, 3, 4 and 5 have only a variable number of EC domains and no TM domain and CD

contacts; rather the isoform containing EC1 domain (possibly isoform A, the full-length cadherin), not captured in immunoblots, is engaged in the cell-cell contact formation. The western blots as well as qRT-PCR performed in our laboratory confirmed the presence of full-length cadherins in MCF-7, though expression is very low (Singaraju et al., unpublished).

Unlike classical cadherins, the cytosolic part of Cdh-23 was not found to associate with β-catenin. However, Cdh-23 was shown to bind to the PDZ4 domain of a scaffolding protein, MAGI-1, a member of the family of membrane-associated guanylate kinases (MAGUK) (Xu et al. 2008). MAGI-1 can bind to β-catenin via its PDZ5 domain, suggesting that MAGI-1 may act as a bridge between cdh-23 and β-catenin (Dobrosotskaya and James 2000). In neuroepithelial, the cytosolic domain of Cdh-23 was found to interact with PDZ1 domain of harmonin B. Harmonin B has three PDZ domains and two coiled-coil (CC) domains. PDZ1 domain of harmonin B is known to interact with the tail of myosin-7a motor protein and connect Cdh-23 to actin cytoskeleton. Cdh-23 C isoform is linked to F-actin bundle network through harmonin A isoform (Takahashi et al. 2016) (Fig. 8.5).

Ex vivo experiments with non-aggregating L929 cells expressing recombinant full-length Cdh-23 were also shown to form aggregates exclusively via homophilic interactions. The aggregation index of Cdh-23 was higher compared with those L929 cells expressing recombinant full-length E-cadherin. Also, the cells aggregated into two separate groups expressing either of the two proteins and did not form mixed aggregates suggesting nonclassical cadherins and classical cadherins don't interact with each other (Siemens et al. 2004). Further, genome-wide association studies (GWAS) of kidney function have identified a strong association of Cdh-23 with cross-sectional estimated glomerular filtration rate (eGFR), and its knockdown in zebrafish embryo resulted in severe oedema, suggesting its important role in normal kidney functions (Gorski et al. 2015).

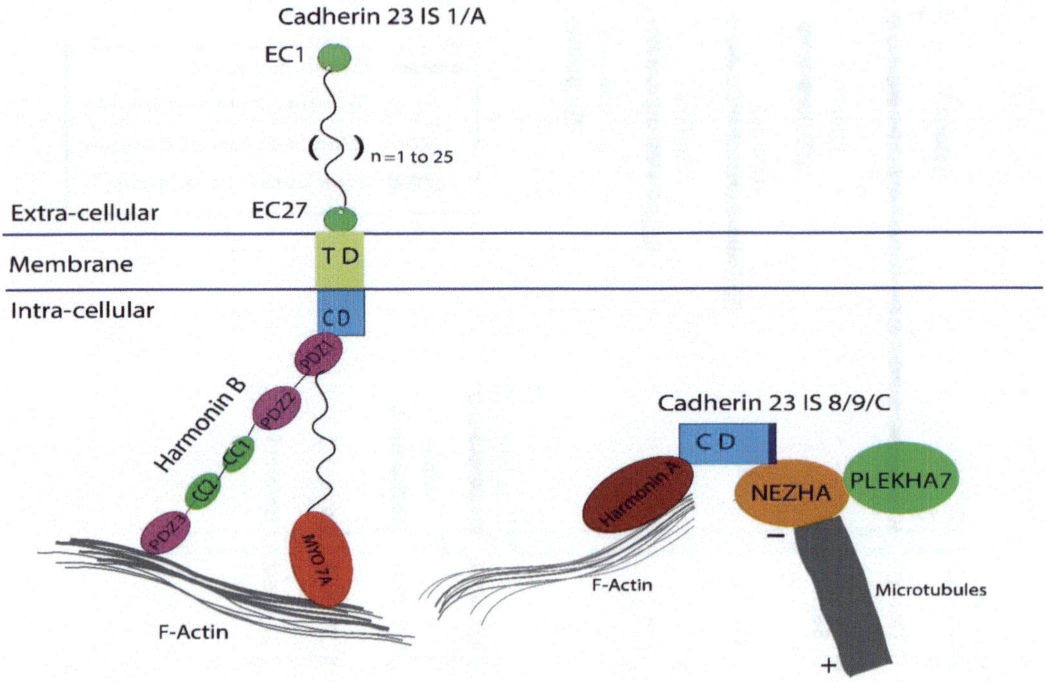

Fig. 8.5 Cadherin-23 isoforms contain PDZ-binding motif in the cytosolic domain through which it interacts strongly with PDZ1 domain of harmonin isoform B. PDZ1 domain of harmonin B is known to interact with the tail of Myosin-7a motor protein and connect cadherin-23 to actin cytoskeleton. Cadherin-23 isoform C is connected to microtubule minus-end binding protein called NEZHA/CAMSAP3, hence indirectly regulating microtubule network. NEZHA is known to interact with protein PLEKHA7 which helps in stabilization of microtubule network. Cadherin-23 C is linked to F-actin bundle network through harmonin A isoform

In neuroepithelium tissue, Cdh-23 interacts heterophilically with Pcdh15 and links the apex of two neighbouring stereocilium. This link is called the "tip-links" which serves as gating spring for mechano-responsive ion channels in hearing (Elledge et al. 2010; Sotomayor et al. 2012). However, tip-links is not a true cell-cell junction. Jaiganesh et al. (2017) have reviewed the molecular and cellular details of this heterophilic structure, and we therefore avoid repeating here (Jaiganesh et al. 2017).

8.3 Conclusion

Since their discovery, the nonclassical cadherins are known for their role in tissue morphogenesis. They usually have varying numbers of EC domains (each domain containing ~110 residues), short transmembrane domains and cytoplasmic domains. Recently, many of them have been explored for their role in either homophilic or heterophilic cell adhesion. The assembly of cadherin-containing AJs is essential for the formation and maintenance of cellular and tissue integrity as well as activation of many intracellular pathways. The nonclassical cadherins also play a very important role in adhesion, migration and patterning of a variety of tissues. However, still little is known about the distribution of nonclassical cadherin on the cell membrane, to understand how these various cadherin-cadherin interactions on cell surface levels are regulated and to determine the biological contexts in which their influence is most important in normal tissue and various diseases including cancer.

References

Apostolopoulou M, Ligon L (2012) Cadherin-23 mediates heterotypic cell-cell adhesion between breast cancer epithelial cells and fibroblasts. PLoS One 7:e33289. https://doi.org/10.1371/journal.pone.0033289

Dobrosotskaya IY, James GL (2000) MAGI-1 interacts with beta-catenin and is associated with cell-cell adhesion structures. Biochem Biophys Res Commun 270:903–909. https://doi.org/10.1006/bbrc.2000.2471

Elledge HM, Kazmierczak P, Clark P, Joseph JS, Kolatkar A, Kuhn P, Muller U (2010) Structure of the N terminus of cadherin 23 reveals a new adhesion mechanism for a subset of cadherin superfamily members. Proc Natl Acad Sci U S A 107:10708–10712. https://doi.org/10.1073/pnas.1006284107

Gillespie PG, Muller U (2009) Mechanotransduction by hair cells: models, molecules, and mechanisms. Cell 139:33–44. https://doi.org/10.1016/j.cell.2009.09.010

Gorski M et al (2015) Genome-wide association study of kidney function decline in individuals of European descent. Kidney Int 87:1017–1029. https://doi.org/10.1038/ki.2014.361

Jaiganesh A, Narui Y, Araya-Secchi R, Sotomayor M (2017) Beyond cell-cell adhesion: sensational Cadherins for hearing and balance. Cold Spring Harb Perspect Biol. https://doi.org/10.1101/cshperspect.a029280

Kazmierczak P, Sakaguchi H, Tokita J, Wilson-Kubalek EM, Milligan RA, Muller U, Kachar B (2007) Cadherin 23 and protocadherin 15 interact to form tip-link filaments in sensory hair cells. Nature 449:87–91. https://doi.org/10.1038/nature06091

Lagziel A, Ahmed ZM, Schultz JM, Morell RJ, Belyantseva IA, Friedman TB (2005) Spatiotemporal pattern and isoforms of cadherin 23 in wild type and waltzer mice during inner ear hair cell development. Dev Biol 280:295–306. https://doi.org/10.1016/j.ydbio.2005.01.015

Michel V et al (2005) Cadherin 23 is a component of the transient lateral links in the developing hair bundles of cochlear sensory cells. Dev Biol 280:281–294. https://doi.org/10.1016/j.ydbio.2005.01.014

Sakaguchi H, Tokita J, Muller U, Kachar B (2009) Tip links in hair cells: molecular composition and role in hearing loss. Curr Opin Otolaryngol Head Neck Surg 17:388–393. https://doi.org/10.1097/MOO.0b013e3283303472

Siemens J, Lillo C, Dumont RA, Reynolds A, Williams DS, Gillespie PG, Müller U (2004) Cadherin 23 is a component of the tip link in hair-cell stereocilia. Nature 428:950. https://doi.org/10.1038/nature02483

Sotomayor M, Weihofen WA, Gaudet R, Corey DP (2012) Structure of a force-conveying cadherin bond essential for inner-ear mechanotransduction. Nature 492:128–132. https://doi.org/10.1038/nature11590

Sotomayor M, Gaudet R, Corey DP (2014) Sorting out a promiscuous superfamily: towards cadherin connectomics. Trends Cell Biol 24:524–536. https://doi.org/10.1016/j.tcb.2014.03.007

Takahashi S, Mui VJ, Rosenberg SK, Homma K, Cheatham MA, Zheng J (2016) Cadherin 23-C regulates microtubule networks by modifying CAMSAP3's function. Sci Rep 6:28706. https://doi.org/10.1038/srep28706

Thul PJ et al (2017) A subcellular map of the human proteome. Science 356. https://doi.org/10.1126/science.aal3321

Uhlen M et al (2015) Proteomics. Tissue-based map of the human proteome. Science 347:1260419. https://doi.org/10.1126/science.1260419

Uhlen M et al (2017) A pathology atlas of the human cancer transcriptome. Science 357. https://doi.org/10.1126/science.aan2507

Xu Z, Peng AW, Oshima K, Heller S (2008) MAGI-1, a candidate stereociliary scaffolding protein, associates with the tip-link component cadherin 23. J Neurosci Off J Soc Neurosci 28:11269–11276. https://doi.org/10.1523/JNEUROSCI.3833-08.2008

Structural-Mechanical and Biochemical Functions of Classical Cadherins at Cellular Junctions: A Review and Some Hypotheses

9

Prince Tiwari, Arpita Mrigwani, Harpreet Kaur, Pallavi Kaila, Rajendra Kumar, and Purnananda Guptasarma

Abstract

This article begins with a general review of cell adhesion molecules (CAMs) and narrows the focus down progressively to the cadherins (calcium binding-dependent CAMs), classifications of subfamilies of the cadherins, type I (E- and N-) cadherins, evolutionary relationships amongst cadherins, structural-mechanical and functional consequences of calcium binding to the cadherins, differential molecular interactions involving the extracellular (ecto) and intracellular (cytoplasmic) domains of the cadherins, multiple adherence-related homophilic and heterophilic interactions and associated functions of E- and N-cadherin in organismal development and disease and cadherin trafficking and membrane rafts. It ends by summarizing multiple perspectives and hypotheses concerning different aspects of cadherin structure, stability and function.

Keywords

Cell adhesion · Epithelial cadherins · Neuronal cadherins · Cadherin interactions · Cadherin trafficking · Calcium binding · Cancer · Metastasis · Development

9.1 Cell Adhesion Molecules (CAMs)

A fundamental determinant of the formation, and maintenance, of diverse three-dimensional assemblies of cells within biological tissues or organs, is the mechanism by which any cell manages to adhere to its neighbouring cells. The surfaces of cells in multicellular organisms display a variety of cell adhesion molecules (CAMs) that work to keep the trillions of cells in the organism tied together into a composite and singular 'whole', mediating and regulating cell-cell interactions.

CAMs bind either to the molecular components of the extracellular matrix within which a cell happens to be embedded or to other CAMs displayed on the surfaces of neighbouring (or juxtaposed) cells. CAMs thus play very important roles in determining cell shape and integrity. Notably, CAMs also play very important roles in cell-cell and environment-cell signalling by linking up with proteins like the catenins, which are associated with the actin-based cellular cytoskeleton. Consequently, CAMs also play

P. Tiwari · A. Mrigwani · H. Kaur · P. Kaila · R. Kumar
P. Guptasarma (✉)
Centre for Protein Science, Design and Engineering, Department of Biological Sciences, Indian Institute of Science Education and Research (IISER) Mohali, Mohali, Punjab, India
e-mail: princetiwari@iisermohali.ac.in; arpitamrigwani@iisermohali.ac.in; pallavi@iisermohali.ac.in; rajendrabits@iisermohali.ac.in; guptasarma@iisermohali.ac.in; http://www.guptasarmalab.com

K. Chattopadhyay, S. C. Basu (eds.), *Biochemical and Biophysical Roles of Cell Surface Molecules*, Advances in Experimental Medicine and Biology 1112, https://doi.org/10.1007/978-981-13-3065-0_9

roles in cellular homeostasis and tissue morphogenesis.

CAMs can be broadly classified into (i) calcium binding-dependent and (ii) calcium binding-independent CAMs. Molecules such as the selectins or the cadherins are examples of calcium binding-dependent CAMs, whereas molecules such as the integrins or the immunoglobulins (including various members of the larger immunoglobulin superfamily) are examples of calcium binding-independent CAMs.

Amongst the calcium binding-independent CAMs, the integrins are CAMs that interact either with components of the extracellular matrix, such as fibronectin, collagen or fibrinogen, or with CAMs on the surfaces of other cells. The other calcium-independent CAM mentioned above, i.e. the immunoglobulin superfamily, is a group of cell surface glycoproteins consisting of immunoglobulin (Ig)-like structural domains consisting of 70–110 amino acids each. These Ig-like domains exhibit both heterophilic and homophilic interactions. To take some examples, ICAMs, or intercellular cell adhesion molecules, engage in heterophilic interactions and bind to other CAMs such as integrins, whereas N-CAMs, or neural cell adhesion molecules, engage in homophilic interactions and participate in cell-cell adhesions through interaction with other N-CAMs (Cruse et al. 2004).

Amongst the calcium binding-dependent CAMs, the selectins are a group of cell surface glycoproteins which bind to fucosylated carbohydrates. Different types of selectins turn out to be expressed on different types of cells, e.g. P-selectin is expressed on the surfaces of platelets and leukocytes; L-selectin on the surfaces of leukocytes, monocytes, neutrophils and eosinophils; and E-selectin on the surfaces of endothelial cells. The functions of the selectins vary considerably. For example, they range from leukocyte trafficking to signal transduction (Lodish et al. 2000).

Like the selectins, the molecules which are the focus of this review, i.e. the cadherins, constitute a distinct group of calcium binding-dependent CAMs. The cadherin superfamily is a multigene family of proteins with diverse structures and functions. Cadherins tend to localize at 'adhesion junctions' which are the dominant structural features visible at cell-cell adhesion interfaces.

9.2 The Cadherins: Calcium Binding-Dependent CAMs

The cadherins mediate cell adhesion in a calcium binding-dependent manner. All cadherins are multidomain proteins. The extracellular domains of the cadherins are a series of domain repeats, displaying considerable mutual sequence as well as structural similarities, suggestive of an origin based on gene duplication. Three calcium ions bind to each of the linker peptides separating any two extracellular cadherin domains. The length of each extracellular (or ecto) domain of a cadherin is about 110 amino acids. Every molecule of a cadherin is a membrane-displayed protein consisting of a cytoplasmic domain, a single-pass transmembrane region and multiple extracellular 'repeat' domains known as 'ectodomains'. In mammalian cadherins, the number of the extracellular repeat domains available for adhesive interactions usually varies from 2 to 34 contiguous domains. The classical cadherins and protocadherins, however, have only about five to seven ectodomains (Suzuki and Hirano 2016). Each cadherin ectodomain has a fold similar to that of the Ig or Ig-like domains, i.e. the typical cadherin ectodomain contains seven β-strands arranged into two β-sheets that associate into a beta-sandwich fold, exactly as is seen in any Ig or Ig-like domain, however with distinct topological differences, i.e. with different contact schemes of the constituent β-strands from that seen in the Ig-like domains.

Unlike the ectodomains, the cytoplasmic domains of the cadherins have only a few conserved motifs, and these are seen amongst specific subfamilies of cadherins involved in binding or interaction(s) with cytoplasmic proteins. For example, β-catenin and catenin p120 (p120ctn) bind to two catenin-binding motifs in the cytoplasmic domains of the classical cadherins. Since these cytoplasmic domains are exceptionally diverse, they form the basis of the categorization of the cadherin superfamily into various subfamilies (Suzuki and Hirano 2016).

9.3 Different Subfamilies of Cadherins

The cadherin superfamily is divided into four major subfamilies. These are (a) classical cadherins, (b) protocadherins, (c) desmosomal cadherins and (d) atypical cadherins (Priest et al. 2017). Below, we provide a brief description of each of these subfamilies, before we bring this review's focus primarily on to the classical cadherins.

(a) *Classical cadherins.* The classical cadherins constitute a major subgroup of the cadherin superfamily. In most higher organisms, the classical cadherins are characterized by the presence of a five domains-long extracellular region, a transmembrane region and a cytoplasmic region which consists of a folded domain that interacts directly with p120 and β-catenin and indirectly with α-catenin, to link up to actin filaments through protein-protein associations. The classical cadherins are further subclassified primarily into types, I, II, III and IV. Type I and type II classical cadherins are present only in vertebrates. They are classified on the basis of the tissue within which they were first identified. In vertebrates, there are 6 type I cadherins and 13 type II classical cadherins. Type III classical cadherins are found both in vertebrates and invertebrates, but not in mammals. Below, we provide a brief overview of these three types of classical cadherins. Type I classical cadherins consist of proteins like epithelial cadherin or E-cadherin which is known as CDH1 and neuronal (or neural) cadherin, known as N-cadherin, or CDH2. The type I classical cadherins display a conserved HAV tripeptide motif and a conserved tryptophan at the second position in the most distal domain, EC1, located farthest from the plasma membrane (Shapiro et al. 1995). The type I cadherins have five main members, which include CDH1 (E-cadherin, epithelial), CDH2 (N-cadherin, neuronal), CDH3 (P-cadherin, placental), CDH4 (R-cadherin, retinal) and CDH15 (M-cadherin, myotubule) (Suzuki and Hirano 2016). Type II classical cadherins consist of proteins like vascular endothelial (VE)-cadherin which is known as CDH5 and kidney (K)-cadherin which is known as CDH6 (Gumbiner 2005; Leckband and Prakasam 2006). Type II classical cadherins are characterized by the presence of two conserved tryptophan residues at the second and fourth positions (Trp2 and Trp4) in the most distal domain, EC1, but they lack the HAV tripeptide which is present in the type I cadherins (Shapiro and Weis 2009). There are currently 13 named type II cadherins, these being CDH5 (VE-cadherin, vascular endothelium), CDH6 (K-cadherin, foetal kidney), CDH7, CDH8, CDH9 (T1-cadherin, testis), CDH10 (T2-cadherin, testis), CDH11 (OB-cadherin, osteoblast), CDH12, CDH18, CDH19, CDH20, CDH22 and CDH24. The type III classical cadherins possess a variable number of ectodomain repeats (Oda et al. 2002; Tanabe et al. 2004). They also possess a conserved region called the primitive classical cadherin domain (PCCD) which lies between the cadherin repeats and the transmembrane helix. For the maturation of E-cadherin in *Drosophila*, proteolytic cleavage of PCCD is required (Oda and Tsukita 1999), indicating that type III and type I cadherins interact. Type IV cadherins have seven ectodomains.

(b) *Protocadherins.* More than 80 members of the cadherin superfamily together constitute another group of cadherins known as protocadherins. These are mainly expressed in the developing and mature vertebrate nervous system, although low levels of expression are also seen in the lungs and the kidney. The protocadherins possess six or seven ectodomain repeats. These are highly conserved amongst the protocadherin subgroup but show low sequence homology with other members of the cadherin superfamily (Hulpiau and Van Roy 2009). As with other cadherins, in addition to ectodomains, the protocadherins have a single-pass transmembrane domain and a very distinct and specific cytoplasmic domain which, however, lacks motifs for catenin binding (Sano et al. 1993; Wu and Maniatis 1999; Nollet et al. 2000; Vanhalst et al. 2005). Based on their genomic organization, protocadherins are further classified into clustered and non-clustered protocadherins.

(c) *Desmosomal cadherins.* Like the classical cadherins, desmosomal cadherins also possess a highly conserved extracellular region consisting of five repeat domains (Boggon et al. 2002; Delva et al. 2009; Shapiro and Weis 2009). The cytoplasmic domain in these cadherins interacts with the β-catenin-related protein, armadillo, with plakoglobin and also with the plakophilins which are associated with the intermediate filaments (Hatzfeld 2007; Carnahan et al. 2010; Al-Amoudi et al. 2011). Desmosomal cadherins known as the desmogleins and desmocollins are highly expressed in epithelial tissues and cardiac muscle (Nollet et al. 2000; Green and Simpson 2007; Hulpiau and Van Roy 2009). Desmosomal cadherins exhibit both homophilic and heterophilic interactions (Green and Simpson 2007; Thomason et al. 2010). Desmosomal adhesion is crucial for the stability of adhesion junctions in epithelial cell sheets and in the regulation of epidermal differentiation (Garrod et al. 2002).

(d) *Atypical cadherins.* The main atypical cadherins are Dachsous (Ds), Fat, and Flamingo (Fmi). These are required for planar cell polarity signalling (Halbleib and Nelson 2006). Unlike classical and desmosomal cadherins, each of which have five extracellular or ecto (EC) domains, Ds and Fat are characterized by the presence of 27 and 34 extracellular repeat domains, respectively. The cytoplasmic domains of Ds and Fat show sequence homology with the catenin-binding motifs present in classical cadherins (Mahoney et al. 1991; Clark et al. 1995). In mammals, Fat1 interacts with Ena/VASP, a family of proteins involved in regulation of actin cytoskeleton assembly and dynamics (Moeller et al. 2004; Tanoue and Takeichi 2004). Fmi-1 is quite unique as it contains a seven-pass transmembrane region and nine extracellular domain repeats (Nakayama et al. 1998), causing it to be one of the most atypical cadherins yet known. Another much-discussed atypical cadherin is cadherin-23, or CDH23, which has 27 extracellular repeat domains and plays a role in hearing involving stereocilia (Siemens et al. 2004).

9.4 Evolutionary Relationships Amongst the Cadherins and Their Ectodomains

Several classifications have been made in respect of the cadherins. One of the earliest classifies them broadly into classical cadherins, protocadherins, desmosomal cadherins and cadherin-related genes or atypical cadherins (Suzuki 1996), even as summarized in the immediate previous section. In embryos undergoing development, the role of the classical cadherins is particularly important. Their evolutionary origins and connections with other cadherins have, therefore, been of some interest.

Invertebrate Cadherins DNA sequencing techniques and sequence comparisons have helped to trace the origins of the cadherins. In organisms like *Branchiostoma floridae* (lancelet), *Nematostella vectensis* (sea anemone) and *Trichoplax adhaerens* (primitive placozoan) which occupy key positions in studies of metazoan evolution, sequencing reveals the presence of 30, 16 and 8 cadherin genes (or cadherin-like genes), respectively, in their genomes. The genome of the sea urchin, *Strongylocentrotus purpuratus*, contains 14 cadherin-like genes. The worm, *Caenorhabditis elegans*, has 12, and the fly, *Drosophila melanogaster*, has 17. In the closest known relative of the metazoans, *Monosiga brevicollis* (a unicellular non-metazoan choanoflagellate), 23 putative cadherin-like genes have been identified (Murray and Zaidel-Bar 2014). Members of cadherin families, lefftyrin, coherin and hedgling, were present in the last common ancestor of choanoflagellates and metazoans. Mainly present in choanoflagellates and sponges, these may have evolved by domain shuffling and lateral gene transfer. These genes are speculated to have adhesive functions in these organisms (Nichols et al. 2012). Cadherins containing extracellular domain repeats linked to Src homology 2 (SH2), Hedgehog N-terminal peptide (N-hh), immunoglobulin (Ig) and von Willebrand type A domains are seen in *M. Brevicollis* and *Amphimedon queenslandica*. Cadherins which are now classified as Fat cadherins are also observed in sponges and sea urchins. The conserved cadherin cytoplasmic domain containing β-catenin binding sites is also observed in *N. Vectensis* (Abedin and King 2008). The function of cadherins in these unicellular organisms is largely unknown, but they are found to be localized in the apical, collar and basal pole of these cells. They play a role in cell shape and polarity and facilitate intracellular processes by taking cues from extracellular environment.

Human Cadherins The human genome encodes 114 cadherins. Although many arise through alternative splicing of mRNA, the presence of such a sizeable repertoire of genes has caused them to be classified as 'cadherin main branch' and 'cadherin-related major branch'. The cadherin-related major branch mainly consists of protocadherins, whereas the cadherin main branch consists of classical (type I) cadherins and atypical (type II) cadherins. Evolution of such a large superfamily of proteins appears to have mainly resulted from whole-genome duplications, individual gene duplications and diversification of duplicated genes. Type I cadherins consist of CDH1/E-cadherin/epithelial cadherin, CDH2/N-cadherin/neuronal cadherins, CDH3/P-cadherin/placental cadherins, CDH4/R-cadherin/retinal cadherin and CDH15/M-cadherin/myotubule cadherins. Each of these consists of the same number (five) of extracellular domains and a highly conserved tryptophan at position 2 (used for adhesion), and cytoplasmic domains are used for association with other proteins of the armadillo family and β-catenin (Hulpiau and Van Roy 2010).

Origins of the Five Ectodomains Bioinformatics-based analyses of DNA and protein sequences from divergent organisms reveal that an ancestral five repeat cadherin gene arose before divergence into paralogs. Repeated duplication of the extracellular domains appears to have led to the formation of a classical cadherin prototype in which introns were inserted because the introns in all cadherin genes are present in exactly the same locations. After divergence of this basic linear gene structure of the cadherin gene into different organisms, mutations could have occurred at fixed rates. Of the five classical cadherins, N-cadherins show the least rate of change because of a selection pressure placed on it due to its presence in the nervous systems. The existence of duplicates of gene paralogs could generate a wider scope for intragenomic recombination. It could also lower selection pressure on copies due to greater redundancy. The somatic morphology of organisms changes dramatically in vertebrates. So, the E-cadherins are placed under much less selection pressure and appear to have evolved faster than N-cadherins (Gallin 1998).

The domains I and II (reckoning from the N-terminal) of E-cadherins only share 25% sequence identity with each other, whereas domains III, IV and V show no significant similarity in their sequences. Conservation is observed at different residue positions amongst different domains. Position A5 is occupied by Glu in all sequences of domains I, II and III, while this position is shared by Gln and Asp in case of domains IV and V. A1 position in EC I of all cadherins is shared by hydrophobic residues, and EC II shares Gly and Ala residues. The position D1 is also conserved for hydrophobic residues for domain I and conserved for hydrophobic and aromatic position for domains II and IV. Domain V is the least conserved domain amongst classical cadherins. The gaps (deletions and insertions) in sequences of ectodomains are almost always found at the borders of strands or helices (Kister et al. 2001). Various studies have suggested that the five and seven extracellular domains of type I and type IV cadherins have independently evolved from a common ancestral cadherin that is represented by type III cadherins. EC1 of type I cadherin and the EC6 of type IV cadherin appear to have evolved from the same extracellular cadherin domain in the common precursor (Nishiguchi et al. 2016).

Relationships with Other Cadherins Phylogenetic studies reveal that the different cadherins, namely, N-, E-, R- and P-, group together more closely than do cadherins from a single species. This indicates that they are paralogs which originated before the divergence of mammals, birds and amphibians. Studies suggest that E- and P-cadherins belong to one paralog group, while N- and R-cadherins belong to another group. In the E-/P-cadherin group, domains I and II are most closely related and so are domains III and IV. In the N−/R-cadherin paralog group, domains I and V are most closely related and so are domains II and IV (Gallin 1998). A total of 72 amino acids at the C-terminal end of the cytoplasmic domains of E-, P- and N-cadherins are highly conserved (Niessen and Gumbiner 1998). Protein sequence similarity searches demonstrate sequence similarities of the outermost extracellular domain, EC1 of N-, R- and P-cadherins to the EC1 of E-cadherins, to be 77–78% (Nollet et al. 2000). In terms of adhesive properties, in simulation studies, R-cadherins appear to have greater homophilic as well as heterophilic binding affinities than E-cadherins (Vendome et al. 2014).

9.5 The Type I Classical Cadherins, E-Cadherin and N-Cadherin

The classical cadherins are a large family of cell surface glycoproteins essential for tissue morphogenesis and development (Takeichi 1995). In both vertebrates and invertebrates, they are characterized by the presence of extracellular (EC) domains, each consisting of 110 amino acid residues folded into a β-sandwich structural motif. The calcium binding regions are highly conserved amongst species (Nollet et al. 2000; Posy et al. 2008). The type I classical cadherins occur only in vertebrates. In this review, our focus is primarily on two type I classical cadherins: epithelial cadherin (E-cadherin) and neuronal or neural cadherin (N-cadherin).

Once an E-cadherin or N-cadherin molecule has begun to be synthesized and emerges from a ribosome, the signal peptide at its N-terminus facilitates the molecule's secretion into the lumen of the endoplasmic reticulum. This causes the five extracellular domains (EC1–EC5) of the cadherin to be serially and progressively transported across the membrane. Transport stops with the transmembrane (single-pass) region crossing the membrane and getting retained and not allowed to be secreted, causing the domain that follows it, i.e. the cytoplasmic domain, to fold on the cytoplasmic side of the membrane. The folded cytoplasmic domain then binds to catenins, at some point. The catenins are linked to the actin cytoskeleton. Prior to the completion of this entire assembly, the budding of the ER into vesicles takes place. These vesicles travel to the cell membrane and fuse with it, and this causes the cadherins to appear on the cell surface, where they can perform their adhesive function (s) involving the extracellular matrix or other cells, with their extracellular domains, or ectodomains, facing the outside of the cell. Prior to their appearance on the cell surface, the signal

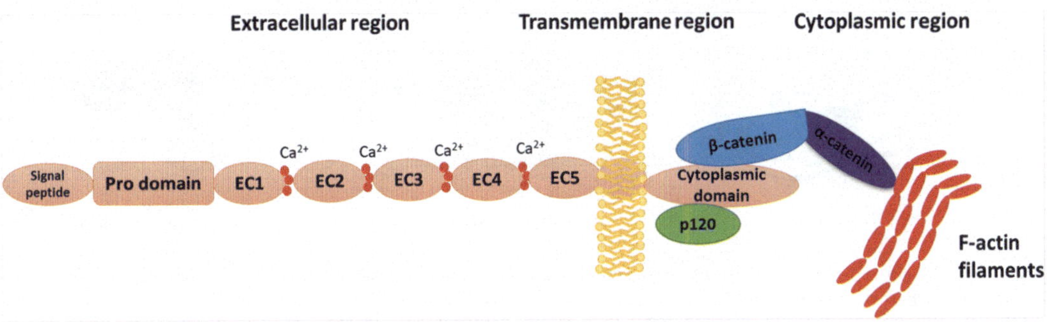

Fig. 9.1 Schematic diagram representing an E- or N-cadherin and its interactions with the cytoskeleton. The entire translated polypeptide is shown; however, the signal peptide and the pro-domain are removed through protease action, prior to adhesive function, and this removal presumably also precedes the interaction with actin for most molecules

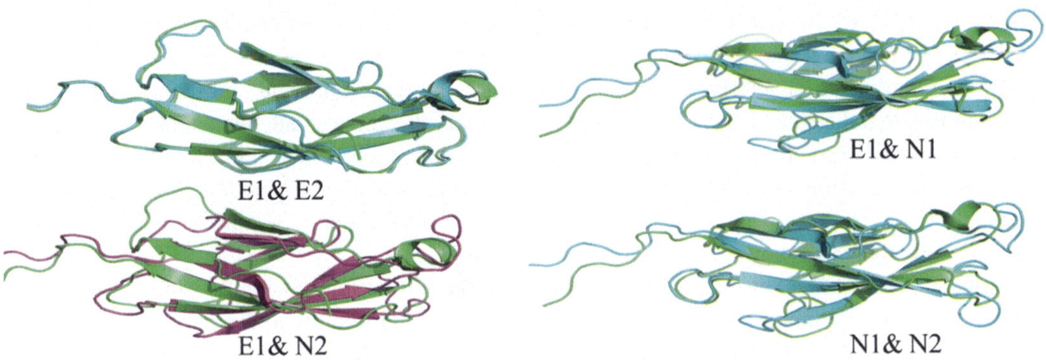

Fig. 9.2 Backbone ribbon diagram representations displaying the structural similarity and superimposability of various pairs of the first two ectodomains, EC1 and EC2, of E-cadherin (E1, E2) and N-cadherin (N1, N2). The figures were generated using the software PYMOL using PDB ID 3Q2V for mouse E-cadherin and PDB ID 3Q2W for mouse N-cadherin. Structure superimposition was also done using PYMOL

peptide and the pro-domain immediately following the signal peptide are required to be excised and removed at some stage, by proteases present in the lumen of the ER (as occurs with most secreted proteins).

Figure 9.1 shows a representation of a classical (E- or N-) cadherin molecule incorporated into the endomembrane system, prior to the proteolytic removal of the signal peptide and the pro-domain, i.e. as the entire translated polypeptide, with a putative interaction having already occurred with the catenins. This diagram must not be taken literally, in that it is rather unlikely, although not impossible, for cadherins to interact with the cytoskeleton prior to the removal of the N-terminal regions of the molecule. Following excision and removal of the signal peptide and the pro-domain and following transport to the cell surface, the outermost section of the cadherin which then comes into contact with similarly processed cadherins on the surfaces of juxtaposed cells is the domain known as EC1.

The E- and N-cadherin ectodomains show sequence homology as well as structural homology with each other (Patel et al. 2003; Chen et al. 2005; Posy et al. 2008). As shown below in Fig. 9.2, the structures of the EC1 and EC2 ectodomains of E- and N-cadherins are perfectly superimposable, both in cases in which these domains are drawn from the same cadherin and

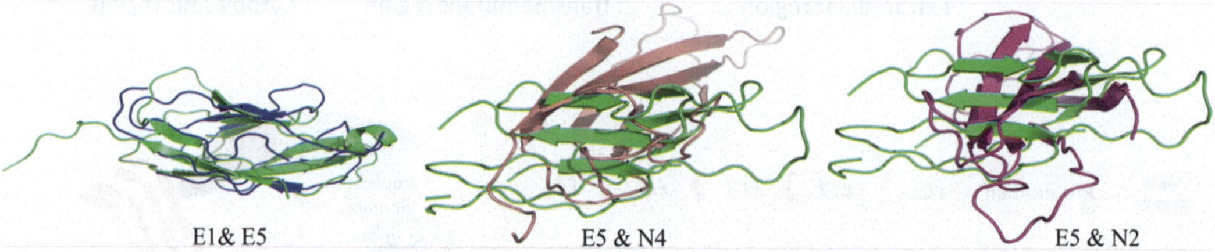

Fig. 9.3 Backbone ribbon diagram representations displaying the lack of structural similarity and superimposability for some of the ectodomains, e.g. EC1, EC2 and EC5, of E-cadherin (E1, E2, E5) or N-cadherin (N2, N4). The figures were generated using the software PYMOL using PDB ID 3Q2V for mouse E-cadherin and PDB ID 3Q2W for mouse N-cadherin. Structure superimposition was also done using PYMOL

also where they are drawn from different cadherins, for comparison.

Notably, however, this structural homology does not extend to all five ectodomains. The structures of EC3, EC4 and EC5 are quite unlike the structures of the EC1 and EC2 domains. As shown in Fig. 9.3, the structures of EC1 and EC5 of E-cadherin (E1, E5) are not as similar to each other as the structures of EC1 and EC2 of E-cadherin (E1, E2) shown superimposed in Fig. 9.2. Figure 9.3 also shows (i) how dissimilar the structures of the other cadherin domains (which are more membrane-proximal) tend to be, in respect of the E- and N-cadherins, as well as (b) how the membrane-distal and membrane-proximal domains tend to be dissimilar, although all of these ectodomains (or EC domains) consist of similar beta-sandwich folds.

9.6 Effects of Binding of Calcium to E- and N-Cadherins

Of all the CAMs, the cadherins, in particular, show a unique dependence on extracellular calcium for their activity, i.e. for the occurrence of homophilic interactions with other CAMs. There are three calcium-binding sites present in each inter-domain region, showing sequential binding to three calcium ions suggestive of a cooperative mechanism of calcium binding. The calcium ions are bound by the negatively charged, aspartate and glutamate side chains and by the backbone oxygen. The inter-domain regions possess three copies of the consensus sequence, DXNDNXP, acting as the master binding motifs for the three calcium ions. Within these, the calcium-binding sequences, PENE (residues 10–13), LDRE (residues 66–69) and DAD (residues 134–136), utilize their aspartate and glutamate residues to bind to calcium, and these sequences are conserved amongst all inter-domain regions in all classical cadherins, and not just the Type I classical cadherins. Figure 9.4 shows a representative inter-domain region and its calcium-binding sites.

The binding of calcium is reported to impart structure and stability to the ectodomains (Prasad and Pedigo 2005), in that the binding of three calcium ions to each inter-domain region causes ‘rigidification’ of that section of the cadherin and facilitates homophilic interactions by aligning the five EC domains into a rigid ‘rod-like’ arrangement. Monovalent cations, e.g. potassium and sodium, can also bind to the cadherins. However, such binding induces no conformational change (or rigidification) in the ectodomains or in their geometric dispositions towards each other. Electron microscopic studies have shown that treatment of full-length (EC1-EC5) epithelial cadherin [or fusion constructs comprising the first two domains (EC1 and EC2) of E-cadherin] with physiological concentrations of calcium induces changes in structure, causing the arrangement of the domains to become more rod-shaped. The absence of calcium, in contrast, makes the structure more compact and flexible, as well as more susceptible to proteolytic cleavage (Takeichi 1990). A higher vulnerability to proteases has also been observed where mutation(s) exist in

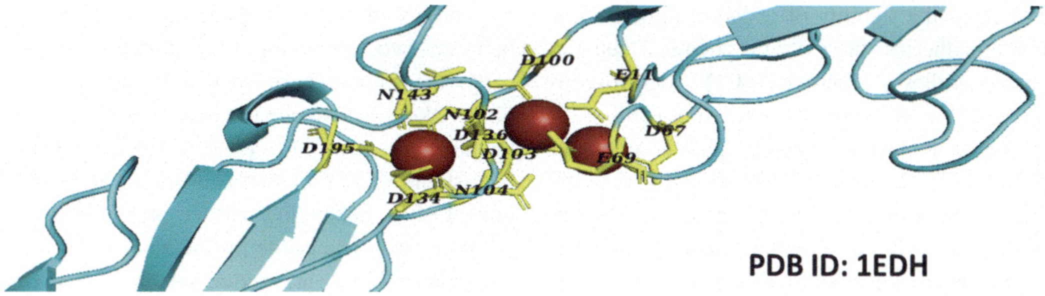

Fig. 9.4 Linker regions separating any two cadherin extracellular (EC) or ecto repeat domains are capable of binding to three calcium ions (shown in red, with residues involved in the binding shown in yellow). The detail shown is specific to the linker separating the EC1 and EC2 domains of E-cadherin). The figure was generated using the software PYMOL using PDB ID 1EDH

certain amino acid residues involved in calcium binding, such as Asp-134 (Ozawa et al. 1990a, b). Circular dichroism studies have shown that the ectodomains undergo a secondary structural alteration, from a randomly coiled conformation to a conformation with greater β-sheet content, upon calcium addition (Prasad and Pedigo 2005). X-ray crystallographic studies reveal that a two-fold symmetric dimer of the N-terminal EC domains of E-cadherin forms in the calcium-bound state, but not in the absence of calcium (Nagar et al. 1996). Molecular dynamics simulations suggest that there is a significant change in the structural dynamics of an EC1-EC2 fusion construct when solvent-exposed calcium ions are removed from a calcium-bound state (Cailliez and Lavery 2005). Thus, various computational and biophysical studies have confirmed that the apo-cadherin (the unbound state) shows significant conformational flexibility and that calcium-bound cadherin ectodomains display significant rigidity of conformation and restriction of conformational flexibility involving inter-domain regions. In terms of function(s), the absence of calcium also appears to abolish the adhesive function of the cadherins and makes them more susceptible to proteases. The increased rigidity of ectodomains upon calcium addition also causes molecules to come into proximity and exhibit interactions with each other to form dimers, either between molecules on the same cell (*cis* dimers) or between molecules on the surfaces of cells which are juxtaposed to each other (*trans* dimers). It could be conceived that binding of calcium cooperatively makes the otherwise flexible (and flippy-floppy) string of extracellular domains into a rodlike unit behaving as a single entity and that the reduction in intra-chain motions and the stabilization of a single conformation facilitate weak intermolecular interactions which would otherwise have been disfavoured by the occurrence of strong chain motions.

The binding of calcium ions to the inter-domain regions occurs with affinities which are rather poor, in comparison with the affinities of many other molecular systems, ranging from micromolar to millimolar values of binding constants (Shapiro et al. 1995; Nagar et al. 1996; Tamura et al. 1998; Boggon et al. 2002; He et al. 2003; Prasad and Pedigo 2005; Abedin and King 2008; Harrison et al. 2010). Such a range of calcium-binding affinities suggests that the cadherins might exhibit a dynamic response to the changing calcium ion concentrations in the extracellular milieu. Such a dynamic response could potentially facilitate the transmission of information about the junctional calcium status into the interior of the cell via interactions of cadherin cytoplasmic domains with the catenins and the cytoskeletal network in the cytoplasm. The binding of calcium ions to the inter-domain regions in ectodomains not only facilitates the rigidification and dimerization of these domains but also appears to stabilize them conformationally. Numerous cellular and biophysical studies have been performed to elucidate the strength as well as the stability of cadherin dimerization. For E-cadherins, thermal and chemical denaturation studies employing CD spectroscopy observed a shift in the structure-melting or unfolding temperature of the protein upon the addition of calcium (10 mM), from 40 °C (without calcium) to 65 °C (with calcium). Thermal studies involving differential scanning calorimetry are also in agreement with these spectroscopic experiments (Prasad and Pedigo 2005) suggestive of a role for calcium in stabilizing some aspect of cadherin conformation.

9.7 Cadherin-Cadherin Interactions in the *Cis* and *Trans* Modes Involving Ectodomains

The structures of the cadherin domains elucidated through X-ray crystallographic studies reveal that homophilic interactions between the cadherin ectodomains involve the N-terminal domain, EC1, i.e. the domain which is most distal to the plasma membrane (Shapiro et al. 1995). When cadherin ectodomains on the same cell's surface interact in a 'side-by-side' manner, they form *cis* dimers. When cadherin ectodomains present on juxtaposed or opposing cells interact, they form *trans* dimers. Currently, it appears that the binding interface is formed by only the EC1 domains of interacting partner cadherins, regardless of whether the interaction happens to be a *cis* or *trans* interaction. Various crystallographic studies suggest that *trans* dimers are formed in two conformations: strand-swapped dimers (S-dimers) and X-shaped dimers (X-dimers) (Shapiro et al. 1995; Boggon et al. 2002; Parisini et al. 2007; Ciatto et al. 2010; Harrison et al. 2010; Vendome et al. 2011). The crystal structures of N-cadherins show naturally formed strand-swap dimers, which are formed during crystallization (Shapiro et al. 1995).

S- and X-trans Dimers A *trans* interaction of the S-dimer variety occurs through the exchange or swapping of the β-strands located at the N-terminus of the EC1 (Harrison et al. 2010), between two interacting EC1 domains. Physically, the formation of such a strand-swapped dimer takes place by the insertion of a tryptophan residue present at the second position from the beta-strand of one partner into the hydrophobic pocket of its interacting partner, and vice versa. All type I classical cadherins have one conserved tryptophan, Trp2, at residue position 2 in the chain, and this tryptophan anchors the exchanged strands between the interacting EC1 domains, whereas for type II cadherins, both Trp2 and Trp4 participate in anchoring the exchanged strands. In addition, in both type I and type II cadherins, the remaining residues at the positively charged N-terminus also participate in the formation of various intermolecular salt bridges that further stabilize the strand-swapped dimers which are initially formed by the insertion of the tryptophan residue(s) into hydrophobic pocket (s) in the interacting partner domain. Notably, after the insertion of cadherins into membranes, the proteolytic cleavage of the signal peptide and 'pro-region' is absolutely essential for such salt bridge interactions to occur (Häussinger et al. 2004), suggesting that this removal could regulate the timing and context of the function of the cadherin and that cadherins remain in an essentially non-functional state until this removal takes place. Mutation of Trp2 in E-cadherin as well as N-cadherin prevents cell aggregation through cell-cell adhesion, suggesting that the strand-swapping mechanism is essential for cell-cell adhesion involving classical cadherins (Tamura et al. 1998). It has been demonstrated that the strand-swap dimerization is driven by the release of some physical strain present in the anchored N-terminal β-strands on cadherin monomers. The ease of the release of the strain thus appears to determine the binding affinity observed, where measurements have been made. Further, it has been observed that alteration in the length of the strand, or mutation of anchoring amino acid residues, can significantly change binding affinities. Various studies have also proposed that the binding specificity in type I cadherins is regulated by individual binding affinities and that there is a conserved Pro5-Pro6 motif which prohibits non-specific high-affinity interactions (e.g. heterophilic interactions) by preventing the formation of hydrogen bonds between the opposing β-strands. It is reported that mutations in the Pro5-Pro6 motif (involving one or both prolines) can result in the formation of not just high-affinity homodimers but also various heterodimers that do not otherwise form (Vendome et al. 2011).

As opposed to S-dimers, the classical cadherin X-dimers are formed by surface interactions between two monomeric cadherins along the inter-domain region between the first two ectodomains EC1-EC2 in a trans fashion that resembles an 'X' shape (Harrison et al. 2010). The crystal structure of two engineered EC1-EC2 domains of human E-cadherin containing some additional N-terminal amino acid residues provides evidence of the formation of X-dimers through the obstruction of the strand-swapped mode of interaction (Nagar et al. 1996; Pertz et al. 1999). Similarly, various biophysical and structural studies have shown the presence of X-dimers in P-cadherins and nonclassical T-cadherin (Ciatto et al. 2010; Vendome et al. 2014). Nuclear magnetic resonance (NMR) studies along with some single-molecule fluorescence resonance energy transfer (FRET) experiments involving E-cadherins have confirmed that the mutation of Trp2 to alanine leads to the formation of X-dimers, through inhibition of the formation of strand-swapped dimers (Sivasankar et al. 2009; Li et al. 2013). Early studies proposed X-dimers to be a transient intermediate in the formation of S-dimers (Sivasankar et al. 2009), but recent biophysical studies suggest that the X-dimer and the S-dimer conformations constantly alternate during interactions of the ectodomains (Manibog et al. 2016), transforming from one into the other in a dynamic fashion. Nothing is yet known about how calcium affects this equilibrium, i.e. whether the concentration of calcium in the vicinity of the cadherin determines the frequency of the interconversion or the lifetime of either the X-dimers state or the S-dimer state.

Despite being involved in cell-cell adhesion, classical cadherins have remarkably high dissociation constants in solution. The dissociation constants (K_d) for the majority of the classical cadherins are in the micromolar range which indicates weak interactions. For instance, the K_d for the full-length ectodomains EC1-EC5 of C-cadherins interacting in homophilic fashion, determined by analytical ultracentrifugation, is 64 μM (Chappuis-Flament et al. 2001). Similarly, K_d values for the EC1-EC2 domain fusions of E-, N-, R-, C- and P-cadherins, obtained from such experiments, are only about 97 μM, 26 μM, 14 μM, 127 μM and 31 μM, respectively (Vendome et al. 2014), i.e. they are all in the micromolar range. In E-cadherins, the mutations that exclude possibilities of strand-swapping show an exceptionally larger K_d value (around 916 μM) suggesting that the monomers forming X-dimers have indeed very weak affinity for each other and that they are less favoured than S-dimers (Harrison et al. 2010).

Through thermodynamic and simulation studies, binding free energies for S-dimers and X-dimers have been calculated. These studies revealed that the S-dimers are more stable and have larger interacting surfaces than X-dimers. It has been proposed that the staggered interface in X-dimers could just be a crystal contact rather than an actual interacting (functional) surface. Since cadherin-cadherin interactions are quite weak (Koch et al. 1997) and are characterized by a small but concentrated interface, it would not be prudent to neglect the evidence of a staggered interface obtained from the crystal structure studies. Through analysis of dimeric interfaces, it has been stated that the specificity of interaction in the S-dimers must depend on subtle thermodynamic or kinetic factors. The strand-swapped interface is identical for E- and C-cadherins but varies a little from that in the N-cadherins (Cailliez and Lavery 2006).

Cis Interactions In addition to the strong *trans* interactions, weak *cis* (lateral) interactions occurring in the other regions of EC1 and some regions of the EC2 domain are also reported. It is not yet known whether the remaining domains, i.e. EC3, EC4 and EC5, play a role in *cis* interactions. The *cis* dimeric interface, proposed from the crystal structure of C-cadherin, is conserved in terms of residue content and geometry in E-cadherin as well as in N-cadherin (Boggon, Murray et al. 2002). The *cis* dimer interface possesses an asymmetric junction between the EC1 ectodomain of one cadherin monomer and the EC2 ectodomain of its interacting partner; mutation of the interfacial residues involved abolishes the *cis* dimer assembly. Various biophysical techniques have suggested that *cis* dimerization interactions are either weak or transient (Häussinger et al. 2004; Harrison et al. 2011).

Monte Carlo simulations suggest that *trans* dimerization could be a prerequisite for *cis* dimerization because *trans* interactions lower the entropic penalty associated with *cis* dimer formation by reducing the conformational flexibility of the interacting cadherin domains (Wu et al. 2010, 2011). Recent studies have shown that both cells expressing E-cadherins, as well as liposomes coated with E-cadherins, tend to form ordered arrangements of 2D cadherin lattices at intercellular or inter-liposome junctions; however, mutants which abolish *cis* interactions fail to form such arrangements. This suggests that *cis* interactions must not only occur to facilitate the formation of ordered cadherin lattices but also that they could play a crucial role in forming cadherin clusters on membranes which stabilize cell-cell adhesions (Harrison et al. 2011).

It is believed that slow mechanisms such as exocytosis and endocytosis, as well as rapid mechanisms such as the lateral diffusion of cadherins to form cadherin clusters, could facilitate increased avidity of interactions between cellular surfaces which could potentially compensate for the poor affinity of cadherin-cadherin interactions and create the required total strength of collective cadherin-cadherin interactions to facilitate substantive cell-cell interactions (Iino et al. 2001; de Beco et al. 2009; Zhang et al. 2009). If this is the case, *cis* interactions must also be of consequence.

However, there are some conflicting studies which challenge the role of *cis* dimers in the formation of cadherin clusters. A recent study of E-cadherins embedded in liposomal bilayers suggests that cadherin clusters could be stabilized by intracellular linkage to the F-actin cytoskeleton, and not really by the formation of *cis* dimers. Further, there is no consensus yet about whether prior *trans* dimerization is required to facilitate *cis* interactions or whether the opposite is true, i.e. whether prior *cis* dimerization is required to occur to facilitate *trans* dimerization. Indeed, the two could be interdependent and operate through cooperative feedback mechanisms. Various computational and super-resolution microscopic studies have shown that glycosylation of ectodomains could also facilitate *cis* dimerization, indicating that the importance of *trans* dimerization prior to the formation of *cis* dimers might apply mainly to cadherins produced without any glycosylation (e.g. in bacterial cells for in vitro studies), whereas glycosylation could facilitate *cis* interactions in vivo. Towards the end of this review, a summary figure regarding possible *cis* and *trans* modes of interactions of cadherin ectodomains is included; this figure incorporates both currently known and hypothesized modes of interactions and also certain new hypotheses regarding such interactions.

9.8 Cadherin-Cytoskeleton Interactions Involving the Cytoplasmic Domain

The classical cadherins possess a cytoplasmic domain which is involved in direct binding to p120-catenin and β-catenin and indirect binding to α-catenin, a member of the vinculin superfamily (Shapiro and Weis 2009). The p120 catenin binds with the juxtamembrane domain (JMD) of the cadherin. It imparts stability to this domain by inhibiting internalization and degradation of the domain (Davis et al. 2003; Xiao et al. 2003). Cadherins are linked with the actin cytoskeleton via salt bridge-based interactions of α-catenin with cadherin-bound β-catenin, on the one side, and F-actin on the other (Gates and Peifer 2005; Kwiatkowski et al. 2010). This suggests that α-catenin could bridge cadherin-bound β-catenin to actin. However, there is no substantial evidence to back up this possibility, and there have been challenges of the mechanism suggesting simultaneous binding of α-catenin to β-catenin and F-actin (Drees et al. 2005; Yamada et al. 2005). Nonetheless, various studies corroborate the proposition that α-catenin linkage is essential for adhesion and junction assembly, without specific reference to any mechanistic details regarding this contention (Pokutta and Weis 2007; Hartsock and Nelson 2008; Kwiatkowski et al. 2010; Yonemura et al. 2010; Taguchi et al. 2011; Yonemura 2011).

The central region of the β-catenin, containing armadillo repeats (each repeat consisting of 40 amino acid residues), binds to the cytoplasmic domain of the cadherin (Hülsken et al. 1994; Funayama et al. 1995). β-catenin just acts as a linker, or intermediary, between α-catenin and the cytoplasmic domain of the E- or N-cadherin. According to a study involving an engineered cadherin construct in which the cytoplasmic domain of the cadherin is replaced by α-catenin, there is an alteration in cell adhesion (compared to the cell adhesion mediated by the whole protein complex) even in the absence of β-catenin (Nagafuchi et al. 1994). It is noteworthy that cadherin expression levels regulate catenin expression post-translationally. Surprisingly, when c-DNAs encoding E-, N-, or P-cadherins were transfected to L-cells, catenin expression levels increased significantly, without affecting the mRNA content (Nagafuchi et al. 1991).

A molecule known as γ-catenin, or plakoglobin, is associated with the desmosomal cadherins (Korman et al. 1989; Witcher et al. 1996). This shares significant structural and functional similarity with β-catenin and armadillo and can, therefore, replace β-catenin in the cadherin-catenin complex (Hülsken et al. 1994); however, γ-catenin associates to form weak complexes that dissociate more readily than β-catenin-cadherin complexes (Haegel et al. 1995). The deletion of the γ-catenin gene is lethal to heart structure formation and results in early death of the embryos, ostensibly due to the disruption of the strong association of γ-catenin with desmosomal cadherins (Bierkamp et al. 1996). The cytoplasmic domain of the classical cadherins is also known to interact with other proteins like tyrosine phosphatases (Brady-Kalnay et al. 1995; Kypta et al. 1996).

Numerous experiments have elucidated the functional significance of the cytoplasmic domain. It has been reported that the deletion or overexpression of the catenin-binding site or the complete cytoplasmic domain abolishes cell-cell adhesion mediated by the cadherins (Ozawa et al. 1990a, b; Nagafuchi et al. 1994). On the other hand, some mutation-based studies have also demonstrated that the presence of the cytoplasmic domain is not essential for cadherin-mediated cell-cell adhesion. In these studies, the cytoplasmic domain is substituted for desmoglein-3, one of the desmosomal cadherins that cannot bind to the catenins. The mutations did not affect the cadherin-mediated cell adhesion in cultured cells indicating that catenin association is not the sole mechanism that regulates cell adhesion (Roh and Stanley 1995). In principle, this assertion is not unreasonable. Since many in vitro experiments demonstrate that the EC1 domains of the cadherins are fully capable of engaging in *cis* and *trans* interactions by themselves, without the presence of the remaining EC domains and even without the presence of the membrane and cytoplasmic domains, it is conceivable that cadherins lacking the cytoplasmic domain can mediate cell-cell adhesion; of course, signalling into the cell would be expected to be significantly affected.

9.9 Adherence-Related Functions of E- and N-Cadherins in Organismal Development and Disease

Cell Sorting and Segregation It is natural to assume that differences in cadherin-cadherin binding affinities and specificities must be somehow used by organisms to facilitate the sorting and segregation of cells expressing these cadherins differentially, into different tissues. Given their extreme structural and conformational similarities, it would be interesting to understand the differences between them which regulate the highly specific homophilic pairing, heterophilic pairing and cell patterning behaviour seen in organisms. We know that cells expressing different cadherins can either be sorted homotypically (forming separate aggregates based on the type of cadherin expressed on these cells) or heterotypically (forming intermixed aggregates of different types of cadherin-expressing cells). However, due to our poor understanding of the relationship between cellular binding specificity and free energies of intermolecular binding for different types of cadherins and cadherin pairs, good models (including molecular models) are not yet available for cadherin-mediated cellular patterning, despite our knowledge of the structures of the domains of some of these cadherins and some structural details of their interactions.

Even so, some recent studies have shown that E- and N-cadherin bearing cells first form distinct homotypic cell aggregates and then interact heterotypically through the interfaces between the two distinct cell aggregates. Studies using analytical ultracentrifugation and surface plasmon resonance have determined the homophilic and heterophilic affinities, respectively, for both full-length E-cadherins and N-cadherins (Katsamba et al. 2009). The free energies for N- and E-cadherins turn out to be comparable, at 6.5 kcal/mol for N-cadherin and 5.3 kcal/mol for E-cadherin. The K_d values from ultracentrifugation studies determined for N- and E-cadherins were 22.6 μM and 160 μM, respectively, i.e. the homo-dimerization affinity for N-cadherins was found to be approximately tenfold greater than that for E-cadherins. Similarly, surface plasmon resonance experiments have provided the K_d for the heterophilic affinity between E- and N-cadherins, and this has been found to be intermediate to that applying to the two homophilic dimerization affinities. Many reports are suggestive of significant heterophilic interactions between E- and N-cadherins (Volk et al. 1987; Niessen and Gumbiner 2002). It may be argued that the small differences in affinity of homophilic and heterophilic interactions can be compensated for (or competed out) with high avidities and that this can play a significant role in determining cell adhesion behaviour and apparent specificity of interaction. Within a population of cells, subpopulations required to engage in differential interactions with other cells can express different numbers of different cadherins, on different facets of the cell's surface, and this detail can get smeared out in studies assuming that all cells of a particular type express the same numbers of the same types of cadherins. Thus, we may believe that based on the dimerization affinities and availabilities of E- and N-cadherins in requisite numbers, the equilibrium between homotypic and heterotypic interactions would be established according to the requirements of cells (and their genetic programming) and would not only facilitate cell sorting, segregation and formation of separate tissue layers but also allow the adhesion of these layers onto one another.

The cadherin superfamily plays a fundamental role in tissue morphogenesis, development and homeostasis, simply because cadherins are crucial for maintaining cellular contacts that can either directly participate in signalling or facilitate signalling involving other molecular interactions which are facilitated by cell-cell contacts. For example, reduced expression levels of epithelial cadherin (E-cadherin) are associated with enhanced invasion and metastasis in many tumours. Through their stable cell-cell adhesive interactions, cadherins mediate 'contact expansion' by reducing the 'interfacial tension' at the junctional interfaces of cells. Another way in which cadherins help to release interfacial tension or cortex tension is via the signal transmission to the actomyosin cytoskeleton through interactions with catenins. Cadherins also resist mechanical forces that attempt to disrupt cell-cell contacts. Also, cadherins are reported to be involved in the signal transduction processes that regulate cell fate (Stephenson et al. 2010; Lorthongpanich et al. 2012; Sarpal et al. 2012), cell polarity

(Wang et al. 2010; Bosveld et al. 2012) and cell proliferation (Nelson and Chen 2003; Kim et al. 2011; Schlegelmilch et al. 2011). One of the intriguing adhesive variants of the E-cadherin, known as uvomorulin, has the ability to adhere cells to each other during early stages of embryogenesis involving the morula. Since the loss in expression levels and functions of E-cadherins is correlated with tumour metastasis and their re-expression is associated with the decrease in proliferative and invasive capacity, they are proposed to function as tumour suppressors and metastasis suppressors.

Interactions with Growth Factors The N-cadherins have been reported to exhibit heterophilic interactions with a wide variety of proteins, including some growth factor receptors and some cell matrix proteins. One such protein is the fibroblast growth factor receptor (FGFR). The regulatory mechanism for the interactions between FGFR and N-cadherin is not known. However, these interactions can potentiate signalling in cancer cells either by the aid of endogenous FGFs, stabilizing the cell surface receptors, or by forming a higher-order complex which does not require FGF for downstream signalling (Nakamura et al. 2001; Nourse et al. 2007). Recent studies have also proposed a direct crosstalk between E-cadherin and EGF receptors (EGFR). EGFR is known to disrupt cell adhesions by destabilizing cadherin-catenin interactions, downregulating cadherin expression and exocytosis. The engagement of E-cadherin in newly formed cell-cell contacts appears to stimulate the rapid activation of EGFR in an EGF-independent manner. Notably, EGFR-initiated signalling pathways enhance cell proliferation and cell survival through MAP kinase, PI3-kinase and Rho GTPases. Future studies on the understanding of the molecular mechanisms underlying the interactions between EGFR and E-cadherin in normal and tumour epithelial cells can provide new insights into the development of suitable therapeutics for cancer treatment (Gavard and Gutkind 2008).

Interactions with Matrix Proteins It has been reported that dimeric E-cadherin interacts in a heterophilic manner with the cell matrix protein, integrin αEβ7, in a calcium-dependent manner (Corps et al. 2001). Since integrin αEβ7 has only one ligand, i.e. E-cadherin, high specificity of binding is ensured. This restricts autoreactive mucosal (cytotoxic) T cells in their specific locations. Both the integrin subunits, αE and β7, participate in the interaction with E-cadherins. In addition, the integrin α2β1 also shows interactions with both E- and N-cadherins but in somewhat different modes of interactions. In adhesion networks involving N-cadherins, the integrin α2β1 interacts with type I collagen and is involved in melanoma cell invasion and metastasis. Independently, E-cadherin/α2β1 integrin adhesion networks are also thought to regulate cell-cell adhesion in a type I collagen-independent manner (Siret et al. 2015).

Interactions with Other Cadherins and Receptors Various in vitro studies and site-directed mutagenesis-based studies have thrown up evidence of heterotypic interactions between N-cadherin and R (retinal)-cadherin. These studies support the S-dimer model for *cis*-heterodimerization of N- and R-cadherin molecules on the surfaces of the same cells. In vitro studies have demonstrated that the two cadherins are co-expressed in neurons and show co-localization at certain neural synapses, implying significance in cell adhesion in the neural retina (Shan et al. 2000).

Some biochemical studies have shown that N-cadherin interacts with the N-terminal domain of the glutamate receptor, GluR2, in both *cis* and *trans* fashions. In hippocampal neurons, N-cadherin and GluR2 form a synaptic complex that stimulates presynaptic development and function and also promotes dendritic spine formation (Saglietti et al. 2007). N-cadherin is also known to show robust interactions with another cadherin known as protocadherin-19 (Pcdh19). Bead aggregation studies examining the interactions of beads bearing proteins have revealed that the heterophilic Pcdh19–N-cadherin complex forms along with homophilic complexes, suggesting the usefulness of this interaction as a switch, converting between distinct binding specificities (Emond et al. 2011).

An atypical E3 ubiquitin ligase, Fbxo45, which facilitates ubiquitin-mediated degradation of proteins, has also been identified as an interacting partner for N-cadherin. The binding interface for this interaction overlaps with the calcium-binding motifs as well as with the dimerization interface, such that these interactions are disrupted by the addition of calcium, implying an 'either-or' mechanism of switching of interactions. N-cadherin proteolysis is also substantially enhanced by RNAi-mediated Fbxo45 gene silencing, leading to the impairment of neuronal differentiation and reduced expression of N-cadherins. Surprisingly, Fbxo45 prevents the calcium depletion-induced proteolysis of N-cadherin and R-cadherin and promotes neuronal differentiation by directly interacting with either N- or R-cadherins (Chung et al. 2014). The wide gamut of interactions of N-cadherin with different proteins suggests that it plays important regulatory and switch-like roles, sensing different situations and responding like a node in a protein interaction network.

Cadherin Gene Organization and Regulation of Expression During Development There are around 114 different types of cadherins encoded by the human genome. Not all of this diversity arises from the presence of a comparable number of genes; rather the diversity owes significantly to alternative RNA splicing mechanisms responsible for producing multiple splice variants from a smaller group of genes. The encoded classical cadherins include E-cadherins, N-cadherins, P-cadherins and cadherin 12 (Type2 cadherin). Several genes encoding classical cadherins have been identified and sequenced.

The gene *cdh1* encodes E-cadherin (CDH1). It is localized in the 16q22.1 region of chromosome 16. This gene has 14 splice variants. One of these encodes a preproprotein which is proteolytically processed to yield the mature glycoprotein. The transcription of the CDH1 gene is directly regulated by methylation of CpG islands present within the gene's promoter. Methylation of these CpG sites by DNA methyltransferase enzymes leads to the downregulation of the CDH1 encoding gene. Also several transcription factors such as Snail, Slug, E12/E47, ZEB-1 and SIP-1 are known to regulate E-cadherin expression through interactions with the enhancer boxes present upstream of the gene (Bolós et al. 2003). A substantial part of the *cdh1* gene consists of introns, the largest amongst which is the second intron. This aberrantly large intron is known to contain *cis* regulatory elements that regulate E-cadherin expression during development (Stemmler et al. 2005). The expression of E-cadherins begins very early during embryonic development, probably at the 1-cell stage, i.e. the zygote (Ogou et al. 1982). The differentiation and polarization of epithelia occur early, during the morula stage, in ontogenic terms, when there is a compaction of the embryo and each cell undergoes polarization along its apicobasal axis, generating an 'epithelial-like' phenotype. E-cadherins are thought to assume a critical part in the compaction of the morula, since functionally interfering antibodies against the E-cadherins have been shown to decompact the morula (Riethmacher et al. 1995). However, the source of E-cadherins at this stage is very likely to be maternally encoded (i.e. using proteins encoded from maternal mRNA), as embryos that are homozygous for an E-cadherin mutation develop normally up to the morula stage and have been shown to compact properly despite the mutations being present in the germline. In any case, most zygotic genes do not express this early during development, and cellular characteristics are essentially maternally encoded. The mutant morula cells become initially polarized, based on their maternally inherited cadherins, but very soon the embryo appears to become severely distorted, owing ostensibly to the mutation in the zygotically encoded version of the E-cadherin protein once this begins to be produced. This establishes that E-cadherin expression plays a very important role in maintaining cell polarity during the early stages of development.

At the stage of implantation, all embryonic cells express E-cadherin. The molecule, however, disappears from some cell layers as the differentiation of cell occurs, into various cell types. The best known example is the epithelial to mesenchymal transition (or EMT) which occurs during the formation of the mesoderm. The cells of the mesoderm migrate into the space between the ectoderm and the endoderm and lose E-cadherin during migration. During invagination of the neural plate, this region also loses E-cadherin. Other regions of the ectoderm continue to express E-cadherin, as do all endodermal cells. In older embryos, all ectoderm-derived cells express E-cadherin, barring certain terminally differentiated, non-proliferating epithelial cells such as lens fibre cells and keratinized epidermal cells, which lose E-cadherin from the membrane. Mutations in the *cdh1* gene leads to various diseases in humans which include cancers of the breast, colo-rectum, ovaries and many more due to increased proliferation, or occurrence of metastasis.

The N-cadherins are encoded by the *cdh2* gene in humans which maps to the 18q21.1 region of the chromosome 18. N-cadherins are known for their role in neural and mesodermal development. N-cadherins first appear in some ectodermal cells at the time of gastrulation. These are the most abundant type of cadherins in the neural tube at the early stages of development; however, during the process of differentiation of the central nervous system, they become regionally localized followed by the complete loss of N-cadherins from the neural retina and some layers of the cerebellum. Also the early neural tube expresses N-cadherin, in the dorsal-most region where neural crest cells are generated. However, the expression of this cadherin is downregulated in neural crest cells migrating from the neural tube; they instead begin to express cadherin-7 (Nakagawa and Takeichi 1998). When dorsoventral migration is concluded, N-cadherin tends to be upregulated in neural crest cells just before their their differentiation into the dorsal root ganglia (DRG) and sympathetic ganglia. After dorsolateral migration, dermal melanocytes express N-cadherin, facilitating contacts with fibroblasts in the skin dermis. Following the formation of the ganglion, N-cadherin appears at various places: on the apical surface of the neural tube, at the basolateral surface of the floorplate, upon the neuronal cells which are localized ventrally or laterally within the neural tube, upon fibrous axonal processes and in the ventral root and sympathetic ganglia.

When organisms turn into adults, N-cadherin is observed in neural tissue, endothelial cells, fibroblasts, the retina, as well as myocytes and osteoblasts (Taneyhill 2008). N-cadherin is also asymmetrically expressed in the chicken embryo; its activity is required during gastrulation to establish the left-right axis. Blocking the function of N-cadherin alters the expression of transcription factors, Snail and Pitx2, both downstream components of the cascade of factors regulating establishment of left-right asymmetry (García-Castro et al. 2000). N-cadherin mutations in the heart lead to the disassembly of the intercalated disc structure in the adult myocardium (Kostetskii et al. 2005). Mutations in the *cdh2* gene are also associated with the arrhythmogenic right ventricular cardiomyopathy (ARVC) which has been identified as the cause for sudden cardiac arrest in young people (Mayosi et al. 2017).

Cadherins and Their Connection to the Epithelial to Mesenchymal Transition (EMT) EMT is a core process in embryonic development, cancer progression and fibrosis (Kalluri and Weinberg 2009). By definition, EMT is the process of conversion of the epithelial cell type into the mesenchymal type under the influence of epigenetic mechanisms. The plasticity of epithelial cells entering the transition is highlighted by loss of ZO-1 and E-cadherin expression and upregulation of the proteins, fibroblast-specific protein 1 (FSP-1) and vimentin (Zeisberg and Kalluri 2004; Liu 2010). Such 'transitional' cells interact less with the extracellular matrix at its basal surface and gain the ability to invade it. Through this plasticity-marking trait, highly proliferative epithelial cells tend to develop the mesodermal tissue by undergoing the transition

(Tsai and Yang 2013). EMT thus appears to be an energy-conserving and resource-scavenging mechanism designed to quickly deliver myofibroblast- or fibroblast-type cells urgently required by an organism during a response to an injury and in organ development. However, this fluidity of state is also associated with pathophysiological evils like fibrosis and tumour formation, and cells undergoing EMT can enter into these states instead of undergoing transdifferentiation (Kalluri 2009).

The transdifferentiation process in EMT starts with the cell-cell contacts at tight, adherence and gap junction followed by a loss of cell polarity. E-cadherin along with other signatures of epithelial phenotype disappears from the cell surface during this transition through repression of expression of E-cadherin at the transcriptional, translational and post-translational levels. A variety of injury-causing factors and growth factors, including transforming growth factor-β, basic fibroblast growth factor (BFGF) and hepatocyte growth factor, are known to induce this process (Chen et al. 2005; Farrell et al. 2014). N-cadherin is produced in some carcinomas that have lost or downregulated E-cadherin. Unlike E-cadherin, N-cadherin exhibits weaker adhesive interactions. Thus there is a correlation between the switch from E- to N-cadherin and invasive cellular behaviour. The switch from E- to N-cadherin and the associated higher degree of invasiveness are linked to higher tumorigenicity of E-cadherin-negative cell lines and also to the poorer prognosis of E-cadherin-lacking tumours. Similarly, reversal of EMT by constitutive E-cadherin expression is shown to inhibit neoplasticity and invasiveness in tumours. Usually, differentiated tumours show E-cadherin expression, and levels of expression are inversely related to the grade of cancer in solid-organ malignancy (Lombaerts et al. 2006; Wheelock et al. 2008). Many signalling pathways converge to cause EMT via inhibition of E-cadherin expression, and these include Wnt, TGF-β and Notch/delta signalling (Son and Moon 2010; Tsai and Yang 2013). In the mammary gland's epithelial cells, integrin-linked kinase activity stimulates Wnt signalling through the Snail and Slug transcription factors, repressing E-cadherin and further enhancing tumorigenicity (Tsai and Yang 2013). Furthermore, hypoxia-inducing factor also directs the expression of Snail and Slug, and TGF-β (which is responsible for extracellular matrix remodelling) induces Snail leading to repression and loss of E-cadherin (Wang et al. 2013). Collectively, all inducers of EMT converge to cause multiple and pleiotropic effects, and the loss of E-cadherin is the most common effect. Below, we focus some more on cadherins and neoplasia/cancers (Thiery et al. 2009).

Cadherins and Their Connection to Cancers Dysfunction of classical cadherins has been suggested to be a hallmark in the origin and progression of neoplastic diseases (Kawanishi et al. 1995). E-cadherin, being a prime marker (and potentially also a prime maintainer) of the epithelial phenotype, has received significant attention in cancer literature (Thiery 2002). There are clinical correlations between aberrant E-cadherin expression and tumour prognosis. There are also observations of E-cadherin dysfunction in tumour progression in in vitro and in vivo models. These combinedly suggest an important role for E-cadherin (Vleminckx et al. 1991; Kowalski et al. 2003). Overall, dysfunction or reduced E-cadherin function has been reported in carcinoma of the breast, nasopharyngeal cavity, pancreas, lung, stomach, GI cavity, kidney and prostate. Considering the source or origin, mechanisms of E-cadherin dysfunction or dysregulation can be genetic, epigenetic, transcriptional and translational or post-translational, i.e. mechanisms of E-cadherin's loss of function can include proteolytic cleavage of ectodomains, proteasomal degradation of E-cadherin upon endocytosis, miRNA-induced downregulation, transcriptional repression involving gene hypermethylation or expression of repressors through signalling, germline and somatic mutations, aberrant TGF-signalling and/or loss of heterozygosity. These mechanisms were observed and reported in a variety of cancers, suggestive of a multilevel association of loss of E-cadherin function with the origin and propagation of neoplasms (Berx and van Roy 2009). Cadherin switching is another recently discussed mechanism associated with the E-cadherin null phenotype of most cancers (Wheelock et al. 2008). Below, we focus on genetic and epigenetic associations, transcription factor involvement and other processes.

Genetic Modifications Loss of heterozygosity (LOH) at the chromosomal level and loss of function (LOF) through mutations at the protein level are two ways in which the expression of E-cadherin is downregulated genetically. LOH is a common genetic aberration found in cancers in which the functional tumour suppressor gene is absent due to cross chromosomal events. The loss of heterozygosity of chromosome 16q21-22, where the E-cadherin gene is located, has been demonstrated in a variety of tumours including carcinoma of the breast, gastric, prostate and oesophageal cancer (Wijnhoven et al. 1999; Cleton-Jansen 2002; Corso et al. 2013). Similarly, progressive accumulation of somatic mutations in E-cadherin can lead to the process of carcinogenesis. Loss-of-function mutations in E-cadherin genes are known in diffusive gastric cancer, but these are rare events; these include in-frame deletion, truncation and splice-site-type mutations (Carneiro et al. 2008). Although rare in nature, E-cadherin mutations are known to cause familial aggregation of the diffusive form of gastric cancer. Most of these cause the occurrence of a stop codon resulting in premature cessation of expression. Polypeptide truncating germline mutations are also known to occur along the entire length of E-cadherin and can be associated with lobular breast carcinoma (Masciari et al. 2007). Whether the loss of E-cadherin is merely an effect is a moot point, since clearly the loss of the protein increases invasiveness, indicating that it can be a cause. There might also be other cause-effect relationships. It would be interesting to see whether the loss of E-cadherin can be a sole triggering event of a causal nature, in cancer. Of course, in cells that have lost E-cadherin, there are so many other changes, including EMT-type changes that it is difficult to establish which constitute 'cause' and which constitute 'effect'.

Epigenetic Modifications Many types of epigenetic modifications regulate gene expression, including methylation, acetylation, sumoylation, phosphorylation and ubiquitylation. Of course, methylation is the easiest to study with the existing technology and, therefore, the best known and most explored (Weinhold 2006). Hypermethylation of the E-cadherin 5′ proximal promoter has been found to result in reduced E-cadherin expression. Methylation-induced downregulation of E-cadherin is seen in many types of carcinoma, with a clear negative correlation between the levels of methylation and E-cadherin expression (Yoshiura et al. 1995). Hypermethylation leads to methyl-CpG binding proteins, MeCP2, and MBP2, interacting with the E-cadherin promoter, resulting in histone deacetylase (HDAC) recruitment, leading to the compaction of chromatin and suppression of transcription of the E-cadherin gene (Bhatt et al. 2013).

Other than hypermethylation, expression of the E-cadherin gene is regulated by Snail, Slug (Snail2), Twist, ZEB1 and ZEB2 which repress E-cadherin expression. Snail binds to E-box elements and recruits HDACs, triggering the cascade leading to chromatin compaction. Elevated Snail expression is common in invasive ductal carcinoma of the breast, and its higher expression is correlated with high-grade and lymph node-metastasized mammary tumours (Blanco et al. 2002). Similarly, Twist (a member of the basic helix-loop-helix family) recruits histone-lysine N-methyltransferase to the E-cadherin promoter to repress E-cadherin expression while inducing N-cadherin expression, giving rise to poor prognosis in cancer (Lamouille et al. 2014).

Other Mechanisms of Cadherin Downregulation Endocytosis-based uptake, shedding of ectodomains through extracellular cleavage of E-cadherin and intracellular cleavage of the linker region between the transmembrane section and the cytoplasmic domain, all singly or in combination, correlate with malignancy. Physiological recycling of E-cadherin involves endocytic pathways mediated by clathrin, caveolae and micropinocytosis, through which E-cadherin is recycled to new sites in cell-cell junctions. Endocytosis and recycling of E-cadherin are significantly increased in cells devoid of stable cell-cell contacts due to either low confluency or depletion of extracellular Ca^{2+} by a chelating agent (Le et al. 1999). Additionally, abnormal phosphorylation of a tyrosine residue in the cytoplasmic domain, induced by over-activation of proto-oncogenes like EGFR, Met and Src, results in internalization and ubiquitin-mediated proteasomal degradation of E-cadherin. Additionally, in the cytoplasmic domain of E-cadherin, where β-catenin binds, there are a number of serine and threonine residues which become putative sites for phosphorylation by a diversity of kinases. Phosphorylation of these sites in E-cadherin may alter its binding to β-catenin. On the other hand, phosphorylation of β-catenin by Src kinase results in disassembly of the cadherin-catenin complex, leading to loss of cell-cell adhesion and migration of β-catenin into the nucleus. Similarly, growth factors like epidermal growth factor and scatter factor can also lead to similar effects (Roura et al. 1999; McEwen et al. 2014). Another catenin, known as P120 catenin, is also known to regulate the expression of E-cadherin, presumably through feedback and sensing mechanisms, by ensuring the stability of the protein. Loss of P120 catenin is seen in a variety of cancers.

The regulation of what happens to the ectodomains also decides the fate of adhesion junctions between cells in cancers (Strumane et al. 2006). The proteolytic degradation of E-cadherin by zinc-dependent matrix metalloproteinases is known. Increased expression of these proteases correlates with the progression of cancer and inflammatory diseases. The ectodomain of E-cadherin near the plasma membrane is cleaved by these metalloproteinases, causing free E-cadherin fragments to be found circulating in the serum of patients with neoplastic diseases. Fascinatingly, soluble fragments of E-cadherin are also found to stimulate the migration of cells grown in collagen matrix under in vitro conditions (Nawrocki-Raby et al. 2003). Serine proteases like kallikrein 6 and 7 have also been found to be overexpressed in pancreatic and squamous cell carcinoma. Kallikrein 6 is known to modulate the protease activity of other proteinases like disintegrin which leads to the shedding of the extracellular domain of E-cadherin, resulting in metastasis of tumour cells (Johnson et al. 2007; Klucky et al. 2007).

9.10 Cadherin Trafficking, Association with Membrane Rafts and Non-association with Organelles

Trafficking of Cadherins Changes in cellular morphology and interactions and rearrangements accompanying physiological changes require changes in cadherin composition and constitution (Kowalczyk and Nanes 2012). Endocytosis, degradation and recycling of cadherins do occur, with proteins constantly being removed from the plasma membrane through endocytosis and recycled back into the membrane through exocytosis. E-cadherin is known to be internalized through clathrin-mediated endocytosis (Le et al. 1999), as well as through growth factor-induced pathways utilizing non-clathrin-mediated endocytosis like Rac1-dependent micropinocytosis (Watanabe et al. 2009). Once E-cadherin is internalized, it enters a Rab5 + ve compartment meant for sorting transmembrane proteins (Zerial and McBride 2001). Vesicles bud off from these compartments, mediated by the GTPase dynamin (Doherty and McMahon 2009). These can either be recycled back to the plasma membrane or marked for lysosomal degradation in a polarized manner (Woichansky et al. 2016). The amount and location of cadherin already present on the cell surface appear to determine whether a cadherin gets degraded after endocytosis or recycled back to the cell surface. The molecular mechanisms governing the recycling of cadherins have not been fully explored, and whatever is known is largely about E-cadherin, with virtually nothing being known about N-cadherin recycling. Several studies suggest that p120 catenin inhibits endocytosis. Several amino acid motifs have been identified as being responsible for cadherin internalization. In certain cases, adaptor proteins mediating endocytosis have been identified (Cadwell et al. 2016). Processes requiring cell migration are dominated by endocytic trafficking, and so trafficking affects healing of wounds after injury, tumour metastasis and angiogenesis, as already mentioned in the section dealing with neoplasia/cancer. This is supported by the finding that endothelial cell migration is inhibited by mutations in DEE endocytic motifs in the molecule, VE-cadherin. Thus identification of all such motifs would be of use in revealing how endocytic signals contribute to adhesion, migration and cell patterning in tissues.

Association of Cadherins with Rafts Lipid rafts are an integral part of the plasma membrane which exist as liquid-ordered regions. Rafts are small in size and are abundant in cholesterol and glycosphingolipids. Despite having somewhat distinct protein and lipid compositions, rafts are not identical in terms of composition of their constituents in all cells and all situations. Various proteins, especially those involved in cell signalling, have been shown to be present in rafts (Pike 2003). N-cadherin present at cell junctions is colocalized with lipid rafts, and disruption of lipid rafts results in the inhibition of cell-cell adhesion, without any modification of the interaction of N-cadherin and catenins to its plasma membrane. This suggests that lipid rafts might be an important site for the dynamic assembly of classical cadherins like N-cadherin at cell junctions; in fact, lipid rafts appear to stabilize cadherin-dependent adhesion complexes (Causeret et al. 2005). The presence of E-cadherin in lipid rafts has been shown to be necessary for the initial interaction of *Listeria monocytogenes* with cells, in order for it to gain entry into host cells (Seveau et al. 2004). Proteins in lipid rafts are sometimes interaction sites for the entry of pathogens, and it is interesting that E-cadherin associated with rafts plays a role in listeria infection.

Non-association of Cadherins with Exosomes and Organelles Hardly anything is known about whether cadherins are present on exosomes, or about whether they affect the fusion of the exosome with other cells, or play any role in cell-cell fusion which is either stimulatory or inhibitory. Similarly, almost nothing is yet known about whether cadherins are present on the membranes covering intracellular organelles such as the nucleus, or lysosomes, or whether they play any causative roles in endocytosis or exocytosis, or phagocytosis, rather than just being the subjects of these processes in terms of their trafficking. Given that the entire membranous pool of the cell is commonly described as the 'endomembrane system', with evidence of the rapid exchange of lipids between the plasma membrane at the cell surface and the membranes covering all intracellular organelles, it is interesting that there is no evidence of the association of any cadherin, including any of the classical cadherins, with the entire endomembrane system, with organelles or with exosomes. Of course, the lack of evidence in this regard cannot be assumed to be the evidence of lack and it might be a good idea to examine whether indeed organelle membranes lack cadherins and other CAMs and also how these are restricted and removed. It may be said, in jest, that the endoplasmic reticulum is associated with cadherins; only that they face the lumen of the ER and not the cytoplasm. What about the other organelle membranes? Can the presence or location of cadherins in these tell us something about the nature of their equilibrium with the plasma membrane?

9.11 Hypotheses, Perspectives and Questions

Do Cadherin Domains Act Like Extendable 'Unfoldable-Refoldable' Springs? Anyone familiar with the studies of the mechanical unfolding of the repeat domains of the muscle protein 'Titin' performed over a decade ago

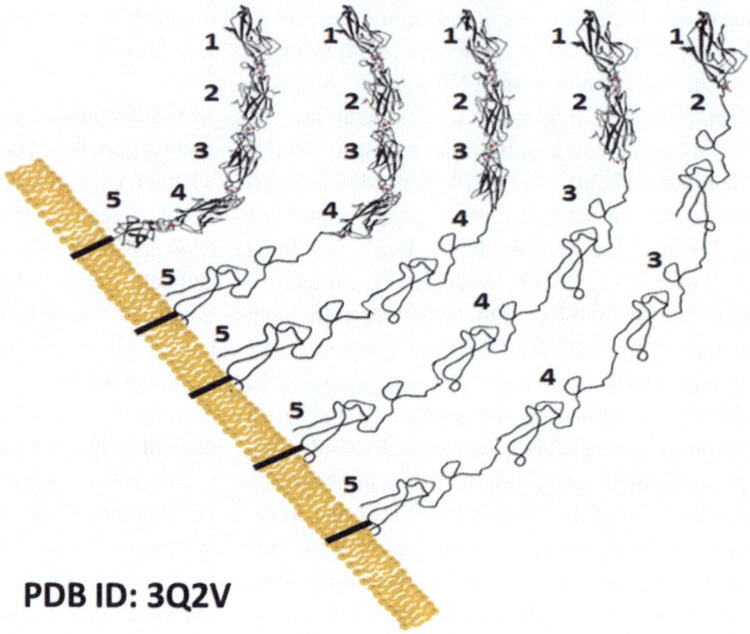

Fig. 9.5 As long as EC1 remains folded and capable of engaging in intercellular contacts, other domains could undergo unfolding in order to undergo lengthening and allow cells to move away and towards each other while remaining in contact through the unfolding and refolding of domains. The structures of the folded domains in this schematic figure were generated using the software UCSF Chimera and the PDB ID 3Q2V for mouse full-length E-cadherin

(in order to examine how many piconewtons of force are required to unfold individual domains, as well as whether domains refold when allowed to do so) would be likely to wonder whether the repeat domains of cadherins undergo mechanical force-induced unfolding as cells move away from each other. It is conceivable that the mechanical forces exerted upon the cadherins by the movements of cells could indeed result in the unfolding of the ectodomains. In particular, with cadherins which have many more than five repeat domains, e.g. cadherin-23, it is even conceivable that cells displaying such cadherins on their surface, which appear not to be in contact, happen to actually still remain in contact (despite having moved away from each other) or be already in contact (long before they have physically touched each other). Their still-folded (and still interacting) outermost EC1 and EC2 domains could very well have 'snaked' away from their surfaces, due to the existence of other unfolded repeat domains in the polypeptide chain, right up to the domains that lie next to the membranes. It is even conceivable that where there are enough of such partially unfolded cadherin polypeptides on a cell's surface, these could trap cells at large distances prior to the actual physical contact and then 'reel in' such cells into coming into close contact through the progressive calcium binding-aided folding of unfolded domains. This concept is schematically illustrated in Fig. 9.5, for a classical cadherin containing five ectodomains, e.g. E-cadherin or N-cadherin.

Are Longer Cadherins Used for Long-Distance Cell Contacts and Looser Adherence Junctions in Cancers? As a corollary to the above, in certain situations, e.g. in cancers in which cadherin expression profiles are altered and cadherins with larger numbers of repeat domains, e.g. cadherin-23, tend to be overexpressed, could the membranes of juxtaposed cells appear to be farther away from each other than in the case of adhesions based on classical cadherins? What implications might such a situation have for the cells in question? Would the adherence junctions be much more accommodating, in terms of allowing greater contact with the extracellular fluids, nutrients, cytokines and other factors?

Are N-cadherin's Outermost Ectodomains Likely to Be More Refoldable than Those of E-cadherin? Although it might seem like a specious argument at first sight, it could be argued that different kinds of cadherins have evolved for the mutual attachments of neuronal and epithelial cells because neurons are required to very quickly 'make or break' synaptic connections with other neurons, whereas contacts between epithelial cells tend to be somewhat more long-lived (and in certain instances, e.g. in tissues, lifelong contacts). The question is whether this is likely to have a biophysical correlate, in terms of the unfolding and refolding characteristics of the EC1 and EC2 domains of the E- and N-cadherins. If one were to make a prediction, one could argue that since neurons often need to rapidly separate away from other neurons in order to join yet other neurons, the mechanical forces involved in such rapid separation could result in the unfolding of the ectodomains of N-cadherin. If such domains were completely incapable of undergoing rapid refolding and no time were available to replenish the relevant regions of the neuron's surface with freshly synthesized and folded cadherins, neurons could become incapable of breaking contacts and making new contacts. In contrast, in epithelial cells, there would presumably be time available to replenish E-cadherins destroyed through the unfolding of ectodomains. Thus, the prediction would be that N-cadherin's ectodomains, especially domains EC1 and EC2, must be far more amenable to undergoing unfolding and refolding than those of E-cadherin. It would be interesting to verify whether this is indeed the case.

Are Homophilic Contacts Between N-cadherins Less Strong than Those Between E-cadherins? For reasons entirely similar to those laid out in the perspective mentioned immediately above, it could be argued that, on balance, contacts between N-cadherins would be weaker, in terms of dissociation constants, than contacts between E-cadherins, because neurons have a greater need, and tendency, to dissociate. On the other hand, it could also be argued that the areas of contacts between neurons tend to be smaller than those between epithelial cells and that, therefore, fewer cadherin molecules could be involved in building cell-cell contacts, with these contacts being much stronger than those between epithelial cells. It would be interesting to more fully examine which of these scenarios are true, given that there is already some evidence reviewed in this article to suggest that the former scenario is true.

Does Avidity Compensate for Differences in Affinity? In general, it might be argued that differences in the affinities of homophilic contacts between cadherins might not be extremely relevant, because it is conceivable that cells overexpress a cadherin that employs weaker homophilic contacts, if necessary, to ensure that there are a much larger number of molecules involved in making contacts, with greater avidity compensating for lower affinity. Of course, there is only a finite amount of area available on the cell surface, and cadherins and other CAMs have to share it with a multitude of other cell surface proteins and receptors. It would be interesting to understand how cells manage these issues of avidity versus affinity, based on their shapes and surface areas.

Do Calcium Channels and Transporters Regulate Cadherin Function and Cell Separation in Development and Cancer? Calcium channels and transporters on the cell membrane can be thought to underlie a region of the cell's surface which is engaged in cadherin-based cell-cell adhesive interactions, much like the underbrush on a forest's floor underlies the tall trees of a forest. Since cadherins have relatively weak (micromolar to millimolar) affinities for calcium, it is likely that there occurs a significant dissociation and reassociation of calcium, allowing cadherins to remain associated for long durations only when the equilibrium concentrations of calcium are sufficiently high to overcome the poor affinity of cadherins for calcium. Under such circumstances, if the replenishment of calcium from the serum were to be restricted, such that a requisitely high, equilibrium concentration of calcium was slow to be re-established, ions transported away from the vicinity of cadherins by channels or transporters could facilitate rapid dissociation of cells in a region of the surface through rapid calcium depletion-aided loss of cadherin-cadherin *trans* contacts. Such a mechanism would be particularly effective if there were insufficient scope for the rapid replenishment of calcium owing to the restriction of ion movement by molecular crowding, calcium trapping or continued calcium removal by channel/transporter action. From a design and engineering viewpoint, one very efficient way to cause rapid cell-cell separation would be to upregulate the presence or activity of calcium channels and transporters, in a 'hit-and-run' mode, rapidly sucking up the available calcium and transporting it into the cell until cadherin-cadherin separation is achieved. Separation would, of course, immediately allow replenishment of calcium from the serum and extracellular fluids, but presumably by this time cells would have already separated. Intriguingly, there is evidence that verapamil which blocks calcium transport also does reduce cellular metastasis, which requires cell-cell separation (Tsuruo et al. 1985). There is also considerable evidence available now to suggest that metastasising cells do have upregulated levels of at least three different calcium channels (Mo and Yang 2018), suggesting that the above scenario is likely to be true. The concept is explained in Fig. 9.6 below.

Do Endogenous Proteases Function in Cell Separation? Cell-cell separations in cell culture experiments are achieved by adding trypsin and EDTA. Presumably, the EDTA chelates calcium away from the cadherins, making them more susceptible to proteolysis by trypsin in a non-calcium-bound state, and this allows trypsin to then selectively act on the cadherins to rapidly cut them away from each other and break up residual cadherin-cadherin contacts. If, indeed, this is how cell-cell separations are achieved in cell culture, it is conceivable that the same could also apply to cells in vivo, i.e. cells could produce and secrete trypsin whenever necessary, in the vicinity of the regions of a cell's surface attempting to dissociate from surrounding cells, to facilitate the dissociation process along with the mechanism outlined above about upregulation of calcium channels and transporters. It must be mentioned here that originally, it was assumed that trypsin is only made by acinar cells in the pancreas; however, for over two decades now, it is known from in situ hybridization, immunohistochemistry and reverse-transcription PCR that trypsin is expressed widely in epithelial cells in the oesophagus, stomach, skin, lung, small intestine, liver, kidney and extrahepatic bile duct and also in neuronal and splenic cells, as well as in the brain. Also, many cell types have receptors for trypsin on their surfaces. So, basically, the hypothesis is that a cell wishing to exit from a tissue, or alter its contacts, could simply produce and secrete trypsin to act in autocrine fashion.

Is There a Division of Labour Amongst Cadherin Domains for Cis (EC3, EC4 and EC5) and Trans (EC1 and EC2) Interactions? While there is now much evidence that EC1 and EC2 engage in adhesive interactions, not much is known about the role of the remaining ectodomains, i.e. EC3, EC4 and EC5, besides some information which is available about interactions of such domains with certain growth factors and some receptors. Our proposal is that EC3, EC4 and EC5 are domains that engage in *cis* interactions amongst cadherins on the same surface, with EC1 and EC2 are engaged primarily in *trans* interactions between cells. We further propose that *cis* and *trans* interactions cooperate and are interdependent. It is already known that EC1 and EC2 form monomers and dimers and nothing larger, either individually or in fusion constructs. We propose that EC3, EC4 and EC5 will turn out to form large multimolecular complexes consisting of folded polypeptide chains, individually and in fusion constructs, suggesting that they have a natural tendency to associate and cluster together and that they could bring cadherins together through *cis* interactions.

Do Cis Cadherin Interactions Contribute to Cell Surface Flatness? Calcium binding straightens up the entire set of five ectodomains

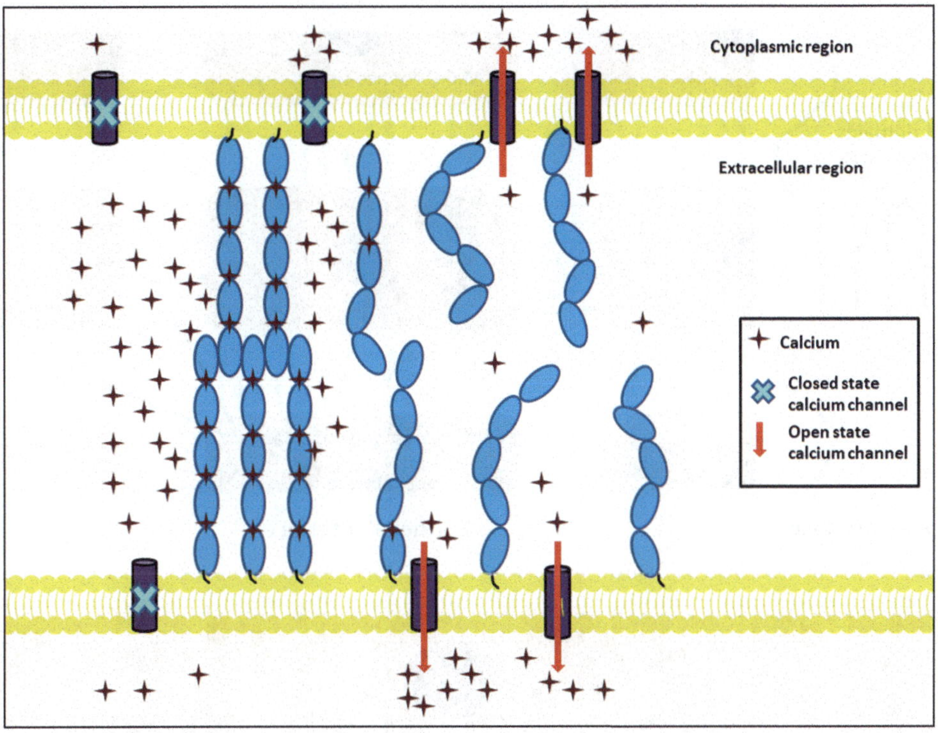

Fig. 9.6 Two adjacent regions of an adherence junction are shown. The one on the left is shown to have fewer calcium channels/transporters. These are also shown as being inactive and/or fewer in number. This allows calcium association-dissociation from the cadherins in the region to occur in such a manner that a high equilibrium concentration of calcium can be maintained in the region, with available calcium ions remaining titrated within the region by binding dissociation with cadherin. As a consequence, cell-cell contacts also remain stable. In contrast, in the region on the right, there are larger numbers of calcium channels transporting calcium into cells and/or higher numbers of functionally active channels/transporters. This results in the available calcium being depleted through transport into the cell and cellular organellar stores. This is proposed to result in a poorer (suboptimal) equilibrium calcium concentration in the region, owing to the slowness of replenishment of calcium from the serum. Consequently, the loss of the requisite calcium concentration in the region leads to the loss of cadherin-cadherin associations and separation of cells. The above is proposed as a mechanism for cell-cell separations during metastasis

into a rodlike rigid shape. Above, we have hypothesized that this can stimulate cis interactions between cadherins displayed on the same cell's surface, involving interactions of the EC3-EC3, EC4-EC4 and EC5-EC5 varieties, especially where a high concentration of calcium-bound cadherins pre-exists on the cell's surface allowing molecules to collide and associate. Here we propose that this is the primary mechanism for causing a cell's surface to be flat, i.e. by causing the formation of two-dimensional lattices of cadherins that then hold the plasma membrane in a flat shape. Of course, this could then be further supported by actin cytoskeletal dynamics involving the catenins. This concept is shown in Fig. 9.7, which shows how a combination of *cis* and *trans* interactions between E-cadherins can lead to the formation of a rigid adherens junction that ensures the flatness of the cell's surface. Of course, such a flat interface region between cells would need to be supported adequately by the formation of intracellular contacts of the cadherins with suitably disposed catenins and the actin cytoskeletal network.

Summary of Concepts Figure 9.7a also shows schematically some differences between E- and N-cadherin-based contacts involving cuboidal

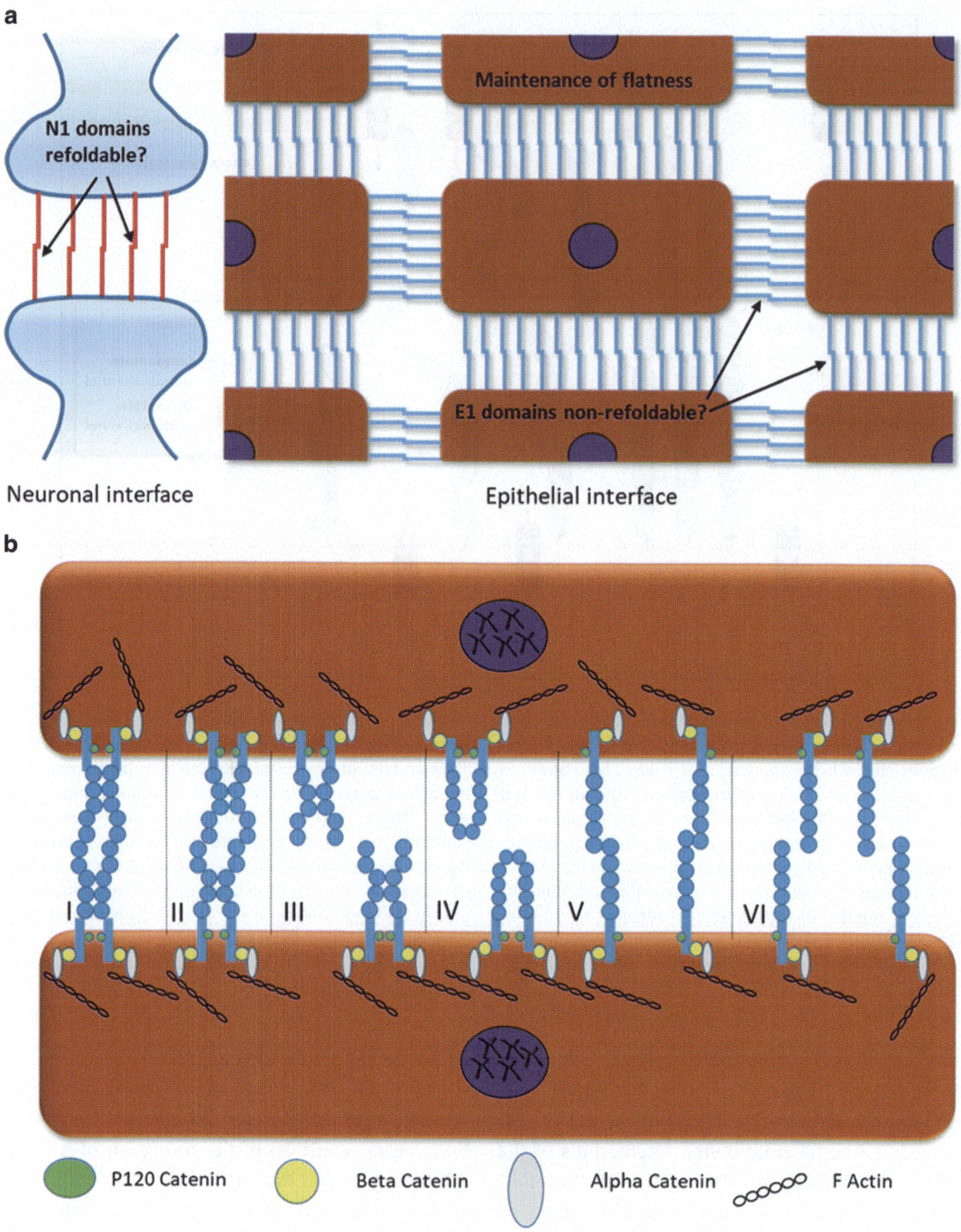

Fig. 9.7 *Panel* (**a**) Likely differential behaviour of N- and E-cadherin at neuronal cell-cell interfaces and epithelial cell-cell interfaces, respectively. It is proposed that the outermost ectodomains of N-cadherin must interact with lower affinity, there must be fewer N-cadherins present, and also the outermost ectodomains of N-cadherin must unfold and refold in a facile manner, to facilitate the rapid breaking and remaking of connections between neurons. On the other hand, it is proposed that the outermost ectodomains of E-cadherin must interact with higher affinity, there must be more E-cadherin molecules present and the outermost ectodomains of E-cadherin may not be amenable to unfolding and refolding (but require replenishment after breaking of contacts), to facilitate long-lived contacts. The figure also emphasizes that high-density E-cadherin contacts could facilitate maintenance of cell surface flatness. *Panel* (**b**) Six likely different modes of existence of the ectodomains of cadherins at cell-cell

epithelial cells and neuronal synapses, respectively. Our hypothesis is that the latter type of contacts involve EC1 and EC2 domains which are more easy to unfold and refold (refer to the concept outlined in Fig. 9.5) because neurons are required to frequently dissociate and reassociate. Thus, the EC1 and EC2 domains of N-cadherin are likely to be amenable to multiple rounds of unfolding and refolding. In contrast, epithelial cells engage in contacts that are far more long-lived, on the average. Therefore, their EC1 and EC2 domains could have evolved in a manner that would not allow them to undergo multiple rounds of unfolding and refolding. Also, these differential needs of neuronal and epithelial cells could be serviced by a lower density of cadherins at the cell-cell interface, as well as a lower strength of the EC1-EC1 trans interaction in the case of neurons. All these possibilities are summarized in Fig. 9.7a, which would be fascinating to explore experimentally. Figure 9.7b serves the purpose of summarizing all the different modes of cadherin-cadherin interactions that have been discussed in this review.

References

Abedin M, King N (2008) The premetazoan ancestry of cadherins. Science 319(5865):946–948

Al-Amoudi A, Castano-Diez D et al (2011) The three-dimensional molecular structure of the desmosomal plaque. Proc Natl Acad Sci U S A 108(16):6480–6485

Berx G, van Roy F (2009) Involvement of members of the cadherin superfamily in cancer. Cold Spring Harb Perspect Biol 1(6):a003129

Bhatt T, Rizvi A et al (2013) Signaling and mechanical roles of E-cadherin. Cell Commun Adhes 20 (6):189–199

Bierkamp C, Mclaughlin KJ et al (1996) Embryonic heart and skin defects in mice lacking plakoglobin. Dev Biol 180(2):780–785

Blanco MJ, Moreno-Bueno G et al (2002) Correlation of snail expression with histological grade and lymph node status in breast carcinomas. Oncogene 21 (20):3241–3246

Boggon TJ, Murray J et al (2002) C-cadherin ectodomain structure and implications for cell adhesion mechanisms. Science 296(5571):1308–1313

Bolós V, Peinado H et al (2003) The transcription factor slug represses E-cadherin expression and induces epithelial to mesenchymal transitions: a comparison with snail and E47 repressors. J Cell Sci 116(3):499–511

Bosveld F, Bonnet I et al (2012) Mechanical control of morphogenesis by Fat/Dachsous/Four-jointed planar cell polarity pathway. Science 336(6082):724–727

Brady-Kalnay SM, Rimm DL et al (1995) Receptor protein tyrosine phosphatase PTPmu associates with cadherins and catenins in vivo. J Cell Biol 130 (4):977–986

Cadwell CM, Su W et al (2016) Cadherin tales: regulation of cadherin function by endocytic membrane trafficking. Traffic 17(12):1262–1271

Cailliez F, Lavery R (2005) Cadherin mechanics and complexation: the importance of calcium binding. Biophys J 89(6):3895–3903

Cailliez F, Lavery R (2006) Dynamics and stability of E-cadherin dimers. Biophys J 91(11):3964–3971

Carnahan RH, Rokas A et al (2010) The molecular evolution of the p120-catenin subfamily and its functional associations. PLoS One 5(12):e15747

Carneiro F, Oliveira C et al (2008) Molecular pathology of familial gastric cancer, with an emphasis on hereditary diffuse gastric cancer. J Clin Pathol 61(1):25–30

Causeret M, Taulet N et al (2005) N-cadherin association with lipid rafts regulates its dynamic assembly at cell-cell junctions in C2C12 myoblasts. Mol Biol Cell 16 (5):2168–2180

Chappuis-Flament S, Wong E et al (2001) Multiple cadherin extracellular repeats mediate homophilic binding and adhesion. J Cell Biol 154(1):231–243

Chen CP, Posy S et al (2005) Specificity of cell-cell adhesion by classical cadherins: critical role for low-affinity dimerization through -strand swapping. Proc Natl Acad Sci U S A 102(24):8531–8536

Chung F-Z, Sahasrabuddhe AA et al (2014) Fbxo45 inhibits calcium-sensitive proteolysis of N-cadherin and promotes neuronal differentiation. J Biol Chem 289(41):28448–28459

Ciatto C, Bahna F et al (2010) T-cadherin structures reveal a novel adhesive binding mechanism. Nat Struct Mol Biol 17(3):339

Fig. 9.7 (continued) interfaces. Mode I shows *cis* interactions involving EC3, EC4 and EC5 and *trans* interactions involving primarily EC1, supported by EC2. Mode II shows an interaction similar to mode I, except that *cis* interactions involve only EC3 and EC4. Mode III shows only *cis* interactions involving EC3 and EC4 with no *trans* interactions. Mode IV also shows only *cis* interactions; however, these involve only EC1 with no *trans* interactions. Mode V shows *trans* interactions involving EC1 with no *cis* interactions. Mode VI shows cadherins with neither *cis* nor *trans* interactions

Clark HF, Brentrup D et al (1995) Dachsous encodes a member of the cadherin superfamily that controls imaginal disc morphogenesis in Drosophila. Genes Dev 9(12):1530–1542
Cleton-Jansen AM (2002) E-cadherin and loss of heterozygosity at chromosome 16 in breast carcinogenesis: different genetic pathways in ductal and lobular breast cancer? Breast Cancer Res 4(1):5–8
Corps E, Carter C et al (2001) Recognition of E-cadherin by integrin alpha(E)beta(7): requirement for cadherin dimerization and implications for cadherin and integrin function. J Biol Chem 276(33):30862–30870
Corso G, Carvalho J et al (2013) Somatic mutations and deletions of the E-cadherin gene predict poor survival of patients with gastric cancer. J Clin Oncol 31 (7):868–875
Cruse JM, Lewis RE et al (2004) Immunology Guidebook. Elsevier, San Diego
Davis MA, Ireton RC et al (2003) A core function for p120-catenin in cadherin turnover. J Cell Biol 163 (3):525–534
de Beco S, Gueudry C et al (2009) Endocytosis is required for E-cadherin redistribution at mature adherens junctions. Proc Natl Acad Sci 106(17):7010–7015
Delva E, Tucker DK et al (2009) The desmosome. Cold Spring Harb Perspect Biol 1(2):a002543–a002543
Doherty GJ, McMahon HT (2009) Mechanisms of endocytosis. Annu Rev Biochem 78:857–902
Drees F, Pokutta S et al (2005) α-catenin is a molecular switch that binds E-cadherin-β-catenin and regulates actin-filament assembly. Cell 123(5):903–915
Emond MR, Biswas S et al (2011) A complex of Protocadherin-19 and N-cadherin mediates a novel mechanism of cell adhesion. J Cell Biol 195 (7):1115–1121
Farrell J, Kelly C et al (2014) HGF induces epithelial-to-mesenchymal transition by modulating the mammalian hippo/MST2 and ISG15 pathways. J Proteome Res 13 (6):2874–2886
Funayama N, Fagotto F et al (1995) Embryonic axis induction by the armadillo repeat domain of beta-catenin: evidence for intracellular signaling. J Cell Biol 128(5):959–968
Gallin WJ (1998) Evolution of the" classical" cadherin family of cell adhesion molecules in vertebrates. Mol Biol Evol 15(9):1099–1107
García-Castro MI, Vielmetter E, Bronner-Fraser M (2000) N-cadherin, a cell adhesion molecule involved in establishment of embryonic left-right asymmetry. Science 288(5468):1047–1051
Garrod DR, Merritt AJ et al (2002) Desmosomal cadherins. Curr Opin Cell Biol 14(5):537–545
Gates J, Peifer M (2005) Can 1000 reviews be wrong? Actin, α-catenin, and adherens junctions. Cell 123 (5):769–772
Gavard J, Gutkind JS (2008) A molecular crosstalk between E-cadherin and EGFR signaling networks. In: EGFR Signaling Networks in Cancer Therapy. Humana Press, Totowa, pp 131–146
Green KJ, Simpson CL (2007) Desmosomes: new perspectives on a classic. J Investig Dermatol 127 (11):2499–2515
Gumbiner BM (2005) Regulation of cadherin-mediated adhesion in morphogenesis. Nat Rev Mol Cell Biol 6 (8):622–634
Haegel H, Larue L et al (1995) Lack of beta-catenin affects mouse development at gastrulation. Development 121 (11):3529–3537
Halbleib JM, Nelson WJ (2006) Cadherins in development: cell adhesion, sorting, and tissue morphogenesis. Genes Dev 20(23):3199–3214
Harrison OJ, Bahna F et al (2010) Two-step adhesive binding by classical cadherins. Nat Struct Mol Biol 17(3):348
Harrison OJ, Jin X et al (2011) The extracellular architecture of adherens junctions revealed by crystal structures of type I cadherins. Structure 19(2):244–256
Hartsock A, Nelson WJ (2008) Adherens and tight junctions: structure, function and connections to the actin cytoskeleton. Biochimica et Biophysica Acta (BBA)-Biomembranes 1778(3):660–669
Hatzfeld M (2007) Plakophilins: multifunctional proteins or just regulators of desmosomal adhesion? Biochimica et Biophysica Acta (BBA)-Mol Cell Res 1773(1):69–77
Häussinger D, Ahrens T et al (2004) Proteolytic E-cadherin activation followed by solution NMR and X-ray crystallography. EMBO J 23(8):1699–1708
He W, Cowin P et al (2003) Untangling desmosomal knots with electron tomography. Science 302 (5642):109–113
Hulpiau P, Van Roy F (2009) Molecular evolution of the cadherin superfamily. Int J Biochem Cell Biol 41 (2):349–369
Hulpiau P, Van Roy F (2010) New insights into the evolution of metazoan cadherins. Mol Biol Evol 28 (1):647–657
Hülsken J, Birchmeier W et al (1994) E-cadherin and APC compete for the interaction with beta-catenin and the cytoskeleton. J Cell Biol 127(6):2061–2069
Iino R, Koyama I et al (2001) Single molecule imaging of green fluorescent proteins in living cells: E-cadherin forms oligomers on the free cell surface. Biophys J 80 (6):2667–2677
Johnson SK, Ramani VC et al (2007) Kallikrein 7 enhances pancreatic cancer cell invasion by shedding E-cadherin. Cancer 109(9):1811–1820
Kalluri R (2009) EMT: when epithelial cells decide to become mesenchymal-like cells. J Clin Invest 119 (6):1417–1419
Kalluri R, Weinberg RA (2009) The basics of epithelial-mesenchymal transition. J Clin Invest 119 (6):1420–1428
Katsamba P, Carroll K et al (2009) Linking molecular affinity and cellular specificity in cadherin-mediated adhesion. Proc Natl Acad Sci 106(28):11594–11599
Kawanishi J, Kato J et al (1995) Loss of E-cadherin-dependent cell-cell adhesion due to mutation of the

beta-catenin gene in a human cancer cell line, HSC-39. Mol Cell Biol 15(3):1175–1181

Kim N-G, Koh E et al (2011) E-cadherin mediates contact inhibition of proliferation through Hippo signaling-pathway components. Proc Natl Acad Sci 108 (29):11930–11935

Kister AE, Roytberg MA et al (2001) The sequence determinants of cadherin molecules. Protein Sci 10 (9):1801–1810

Klucky B, Mueller R et al (2007) Kallikrein 6 induces E-cadherin shedding and promotes cell proliferation, migration, and invasion. Cancer Res 67 (17):8198–8206

Koch AW, Pokutta S et al (1997) Calcium binding and homoassociation of E-cadherin domains. Biochemistry 36(25):7697–7705

Korman NJ, Eyre RW et al (1989) Demonstration of an adhering-junction molecule (plakoglobin) in the autoantigens of pemphigus foliaceus and pemphigus vulgaris. N Engl J Med 321(10):631–635

Kostetskii I, Li J et al (2005) Induced deletion of the N-cadherin gene in the heart leads to dissolution of the intercalated disc structure. Circ Res 96(3):346–354

Kowalczyk AP, Nanes BA (2012) Adherens junction turnover: regulating adhesion through cadherin endocytosis, degradation, and recycling. Adherens Junctions: from molecular mechanisms to tissue development and disease, Springer:197–222

Kowalski PJ, Rubin MA et al (2003) E-cadherin expression in primary carcinomas of the breast and its distant metastases. Breast Cancer Res 5(6):R217–R222

Kwiatkowski AV, Maiden SL et al (2010) In vitro and in vivo reconstitution of the cadherin–catenin–actin complex from Caenorhabditis elegans. Proc Natl Acad Sci 107(33):14591–14596

Kypta RM, Su H et al (1996) Association between a transmembrane protein tyrosine phosphatase and the cadherin-catenin complex. J Cell Biol 134 (6):1519–1529

Lamouille S, Xu J et al (2014) Molecular mechanisms of epithelial-mesenchymal transition. Nat Rev Mol Cell Biol 15(3):178–196

Le TL, Yap AS et al (1999) Recycling of E-cadherin. Potential Mech Regul Cadherin Dyn 146(1):219–232

Leckband D, Prakasam A (2006) Mechanism and dynamics of cadherin adhesion. Annu Rev Biomed Eng 8:259–287

Li Y, Altorelli NL et al (2013) Mechanism of E-cadherin dimerization probed by NMR relaxation dispersion. Proc Natl Acad Sci 110(41):16462–16467

Liu Y (2010) New insights into epithelial-mesenchymal transition in kidney fibrosis. J Am Soc Nephrol 21 (2):212–222

Lodish H, Berk A et al (2000) Molecular cell biology, 4th edn. Bookshelf, National Center for Biotechnology Information

Lombaerts M, Van Wezel T et al (2006) E-cadherin transcriptional downregulation by promoter methylation but not mutation is related to epithelial-to-mesenchymal transition in breast cancer cell lines. Br J Cancer 94(5):661

Lorthongpanich C, Doris TPY et al (2012) Developmental fate and lineage commitment of singled mouse blastomeres. Development 139(20):3722–3731

Mahoney PA, Weber U et al (1991) The fat tumor suppressor gene in Drosophila encodes a novel member of the cadherin gene superfamily. Cell 67(5):853–868

Manibog K, Sankar K et al (2016) Molecular determinants of cadherin ideal bond formation: conformation-dependent unbinding on a multidimensional landscape. Proc Natl Acad Sci 113(39):E5711–E5720

Masciari S, Larsson N et al (2007) Germline E-cadherin mutations in familial lobular breast cancer. J Med Genet 44(11):726–731

Mayosi BM, Fish M et al (2017) Identification of cadherin 2 (CDH2) mutations in Arrhythmogenic right ventricular cardiomyopathy. Circ Cardiovasc Genet 10(2): e001605

McEwen AE, Maher MT et al (2014) E-cadherin phosphorylation occurs during its biosynthesis to promote its cell surface stability and adhesion. Mol Biol Cell 25 (16):2365–2374

Mo P, Yang S (2018) The store-operated calcium channels in cancer metastasis: from cell migration, invasion to metastatic colonization. Front Biosci (Landmark Ed) 23:1241–1256

Moeller MJ, Soofi A et al (2004) Protocadherin FAT1 binds Ena/VASP proteins and is necessary for actin dynamics and cell polarization. EMBO J 23 (19):3769–3779

Murray PS, Zaidel-Bar R (2014) Pre-metazoan origins and evolution of the cadherin adhesome. Biol open 3 (12):1183–1195

Nagafuchi A, Ishihara S et al (1994) The roles of catenins in the cadherin-mediated cell adhesion: functional analysis of E-cadherin-alpha catenin fusion molecules. J Cell Biol 127(1):235–245

Nagafuchi A, Takeichi M et al (1991) The 102 kd cadherin-associated protein: similarity to vinculin and posttranscriptional regulation of expression. Cell 65 (5):849–857

Nagar B, Overduin M et al (1996) Structural basis of calcium-induced E-cadherin rigidification and dimerization. Nature 380(6572):360

Nakagawa S, Takeichi M (1998) Neural crest emigration from the neural tube depends on regulated cadherin expression. Development 125(15):2963–2971

Nakamura T, Mochizuki Y et al (2001) Signals via FGF receptor 2 regulate migration of endothelial cells. Biochem Biophys Res Commun 289(4):801–806

Nakayama M, Nakajima D et al (1998) Identification of high-molecular-weight proteins with multiple EGF-like motifs by motif-trap screening. Genomics 51(1):27–34

Nawrocki-Raby B, Gilles C et al (2003) Upregulation of MMPs by soluble E-cadherin in human lung tumor cells. Int J Cancer 105(6):790–795

Nelson CM, Chen CS (2003) VE-cadherin simultaneously stimulates and inhibits cell proliferation by altering cytoskeletal structure and tension. J Cell Sci 116 (Pt 17):3571–3581

Nichols SA, Roberts BW et al (2012) Origin of metazoan cadherin diversity and the antiquity of the classical cadherin/β-catenin complex. Proc Natl Acad Sci 109 (32):13046–13051

Niessen CM, Gumbiner BM (1998) The juxtamembrane region of the cadherin cytoplasmic tail supports lateral clustering, adhesive strengthening, and interaction with p120ctn. J Cell Biol 141(3):779–789

Niessen CM, Gumbiner BM (2002) Cadherin-mediated cell sorting not determined by binding or adhesion specificity. J Cell Biol 156(2):389–400

Nishiguchi S, Yagi A et al (2016) Divergence of structural strategies for homophilic E-cadherin binding among bilaterians. J Cell Sci 129(17):3309–3319

Nollet F, Kools P et al (2000) Phylogenetic analysis of the cadherin superfamily allows identification of six major subfamilies besides several solitary members. J Mol Biol 299(3):551–572

Nourse MB, Rolle MW et al (2007) Selective control of endothelial cell proliferation with a synthetic dimerizer of FGF receptor-1. Lab Investig 87(8):828

Oda H, Tsukita S (1999) Nonchordate classic cadherins have a structurally and functionally unique domain that is absent from chordate classic cadherins. Dev Biol 216 (1):406–422

Oda H, Wada H et al (2002) A novel amphioxus cadherin that localizes to epithelial adherens junctions has an unusual domain organization with implications for chordate phylogeny. Evol Dev 4(6):426–434

Ogou S-I, Okada T et al (1982) Cleavage stage mouse embryos share a common cell adhesion system with teratocarcinoma cells. Dev Biol 92(2):521–528

Ozawa M, Engel J et al (1990a) Single amino acid substitutions in one Ca2+ binding site of uvomorulin abolish the adhesive function. Cell 63(5):1033–1038

Ozawa M, Ringwald M et al (1990b) Uvomorulin-catenin complex formation is regulated by a specific domain in the cytoplasmic region of the cell adhesion molecule. Proc Natl Acad Sci 87(11):4246–4250

Parisini E, Higgins JM et al (2007) The crystal structure of human E-cadherin domains 1 and 2, and comparison with other cadherins in the context of adhesion mechanism. J Mol Biol 373(2):401–411

Patel SD, Chen CP et al (2003) Cadherin-mediated cell–cell adhesion: sticking together as a family. Curr Opin Struct Biol 13(6):690–698

Pertz O, Bozic D et al (1999) A new crystal structure, Ca2 + dependence and mutational analysis reveal molecular details of E-cadherin homoassociation. EMBO J 18 (7):1738–1747

Pike LJ (2003) Lipid rafts bringing order to chaos. J Lipid Res 44(4):655–667

Pokutta S, Weis WI (2007) Structure and mechanism of cadherins and catenins in cell-cell contacts. Annu Rev Cell Dev Biol 23:237–261

Posy S, Shapiro L et al (2008) Sequence and structural determinants of strand swapping in cadherin domains: do all cadherins bind through the same adhesive interface? J Mol Biol 378(4):954–968

Prasad A, Pedigo S (2005) Calcium-dependent stability studies of domains 1 and 2 of epithelial cadherin. Biochemistry 44(42):13692–13701

Priest AV, Shafraz O et al (2017) Biophysical basis of cadherin mediated cell-cell adhesion. Exp Cell Res 358 (1):10–13

Riethmacher D, Brinkmann V et al (1995) A targeted mutation in the mouse E-cadherin gene results in defective preimplantation development. Proc Natl Acad Sci 92(3):855–859

Roh J-Y, Stanley JR (1995) Plakoglobin binding by human Dsg3 (pemphigus vulgaris antigen) in keratinocytes requires the cadherin-like intracytoplasmic segment. J Investig Dermatol 104 (5):720–724

Roura S, Miravet S et al (1999) Regulation of E-cadherin/catenin association by tyrosine phosphorylation. J Biol Chem 274(51):36734–36740

Saglietti L, Dequidt C et al (2007) Extracellular interactions between GluR2 and N-cadherin in spine regulation. Neuron 54(3):461–477

Sano K, Tanihara H et al (1993) Protocadherins: a large family of cadherin-related molecules in central nervous system. EMBO J 12(6):2249–2256

Sarpal R, Pellikka M et al (2012) Mutational analysis supports a core role for Drosophila α-catenin in adherens junction function. J Cell Sci 125(1):233–245

Schlegelmilch K, Mohseni M et al (2011) Yap1 acts downstream of α-catenin to control epidermal proliferation. Cell 144(5):782–795

Seveau S, Bierne H et al (2004) Role of lipid rafts in E-cadherin–and HGF-R/Met–mediated entry of Listeria monocytogenes into host cells. J Cell Biol 166(5):743–753

Shan W-S, Tanaka H et al (2000) Functional cis-heterodimers of N-and R-cadherins. J Cell Biol 148(3):579–590

Shapiro L, Fannon AM et al (1995) Structural basis of cell-cell adhesion by cadherins. Nature 374(6520):327

Shapiro L, Weis WI (2009) Structure and biochemistry of cadherins and catenins. Cold Spring Harb Perspect Biol 1(3):a003053

Siemens J, Lillo C et al (2004) Cadherin 23 is a component of the tip link in hair-cell stereocilia. Nature 428 (6986):950–955

Siret C, Terciolo C et al (2015) Interplay between cadherins and α2β1 integrin differentially regulates melanoma cell invasion. Br J Cancer 113(10):1445

Sivasankar S, Zhang Y et al (2009) Characterizing the initial encounter complex in cadherin adhesion. Structure 17(8):1075–1081

Son H, Moon A (2010) Epithelial-mesenchymal transition and cell invasion. Toxicol Res 26(4):245

Stemmler MP, Hecht A et al (2005) E-cadherin intron 2 contains cis-regulatory elements essential for gene expression. Development 132(5):965–976
Stephenson RO, Yamanaka Y et al (2010) Disorganized epithelial polarity and excess trophectoderm cell fate in preimplantation embryos lacking E-cadherin. Development 137(20):3383–3391
Strumane K, Bonnomet A et al (2006) E-cadherin regulates human Nanos1, which interacts with p120ctn and induces tumor cell migration and invasion. Cancer Res 66(20):10007–10015
Suzuki ST (1996) Structural and functional diversity of cadherin superfamily: are new members of cadherin superfamily involved in signal transduction pathway? J Cell Biochem 61(4):531–542
Suzuki ST, Hirano S (2016) The cadherin superfamily: key regulators of animal development and physiology. Springer, Tokyo
Taguchi K, Ishiuchi T et al (2011) Mechanosensitive EPLIN-dependent remodeling of adherens junctions regulates epithelial reshaping. J Cell Biol: JCB 201104124
Takeichi M (1990) Cadherins: a molecular family important in selective cell-cell adhesion. Annu Rev Biochem 59(1):237–252
Takeichi M (1995) Morphogenetic roles of classic cadherins. Curr Opin Cell Biol 7(5):619–627
Tamura K, Shan W-S et al (1998) Structure-function analysis of cell adhesion by neural (N-) cadherin. Neuron 20(6):1153–1163
Tanabe K, Takeichi M et al (2004) Identification of a nonchordate-type classic cadherin in vertebrates: chicken Hz-cadherin is expressed in horizontal cells of the neural retina and contains a nonchordate-specific domain complex. Dev Dyn 229(4):899–906
Taneyhill LA (2008) To adhere or not to adhere: the role of Cadherins in neural crest cell development. Cell Adhes Migr 2(4):223–230
Tanoue T, Takeichi M (2004) Mammalian Fat1 cadherin regulates actin dynamics and cell–cell contact. J Cell Biol 165(4):517–528
Thiery JP (2002) Epithelial-mesenchymal transitions in tumour progression. Nat Rev Cancer 2(6):442–454
Thiery JP, Acloque H et al (2009) Epithelial-mesenchymal transitions in development and disease. Cell 139 (5):871–890
Thomason HA, Scothern A et al (2010) Desmosomes: adhesive strength and signalling in health and disease. Biochem J 429(3):419–433
Tsai JH, Yang J (2013) Epithelial-mesenchymal plasticity in carcinoma metastasis. Genes Dev 27 (20):2192–2206
Tsuruo T, Iida H et al (1985) Inhibition of spontaneous and experimental tumor metastasis by the calcium antagonist verapamil. Cancer Chemother Pharmacol 14 (1):30–33
Vanhalst K, Kools P et al (2005) δ-Protocadherins: a gene family expressed differentially in the mouse brain. Cell Mol Life Sci CMLS 62(11):1247–1259
Vendome J, Felsovalyi K et al (2014) Structural and energetic determinants of adhesive binding specificity in type I cadherins. Proc Natl Acad Sci 111(40):E4175–E4184
Vendome J, Posy S et al (2011) Molecular design principles underlying β-strand swapping in the adhesive dimerization of cadherins. Nat Struct Mol Biol 18 (6):693
Vleminckx K, Vakaet L Jr et al (1991) Genetic manipulation of E-cadherin expression by epithelial tumor cells reveals an invasion suppressor role. Cell 66 (1):107–119
Volk T, Cohen O et al (1987) Formation of heterotypic adherens-type junctions between L-CAM-containing liver cells and A-CAM-containing lens cells. Cell 50 (6):987–994
Wang Y, Kaiser MS et al (2010) Moesin1 and Ve-cadherin are required in endothelial cells during in vivo tubulogenesis. Development 137(18):3119–3128
Wang Y, Shi J et al (2013) The role of snail in EMT and tumorigenesis. Curr Cancer Drug Targets 13 (9):963–972
Watanabe T, Sato K et al (2009) Cadherin-mediated intercellular adhesion and signaling cascades involving small GTPases. Cold Spring Harb Perspect Biol 1(3): a003020
Weinhold B (2006) Epigenetics: the science of change. Environ Health Perspect 114(3):A160–A167
Wheelock MJ, Shintani Y et al (2008) Cadherin switching. J Cell Sci 121(Pt 6):727–735
Wijnhoven BP, de Both NJ et al (1999) E-cadherin gene mutations are rare in adenocarcinomas of the oesophagus. Br J Cancer 80(10):1652–1657
Witcher LL, Collins R et al (1996) Desmosomal cadherin binding domains of plakoglobin. J Biol Chem 271 (18):10904–10909
Woichansky I, Beretta CA et al (2016) Three mechanisms control E-cadherin localization to the zonula adherens. Nat Commun 7:10834
Wu Q, Maniatis T (1999) A striking organization of a large family of human neural cadherin-like cell adhesion genes. Cell 97(6):779–790
Wu Y, Jin X et al (2010) Cooperativity between trans and cis interactions in cadherin-mediated junction formation. Proc Natl Acad Sci 107(41):17592–17597
Wu Y, Vendome J et al (2011) Transforming binding affinities from three dimensions to two with application to cadherin clustering. Nature 475(7357):510
Xiao K, Allison DF et al (2003) Cellular levels of p120 catenin function as a set point for cadherin expression levels in microvascular endothelial cells. J Cell Biol 163(3):535–545
Yamada S, Pokutta S et al (2005) Deconstructing the cadherin-catenin-actin complex. Cell 123(5):889–901
Yonemura S (2011) A mechanism of mechanotransduction at the cell-cell interface. BioEssays 33(10):732–736

Yonemura S, Wada Y et al (2010) α-Catenin as a tension transducer that induces adherens junction development. Nat Cell Biol **12**(6):533

Yoshiura K, Kanai Y et al (1995) Silencing of the E-cadherin invasion-suppressor gene by CpG methylation in human carcinomas. Proc Natl Acad Sci U S A 92(16):7416–7419

Zeisberg M, Kalluri R (2004) The role of epithelial-to-mesenchymal transition in renal fibrosis. J Mol Med (Berl) 82(3):175–181

Zerial M, McBride H (2001) Rab proteins as membrane organizers. Nat Rev Mol Cell Biol 2(2):107

Zhang Y, Sivasankar S et al (2009) Resolving cadherin interactions and binding cooperativity at the single-molecule level. Proc Natl Acad Sci 106(1):109–114

Part III

G-Protein Coupled Receptors: From Structure to Function

Identification of Sphingolipid-binding Motif in G Protein-coupled Receptors

10

Sandeep Shrivastava, Md. Jafurulla, Shrish Tiwari, and Amitabha Chattopadhyay

Abstract

Sphingolipids correspond to a major class of lipids which serve as indispensable structural components of membranes and play an important role in various cellular functions. They constitute ~10–20% of total membrane lipids and are known to form segregated domains in biological membranes. Sphingolipids have been shown to play a vital role in the function of various G protein-coupled receptors (GPCRs). We report here the presence of sphingolipid-binding motif (SBM) in representative GPCRs such as cholecystokinin, oxytocin and secretin receptors, and subtypes of human serotonin receptors. We previously reported the importance of sphingolipids in the function of the serotonin$_{1A}$ receptor, a representative member of the GPCR superfamily, involved in behavioral, cognitive, and developmental functions. In this work, we show that the serotonin$_{1A}$ receptor contains a putative SBM, corresponding to amino acids 205 to 213. In addition, our analysis shows that SBM is an intrinsic characteristic feature of the serotonin$_{1A}$ receptor and is conserved throughout the course of natural evolution. Our results represent the first report on the presence of SBM in serotonin$_{1A}$ receptors and provide novel insight on the molecular mechanism of GPCR-sphingolipid interaction.

Keywords

SBM · Sphingolipids · GPCR · Serotonin$_{1A}$ receptor · CRAC

Abbreviations

CRAC	Cholesterol recognition/interaction amino acid consensus
GPCR	G protein-coupled receptor
SBM	Sphingolipid-binding motif

10.1 Introduction

Sphingolipids are indispensable constituents of cellular membranes of eukaryotes and account for ~10–20% of the entire lipids associated with membranes (Holthuis et al. 2001). They are believed to form laterally segregated domains with cholesterol (also referred as 'lipid rafts') (Brown 1998; Masserini and Ravasi 2001; Jacobson et al. 2007). The concept of these membrane domains assumes relevance as they have been implicated in crucial physiological functions such as cellular sorting, trafficking (Simons and

S. Shrivastava · M. Jafurulla · S. Tiwari
A. Chattopadhyay (✉)
CSIR-Centre for Cellular and Molecular Biology, Hyderabad, India
e-mail: sandeeps@ccmb.res.in; jafri@ccmb.res.in; shrish@ccmb.res.in; amit@ccmb.res.in

K. Chattopadhyay, S. C. Basu (eds.), *Biochemical and Biophysical Roles of Cell Surface Molecules*, Advances in Experimental Medicine and Biology 1112, https://doi.org/10.1007/978-981-13-3065-0_10

van Meer 1988), cellular signaling (Simons and Toomre 2000), and the entry of pathogens into host cells (Riethmüller et al. 2006; Pucadyil and Chattopadhyay 2007; Vieira et al. 2010; Chattopadhyay and Jafurulla 2012; Kumar et al. 2016).

The G protein-coupled receptor (GPCR) superfamily constitutes an important class of proteins employed in signal transduction (Rosenbaum et al. 2009; Chattopadhyay 2014). GPCRs are associated with an array of physiological processes and have therefore emerged as major drug targets (Heilker et al. 2009; Chattopadhyay 2014; Cooke et al. 2015; Jacobson 2015). An estimate of ~50% of present clinically recommended drugs act as either agonists or antagonists to GPCRs (Jacobson 2015). The serotonin$_{1A}$ receptor represents an important G protein-coupled neurotransmitter receptor and plays a vital role in several neurological functions such as behavior, cognition, anxiety, depression, and learning (Pucadyil et al. 2005; Müller et al. 2007; Kalipatnapu and Chattopadhyay 2007; Fiorino et al. 2014). It therefore naturally has emerged as an important target in the development of therapeutics against neurological disorders (Kaufman et al. 2016).

GPCRs are transmembrane proteins, and a large part of the receptor remains embedded in the membrane. Any change in the membrane lipid milieu therefore could influence the structure and function of the receptor. For example, it is estimated from molecular dynamics simulations that the lipid-protein interface accounts for ~38% of the entire surface area of rhodopsin (Huber et al. 2004). Keeping in mind the increasing relevance of the serotonin$_{1A}$ receptor in pharmacology and drug development, interaction of surrounding membrane lipids with the receptor assumes significance. In this context, previous work from our laboratory has demonstrated the requirement of membrane cholesterol and sphingolipids (reviewed in Pucadyil and Chattopadhyay 2006; Paila and Chattopadhyay 2010; Jafurulla and Chattopadhyay 2013, 2015) in the function of the serotonin$_{1A}$ receptor. The influence of sphingolipids on the structure and function of a variety of membrane proteins has been attributed to specific interaction (Snook et al. 2006).

A characteristic amino acid sequence which could represent the conserved sphingomyelin-binding motif in proteins has been previously proposed (Contreras et al. 2012). These authors identified a specific binding motif for sphingomyelin, termed the sphingolipid-binding motif (SBM), in the transmembrane protein p24 (a COPI machinery protein) and demonstrated the headgroup and acyl chain specificity of sphingomyelin in its interaction with this motif. Importantly, these authors further showed that only sphingomyelin with an appropriate dynamic volume would fit into the cavity formed by SBM residues in the transmembrane domain of p24. In spite of the importance of sphingolipids in the structure and function of several membrane proteins, including GPCRs and ion channels (Alves et al. 2005; Harikumar et al. 2005; Sjögren and Svenningsson 2007; Fantini and Barrantes 2009; Slotte 2013; Jafurulla and Chattopadhyay 2015; Jafurulla et al. 2017), limited information is available on specific binding motifs involved in GPCR-sphingolipid interaction. Exploring the possibility of specific interaction between GPCRs and sphingolipids therefore assumes relevance. In the present study, we identified the presence of SBM in representative GPCRs such as cholecystokinin, oxytocin and secretin receptors, and subtypes of human serotonin receptors. Earlier results from our laboratory have shown that membrane sphingomyelin is important for regulating the function of the serotonin$_{1A}$ receptor (Jafurulla et al. 2008; Singh and Chattopadhyay 2012). We therefore explored the presence of SBM in the human serotonin$_{1A}$ receptor and its conservation over the course of natural evolution. Our results show that human serotonin$_{1A}$ receptors contain a putative SBM in transmembrane helix V. In addition, sequence analysis of the serotonin$_{1A}$ receptor from various species across diverse phyla shows that SBM is an inherent characteristic attribute of the receptor and is conserved throughout the course of natural evolution.

10.2 Methods

10.2.1 Identification of SBM in Representative Human GPCRs

The conserved signature sequence for SBM is (I/L/T/V)XX(I/L/T/V)(I/L/T/V)XX(I/L/T/V)(F/W/Y), where X represents any of the 20 naturally occurring amino acids (Contreras et al. 2012; Björkholm et al. 2014). The putative SBMs were identified by visual inspection in cholecystokinin, oxytocin, serotonin, and secretin receptors with the help of regular sequence of SBM (see Fig. 10.1a). The amino acid sequences of GPCRs were obtained from NCBI database. The positions of starting amino acid residues in the corresponding sequences are marked in parentheses.

10.2.2 Sequence Alignments of the Predicted SBM in Human Serotonin Receptor Subtypes

Multiple sequence alignment for various serotonin receptor subtypes was performed with ClustalW (Larkin et al. 2007), using the human serotonin$_{1A}$ receptor sequence as a reference. The positions of starting amino acid residues in various serotonin receptor subtypes are marked in parentheses (see Fig. 10.1b).

10.2.3 Sequence Alignment of the Serotonin$_{1A}$ Receptor and Identification of SBM

The transmembrane helices of the serotonin$_{1A}$ receptor were predicted using the program TMHMM2 (Krogh et al. 2001; see Fig. 10.2). The putative SBM in the serotonin$_{1A}$ receptor is identified as described above. The conservation of sphingolipid-binding motif (SBM) in the serotonin$_{1A}$ receptor during evolution was analyzed by examining amino acid sequences of the receptor over various phyla obtained from NCBI and ExPASy databases (see Fig. 10.3). Partial, duplicate, and other non-specific sequences were removed from the set of sequences obtained. Initial alignment of sequences was carried out using ClustalW. After eliminating the relatively divergent parts of the receptor, the sequence was realigned using the same program. The putative SBMs in the serotonin$_{1A}$ receptor were identified by visual inspection. The amino acid sequences used for the analysis belong to diverse taxa that include insects, fish and other marine species, amphibians, and extending up to mammals. The quality of alignment shown in Figs. 10.1b and 10.3 was computed in Jalview, the software used to view the alignment.

10.3 Results and Discussion

Sphingolipids have been shown to modulate the function and organization of important classes of membrane proteins such as GPCRs (Jafurulla and Chattopadhyay 2015). In the overall context of sphingolipid sensitivity of GPCR function (Sjögren and Svenningsson 2007; Fantini and Barrantes 2009; Paila et al. 2010; Singh et al. 2012; Jafurulla and Chattopadhyay 2015; Jafurulla et al. 2017), we examined whether the sequence of some of the representative GPCRs includes any SBM(s). We identified the presence of SBM in representative GPCRs such as cholecystokinin, oxytocin, serotonin$_{1A}$, and secretin receptors (see Fig. 10.1a). Interestingly, the function of some of these receptors (e.g., the cholecystokinin receptor) has been shown to be modulated by sphingolipids (Harikumar et al. 2005). Sequence analysis of these receptors revealed that while the secretin receptor contains two SBMs (residues 31–39 and 182–190), the cholecystokinin, the oxytocin, and the serotonin$_{1A}$ receptor sequences display only one motif. It is noteworthy that while the characteristic SBM identified in oxytocin and one of the

a

Cholecystokinin receptor (AAA35660)	(223) LLLLF**FIPGVVMAV**AYGLI
Oxytocin receptor (NP_000907)	(78) MKHLS**IADLVVAVF**QVLPQ
Serotonin$_{1A}$ receptor (BAA94488)	(200) TFGAF**YIPLLLMLV**LYGRI
Secretin receptor (AAA64949)	(26) GALPR**LCDVLQVLW**EEQDQ
	(177) IHMHL**FVSFILRAL**SNFIK

b

Serotonin$_{1A}$ (BAA94488)	(202) GAF**YIPLLLMLV**LYG
Serotonin$_{1D}$ (P28221)	(204) GAF**YIPSVLLII**LYG
Serotonin$_{1E}$ (P28566)	(189) GAF**YIPLTLILI**LYY
Serotonin$_{2A_1}$ (NP_000612)	(241) VSF**FIPLTIMVI**TYF
Serotonin$_{2A_2}$ (NP_001159419)	(157) VSF**FIPLTIMVI**TYF
Serotonin$_{2C}$ (P28335)	(221) VAF**FIPLTIMVI**TYC
Serotonin$_4$ (CAA70774)	(199) VAF**YIPFLLMVL**AYY

Quality

Fig. 10.1 Sequence alignment of the predicted sphingolipid-binding motifs in representative human G protein-coupled receptors and serotonin receptor subtypes. (**a**) SBM(s) in cholecystokinin, oxytocin, serotonin$_{1A,}$ and secretin receptors are shown (highlighted in blue). The numbers corresponding to the starting amino acid position in the respective sequences are mentioned in parentheses. (**b**) Multiple alignments of human serotonin receptor subtypes were performed with ClustalW using human serotonin$_{1A}$ receptor sequence as reference. The putative SBM is shown in blue. The positions of amino acid residues are marked in parentheses for various serotonin receptor subtypes. A graphical representation displaying the quality of alignment for serotonin receptor subtypes, with lighter shades representing higher quality. The amino acid sequences of the receptors were obtained from NCBI database, and the protein accession numbers are indicated in parentheses. See text and Methods for more details

motifs in secretin receptors (residues 31–39) show orientation of amino acids similar to what was previously reported (Contreras et al. 2012; Björkholm et al. 2014), SBM of the cholecystokinin, the serotonin$_{1A,}$ and the second motif (residues 182–190) in the secretin receptor was found to be oriented in opposite direction.

In view of the fact that the various subtypes of serotonin receptors share considerable sequence homology (Hoyer et al. 2002), we examined the occurrence of SBM in subtypes of human serotonin receptors. Figure 10.1b shows the sequence alignment of subtypes of human serotonin receptors. Multiple alignments were carried out with ClustalW using the serotonin$_{1A}$ receptor sequence as reference. Interestingly, SBM was found to be conserved in most of the serotonin receptor subtypes analyzed (highlighted in blue in Fig. 10.1b).

Previous work from our laboratory has demonstrated the requirement of sphingomyelin (Jafurulla et al. 2008; Singh and Chattopadhyay 2012) in regulating the function of the serotonin$_{1A}$ receptor. The putative SBM identified in the human serotonin$_{1A}$ receptor is present in transmembrane helix V and comprises of Tyr-205, Iso-206, Leu-209, Leu-210, and Val-213 (see Figs. 10.1 and 10.2). Within SBM, the β-branched residue Ile is found in position 206 in serotonin$_{1A}$ receptors (see Fig. 10.1). These β-branched residues were shown to contribute to interactions between transmembrane helices (Senes et al. 2000) by providing additional rigidity and thereby enhancing London dispersion forces.

It is interesting to note that SBM in the serotonin$_{1A}$ receptor exhibits partial overlap (specifically, in amino acid residues 209–213) with the cholesterol recognition/interaction amino acid

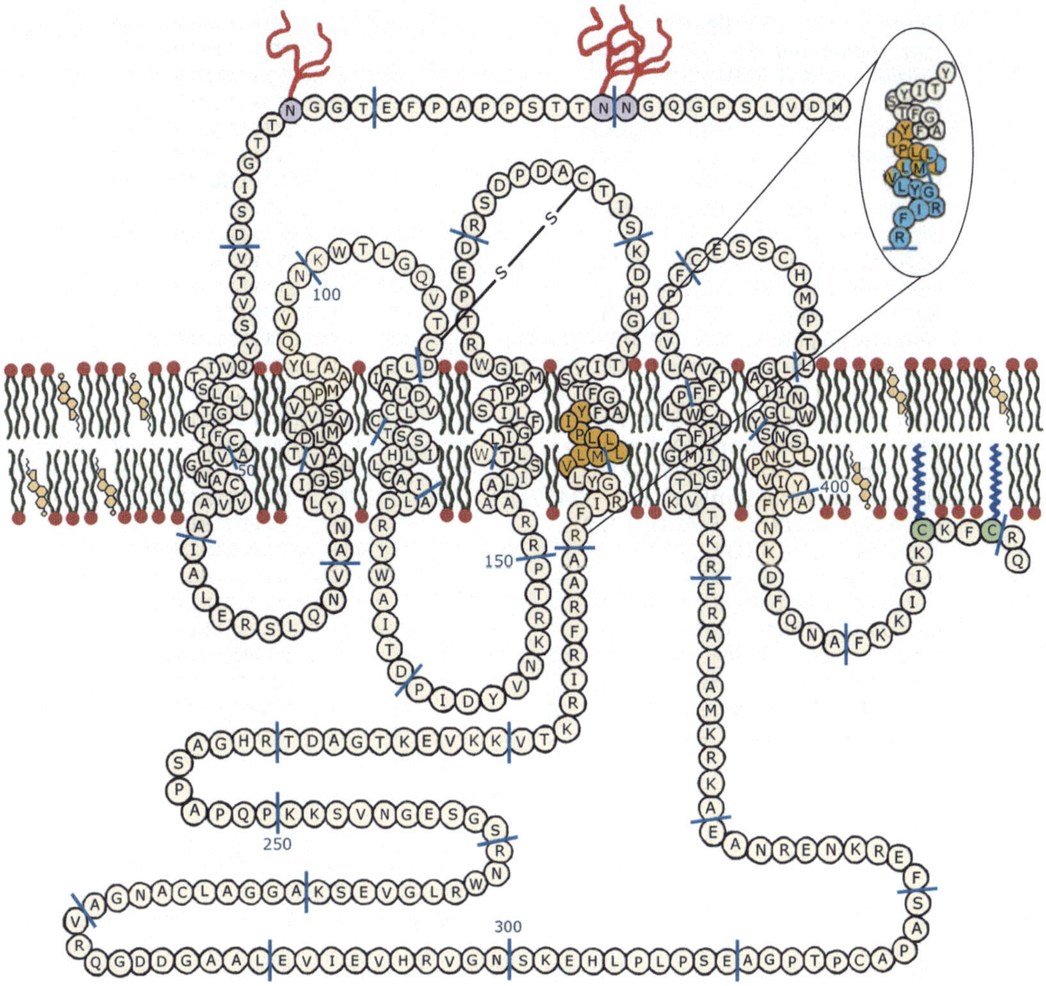

Fig. 10.2 A schematic representation of the membrane-embedded human serotonin$_{1A}$ receptor showing its topological and other structural features. The membrane is shown as a bilayer of phospholipids and cholesterol, representative of typical eukaryotic membranes. The transmembrane helices of the receptor were predicted using the program TMHMM2 (see Methods for details). Seven transmembrane stretches, each composed of ~22 amino acids, are depicted as putative α-helices. The putative SBM in the transmembrane helix V is highlighted (in yellow). The amino acids in the receptor sequence are shown as circles. An enlarged representation of transmembrane helix V of the human serotonin$_{1A}$ receptor showing the overlap of putative SBM highlighted in yellow and CRAC highlighted in cyan in the transmembrane helix V is shown. The common residues corresponding to both SBM and CRAC are shown in a combination of yellow and cyan. (Adapted and modified from Paila et al. 2009)

consensus (CRAC) motif identified earlier by us in transmembrane helix V (Jafurulla et al. 2011; see Fig. 10.2). This observation assumes significance in light of our previous results that showed both cholesterol (Pucadyil and Chattopadhyay 2006; Paila and Chattopadhyay 2010; Jafurulla and Chattopadhyay 2013) and sphingolipids (Jafurulla and Chattopadhyay 2013, 2015) to be essential for the function of the serotonin$_{1A}$ receptor. Such overlapping interaction sites for cholesterol and sphingomyelin in the serotonin$_{1A}$ receptor could help in understanding the mode of regulation of receptor function by these important membrane lipids. Previous results from our laboratory have shown the presence of a sphingolipid-binding domain (SBD) in the

Homo sapiens (BAA94488)	(200) TFGAFYIPLLLMLVLYGRI
Pan troglodytes (NP_001129094)	(200) TFGAFYIPLLLMLVLYGRI
Gorilla gorilla (BAA94490)	(200) TFGAFYIPLLLMLVLYGRI
Pongo pygmaeus (Q9N296)	(200) TFGAFYIPLLLMLVLYGRI
Macaca mulatta (NP_001185629)	(199) TFGAFYIPLLLMLVLYGRI
Canis familiaris (NP_001012397)	(200) TFGAFYIPLLLMLVLYGRI
Vulpes vulpes (AAP12466)	(200) TFGAFYIPLLLMLVLYGRI
Equus caballus (NP_001075251)	(200) TFGAFYIPLLLMLVLYGRI
Bos taurus (DAA17967)	(200) TFGAFYIPLLLMLVLYGRI
Oryctolagus cuniculus (NP_001185864)	(200) TFGAFYIPLLLMLVLYGRI
Mus musculus (NP_032334)	(200) TFGAFYIPLLLMLVLYGRI
Rattus norvegicus (NP_036717)	(200) TFGAFYIPLLLMLVLYGRI
Monodelphis domestica (XP_007486441)	(211) TFGAFYIPLLLMLVLYGRI
Ornithorhynchus anatinus (XP_016083033)	(238) TFGAFYIPIVLMLVLYGRI
Taeniopygia guttata (XP_002188876)	(194) TFGAFYIPLLLMLVLYGRI
Xenopus laevis (NP_001079299)	(194) TFGAFYIPLILMLVLYGKI
Aspidoscelis uniparens (ACE88682)	(62) TFGAFYIPLLLMLVLYGRI
Aspidoscelis inornata (ABY65236)	(73) TFGAFYIPLLLMLVLYGRI
Oreochromis mossambicus (AAP83427)	(202) TFGAFYIPLTLMLVLYGRI
Haplochromis burtoni (ADG21972)	(189) TFGAFYIPLTLMLVLYGRI
Fugu rubripes 1A (O42385)	(209) TFGAFYIPLILMLVLYGRI
Fugu rubripes 1B (O42384)	(202) TFGAFYIPLTLMLVLYGRI
Danio rerio 1a (NP_001116793)	(192) TFGAFYIPLILMLVLYGRI
Danio rerio 1b (NP_001139238)	(194) TFCAFYIPLILMLVLYGRI
Opsanus beta (ACN86308)	(210) TFGAFYIPLVLMLVLYGRI
Gryllus bimaculatus (BAJ83479)	(102) TCSTFYVPLLVILVLYWKI
Caenorhabditis elegans (NP_741730)	(208) TIISFYAPTFIMVILNIKI
Clonorchis sinensis (GAA36232)	(187) SSTSFYVPFVVCVTLYSRI

Quality

Fig. 10.3 Multiple alignment of putative SBM in the serotonin$_{1A}$ receptor over various phyla across natural evolution. The conserved signature sequence for SBM (highlighted in blue) is (I/L/T/V)XX(I/L/T/V)(I/L/T/V)XX(I/L/T/V)(F/W/Y) residues, where X represents any of the 20 naturally occurring amino acids. As shown, the putative SBM is conserved from fish to humans. The numbers corresponding to the starting amino acid position in the respective sequences are mentioned in parentheses. Amino acid sequences of serotonin$_{1A}$ receptors are from NCBI and ExPASy databases, and the protein accession numbers are indicated in parentheses. A graphical representation displaying the quality of alignment with lighter shades representing higher quality is shown below. See text and Methods for more details

serotonin$_{1A}$ receptor (Chattopadhyay et al. 2012) that showed an overlap with the CRAC motif in transmembrane helix II. We recently demonstrated using coarse-grain molecular dynamic simulations that GM_1 (monosialotetrahexosylganglioside) predominantly interacts at the extracellular loop 1 specifically at the proposed SBD site of the serotonin$_{1A}$ receptor in a cholesterol-dependent manner (Prasanna et al. 2016). These results provided better understanding of the importance of such overlap of SBD and CRAC motifs. With this background, we plan to explore in our future studies the possible role of SBM (identified in the present study) in the interaction of sphingomyelin with the serotonin$_{1A}$ receptor and dependency of any such interaction on membrane cholesterol.

We further explored the conservation of SBM in the serotonin$_{1A}$ receptor over natural evolution. For this, we examined the amino acid sequences of the serotonin$_{1A}$ receptor from various organisms across diverse phyla (see Fig. 10.3;

the position of the first amino acid in the alignment is denoted in parentheses). We analyzed the amino acid sequences from various species that belong to diverse taxa which include insects, fish, amphibians, and extending up to mammals. Figure 10.3a shows multiple sequence alignment of the serotonin$_{1A}$ receptor from diverse phyla in the region of the putative SBM with the conserved amino acid residues highlighted. It is evident from this sequence alignment that SBM (Tyr-205, Iso-206, Leu-209, Leu-210, and Val-213) identified in the serotonin$_{1A}$ receptor of *Homo sapiens* (see Figs. 10.1 and 10.2) is conserved in most species included in the study. Initial sequence alignment performed using ClustalW demonstrated that SBM is conserved in the majority of species. Realignment with ClustalW (after deleting the comparatively diverse fractions from the sequence of the serotonin$_{1A}$ receptor) displayed conservation of SBM across various phyla studied (see Fig. 10.3). These results therefore show that SBM is conserved over natural evolution and corresponds to an inherent characteristic attribute of the serotonin$_{1A}$ receptor.

Sphingolipids are enriched in neural tissue and play an important role in the metabolism, survival, and regeneration of the nervous system (van Echten-Deckert and Herget 2006; Piccinini et al. 2010). Regulation of neuronal sphingolipid metabolism has been shown to be crucial, with any deregulation resulting in severe neurodegenerative diseases (Zeidan and Hannun 2007; Piccinini et al. 2010; Prinetti et al. 2011). For example, sphingolipids have been shown to be critical players in the pathogenesis of Alzheimer's disease (Ariga et al. 2008; van Echten-Deckert and Walter 2012) and Parkinson's disease (Wu et al. 2012). In particular, deregulation of sphingomyelin content has been reported in neurological disorders such as Alzheimer's disease, schizophrenia, Parkinson's disease, and Niemann-Pick disease (Bienias et al. 2016). On the other hand, it is interesting to note that imbalance in serotonergic signaling is implicated in Alzheimer's disease, schizophrenia, Parkinson's disease, anxiety, and depression (Tan et al. 2011; Wirth et al. 2017). Importantly, signaling mediated by the serotonin$_{1A}$ receptor is shown to be a crucial target in the pharmacotherapy of schizophrenia and Parkinson's disease (Sumiyoshi et al. 2007; Haleem 2015). In view of the reported role of serotonin$_{1A}$ receptors in neurological disorders associated with deregulation of sphingomyelin, along with our previous results on role of sphingomyelin in the function of the receptor (Jafurulla et al. 2008; Singh and Chattopadhyay 2012), SBM identified in serotonin$_{1A}$ receptors assumes greater relevance.

Recent advancements in GPCR-lipid interaction have improved our overall understanding of GPCR function in the context of human physiology. Although pharmacological and signaling aspects of GPCRs have been studied in detail, features highlighting their interaction with membrane lipids such as sphingolipids and cholesterol are recently beginning to be addressed. Our present results, along with previous observations by us and others, could therefore help to understand the molecular basis of the observed role of sphingolipids in the function of GPCRs in general and the serotonin$_{1A}$ receptor in particular. In summary, these results could prove to be useful in understanding malfunctioning of GPCRs in neurodegenerative disorders involving sphingolipids.

Acknowledgments This work was supported by the Science and Engineering Research Board (Govt. of India) project (EMR/2016/002294). A.C. gratefully acknowledges J.C. Bose Fellowship from the Department of Science and Technology, Govt. of India. A.C. is an Adjunct Professor of Tata Institute of Fundamental Research (Mumbai), RMIT University (Melbourne, Australia), Indian Institute of Technology (Kanpur), and Indian Institute of Science Education and Research (Mohali). We thank members of the Chattopadhyay laboratory for critically reading the manuscript.

References

Alves ID, Salamon Z, Hruby VJ, Tollin G (2005) Ligand modulation of lateral segregation of a G-protein-coupled receptor into lipid microdomains in sphingomyelin/phosphatidylcholine solid-supported bilayers. Biochemistry 44:9168–9178

Ariga T, McDonald MP, Yu RK (2008) Role of ganglioside metabolism in the pathogenesis of Alzheimer's disease-a review. J Lipid Res 49:1157–1175

Bienias K, Fiedorowicz A, Sadowska A, Prokopiuk S, Car H (2016) Regulation of sphingomyelin metabolism. Pharmacol Rep 68:570–581
Björkholm P, Ernst AM, Hacke M, Wieland F, Brügger B, von Heijne G (2014) Identification of novel sphingolipid-binding motifs in mammalian membrane proteins. Biochim Biophys Acta 1838:2066–2070
Brown RE (1998) Sphingolipid organization in biomembranes: what physical studies of model membranes reveal. J Cell Sci 111:1–9
Chattopadhyay A (2014) GPCRs: lipid-dependent membrane receptors that act as drug targets. Adv Biol 2014:143023
Chattopadhyay A, Jafurulla M (2012) Role of membrane cholesterol in leishmanial infection. Adv Exp Med Biol 749:201–213
Chattopadhyay A, Paila YD, Shrivastava S, Tiwari S, Singh P, Fantini J (2012) Sphingolipid binding domain in the serotonin$_{1A}$ receptor. Adv Exp Med Biol 749:279–293
Contreras F-X, Ernst AM, Haberkant P, Björkholm P, Lindahl E, Gönen B, Tischer C, Elofsson A, von Heijne G, Thiele C, Pepperkok R, Wieland F, Brügger B (2012) Molecular recognition of a single sphingolipid species by a protein's transmembrane domain. Nature 481:525–529
Cooke RM, Brown AJH, Marshall FH, Mason JS (2015) Structures of G protein-coupled receptors reveal new opportunities for drug discovery. Drug Discov Today 20:1355–1364
Fantini J, Barrantes FJ (2009) Sphingolipid/cholesterol regulation of neurotransmitter receptor conformation and function. Biochim Biophys Acta 1788:2345–2361
Fiorino F, Severino B, Magli E, Ciano A, Caliendo G, Santagada V, Frecentese F, Perissutti E (2014) 5-HT$_{1A}$ receptor: an old target as a new attractive tool in drug discovery from central nervous system to cancer. J Med Chem 57:4407–4426
Haleem DJ (2015) 5-HT1A receptor-dependent control of nigrostriatal dopamine neurotransmission in the pharmacotherapy of Parkinson's disease and schizophrenia. Behav Pharmacol 26:45–58
Harikumar KG, Puri V, Singh RD, Hanada K, Pagano RE, Miller LJ (2005) Differential effects of modification of membrane cholesterol and sphingolipids on the conformation, function, and trafficking of the G protein-coupled cholecystokinin receptor. J Biol Chem 280:2176–2185
Heilker R, Wolff M, Tautermann CS, Bieler M (2009) G-protein-coupled receptor-focused drug discovery using a target class platform approach. Drug Discov Today 14:231–240
Holthuis JCM, Pomorski T, Raggers RJ, Sprong H, van Meer G (2001) The organizing potential of sphingolipids in intracellular membrane transport. Physiol Rev 81:1689–1723
Hoyer D, Hannon JP, Martin GR (2002) Molecular, pharmacological and functional diversity of 5-HT receptors. Pharmacol Biochem Behav 71:533–554
Huber T, Botelho AV, Beyer K, Brown MF (2004) Membrane model for the G-protein-coupled receptor rhodopsin: hydrophobic interface and dynamical structure. Biophys J 86:2078–2100
Jacobson K, Mouritsen OG, Anderson RGW (2007) Lipid rafts: at a crossroad between cell biology and physics. Nat Cell Biol 9:7–14
Jacobson KA (2015) New paradigms in GPCR drug discovery. Biochem Pharmacol 98:541–555
Jafurulla M, Bandari S, Pucadyil TJ, Chattopadhyay A (2017) Sphingolipids modulate the function of human serotonin$_{1A}$ receptors: insights from sphingolipid-deficient cells. Biochim Biophys Acta 1859:598–604
Jafurulla M, Chattopadhyay A (2013) Membrane lipids in the function of serotonin and adrenergic receptors. Curr Med Chem 20:47–55
Jafurulla M, Chattopadhyay A (2015) Sphingolipids in the function of G protein-coupled receptors. Eur J Pharmacol 763:241–246
Jafurulla M, Pucadyil TJ, Chattopadhyay A (2008) Effect of sphingomyelinase treatment on ligand binding activity of human serotonin$_{1A}$ receptors. Biochim Biophys Acta 1778:2022–2025
Jafurulla M, Tiwari S, Chattopadhyay A (2011) Identification of cholesterol recognition amino acid consensus (CRAC) motif in G-protein coupled receptors. Biochem Biophys Res Commun 404:569–573
Kalipatnapu S, Chattopadhyay A (2007) Membrane organization and function of the serotonin$_{1A}$ receptor. Cell Mol Neurobiol 27:1097–1116
Kaufman J, DeLorenzo C, Choudhury S, Parsey RV (2016) The 5-HT$_{1A}$ receptor in major depressive disorder. Eur Neuropsychopharmacol 26:397–410
Krogh A, Larsson B, von Heijne G, Sonnhammer ELL (2001) Predicting transmembrane protein topology with a hidden Markov model: application to complete genomes. J Mol Biol 305:567–580
Kumar GA, Jafurulla M, Chattopadhyay A (2016) The membrane as the gatekeeper of infection: cholesterol in host-pathogen interaction. Chem Phys Lipids 199:179–185
Larkin MA, Blackshields G, Brown NP, Chenna R, McGettigan PA, McWilliam H, Valentin F, Wallace IM, Wilm A, Lopez R, Thompson JD, Gibson TJ, Higgins DG (2007) Clustal W and Clustal X version 2.0. Bioinformatics 23:2947–2948
Masserini M, Ravasi D (2001) Role of sphingolipids in the biogenesis of membrane domains. Biochim Biophys Acta 1532:149–161
Müller CP, Carey RJ, Huston JP, De Souza Silva MA (2007) Serotonin and psychostimulant addiction: focus on 5-HT$_{1A}$-receptors. Prog Neurobiol 81:133–178
Paila YD, Chattopadhyay A (2010) Membrane cholesterol in the function and organization of G-protein coupled receptors. Subcell Biochem 51:439–466
Paila YD, Ganguly S, Chattopadhyay A (2010) Metabolic depletion of sphingolipids impairs ligand binding and

signaling of human serotonin$_{1A}$ receptors. Biochemistry 49:2389–2397
Paila YD, Tiwari S, Chattopadhyay A (2009) Are specific nonannular cholesterol binding sites present in G-protein coupled receptors? Biochim Biophys Acta 1788:295–302
Piccinini M, Scandroglio F, Prioni S, Buccinnà B, Loberto N, Aureli M, Chigorno V, Lupino E, DeMarco G, Lomartire A, Rinaudo MT, Sonnino S, Prinetti A (2010) Deregulated sphingolipid metabolism and membrane organization in neurodegenerative disorders. Mol Neurobiol 41:314–340
Prasanna X, Jafurulla M, Sengupta D, Chattopadhyay A (2016) The ganglioside GM1 interacts with the serotonin$_{1A}$ receptor via the sphingolipid binding domain. Biochim Biophys Acta 1858:2818–2826
Prinetti A, Prioni S, Chiricozzi E, Schuchman EH, Chigorno V, Sonnino S (2011) Secondary alterations of sphingolipid metabolism in lysosomal storage diseases. Neurochem Res 36:1654–1668
Pucadyil TJ, Chattopadhyay A (2006) Role of cholesterol in the function and organization of G-protein coupled receptors. Prog Lipid Res 45:295–333
Pucadyil TJ, Chattopadhyay A (2007) Cholesterol: a potential therapeutic target in *Leishmania* infection? Trends Parasitol 23:49–53
Pucadyil TJ, Kalipatnapu S, Chattopadhyay A (2005) The serotonin$_{1A}$ receptor: a representative member of the serotonin receptor family. Cell Mol Neurobiol 25:553–580
Riethmüller J, Riehle A, Grassmé H, Gulbins E (2006) Membrane rafts in host-pathogen interactions. Biochim Biophys Acta 1758:2139–2147
Rosenbaum DM, Rasmussen SGF, Kobilka BK (2009) The structure and function of G-protein-coupled receptors. Nature 459:356–363
Senes A, Gerstein M, Engelman DM (2000) Statistical analysis of amino acid patterns in transmembrane helices: the GxxxG motif occurs frequently and in association with β-branched residues at neighboring positions. J Mol Biol 296:921–936
Simons K, Toomre D (2000) Lipid rafts and signal transduction. Nat Rev Mol Cell Biol 1:31–39
Simons K, van Meer G (1988) Lipid sorting in epithelial cells. Biochemistry 27:6197–6202
Singh P, Chattopadhyay A (2012) Removal of sphingomyelin headgroup inhibits the ligand binding function of hippocampal serotonin$_{1A}$ receptors. Biochem Biophys Res Commun 419:321–325
Singh P, Paila YD, Chattopadhyay A (2012) Role of glycosphingolipids in the function of human serotonin$_{1A}$ receptors. J Neurochem 123:716–724
Sjögren B, Svenningsson P (2007) Depletion of the lipid raft constituents, sphingomyelin and ganglioside, decreases serotonin binding at human 5-HT$_{7(a)}$ receptors in HeLa cells. Acta Physiol 190:47–53
Slotte JP (2013) Biological functions of sphingomyelins. Prog Lipid Res 52:424–437
Snook CF, Jones JA, Hannun YA (2006) Sphingolipid-binding proteins. Biochim Biophys Acta 1761:927–946
Sumiyoshi T, Park S, Jayathilake K, Roy A, Ertugrul A, Meltzer HY (2007) Effect of buspirone, a serotonin$_{1A}$ partial agonist, on cognitive function in schizophrenia: a randomized, double-blind, placebo-controlled study. Schizophr Res 95:158–168
Tan SKH, Hartung H, Sharp T, Temel Y (2011) Serotonin-dependent depression in Parkinson's disease: a role for the subthalamic nucleus. Neuropharmacology 61:387–399
van Echten-Deckert G, Herget T (2006) Sphingolipid metabolism in neural cells. Biochim Biophys Acta 1758:1978–1994
van Echten-Deckert G, Walter J (2012) Sphingolipids: critical players in Alzheimer's disease. Prog Lipid Res 51:378–393
Vieira FS, Corrêa G, Einicker-Lamas M, Coutinho-Silva R (2010) Host-cell lipid rafts: a safe door for micro-organisms? Biol Cell 102:391–407
Wirth A, Holst K, Ponimaskin E (2017) How serotonin receptors regulate morphogenic signalling in neurons. Prog Neurobiol 151:35–56
Wu G, Lu Z-H, Kulkarni N, Ledeen RW (2012) Deficiency of ganglioside GM1 correlates with Parkinson's disease in mice and humans. J Neurosci Res 90:1997–2008
Zeidan YH, Hannun YA (2007) Translational aspects of sphingolipid metabolism. Trends Mol Med 13:327–336

Molecular Signatures of Cholesterol Interaction with Serotonin Receptors

11

Madhura Mohole, Xavier Prasanna, Durba Sengupta, and Amitabha Chattopadhyay

Abstract

The interaction of G protein-coupled receptors (GPCRs) with cholesterol is a hallmark of their function, organization, and structural dynamics. Several cholesterol interaction sites, such as the cholesterol recognition amino acid consensus (CRAC) and cholesterol consensus motif (CCM), have been mapped from crystallography, bioinformatics, and simulation studies. In this article, we characterize common descriptors for cholesterol interaction sites in the serotonin$_{1A}$ receptor from a series of coarse-grain simulations. We have identified a novel interaction mode for cholesterol in which the cholesterol polar headgroup interacts with aromatic amino acid residues, such as tryptophan and tyrosine. The cholesterol rings interact with both aromatic residues and nonpolar residues, thereby constituting a signature aromatic interaction site. In addition, we report a similar binding mode in the crystal structures of the serotonin$_{2B}$ receptor, suggesting that this binding mode could be a general feature of the serotonin receptor family. Interestingly, this signature aromatic interaction site is present along with one of the CRAC motifs in the serotonin$_{1A}$ receptor. Our results represent an important step toward mapping out the diversity of cholesterol-GPCR interaction sites.

Keywords

GPCR · MARTINI coarse-grain simulation · Signature aromatic cholesterol interaction site · Serotonin receptors · Cholesterol interaction site · Cholesterol occupancy

11.1 Introduction

G protein-coupled receptors (GPCRs) represent a unique class of membrane proteins that mediate multiple physiological processes (Pierce et al. 2002; Rosenbaum et al. 2009). These receptors are comprised of a conserved topology of seven transmembrane helices interconnected by extracellular and intracellular loops (Venkatakrishnan et al. 2013). Ligands such as neurotransmitters and hormones bind to these receptors and transduce signals inside the cell (Pierce et al. 2002; Rosenbaum et al. 2009). An important feature of GPCR functional dynamics is the effect of membrane composition on receptor structure and

M. Mohole · D. Sengupta (✉)
CSIR-National Chemical Laboratory, Pune, India

Academy of Scientific and Innovative Research (AcSIR), Ghaziabad, India
e-mail: m.mohole@ncl.res.in; d.sengupta@ncl.res.in

X. Prasanna
CSIR-National Chemical Laboratory, Pune, India

A. Chattopadhyay (✉)
CSIR-Centre for Cellular and Molecular Biology, Hyderabad, India
e-mail: amit@ccmb.res.in

K. Chattopadhyay, S. C. Basu (eds.), *Biochemical and Biophysical Roles of Cell Surface Molecules*, Advances in Experimental Medicine and Biology 1112, https://doi.org/10.1007/978-981-13-3065-0_11

function (Chattopadhyay 2014). For instance, membrane cholesterol has been shown to modulate the structure and function of several GPCRs in a context-dependent fashion (Burger et al. 2000; Pucadyil and Chattopadhyay 2006; Paila and Chattopadhyay 2010; Oates and Watts 2011). To exert its effect on receptor function, cholesterol has been suggested to act by both direct interactions with the receptor and/or indirectly by modulating membrane properties (Paila and Chattopadhyay 2009). Interestingly, the crystal structures of several GPCRs, such as β_2-adrenergic receptor (Hanson et al. 2008), adenosine$_{2A}$ receptor (Liu et al. 2012), and serotonin$_{2B}$ receptor (Wacker et al. 2013), have been able to resolve bound cholesterol molecules, suggesting a direct effect. Similarly, direct interactions of cholesterol have been observed in simulations of several GPCRs (Lee and Lyman 2012; Cang et al. 2013; Prasanna et al. 2014).

One of the most comprehensive cholesterol dependencies has been reported in the serotonin$_{1A}$ receptor, a representative GPCR that is abundant in the hippocampal region of the brain (Chattopadhyay 2014). It is activated by serotonin (a neurotransmitter) and plays a key role in physiological responses such as anxiety, depression, and mood (Pucadyil et al. 2005; Kalipatnapu and Chattopadhyay 2007; Lacivita et al. 2008; Fiorino et al. 2014). We have previously shown that depletion of cholesterol affects the stability, function, and dynamics of the serotonin$_{1A}$ receptor (Pucadyil and Chattopadhyay 2004, 2007; Saxena and Chattopadhyay 2011, 2012; Jafurulla and Chattopadhyay 2013). In addition, cholesterol was observed to modulate receptor organization by modulating the population of the different oligomers (Ganguly et al. 2011; Paila et al. 2011a; Chakraborty et al. 2018). To dissect the molecular basis of these cholesterol effects, molecular dynamics simulations were performed using both atomistic and coarse-grain force fields (Sengupta and Chattopadhyay 2012; Patra et al. 2015; Prasanna et al. 2016). We were able to distinguish direct interactions of cholesterol with the receptor at certain "hot spots". From a series of coarse-grain and atomistic simulations, we have identified multiple cholesterol interaction sites with an occupancy time of nanoseconds to microseconds (Sengupta and Chattopadhyay 2012; Patra et al. 2015; Prasanna et al. 2016). Indirect effects related to membrane thickness changes in the bulk membrane as well as around the receptor have been observed, but their significance is yet to be fully analyzed (Prasanna et al. 2016).

In order to understand the nature of direct GPCR-cholesterol interactions, several cholesterol binding motifs have been reported in GPCRs (Rouviere et al. 2017). These include the cholesterol consensus motif (CCM) (Hanson et al. 2008) and the cholesterol recognition amino acid consensus (CRAC) motif (Jafurulla et al. 2011). The presence of the CCM site was earlier reported in crystal structure of β_2-adrenergic receptor (Hanson et al. 2008). The motif is located between transmembrane helices II and IV and consists of a charged residue R or K followed by W-Y, I-L-V and F-Y residues. An important motif that we previously characterized in GPCRs is the CRAC motif, comprising of hydrophobic and aromatic residues followed by a charged residue (arginine or lysine) (Jafurulla et al. 2011). The CRAC motif is defined by the presence of the pattern –L/V-$(X)_{1-5}$-Y-$(X)_{1-5}$-R/K-, in which $(X)_{1-5}$ represents between one and five residues of any amino acid. This motif is present on the transmembrane region of the helix V of the serotonin$_{1A}$ receptor (along with transmembrane regions of helix II and VII). A mirror image of the CRAC motif, known as the CARC motif, has also been reported. In this motif, the terminal residues of the CRAC motif are reversed, with K/R on the N-terminus and L/V on the C-terminus of the helix (Baier et al. 2011). The CRAC and CARC sites have been mainly identified from bioinformatics and molecular dynamics studies and are yet to be experimentally validated. In all reported cholesterol sites, the hydroxyl group of cholesterol interacts with a charged amino acid residue. Despite these structural and sequence cholesterol motifs, not all crystallographically resolved sites or those identified from molecular dynamics simulations have been characterized. For instance, we have identified several cholesterol interaction sites in the serotonin$_{1A}$ receptor (Sengupta and Chattopadhyay 2012; Prasanna et al. 2016), only one of which correlates to a CRAC site. Other sites that were observed in simulations remain uncharacterized, and the need arises to determine the molecular signatures of these sites.

In this work, we have identified common descriptors of cholesterol dynamics in the serotonin$_{1A}$ receptor using coarse-grain simulations. The work extends from our previous work in which we have analyzed the effect of cholesterol on receptor dimerization (Prasanna et al. 2016). We have performed ten simulations of the receptor embedded in POPC/cholesterol bilayers with 30% cholesterol, as a representative model membrane for eukaryotic cell membranes. In order to extract common features of cholesterol binding, we have calculated the interaction propensities of different parts of the cholesterol molecule (such as hydroxyl group and sterol ring) and have determined a pattern of amino acids that are preferred by each coarse-grain bead of cholesterol molecule. A new interaction mode in which the cholesterol headgroup interacts with an aromatic residue is discussed. These results would contribute to our overall understanding of cholesterol-GPCR interaction.

11.2 Methods

System Setup: The homology model of the serotonin$_{1A}$ receptor was taken from our previous work (Paila et al. 2011b; Sengupta and Chattopadhyay 2012; Prasanna et al. 2016). This model was built based on the crystal structure of the β_2-adrenergic receptor (PDB: 2RH1; Cherezov et al. 2007) and template sequence of the serotonin$_{1A}$ receptor. The receptor monomer was embedded in a pre-equilibrated POPC/cholesterol bilayer with 30% cholesterol. The average cholesterol concentration in a eukaryotic cell is about 30% (Van Meer and de Kroon 2011). Simulations were performed with the MARTINI coarse-grain force field (Marrink et al. 2007; Monticelli et al. 2008) using the GROMACS simulation package (Van Der Spoel et al. 2005). Ten replicates of this system were simulated for 10 μs each. The temperature coupling was carried out using v-rescale thermostat at 300 K for each component of the system separately, with a coupling constant of 0.1 ps (Bussi et al. 2007). Pressure was maintained at 1 bar semi-isotropically in the plane of the bilayer and perpendicular to the bilayer using the Berendsen barostat, with a coupling constant of 0.5 ps and a compressibility of 3×10^{-5} bar^{-1} (Berendsen et al. 1984). A time step of 20 fs was used (Marrink et al. 2007).

Analysis: The maximum occupancy time was calculated as the maximum time a cholesterol molecule is bound at a specific location during the course of simulation (Sengupta and Chattopadhyay 2012). The MARTINI model for cholesterol consists of eight beads, with a ROH bead mapping to the hydroxyl group, R1–R5 beads modeling the rings, and C1 and C2 representing the aliphatic tail. For each of these beads, the residue-wise-specific occupancy was calculated using a cutoff of 0.6 nm. For each of the 20 amino acids, occupancy was calculated as the maximum occupancy time for cholesterol around a given residue type and normalized to the number of instances it appears in the receptor. The occupancy time for each amino acid was averaged and normalized across ten sets of simulations. The maximum occupancy time for cholesterol around transmembrane region of each helix was calculated as described earlier (Prasanna et al. 2016). The transmembrane domain was divided into the upper (extracellular) or lower (intracellular) leaflets depending on its location in the bilayer. Representative images of the cholesterol binding modes were rendered with VMD (Humphrey et al. 1996).

11.3 Results

A series of coarse-grain simulations were performed with the serotonin$_{1A}$ receptor monomer embedded in POPC/cholesterol bilayers with 30% cholesterol. To map out the common descriptors of cholesterol interaction sites, we calculated the interactions of cholesterol beads with each amino acid residue. As described in the Methods section, the coarse-grain model for cholesterol consists of eight beads, with a ROH bead mapping to the hydroxyl group, R1–R5 beads modeling the rings, and C1 and C2 representing the aliphatic tail. From the residue-wise occupancies, we estimated the binding propensity of cholesterol to each amino acid residue type. Using this approach, we have been able to identify the molecular signatures of a novel cholesterol binding mode.

11.3.1 Common Descriptors for Cholesterol Interaction Sites

The maximum occupancy time of cholesterol (considering each coarse-grain bead of cholesterol individually) was calculated for each residue in the serotonin$_{1A}$ receptor. The values were averaged for each residue type and normalized by the frequency of that particular amino acid. The normalized occupancy time for each of the cholesterol beads is shown in Fig. 11.1. The polar ROH bead of cholesterol (representing the hydroxyl group) is observed to mainly interact with the aromatic amino acids Y, W, and F (maximum occupancy $\geq$60%). The positively charged amino acid R exhibits a lower interaction (60% $\geq$ maximum occupancy $\geq$50%). This is surprising since all known cholesterol binding motifs involve a positively charged residue such as R or K interacting with the polar headgroup. Further, the ROH bead is observed to interact with almost all residues (but with a low occupancy), indicating high dynamics.

The coarse-grain beads, R1–R5, represent the sterol rings in cholesterol. These beads show a high interaction with the residues L and M. However the ring beads closer to the polar headgroup (R1 and R2) also show high interactions with the aromatic residues similar to the ROH bead. Interestingly, the R2 and R4 beads map to the "smooth" face of the cholesterol but are nonetheless seen to interact strongly with the receptor. We believe that it represents a limitation of the coarse-grain model since the methyl groups on the cholesterol ring are difficult to represent in such a mapping. The hydrocarbon tail of the cholesterol molecule is mapped to two coarse-grain beads, C1 and C2. These beads show a strong interaction with several residues including the aromatic residues and the nonpolar residues, L and I.

Based on the residue-wise occupancies, we propose a new descriptor for cholesterol interactions in the serotonin$_{1A}$ receptor. A schematic representation of these amino acid descriptors is shown in Fig. 11.2. The residues showing a high occupancy time for each cholesterol bead ($\geq$ 50%) are considered, and the height of the residue in the schematic depicts the relative value of the occupancy times. The most common residues that interact with the polar group of cholesterol are the aromatic residues, while the remaining beads show increased association with nonpolar residues. We would like to point out that these residue descriptors represent a spatial pattern of these residues, rather than sequence motif. The signature aromatic binding mode is easily distinguished in the figure and represents a previously uncharacterized interaction site that could have implications for cholesterol-receptor interactions.

11.3.2 The Signature Aromatic Binding Mode in the Serotonin$_{1A}$ Receptor

In the next step, we evaluated cholesterol interactions at the receptor in order to map these descriptors to individual cholesterol interaction sites. To identify the individual sites with high cholesterol occupancy, the maximum occupancy time was calculated for each helix in the upper and lower leaflets, considering cholesterol as a whole (Fig. 11.3). The values match well to our previous reports (Sengupta and Chattopadhyay 2012; Prasanna et al. 2016), despite the high stochasticity in the interaction sites. As expected, a large occupancy is observed at transmembrane helix V, which contains the CRAC site. However, multiple other sites are mapped as well, similar to our previous work (Sengupta and Chattopadhyay 2012; Prasanna et al. 2016). In the upper leaflet, we observe the highest occupancy time for the cholesterol around transmembrane helices IV and

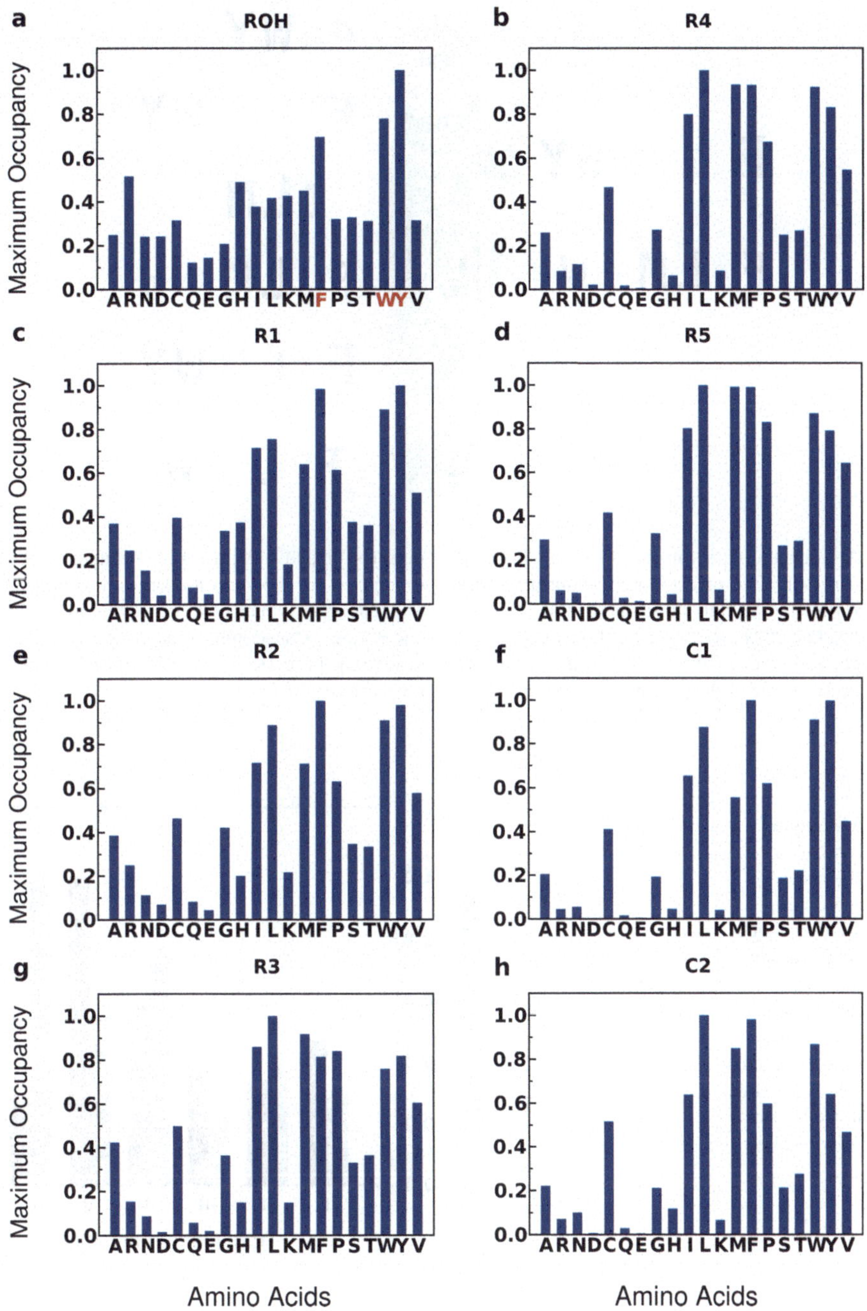

Fig. 11.1 Maximum occupancy time of each cholesterol bead at each residue type in the serotonin$_{1A}$ receptor, averaged and normalized over simulation time and number of residues. A value of 1 indicates that the cholesterol bead is always present at that site, and a value of 0 indicates that there are no interactions during the simulation. See Methods for more details

VI. In the lower leaflet, the maximum cholesterol occupancy was observed around transmembrane helices I, IV, and V.

We mapped four of these high occupancy cholesterol sites to the signature aromatic binding mode. A schematic representation of these four

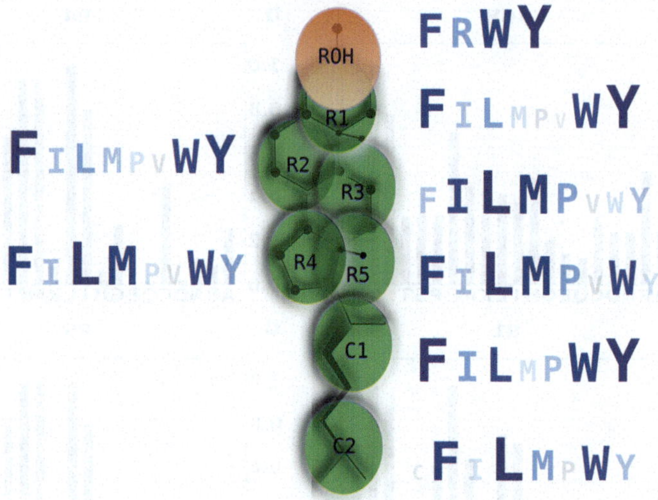

Fig. 11.2 A schematic representation of the amino acid preferences for each bead of cholesterol molecule. Only the residues showing a high occupancy time for each cholesterol bead (≥ 50%) are considered (see Fig. 11.1), and the height and color of the residue in the schematic depict the relative order of the occupancies. For instance, residues W and Y with a high occupancy at cholesterol headgroup are shown as the largest and darkest alphabet. The cholesterol molecule is shown in green with the underlying coarse-grain mapping scheme, and the headgroup bead (ROH) is colored orange. See Methods for more details

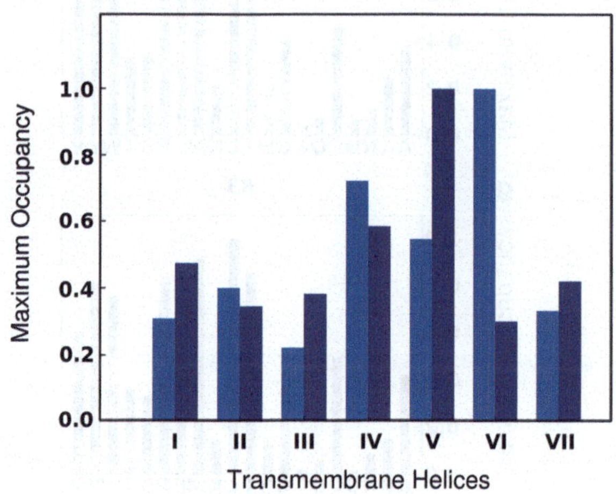

Fig. 11.3 Maximum occupancy time of cholesterol molecule around the serotonin$_{1A}$ receptor that are averaged and normalized over ten sets of simulations. The occupancy time is shown for the transmembrane helices corresponding to the extracellular or upper (light blue) and intracellular or lower (dark blue) leaflets separately

sites is shown in Fig. 11.4. The top panel represents the binding mode present in the upper leaflet of the membrane (transmembrane helices IV and VI) and the lower panel the binding mode in the lower leaflet (transmembrane helices I and V). The four sites are quite varied in their sequence and structure, although all interact with the cholesterol headgroup via an aromatic residue. Interestingly, in transmembrane helix V, the signature aromatic site lies next to the previously identified CRAC site. The fast cholesterol dynamics coupled with site hopping had previously made it difficult to identify the signature aromatic residue binding mode. At the site

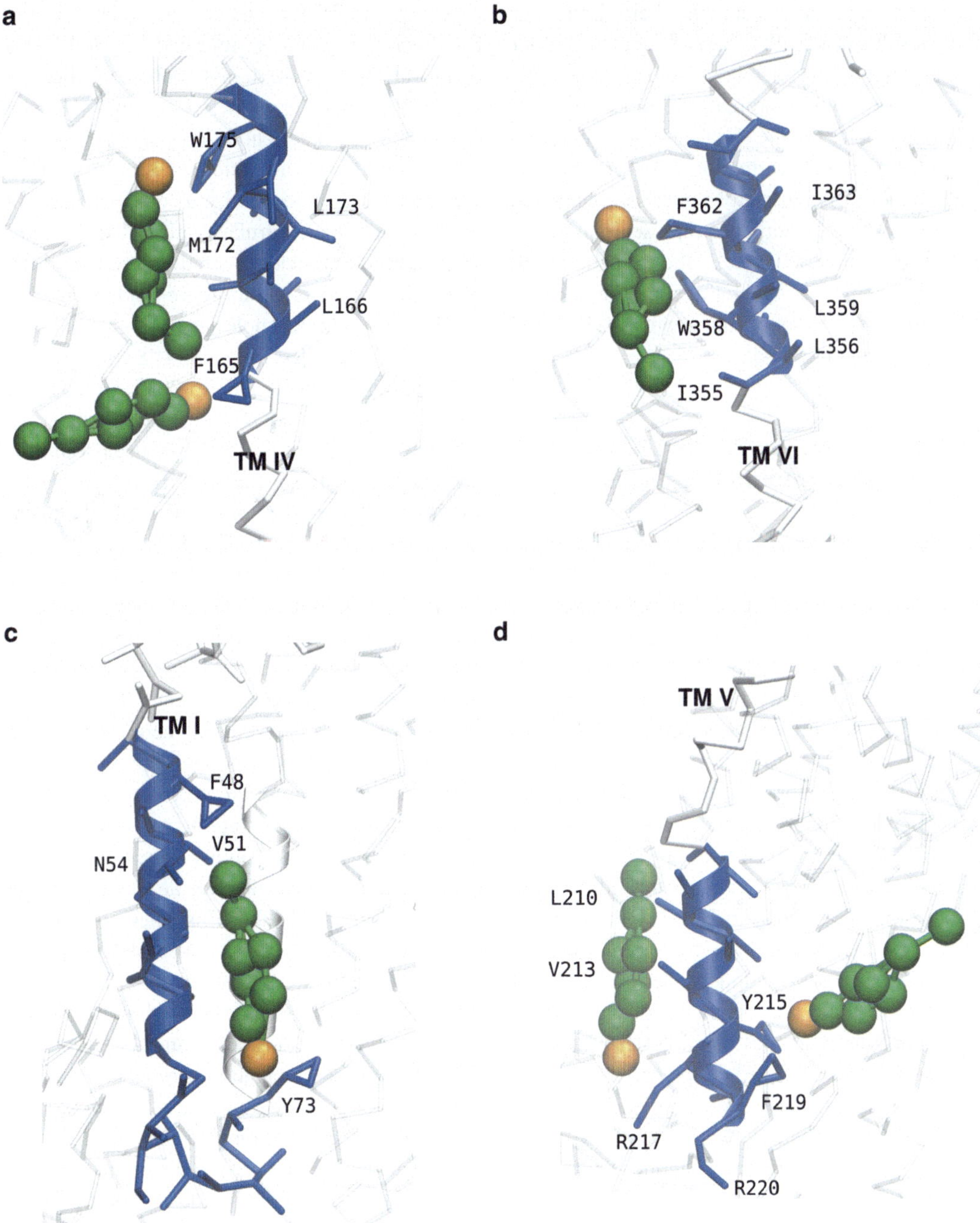

Fig. 11.4 Representative snapshots of cholesterol interaction with the serotonin$_{1A}$ receptor at the signature aromatic binding modes observed in simulations. The top panel represents the binding mode present in the upper leaflet of the membrane (transmembrane helices IV and VI) and the lower panel the binding mode in the lower leaflet (transmembrane helices I and V). The transmembrane helix and the main interacting residues are shown in blue and the remaining part of the receptor in gray. The cholesterol molecules are shown in green, and the headgroup bead (ROH) is colored orange

corresponding to transmembrane helix I, the aromatic residue is actually a part of the extracellular loop and not of transmembrane helix I. Due to the presence of multiple aromatic residues in transmembrane helix IV, the cholesterol molecule can adopt horizontal conformations. As a result, this binding mode of cholesterol cannot be mapped to a single sequence motif but represents a spatial pattern of residues. Overall, the signature aromatic binding mode occurs at multiple sites on the serotonin$_{1A}$ receptor and represents a novel cholesterol binding mode.

11.3.3 A Common Binding Mode for Serotonin Receptors

To analyze whether the signature aromatic binding mode is a general cholesterol binding mode, we analyzed the crystal structures of the serotonin receptor family. Two crystal structures of the serotonin$_{2B}$ receptor (PDB ID: 4IB4 and 5TVN) were resolved along with bound cholesterol (see Fig. 11.5). In both these structures, a cholesterol molecule is observed to interact at transmembrane helix I of the receptor. However, hydroxyl group is oriented toward the residue Y399 on helix VIII. In a similar site in the serotonin$_{1A}$ receptor, the aromatic residue was present on intracellular loop 1. The aliphatic amino acids L and I on transmembrane helix I interact with the sterol rings of cholesterol. The two binding modes in the crystal structures differ in the orientation of cholesterol, confirming the dynamics of cholesterol at these interaction sites. Taken together, the data suggests that cholesterol could interact with the serotonin receptor family by a signature aromatic binding mode. This signature binding mode does not represent an exclusive site but is present together with previously identified motifs such as the CRAC motif.

11.4 Discussion

The identification of cholesterol interaction sites has received attention due to the central role of cholesterol in the function and structural dynamics of GPCRs. The crystal structure of the β_2-adrenergic receptor revealed a cholesterol bound at the cleft of transmembrane helices II and IV that was named the CCM motif (Hanson et al. 2008). Another cholesterol binding site, the CRAC motif was identified in the serotonin$_{1A}$ receptor based on similarity to other cholesterol binding proteins (Jafurulla et al. 2011). In addition, molecular dynamics simulations using both atomistic and coarse-grain force fields were able to identify multiple cholesterol interaction sites on several GPCRs (Lee and Lyman 2012; Sengupta and Chattopadhyay 2012; Cang et al. 2013; Prasanna et al. 2014, 2016; Patra et al. 2015). Although cholesterol has been found to be associated with several GPCRs in their crystal structures, the mapping of the location of cholesterol molecules in these crystal structures to proposed cholesterol binding motifs has proved to be less than straightforward. We identify here a signature aromatic cholesterol interaction mode, in which the headgroup of the cholesterol molecule interacts with an aromatic residue. This interaction mode is observed in serotonin$_{1A}$ and serotonin$_{2B}$ receptors and could represent a common signature in the serotonin receptor family.

An interesting observation from our work is that cholesterol is highly dynamic when interacting at the signature aromatic binding mode. It is possible that this interaction mode represents a site with less favorable interaction energy but is entropically favored due to the high site dynamics. In one of the binding modes identified here (on transmembrane helix IV), cholesterol adopts an orientation parallel to the membrane surface, reminiscent of a site identified in the β_2-adrenergic receptor (Prasanna et al. 2014). Further analysis is required to determine whether this signature interaction mode is specific to the serotonin receptor family or could be generalized to other GPCRs. In general, this binding mode of cholesterol cannot be mapped to a single sequence motif but represents a spatial pattern of residues. To comprehensively map these cholesterol determinants into a spatial 3D motif, we would require to identify more such sites from the serotonin receptor family and other related GPCRs.

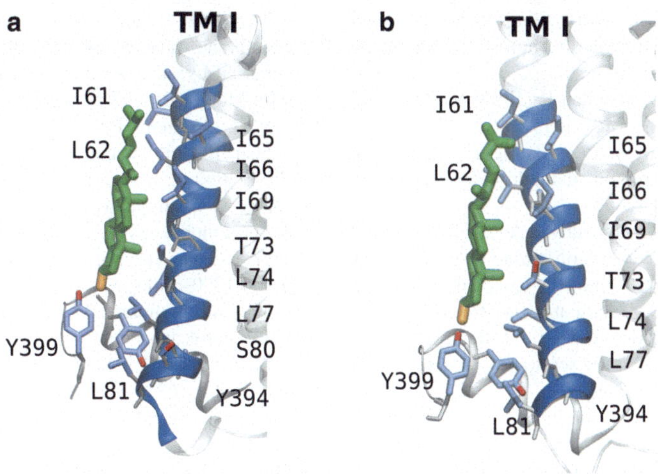

Fig. 11.5 Representative snapshots of cholesterol interaction with the serotonin$_{2B}$ receptor at the signature aromatic binding modes as reported in crystal structures. The snapshots correspond to PDB IDs (**a**) 4IB4 and (**b**) 5TVN. The transmembrane helix and the main interacting residues are shown in blue and the remaining part of the receptor in gray. The cholesterol molecules are shown in green, and the hydroxyl group (-OH) is colored orange

In conclusion, we propose a new signature cholesterol binding mode in which the cholesterol headgroup interacts with an aromatic residue. Increased cholesterol dynamics is observed at this site and could be related to reduced interaction energy but an entropically favorable site. The molecular signatures of cholesterol interaction with receptors represent an important step in our overall understanding of GPCR function in health and disease.

Acknowledgments This work was supported by the Science and Engineering Research Board (Govt. of India) project (EMR/2016/002294) to A.C. and D.S. A.C. gratefully acknowledges support from J.C. Bose Fellowship (Department of Science and Technology, Govt. of India). M.M. thanks the Department of Biotechnology, Govt. of India, for the award of a Junior Research Fellowship. A.C. is an Adjunct Professor of Tata Institute of Fundamental Research (Mumbai), RMIT University (Melbourne, Australia), Indian Institute of Technology (Kanpur), and Indian Institute of Science Education and Research (Mohali). We acknowledge the CSIR Fourth Paradigm Institute (Bangalore) for computational time. We thank Sreetama Pal for help and members of the Chattopadhyay laboratory for their comments.

References

Baier CJ, Fantini J, Barrantes FJ (2011) Disclosure of cholesterol recognition motifs in transmembrane domains of the human nicotinic acetylcholine receptor. Sci Rep 1:69

Berendsen HJC, Postma JPM, van Gunsteren WF, DiNola A, Haak JR (1984) Molecular dynamics with coupling to an external bath. J Chem Phys 81:3684–3690

Burger K, Gimpl G, Fahrenholz F (2000) Regulation of receptor function by cholesterol. Cell Mol Life Sci 57:1577–1592

Bussi G, Donadio D, Parrinello M (2007) Canonical sampling through velocity rescaling. J Chem Phys 126:014101

Cang X, Du Y, Mao Y, Wang Y, Yang H, Jiang H (2013) Mapping the functional binding sites of cholesterol in β_2-adrenergic receptor by long-time molecular dynamics simulations. J Phys Chem B 117:1085–1094

Chakraborty H, Jafurulla M, Clayton AHA, Chattopadhyay A (2018) Exploring oligomeric state of the serotonin$_{1A}$ receptor utilizing photobleaching image correlation spectroscopy: implications for receptor function. Faraday Discuss. (in press 207:409

Chattopadhyay A (2014) GPCRs: lipid-dependent membrane receptors that act as drug targets. Adv Biol 2014:143023

Cherezov V, Rosenbaum DM, Hanson MA, Rasmussen SGF, Thian FS, Kobilka TS, Choi H-J, Kuhn P, Weis WI, Kobilka BK, Stevens RC (2007) High-resolution crystal structure of an engineered human β_2-adrenergic G protein-coupled receptor. Science 318:1258–1265

Fiorino F, Severino B, Magli E, Ciano A, Caliendo G, Santagada V, Frecentese F, Perissutti E (2014) 5-HT$_{1A}$ receptor: an old target as a new attractive tool in drug discovery from central nervous system to cancer. J Med Chem 57:4407–4426

Ganguly S, Clayton AHA, Chattopadhyay A (2011) Organization of higher-order oligomers of the serotonin$_{1A}$ receptor explored utilizing homo-FRET in live cells. Biophys J 100:361–368

Hanson MA, Cherezov V, Griffith MT, Roth CB, Jaakola V-P, Chien EYT, Velasquez J, Kuhn P, Stevens RC

(2008) A specific cholesterol binding site is established by the 2.8 Å structure of the human β_2-adrenergic receptor. Structure 16:897–905
Humphrey W, Dalke A, Schulten K (1996) VMD: visual molecular dynamics. J Mol Graph 14:33–38
Jafurulla M, Chattopadhyay A (2013) Membrane lipids in the function of serotonin and adrenergic receptors. Curr Med Chem 20:47–55
Jafurulla M, Tiwari S, Chattopadhyay A (2011) Identification of cholesterol recognition amino acid consensus (CRAC) motif in G-protein coupled receptors. Biochem Biophys Res Commun 404:569–573
Kalipatnapu S, Chattopadhyay A (2007) Membrane organization and function of the serotonin$_{1A}$ receptor. Cell Mol Neurobiol 27:1097–1116
Lacivita E, Leopoldo M, Berardi F, Perrone R (2008) 5-HT$_{1A}$ receptor, an old target for new therapeutic agents. Curr Top Med Chem 8:1024–1034
Lee JY, Lyman E (2012) Predictions for cholesterol interaction sites on the A$_{2A}$ adenosine receptor. J Am Chem Soc 134:16512–16515
Liu W, Chun E, Thompson AA, Chubukov P, Xu F, Katritch V, Han GW, Roth CB, Heitman LH, IJzerman AP, Cherezov V, Stevens RC (2012) Structural basis for allosteric regulation of GPCRs by sodium ions. Science 337:232–236
Marrink SJ, Risselada HJ, Yefimov S, Tieleman DP, de Vries AH (2007) The MARTINI force field: coarse grained model for biomolecular simulations. J Phys Chem B 111:7812–7824
Monticelli L, Kandasamy SK, Periole X, Larson RG, Tieleman DP, Marrink S-J (2008) The MARTINI coarse-grained force field: extension to proteins. J Chem Theory Comput 4:819–834
Oates J, Watts A (2011) Uncovering the intimate relationship between lipids, cholesterol and GPCR activation. Curr Opin Struct Biol 21:802–807
Paila YD, Chattopadhyay A (2009) The function of G-protein coupled receptors and membrane cholesterol: specific or general interaction? Glycoconj J 26:711–720
Paila YD, Chattopadhyay A (2010) Membrane cholesterol in the function and organization of G-protein coupled receptors. Subcell Biochem 51:439–466
Paila YD, Kombrabail M, Krishnamoorthy G, Chattopadhyay A (2011a) Oligomerization of the serotonin$_{1A}$receptor in live cells: a time-resolved fluorescence anisotropy approach. J Phys Chem B 115:11439–11447
Paila YD, Tiwari S, Sengupta D, Chattopadhyay A (2011b) Molecular modeling of the human serotonin$_{1A}$ receptor: role of membrane cholesterol in ligand binding of the receptor. Mol BioSyst 7:224–234
Patra SM, Chakraborty S, Shahane G, Prasanna X, Sengupta D, Maiti PK, Chattopadhyay A (2015) Differential dynamics of the serotonin$_{1A}$ receptor in membrane bilayers of varying cholesterol content revealed by all atom molecular dynamics simulation. Mol Membr Biol 32:127–137
Pierce KL, Premont RT, Lefkowitz RJ (2002) Seven-transmembrane receptors. Nat Rev Mol Cell Biol 3:639–650
Prasanna X, Chattopadhyay A, Sengupta D (2014) Cholesterol modulates the dimer interface of the β_2-adrenergic receptor via cholesterol occupancy sites. Biophys J 106:1290–1300
Prasanna X, Sengupta D, Chattopadhyay A (2016) Cholesterol-dependent conformational plasticity in GPCR dimers. Sci Rep 6:31858
Pucadyil TJ, Chattopadhyay A (2004) Cholesterol modulates ligand binding and G-protein coupling to serotonin$_{1A}$ receptors from bovine hippocampus. Biochim Biophys Acta 1663:188–200
Pucadyil TJ, Chattopadhyay A (2006) Role of cholesterol in the function and organization of G-protein coupled receptors. Prog Lipid Res 45:295–333
Pucadyil TJ, Chattopadhyay A (2007) Cholesterol depletion induces dynamic confinement of the G-protein coupled serotonin$_{1A}$ receptor in the plasma membrane of living cells. Biochim Biophys Acta 1768:655–668
Pucadyil TJ, Kalipatnapu S, Chattopadhyay A (2005) The serotonin$_{1A}$ receptor: a representative member of the serotonin receptor family. Cell Mol Neurobiol 25:553–580
Rosenbaum DM, Rasmussen SGF, Kobilka BK (2009) The structure and function of G-protein-coupled receptors. Nature 459:356–363
Rouviere E, Arnarez C, Yang L, Lyman E (2017) Identification of two new cholesterol interaction sites on the A$_{2A}$ adenosine receptor. Biophys J 113:2415–2424
Saxena R, Chattopadhyay A (2011) Membrane organization and dynamics of the serotonin$_{1A}$ receptor in live cells. J Neurochem 116:726–733
Saxena R, Chattopadhyay A (2012) Membrane cholesterol stabilizes the human serotonin$_{1A}$ receptor. Biochim Biophys Acta 1818:2936–2942
Sengupta D, Chattopadhyay A (2012) Identification of cholesterol binding sites in the serotonin$_{1A}$ receptor. J Phys Chem B 116:12991–12996
Van Der Spoel D, Lindahl E, Hess B, Groenhof G, Mark AE, Berendsen HJC (2005) GROMACS: fast, flexible, and free. J Comput Chem 26:1701–1718
Van Meer G, de Kroon AI (2011) Lipid map of the mammalian cell. J Cell Sci 124:5–8
Venkatakrishnan AJ, Deupi X, Lebon G, Tate CG, Schertler GF, Babu MM (2013) Molecular signatures of G-protein-coupled receptors. Nature 494:185–194
Wacker D et al (2013) Structural features for functional selectivity at serotonin receptors. Science 340:615–619

Part IV

Cell Surface Macromolecules in Neurobiology

Group I Metabotropic Glutamate Receptors (mGluRs): Ins and Outs

12

Prabhat Kumar Mahato, Namrata Ramsakha, Prachi Ojha, Ravinder Gulia, Rohan Sharma, and Samarjit Bhattacharyya

Abstract

Glutamate is a nonessential amino acid, known to act as a major excitatory neurotransmitter in the central nervous system. Glutamate transduces its signal by activating two types of receptors, viz., ionotropic glutamate receptors and metabotropic glutamate receptors (mGluRs). mGluR1 and mGluR5 are members of the group I mGluR family, and they belong to the G-protein-coupled receptor (GPCR) family. These receptors are involved in various forms of synaptic plasticity including learning and memory. Similar to many other GPCRs, trafficking plays a critical role in controlling the spatiotemporal localization of these receptors on the cell surface, which is critical for the normal ligand/receptor interaction. Improper targeting of GPCRs results in aberrant signaling, which often leads to various diseases. Trafficking also regulates the activity of these receptors. Thus, inappropriate trafficking of these receptors might have pathological consequences. Group I mGluRs have been implicated in various neuropsychiatric disorders like Fragile X syndrome, autism, etc. In this review, we discuss the current understanding of group I mGluR trafficking in the central nervous system and its physiological importance.

Keywords

Endocytosis · Trafficking · Desensitization · Receptor recycling · GPCR · Metabotropic glutamate receptors · Neurotransmitter receptors

12.1 Introduction

Glutamate is a major excitatory neurotransmitter in the central nervous system (CNS), and in addition to the ionotropic glutamate receptors, it also acts through metabotropic glutamate receptors (mGluRs) present at the postsynaptic membrane (Pin and Duvoisin 1995). Metabotropic glutamate receptors (mGluRs) are members of class C G-protein-coupled receptor (GPCR) family (Bhattacharyya 2016; Conn and Pin 1997; Pin and Duvoisin 1995). mGluRs have been subdivided into three classes based on the sequence similarity, the pharmacology, and the second messenger pathways that they are coupled to (Dhami and Ferguson 2006). Among these three groups, group I has two members,

Author contributed equally with all other contributors. Prabhat Kumar Mahato, Namrata Ramsakha, Prachi Ojha, Ravinder Gulia and Rohan Sharma

P. K. Mahato · N. Ramsakha · P. Ojha · R. Gulia
R. Sharma · S. Bhattacharyya (✉)
Department of Biological Sciences, Indian Institute of Science Education and Research (IISER) Mohali, SAS Nagar, Punjab, India
e-mail: samarjit@iisermohali.ac.in

K. Chattopadhyay, S. C. Basu (eds.), *Biochemical and Biophysical Roles of Cell Surface Molecules*, Advances in Experimental Medicine and Biology 1112, https://doi.org/10.1007/978-981-13-3065-0_12

mGluR1 and mGluR5, and they are primarily localized at the postsynaptic sites (Baude et al. 1993; Shigemoto et al. 1993). These receptors are predominantly positively coupled to $G_{\alpha q/11}$, leading to the generation of diacylglycerol (DAG) and inositol 1,4,5-trisphosphate (IP_3), which ultimately activates protein kinase C (PKC) (Conn and Pin 1997; Kim et al. 2008). In addition to neurons, glial cells also express mGluRs, where they play various important roles including glia-neuron communication, neuroprotection, etc. (Winder and Conn 1996; Yao et al. 2005). Group I mGluRs play crucial roles in multiple forms of synaptic plasticity, including learning and memory (Citri and Malenka 2008; Gladding et al. 2009). Furthermore, these receptors have also been implicated in various neuropsychiatric disorders like Fragile X syndrome, schizophrenia, autism, etc. (Bear et al. 2004; Dolen et al. 2007; Ronesi and Huber 2008).

Work done in the last few years has established that many GPCRs are quite mobile and intracellular trafficking of these receptors not only controls the spatiotemporal localization of the receptor; it also plays an important role in the regulation of the activity of these receptors (Drake et al. 2006; Ferguson 2001; Hanyaloglu and von Zastrow 2008). The accurate location of these receptors at specific region of the neuron is a must for the normal signaling of these receptors, and intracellular trafficking plays vital role in controlling this localization. Inaccurate targeting of the receptor might result in improper signaling with pathological consequences. Many GPCRs, subsequent to the activation of the second messenger pathway, get desensitized. Desensitization is a negative feedback mechanism that protects the receptor from chronic overstimulation (Ferguson 2001; Kelly et al. 2008; Krupnick and Benovic 1998). Uncoupling of the receptor from the G-protein involved leads to the desensitization of the receptor. The cellular and molecular mechanisms of the desensitization process of several GPCRs have been studied in detail. Subsequent to desensitization, many GPCRs undergo internalization via various pathways, and this internalization plays an important role in the resensitization/downregulation of those receptors (Drake et al. 2006; Gaborik and Hunyady 2004). Despite this obvious significance, currently the understanding of the protein machineries that control these trafficking events and the physiological significance of these events is very limited. Work done in the last 10–15 years has also taught us that GPCRs are unique and a widely studied GPCR, like β-adrenergic receptor, may not serve as a model system. If one needs to understand a GPCR, that particular receptor has to be studied. Similar to many other GPCRs, activation of group I mGluRs also triggers a variety of cellular responses, viz., receptor desensitization and internalization that might lead to resensitization of the receptor (Bhattacharyya 2016; Mahato et al. 2015; Pandey et al. 2014). In this review, we have focused on the regulation of group I mGluRs in the CNS and its physiological significance.

12.2 Group I mGluR Signaling and Desensitization

Group I mGluRs show differential expression throughout the CNS. For example, mGluR1 is highly expressed in the olfactory bulb, cerebellum, thalamus, CA3 region of the hippocampus, lateral septum, substantia nigra, and globus pallidus (Shigemoto et al. 1992). On the other hand, mGluR5 is present in the CA1 and CA3 region of the hippocampus, cerebral cortex, striatum, nucleus accumbens, lateral septal nucleus, and granule cells of the olfactory bulb (Shigemoto et al. 1993). Suggestions have been made that expression of group I mGluRs varies with the brain development. mGluR5a expression shows a gradual increase in the developing cortex and reaches the peak during the second postnatal week in rodents and then gradually decreases, whereas mGluR5b expression pattern shows increase postnatally, and in adults, this is the most prominent form of mGluR5 (Catania et al. 1994; Minakami et al. 1995; Romano et al. 1996). On the other hand, expression of mGluR1 increases steadily in the hippocampus and neocortex during development (Catania et al. 1994). This differential expression of the two subtypes of group I mGluRs shows a strong correlation with the differential regulation of hippocampal neurons (Mannaioni et al. 2001). Many other studies have also suggested that the distribution of group I mGluRs in the brain regions correlates well with their functions. In the CNS, these receptors are predominantly localized at the peri-synaptic zone of the postsynaptic neurons (Lujan et al. 1996; Luscher and Huber 2010).

Group I mGluRs are positively coupled to the inositol 1,4,5-trisphosphate (IP_3)-diacylglycerol (DAG) pathway through $G_{\alpha q/11}$ and phospholipase C (PLC) (Abdul-Ghani et al. 1996; Dhami and Ferguson 2006). Upon ligand binding, the receptor undergoes a conformational change, which in turn activates the G-protein involved, and, subsequently, the activation of phospholipase C (PLC). PLC cleaves phosphatidylinositol 4, 5-bisphosphate (PIP_2) into inositol 1,4,5-trisphosphate (IP_3) and diacylglycerol (DAG). IP_3 binds to IP_3 receptors at the endoplasmic reticulum (ER) membrane leading to the release of Ca^{2+} from the intracellular stores. DAG and Ca^{2+} together activate protein kinase C (PKC). Subsequently, activated PKC phosphorylates its substrates, which in turn regulates a variety of physiological processes (Gerber et al. 2007; Niswender and Conn 2010; Wang and Zhuo 2012). These receptors could also couple to other types of G-proteins apart from $G_{\alpha q/11}$, in different cell types under different circumstances (Aramori and Nakanishi 1992; Gerber et al. 2007). In hippocampal cells and cultured cortical glial cells, activation of the group I mGluRs activates the MAP kinase pathway, which has been reported to be involved in the mGluR-dependent synaptic plasticity in pyramidal hippocampal neurons (Gallagher et al. 2004; Gerber et al. 2007; Peavy and Conn 1998).

Like many other GPCRs, group I mGluRs get desensitized upon agonist stimulation (Dhami and Ferguson 2006; Francesconi and Duvoisin 2000; Gereau and Heinemann 1998). Prolonged or repeated ligand exposure results in decreased sensitivity of the receptor toward the ligand, and this phenomenon is called "desensitization" of the receptor. Desensitization is believed to be an important physiological feedback mechanism adopted by the cells to prevent themselves from chronic or acute receptor overstimulation. Additionally, GPCR desensitization also filters information from multiple receptor inputs into an integrated and meaningful biological signal (Ferguson 2001; Kelly et al. 2008; Krupnick and Benovic 1998). One of the posttranslational modifications that plays a crucial role in the desensitization of group I mGluRs is phosphorylation. For example, PKC plays a crucial role in the desensitization of group I mGluRs (Francesconi and Duvoisin 2000; Gereau and Heinemann 1998). PKC-dependent phosphorylation of mGluR1a and mGluR1b leads to the desensitization of respective receptors. PKC has also been reported to phosphorylate multiple serine/threonine residues present in the intracellular carboxy-terminal tail of mGluR5 and initiates the desensitization process (Gereau and Heinemann 1998). In contrast to the above results, another protein kinase, viz., PKA, has been reported to inhibit the desensitization of mGluR1. Phosphorylation by PKA results in the uncoupling of adapter proteins, which in turn inhibits the internalization of the receptors (Francesconi and Duvoisin 2000). The G-protein-coupled receptor kinases (GRKs) also play crucial roles in the desensitization of group I mGluRs. For example, GRK4 has been implicated in mGluR1 desensitization in cerebellar Purkinje cells, whereas desensitization of mGluR5 seems to be GRK4-independent (Iacovelli et al. 2003; Sorensen and Conn 2003). The other GRK, viz., GRK2, desensitizes both mGluR1 and mGluR5 in phosphorylation-dependent as well as phosphorylation-independent mechanisms (Dale et al. 2000; Dhami et al. 2002; Ferguson et al. 1996; Ribeiro et al. 2009). The phosphorylation of various residues by GRKs is recognized by arrestin group of proteins, which uncouple the G-proteins from the receptor, leading to the desensitization of the receptor (Lefkowitz and Shenoy 2005). Furthermore, the interaction of group I mGluRs with Huntington binding protein optineurin restrains their coupling with G-proteins to PLC/IP_3 pathway resulting in phosphorylation-independent desensitization of group I mGluRs (Dhami and Ferguson 2006).

The mGluR activity can also be attenuated at the level of G-proteins by the regulators of G-protein signaling (RGS). RGS proteins can cease G-protein signaling following agonist stimulation by acting as GTPase-activating proteins (GAPs). They catalyze the hydrolysis of GTP bound to the G_α subunit of heterotrimeric G-proteins, leading to their inactivation. Among them, RGS2 and RGS4 associate with $G_{\alpha q/11}$ proteins to attenuate group I mGluR-mediated PLC/IP_3 signaling pathway (Dhami and Ferguson 2006).

12.3 Trafficking of Group I mGluRs: Agonist-Dependent and Agonist-Independent

Proper spatiotemporal localization of G-protein-coupled receptors (GPCRs) is necessary for normal interaction of ligand to the receptor and appropriate signaling through the receptor. Trafficking plays a crucial role in controlling the localization of the receptor in a specific region of the cell. It also regulates the activity of the receptor. Improper trafficking of the receptor could result in abnormal signaling which often has serious pathological consequences. GPCRs show remarkable variability in the purpose of endocytosis, the endocytic mechanisms, and the fate of the internalized receptor subsequent to the endocytosis. All these processes depend on the type of the receptor, the ligand, and the system. GPCRs internalize by three principal pathways: (a) arrestin and dynamin-dependent endocytosis; (b) arrestin-independent and dynamin-dependent endocytosis; and (c) arrestin and dynamin-independent endocytosis (Drake et al. 2006). Arrestin and dynamin-dependent endocytic pathway was initially thought to be the model pathway for endocytosis of GPCRs. According to this model, arrestin protein interacts with the receptor and uncouples the receptor from the G-protein involved. Arrestin also acts as a scaffold protein and recruits AP2 adaptor protein and clathrin to form clathrin-coated pits. Subsequently, the Rab GTPase, dynamin interacts with the complex and induces the neck formation of the coated pits, and clathrin-coated vesicles are formed (Claing et al. 2002; Drake et al. 2006; Luttrell and Lefkowitz 2002). Following this, the receptors either recycle back to the cell surface or enter the lysosome for degradation (Drake et al. 2006; Ferguson 2001; Gaborik and Hunyady 2004). It has been reported that some GPCRs internalize via arrestin-independent and dynamin-dependent pathway. For example, overexpression of dominant-negative dynamin inhibited $5\text{-}HT_{2A}$ receptor (serotonin 2A receptor) endocytosis, but overexpression of dominant-negative arrestin did not affect this process (Bhatnagar et al. 2001). Arrestin and dynamin-independent endocytosis has also been reported in case of angiotensin II AT1A receptor (Zhang et al. 1996). Moreover, a particular receptor might get internalized via different mechanisms depending on the phosphorylation status of the receptor and the cellular background. For example, β_1-adrenergic receptor internalizes via clathrin-coated pits upon GRK-mediated phosphorylation, but PKA-dependent phosphorylation of the receptor leads to the caveolin-mediated internalization of the same receptor (Rapacciuolo et al. 2003). Many GPCRs get internalized in a phosphorylation-independent manner as well (Black et al. 2016). Ubiquitination of some GPCRs seems to be critical for the endocytosis of those receptors (Hicke and Riezman 1996; Marchese and Benovic 2001; Shenoy et al. 2001; Terrell et al. 1998). Work done in the last few years has suggested that GPCRs are unique and a widely studied particular receptor may not serve as a model system.

Due to the intense research in the past few decades, our understanding of the biological importance of GPCR endocytosis has expanded rapidly. Initially, the endocytosis of GPCRs was believed to be the primary mechanism for the desensitization of the receptors, since internalization physically separates the receptors from the G-proteins (Ferguson 2001; Sibley and Lefkowitz 1985). However, this hypothesis was challenged by the following observations: (1) for many receptors, the receptor desensitization proceeds more rapidly than the endocytosis of the receptor, and (2) desensitization profile of the β_2-adrenergic receptor remained unaltered even after blocking the endocytosis by various pharmacological and chemical inhibitors (Ferguson 2001; Pippig et al. 1995; Yu et al. 1993). These observations together suggested that for many receptors, internalization is vital for the resensitization and downregulation of the receptor. The idea that internalization is a prerequisite for the receptor resensitization came from several observations: (1) β_2-adrenergic receptors isolated from the light endosomal fractions were less phosphorylated than the receptors isolated from the plasma membrane fraction (Sibley and Lefkowitz 1985), (2) the endosomal fractions were found to be enriched in GPCR-specific phosphatase activity (Pitcher et al. 1995; Sibley and Lefkowitz 1985), (3) resensitization of the GPCRs was blocked upon inhibition of the endocytosis using pharmacological and chemical inhibitors (Garland et al. 1996; Pippig et al. 1995), and (4) internalization-defective mutants were not able to resensitize although their signaling and desensitization properties were intact (Barak et al. 1994). Thus, for many GPCRs, endocytosis is necessary for their resensitization process. For many receptors, internalization is necessary for the downregulation of those receptors (Drake et al. 2006; Gaborik and Hunyady 2004).

Group I mGluRs also undergo ligand-mediated internalization following the desensitization step (Bhattacharyya 2016; Gulia et al. 2017; Mahato et al. 2015; Pandey et al. 2014). Both members of group I mGluRs, viz., mGluR1 and mGluR5, show rapid internalization upon ligand exposure. The ligand-mediated internalization of these receptors is arrestin and dynamin-dependent (Mundell et al. 2001). Moreover, the internalization of these receptors was observed to be dependent on the ubiquitination of the receptor (Gulia et al. 2017). K63-linked polyubiquitination seems to be involved in the internalization of these receptors. Importantly, the lysine residue at the 1112 position of the C-terminal tail of mGluR1 has been reported to be critical for the endocytosis of the receptor. The E3 ubiquitin ligase, Siah-1A is involved in this process. Acute knockdown of this ligase resulted in complete inhibition of the ligand-mediated internalization of mGluR1. Interestingly, the cells in which Siah-1A was knocked down showed enhanced mGluR-mediated AMPAR endocytosis, which is the cellular correlate for mGluR-LTD. Understanding the mechanisms of mGluR-mediated AMPAR trafficking is currently a major attractive area of study because mGluR-LTD has been reported to be altered in various neuropsychiatric disorders, such as autism and Fragile X syndrome. Both second messenger-dependent protein kinases and GRK-mediated phosphorylation have been implicated in the endocytosis of group I mGluRs. The internalization of group I mGluRs has been reported to be PKC-dependent. In case of mGluR1, the region that is responsible for the PKC-dependent internalization is different from the region that is involved in the GRK-mediated internalization of the receptor (Mundell et al. 2003). In the basal state, mGluR5 is associated with calmodulin at the cell surface. Ligand-dependent activation of mGluR5 triggers the rise in the intracellular Ca^{2+} level and activation of PKC. Subsequently, PKC phosphorylates S901 residue at the C-terminus of the receptor, which in turn results in the inhibition of the binding of calmodulin with the receptor and subsequent internalization of the receptor (Lee et al. 2008). Another second messenger-dependent kinase PKA may also modulate the mGluR trafficking by controlling the GRK activity (Mundell et al. 2004). As stated earlier, GRK-mediated phosphorylation of the receptor also plays an important role in the trafficking of group I mGluRs. Acute knockdown of GRK4 in cerebellar Purkinje cells inhibited the ligand-mediated internalization of mGluR1 (Iacovelli et al. 2003; Sallese et al. 2000). GRK4 modulates the trafficking of mGluR1 in a kinase-dependent manner, whereas another GRK, viz., GRK2, acts in a kinase-independent way in this process (Iacovelli et al. 2003). Group I mGluRs undergo agonist-independent (constitutive) internalization as well (Dale et al. 2001; Pula et al. 2004; Trivedi and Bhattacharyya 2012). Conflicting observations exist regarding the mechanism of constitutive endocytosis of group I mGluRs. Some studies have shown that the ligand-independent endocytosis of mGluR1a is β-arrestin and dynamin-independent, whereas few other studies have suggested that it is both β-arrestin and clathrin-dependent (Dale et al. 2001; Francesconi et al. 2009; Pula et al. 2004). Constitutive internalization of group I mGluRs occurs via caveolar/raft pathway (Francesconi et al. 2009). The constitutive endocytosis of mGluR1a appears to be independent of the GRK-mediated phosphorylation of the receptor (Dale et al. 2001). Small GTP-binding protein Ral and phospholipase D2 (PLD2)-dependent pathway has also been shown as an alternative mechanism to regulate constitutive mGluR endocytosis (Bhattacharya et al. 2004). Application of inverse agonists has been reported to inhibit constitutive internalization of mGluR1a, but not mGluR5a (Fourgeaud et al. 2003; Pula et al. 2004). It is possible that mGluR1 inverse agonist is able to promote an inactive conformation of the receptor which is no longer capable of interacting with the components of endocytic machinery, but mGluR5 inverse agonist fails to do so. Thus, looks like, distinct mechanisms regulate constitutive internalization of mGluR1 and mGluR5.

As stated before, subsequent to internalization GPCRs can have various subcellular fates. Some GPCRs recycle back to the cell surface subsequent to internalization, whereas others may enter lysosome for degradation. The recycling seems to be a mechanism for resensitization for many receptors, whereas the lysosomal degradation leads to the downregulation of the receptor. The route taken by the receptor following endocytosis is dictated by the type of the receptor, the type of the ligand, and the type of the cellular background. Majority of the internalized mGluR1 and mGluR5 enters the recycling compartment and recycle back to the cell surface following similar kinetics (Mahato et al. 2015; Pandey et al. 2014). The recycling of mGluR1/mGluR5 subsequent to the ligand-mediated internalization suggests that the internalization might be necessary for the resensitization of these receptors. The process of recycling also favors the cells bio-energetically compared to the de novo synthesis of the receptors, which is a high-energy consuming process. Interestingly, the recycling of both mGluR1 and mGluR5 depends on the pH of the endosomes, and dissipation of the pH of the endosomal compartments results in the inhibition in the recycling of both the receptors (Mahato et al. 2015; Pandey et al. 2014). pH-dependent recycling of group I mGluRs could be because of either of the two possibilities or both: (1) the internalized receptor attains such a confirmation under the acidic pH of the endosomal compartments that exposes the previously modified residue(s) during the desensitization process, and the enzymes present in the endosomes then remodify those residue(s) of the receptor, resulting in the resensitization of the receptor, and (2) activity of the modifying enzymes is dependent on the specific pH of the endosomal compartments, and any change in the pH results in the inactivation of the enzymes. The recycling of mGluR1 has been reported to be dependent on the activity of protein phosphatase 2A (PP2A) (Pandey et al. 2014). On the other hand, the recycling of mGluR5 completely depends on the activity of PP2A and partially depends on the activity of protein phosphatase 2B (PP2B) (Mahato et al. 2015) (Fig. 12.1). It will be important to investigate the substrates for PP2A and PP2B in future. It is possible that PP2A and PP2B dephosphorylate the phosphorylated residues present at the third intracellular loop and carboxy-terminal tail of these receptors. Alternatively, they might also dephosphorylate some other substrate(s) involved in the recycling of mGluR1/mGluR5. It has been observed that subsequent to constitutive internalization also, mGluR5 enters the recycling compartment, similar to the ligand-dependent internalization and recycles back to the cell surface (Trivedi and Bhattacharyya 2012). The mechanism of the recycling of constitutively internalized receptors is currently not understood.

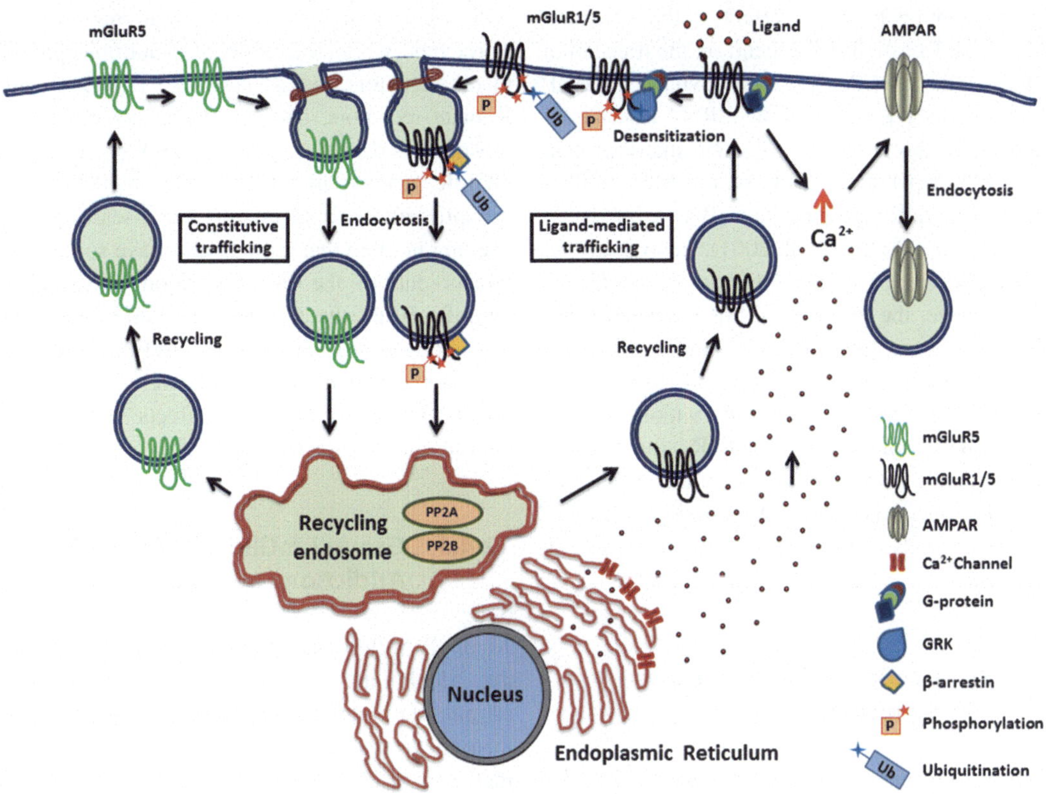

Fig. 12.1 Agonist-dependent and agonist-independent trafficking of group I mGluRs. Upon binding with the ligand, group I mGluRs internalize and recycle back to the cell surface. Internalization of these receptors depends on the phosphorylation of some critical residues as well as on the ubiquitination of the receptor. The recycling of mGluR1 and mGluR5 is PP2A-dependent, whereas recycling of mGluR5 is partially dependent on the activity of PP2B. Activation of group I mGluRs induces internalization of AMPARs, which is the cellular mechanism for mGluR-LTD. mGluR5 also undergoes ligand-independent (constitutive) internalization and subsequently recycles back to the cell surface

12.4 Role of Group I mGluRs in Few Diseases

12.4.1 Group I mGluRs in Fragile X Syndrome and Autism

Autism is a neurodevelopmental disorder characterized by repetitive behavior and withdrawal of the patients from the society. Patients are mostly preoccupied with themselves, and it is more common in males (Harris et al. 2008). One of the genetic causes of autism is the Fragile X syndrome (Brown et al. 1982a, b). Patients with Fragile X syndrome show macrocephaly and macroorchidism and have prominent jaws, ears, and forehead (Penagarikano et al. 2007). Fragile X syndrome is the most common inherited form of mental retardation. The cause of Fragile X syndrome is the lack of the protein product of the fragile X mental retardation 1 (*Fmr1*) gene, FMRP (Bassell and Warren 2008; Bear et al. 2004; Ronesi and Huber 2008). In recent times, a "metabotropic glutamate receptor" theory of Fragile X pathogenesis has got prominence. This theory was based on few observations: (a) activation of group I mGluRs results in the stimulation of proteins present at the synapse;

(b) many of the long-term effects of the activation of mGluR1 or mGluR5 depend on the translation of synaptic mRNA; (c) at the synapse, FMRP functions as a repressor of the mRNA translation; and (d) in the absence of FMRP, many protein synthesis-dependent effects of the activation of group I mGluRs are amplified (Bear et al. 2004; Dolen et al. 2007; Li et al. 2001). Importantly, in the hippocampus of the *Fmr1* knockout (*Fmr1* KO) mouse, the group I mGluR-mediated long-term depression (mGluR-LTD), a form of synaptic plasticity, is enhanced (Hou et al. 2006; Huber et al. 2002). Together, these observations suggest that FMRP and group I mGluRs functionally work against each other. Thus, in the absence of FMRP, unchecked mGluR-dependent protein synthesis leads to the pathogenesis of Fragile X syndrome.

12.4.2 Group I mGluRs in Down Syndrome

Down syndrome is a neurodegenerative disorder that is caused by the trisomy of chromosome 21. It is one of the common congenital disorders that cause intellectual disability. Adult patients with Down syndrome manifest senile plaques and neurofibrillary tangles that are hallmark of Alzheimer's disease (Oka and Takashima 1997; Wisniewski et al. 1985). The amyloid precursor protein (APP) is also located at chromosome 21. APP overexpression is thought to be one of the causative agents that lead to the pathogenesis of Down syndrome (Goldgaber et al. 1987; Neve et al. 1988). It has been observed that in Down syndrome patients, mGluR5 is overexpressed in various brain regions like hippocampus, cerebral cortex, etc. (Oka and Takashima 1999). Generation of DAG and activation of protein kinase C (PKC) accelerate the breakdown of APP and thus formation of soluble and non-amyloidogenic peptides (APPs) (Buxbaum et al. 1992; Caporaso et al. 1992; Nitsch et al. 1992; Slack et al. 1993). As stated before, activation of group I mGluRs leads to the formation of DAG and activation of PKC. Thus, activation of group I mGluRs by specific agonists increases the soluble form of APP (APPs) in both hippocampal region and astrocytes, while activation of ionotropic glutamate receptors, viz., NMDA, AMPA, and Kainate receptors, does not have any effect on this process (Lee et al. 1995). Therefore, selective mGluR agonists might be useful in facilitating synaptic efficacy and treating the disease, while keeping in mind that it might also have some side effects due to the over-activation of group I mGluRs. Thus, the dose and duration of the agonist application need to be precisely determined so that the patient suffers relatively lesser side effects, while the beneficial effects in terms of treating the disease are maximum.

12.4.3 Group I mGluRs in Alcohol Addiction

Alcoholism is a chronic neuropsychiatric disorder. Alcohol acts as a noncompetitive antagonist at multiple glutamate receptors and alters glutamate neurotransmission in the limbic system, including the region nucleus accumbens (Lovinger 1996; Minami et al. 1998). Nucleus accumbens regulates the motivation to drink and other properties of alcoholism. Among other neurotransmitter receptors, group I mGluRs also play critical roles in the alcohol behavior (Schumann et al. 2008). Blocking of mGluR5 in nucleus accumbens and basolateral amygdala has been shown to increase the sensitivity to acute alcohol administration in rodents (Besheer et al. 2010; Cozzoli et al. 2009, 2012; Sinclair et al. 2012). Furthermore, multiple studies have shown that the deletion or blocking of mGluR5 attenuates alcohol drinking, withdrawal, and cue-induced alcohol-seeking (Adams et al. 2008; Bird et al. 2008; Blednov and Harris 2008; Sidhpura et al. 2010). The variation in the mGluR5 gene also shows an association with alcoholism. A population of mGluR5 expressed on dopamine D1 receptor neurons also seems to play a critical role in the deprivation-induced alcohol drinking behavior (Parkitna et al. 2013). The Homer2 member of the Homer family, present at the postsynaptic density, has emerged as a key player in alcohol-induced neuroplasticity in vivo (Szumlinski et al.

2008). Through an Ena/VASP1 Homology (EVH1) domain, Homer2 interacts with group I mGluRs (Tu et al. 1998). Deletion of *Homer2* in mice or *Drosophila* results in an alcohol-avoiding behavior (Szumlinski et al. 2005; Urizar et al. 2007). Interaction of Homer2 to mGluR5 and its downstream signaling partner PI3K has been implicated in excess binge drinking. Disruption of mGluR5 binding with Homer2 results in 50% reduction in the binge alcohol consumption (Cozzoli et al. 2009). Knockdown of Homer2 at the nucleus accumbens region reduces the alcohol consumption. Together, these data point toward an important role for mGluR5-Homer2-PI3K signaling in the nucleus accumbens region that regulates binge-like alcohol consumption. Chronic ethanol exposure has been found to increase the expression of the other member of group I mGluR family, mGluR1, in the nucleus accumbens and central amygdala as well (Obara et al. 2009).

In all the above diseases, as well as many other diseases that we have not discussed in this review, group I mGluR signaling is altered. As stated before, the normal signaling of group I mGluRs depends on the precise localization as well as on the accurate regulation of the activity of the receptor, both of which are controlled by the trafficking of these receptors. Therefore, regulation of mGluR trafficking would provide a powerful means to modulate the long-term synaptic plasticity and other synaptic functions in which mGluRs are involved, in both normal and diseased brain.

12.5 Final Words

Group I mGluRs are major players in regulating various physiological processes in response to external stimuli in the central nervous system. A large number of reports are testaments, there is considerable diversity observed in the group I mGluR-specific signaling, desensitization, internalization, and resensitization processes. In fact, within the same group, the two members, mGluR1 and mGluR5, show diversity in their regulation in the brain. Different subsets of interacting regulatory proteins and the effector molecules might add further complexity in the signaling and regulation of group I mGluRs. Although, we have gained much information about the group I mGluR signaling and regulation in the last few years, it appears that we have just begun to understand the biological complexity that exists in the group I mGluR signaling and regulation. The advent of new technologies and continued advancement in microscopy and in vivo experiments will help us to advance our understanding of the role of group I mGluRs in normal and diseased brain. Thus, continued understanding of the group I mGluR signaling, regulation, and trafficking will provide us important insights to unravel the complexity that exists in nature and also to develop novel therapeutic strategies, in order to cure various diseases that arise due to aberrant group I mGluR signaling and regulation.

References

Abdul-Ghani MA, Valiante TA, Carlen PL, Pennefather PS (1996) Metabotropic glutamate receptors coupled to IP3 production mediate inhibition of IAHP in rat dentate granule neurons. J Neurophysiol 76:2691–2700

Adams CL, Cowen MS, Short JL, Lawrence AJ (2008) Combined antagonism of glutamate mGlu5 and adenosine A2A receptors interact to regulate alcohol-seeking in rats. Int J Neuropsychopharmacol 11:229–241. https://doi.org/10.1017/S1461145707007845

Aramori I, Nakanishi S (1992) Signal transduction and pharmacological characteristics of a metabotropic glutamate receptor, mGluR1, in transfected CHO cells. Neuron 8:757–765

Barak LS, Tiberi M, Freedman NJ, Kwatra MM, Lefkowitz RJ, Caron MG (1994) A highly conserved tyrosine residue in G protein-coupled receptors is required for agonist-mediated beta 2-adrenergic receptor sequestration. J Biol Chem 269:2790–2795

Bassell GJ, Warren ST (2008) Fragile X syndrome: loss of local mRNA regulation alters synaptic development and function. Neuron 60:201–214. https://doi.org/10.1016/j.neuron.2008.10.004

Baude A, Nusser Z, Roberts JD, Mulvihill E, McIlhinney RA, Somogyi P (1993) The metabotropic glutamate receptor (mGluR1 alpha) is concentrated at perisynaptic membrane of neuronal subpopulations as detected by immunogold reaction. Neuron 11:771–787

Bear MF, Huber KM, Warren ST (2004) The mGluR theory of fragile X mental retardation. Trends Neurosci

27:370–377. https://doi.org/10.1016/j.tins.2004.04.009

Besheer J, Grondin JJ, Cannady R, Sharko AC, Faccidomo S, Hodge CW (2010) Metabotropic glutamate receptor 5 activity in the nucleus accumbens is required for the maintenance of ethanol self-administration in a rat genetic model of high alcohol intake. Biol Psychiatry 67:812–822. https://doi.org/10.1016/j.biopsych.2009.09.016

Bhatnagar A, Willins DL, Gray JA, Woods J, Benovic JL, Roth BL (2001) The dynamin-dependent, arrestin-independent internalization of 5-hydroxytryptamine 2A (5-HT2A) serotonin receptors reveals differential sorting of arrestins and 5-HT2A receptors during endocytosis. J Biol Chem 276:8269–8277. https://doi.org/10.1074/jbc.M006968200

Bhattacharya M, Babwah AV, Godin C, Anborgh PH, Dale LB, Poulter MO, Ferguson SS (2004) Ral and phospholipase D2-dependent pathway for constitutive metabotropic glutamate receptor endocytosis. J Neurosci: Off J Soc Neurosci 24:8752–8761. https://doi.org/10.1523/JNEUROSCI.3155-04.2004

Bhattacharyya S (2016) Inside story of Group I Metabotropic Glutamate Receptors (mGluRs). Int J Biochem Cell Biol 77:205–212. https://doi.org/10.1016/j.biocel.2016.03.003

Bird MK, Kirchhoff J, Djouma E, Lawrence AJ (2008) Metabotropic glutamate 5 receptors regulate sensitivity to ethanol in mice. Int J Neuropsychopharmacol 11:765–774. https://doi.org/10.1017/S1461145708008572

Black JB, Premont RT, Daaka Y (2016) Feedback regulation of G protein-coupled receptor signaling by GRKs and arrestins. Semin Cell Dev Biol 50:95–104. https://doi.org/10.1016/j.semcdb.2015.12.015

Blednov YA, Harris RA (2008) Metabotropic glutamate receptor 5 (mGluR5) regulation of ethanol sedation, dependence and consumption: relationship to acamprosate actions. Int J Neuropsychopharmacol 11:775–793. https://doi.org/10.1017/S1461145708008584

Brown WT, Friedman E, Jenkins EC, Brooks J, Wisniewski K, Raguthu S, French JH (1982a) Association of fragile X syndrome with autism. Lancet 1:100

Brown WT, Jenkins EC, Friedman E, Brooks J, Wisniewski K, Raguthu S, French J (1982b) Autism is associated with the fragile-X syndrome. J Autism Dev Disord 12:303–308

Buxbaum JD, Oishi M, Chen HI, Pinkas-Kramarski R, Jaffe EA, Gandy SE, Greengard P (1992) Cholinergic agonists and interleukin 1 regulate processing and secretion of the Alzheimer beta/A4 amyloid protein precursor. Proc Natl Acad Sci U S A 89:10075–10078

Caporaso GL, Gandy SE, Buxbaum JD, Ramabhadran TV, Greengard P (1992) Protein phosphorylation regulates secretion of Alzheimer beta/A4 amyloid precursor protein. Proc Natl Acad Sci U S A 89:3055–3059

Catania MV, Landwehrmeyer GB, Testa CM, Standaert DG, Penney JB Jr, Young AB (1994) Metabotropic glutamate receptors are differentially regulated during development. Neuroscience 61:481–495

Citri A, Malenka RC (2008) Synaptic plasticity: multiple forms, functions, and mechanisms. Neuropsychopharmacol: Off Publ Am Coll Neuropsychopharmacol 33:18–41. https://doi.org/10.1038/sj.npp.1301559

Claing A, Laporte SA, Caron MG, Lefkowitz RJ (2002) Endocytosis of G protein-coupled receptors: roles of G protein-coupled receptor kinases and beta-arrestin proteins. Progress in neurobiology 66:61–79

Conn PJ, Pin JP (1997) Pharmacology and functions of metabotropic glutamate receptors. Annu Rev Pharmacol Toxicol 37:205–237

Cozzoli DK et al (2012) Nucleus accumbens mGluR5-associated signaling regulates binge alcohol drinking under drinking-in-the-dark procedures. Alcohol Clin Exp Res 36:1623–1633. https://doi.org/10.1111/j.1530-0277.2012.01776.x

Cozzoli DK et al (2009) Binge drinking upregulates accumbens mGluR5-Homer2-PI3K signaling: functional implications for alcoholism. J Neurosci: Off J Soc Neurosci 29:8655–8668. https://doi.org/10.1523/JNEUROSCI.5900-08.2009

Dale LB, Bhattacharya M, Anborgh PH, Murdoch B, Bhatia M, Nakanishi S, Ferguson SS (2000) G protein-coupled receptor kinase-mediated desensitization of metabotropic glutamate receptor 1A protects against cell death. J Biol Chem 275:38213–38220

Dale LB, Bhattacharya M, Seachrist JL, Anborgh PH, Ferguson SS (2001) Agonist-stimulated and tonic internalization of metabotropic glutamate receptor 1a in human embryonic kidney 293 cells: agonist-stimulated endocytosis is beta-arrestin1 isoform-specific. Mol Pharmacol 60:1243–1253

Dhami GK, Anborgh PH, Dale LB, Sterne-Marr R, Ferguson SS (2002) Phosphorylation-independent regulation of metabotropic glutamate receptor signaling by G protein-coupled receptor kinase 2. J Biol Chem 277:25266–25272. https://doi.org/10.1074/jbc.M203593200

Dhami GK, Ferguson SS (2006) Regulation of metabotropic glutamate receptor signaling, desensitization and endocytosis. Pharmacol Ther 111:260–271

Dolen G, Osterweil E, Rao BS, Smith GB, Auerbach BD, Chattarji S, Bear MF (2007) Correction of fragile X syndrome in mice. Neuron 56:955–962

Drake MT, Shenoy SK, Lefkowitz RJ (2006) Trafficking of G protein-coupled receptors. Circ Res 99:570–582

Ferguson SS (2001) Evolving concepts in G protein-coupled receptor endocytosis: the role in receptor desensitization and signaling. Pharmacol Rev 53:1–24

Ferguson SS, Barak LS, Zhang J, Caron MG (1996) G-protein-coupled receptor regulation: role of G-protein-coupled receptor kinases and arrestins. Can J Physiol Pharmacol 74:1095–1110

Fourgeaud L, Bessis AS, Rossignol F, Pin JP, Olivo-Marin JC, Hemar A (2003) The metabotropic glutamate receptor mGluR5 is endocytosed by a clathrin-independent pathway. J Biol Chem 278:12222–12230

Francesconi A, Duvoisin RM (2000) Opposing effects of protein kinase C and protein kinase A on metabotropic glutamate receptor signaling: selective desensitization of the inositol trisphosphate/Ca2+ pathway by phosphorylation of the receptor-G protein-coupling domain. Proc Natl Acad Sci U S A 97:6185–6190

Francesconi A, Kumari R, Zukin RS (2009) Regulation of group I metabotropic glutamate receptor trafficking and signaling by the caveolar/lipid raft pathway. J Neurosci: Off J Soc Neurosci 29:3590–3602

Gaborik Z, Hunyady L (2004) Intracellular trafficking of hormone receptors. Trends Endocrinol Metabol: TEM 15:286–293. https://doi.org/10.1016/j.tem.2004.06.009

Gallagher SM, Daly CA, Bear MF, Huber KM (2004) Extracellular signal-regulated protein kinase activation is required for metabotropic glutamate receptor-dependent long-term depression in hippocampal area CA1. J Neurosci: Off J Soc Neurosci 24:4859–4864

Garland AM, Grady EF, Lovett M, Vigna SR, Frucht MM, Krause JE, Bunnett NW (1996) Mechanisms of desensitization and resensitization of G protein-coupled neurokinin1 and neurokinin2 receptors. Mol Pharmacol 49:438–446

Gerber U, Gee CE, Benquet P (2007) Metabotropic glutamate receptors: intracellular signaling pathways. Curr Opin Pharmacol 7:56–61. https://doi.org/10.1016/j.coph.2006.08.008

Gereau RW, Heinemann SF (1998) Role of protein kinase C phosphorylation in rapid desensitization of metabotropic glutamate receptor 5. Neuron 20:143–151

Gladding CM, Fitzjohn SM, Molnar E (2009) Metabotropic glutamate receptor-mediated long-term depression: molecular mechanisms. Pharmacol Rev 61:395–412

Goldgaber D, Lerman MI, McBride OW, Saffiotti U, Gajdusek DC (1987) Characterization and chromosomal localization of a cDNA encoding brain amyloid of Alzheimer's disease. Science 235:877–880

Gulia R, Sharma R, Bhattacharyya S (2017) A critical role for ubiquitination in the endocytosis of glutamate receptors. J Biol Chem 292:1426–1437. https://doi.org/10.1074/jbc.M116.752105

Hanyaloglu AC, von Zastrow M (2008) Regulation of GPCRs by endocytic membrane trafficking and its potential implications. Annu Rev Pharmacol Toxicol 48:537–568. https://doi.org/10.1146/annurev.pharmtox.48.113006.094830

Harris SW et al (2008) Autism profiles of males with fragile X syndrome. Am J Ment Retard: AJMR 113:427–438. https://doi.org/10.1352/2008.113:427-438

Hicke L, Riezman H (1996) Ubiquitination of a yeast plasma membrane receptor signals its ligand-stimulated endocytosis. Cell 84:277–287

Hou L, Antion MD, Hu D, Spencer CM, Paylor R, Klann E (2006) Dynamic translational and proteasomal regulation of fragile X mental retardation protein controls mGluR-dependent long-term depression. Neuron 51:441–454

Huber KM, Gallagher SM, Warren ST, Bear MF (2002) Altered synaptic plasticity in a mouse model of fragile X mental retardation. Proc Natl Acad Sci U S A 99:7746–7750

Iacovelli L et al (2003) Role of G protein-coupled receptor kinase 4 and beta-arrestin 1 in agonist-stimulated metabotropic glutamate receptor 1 internalization and activation of mitogen-activated protein kinases. J Biol Chem 278:12433–12442. https://doi.org/10.1074/jbc.M203992200

Kelly E, Bailey CP, Henderson G (2008) Agonist-selective mechanisms of GPCR desensitization. Br J Pharmacol 153(Suppl 1):S379–S388

Kim CH, Lee J, Lee JY, Roche KW (2008) Metabotropic glutamate receptors: phosphorylation and receptor signaling. J Neurosci Res 86:1–10

Krupnick JG, Benovic JL (1998) The role of receptor kinases and arrestins in G protein-coupled receptor regulation. Annu Rev Pharmacol Toxicol 38:289–319

Lee JH et al (2008) Calmodulin dynamically regulates the trafficking of the metabotropic glutamate receptor mGluR5. Proc Natl Acad Sci U S A 105:12575–12580

Lee RK, Wurtman RJ, Cox AJ, Nitsch RM (1995) Amyloid precursor protein processing is stimulated by metabotropic glutamate receptors. Proc Natl Acad Sci U S A 92:8083–8087

Lefkowitz RJ, Shenoy SK (2005) Transduction of receptor signals by beta-arrestins. Science 308:512–517. https://doi.org/10.1126/science.1109237

Li Z, Zhang Y, Ku L, Wilkinson KD, Warren ST, Feng Y (2001) The fragile X mental retardation protein inhibits translation via interacting with mRNA. Nucleic Acids Res 29:2276–2283

Lovinger DM (1996) Interactions between ethanol and agents that act on the NMDA-type glutamate receptor. Alcohol Clin Exp Res 20:187A–191A

Lujan R, Nusser Z, Roberts JD, Shigemoto R, Somogyi P (1996) Perisynaptic location of metabotropic glutamate receptors mGluR1 and mGluR5 on dendrites and dendritic spines in the rat hippocampus. Eur J Neurosci 8:1488–1500

Luscher C, Huber KM (2010) Group 1 mGluR-dependent synaptic long-term depression: mechanisms and implications for circuitry and disease. Neuron 65:445–459. https://doi.org/10.1016/j.neuron.2010.01.016

Luttrell LM, Lefkowitz RJ (2002) The role of beta-arrestins in the termination and transduction of G-protein-coupled receptor signals. J Cell Sci 115:455–465

Mahato PK, Pandey S, Bhattacharyya S (2015) Differential effects of protein phosphatases in the recycling of metabotropic glutamate receptor 5. Neuroscience 306:138–150

Mannaioni G, Marino MJ, Valenti O, Traynelis SF, Conn PJ (2001) Metabotropic glutamate receptors 1 and 5 differentially regulate CA1 pyramidal cell function. J Neurosci: Off J Soc Neurosci 21:5925–5934

Marchese A, Benovic JL (2001) Agonist-promoted ubiquitination of the G protein-coupled receptor CXCR4 mediates lysosomal sorting. J Biol Chem 276:45509–45512. https://doi.org/10.1074/jbc.C100527200

Minakami R, Iida K, Hirakawa N, Sugiyama H (1995) The expression of two splice variants of metabotropic glutamate receptor subtype 5 in the rat brain and neuronal cells during development. J Neurochem 65:1536–1542

Minami K, Gereau RW, Minami M, Heinemann SF, Harris RA (1998) Effects of ethanol and anesthetics on type 1 and 5 metabotropic glutamate receptors expressed in Xenopus laevis oocytes. Mol Pharmacol 53:148–156

Mundell SJ, Matharu AL, Pula G, Roberts PJ, Kelly E (2001) Agonist-induced internalization of the metabotropic glutamate receptor 1a is arrestin- and dynamin-dependent. J Neurochem 78:546–551

Mundell SJ, Pula G, Carswell K, Roberts PJ, Kelly E (2003) Agonist-induced internalization of metabotropic glutamate receptor 1A: structural determinants for protein kinase C- and G protein-coupled receptor kinase-mediated internalization. J Neurochem 84:294–304

Mundell SJ, Pula G, More JC, Jane DE, Roberts PJ, Kelly E (2004) Activation of cyclic AMP-dependent protein kinase inhibits the desensitization and internalization of metabotropic glutamate receptors 1a and 1b. Mol Pharmacol 65:1507–1516. https://doi.org/10.1124/mol.65.6.1507

Neve RL, Finch EA, Dawes LR (1988) Expression of the Alzheimer amyloid precursor gene transcripts in the human brain. Neuron 1:669–677

Niswender CM, Conn PJ (2010) Metabotropic glutamate receptors: physiology, pharmacology, and disease. Annu Rev Pharmacol Toxicol 50:295–322

Nitsch RM, Slack BE, Wurtman RJ, Growdon JH (1992) Release of Alzheimer amyloid precursor derivatives stimulated by activation of muscarinic acetylcholine receptors. Science 258:304–307

Obara I et al (2009) Differential effects of chronic ethanol consumption and withdrawal on homer/glutamate receptor expression in subregions of the accumbens and amygdala of P rats. Alcohol Clin Exp Res 33:1924–1934. https://doi.org/10.1111/j.1530-0277.2009.01030.x

Oka A, Takashima S (1997) Induction of cyclo-oxygenase 2 in brains of patients with Down's syndrome and dementia of Alzheimer type: specific localization in affected neurones and axons. Neuroreport 8:1161–1164

Oka A, Takashima S (1999) The up-regulation of metabotropic glutamate receptor 5 (mGluR5) in Down's syndrome brains. Acta Neuropathol 97:275–278

Pandey S, Mahato PK, Bhattacharyya S (2014) Metabotropic glutamate receptor 1 recycles to the cell surface in protein phosphatase 2A-dependent manner in non-neuronal and neuronal cell lines. J Neurochem 131:602–614. https://doi.org/10.1111/jnc.12930

Parkitna JR et al (2013) Novelty-seeking behaviors and the escalation of alcohol drinking after abstinence in mice are controlled by metabotropic glutamate receptor 5 on neurons expressing dopamine d1 receptors. Biol Psychiatry 73:263–270. https://doi.org/10.1016/j.biopsych.2012.07.019

Peavy RD, Conn PJ (1998) Phosphorylation of mitogen-activated protein kinase in cultured rat cortical glia by stimulation of metabotropic glutamate receptors. Journal of neurochemistry 71:603–612

Penagarikano O, Mulle JG, Warren ST (2007) The pathophysiology of fragile x syndrome. Annu Rev Genomics Hum Genet 8:109–129. https://doi.org/10.1146/annurev.genom.8.080706.092249

Pin JP, Duvoisin R (1995) The metabotropic glutamate receptors: structure and functions. Neuropharmacology 34:1–26

Pippig S, Andexinger S, Lohse MJ (1995) Sequestration and recycling of beta 2-adrenergic receptors permit receptor resensitization. Mol Pharmacol 47:666–676

Pitcher JA, Payne ES, Csortos C, DePaoli-Roach AA, Lefkowitz RJ (1995) The G-protein-coupled receptor phosphatase: a protein phosphatase type 2A with a distinct subcellular distribution and substrate specificity. Proc Natl Acad Sci U S A 92:8343–8347

Pula G, Mundell SJ, Roberts PJ, Kelly E (2004) Agonist-independent internalization of metabotropic glutamate receptor 1a is arrestin- and clathrin-dependent and is suppressed by receptor inverse agonists. J Neurochem 89:1009–1020. https://doi.org/10.1111/j.1471-4159.2004.02387.x

Rapacciuolo A, Suvarna S, Barki-Harrington L, Luttrell LM, Cong M, Lefkowitz RJ, Rockman HA (2003) Protein kinase A and G protein-coupled receptor kinase phosphorylation mediates beta-1 adrenergic receptor endocytosis through different pathways. J Biol Chem 278:35403–35411. https://doi.org/10.1074/jbc.M305675200

Ribeiro FM, Ferreira LT, Paquet M, Cregan T, Ding Q, Gros R, Ferguson SS (2009) Phosphorylation-independent regulation of metabotropic glutamate receptor 5 desensitization and internalization by G protein-coupled receptor kinase 2 in neurons. J Biol Chem 284:23444–23453. https://doi.org/10.1074/jbc.M109.000778

Romano C, Van den Pol AN, OMalley KL (1996) Enhanced early developmental expression of the metabotropic glutamate receptor mGluR5 in rat brain: protein, mRNA splice variants, and regional distribution. J Comp Neurol 367:403–412

Ronesi JA, Huber KM (2008) Metabotropic glutamate receptors and fragile x mental retardation protein: partners in translational regulation at the synapse. Sci Signal 1:pe6

Sallese M et al (2000) The G-protein-coupled receptor kinase GRK4 mediates homologous desensitization of metabotropic glutamate receptor 1. Faseb J 14:2569–2580

Schumann G et al (2008) Systematic analysis of glutamatergic neurotransmission genes in alcohol dependence and adolescent risky drinking behavior. Arch Gen Psychiatry 65:826–838. https://doi.org/10.1001/archpsyc.65.7.826

Shenoy SK, McDonald PH, Kohout TA, Lefkowitz RJ (2001) Regulation of receptor fate by ubiquitination of activated beta 2-adrenergic receptor and beta-arrestin. Science 294:1307–1313. https://doi.org/10.1126/science.1063866

Shigemoto R, Nakanishi S, Mizuno N (1992) Distribution of the mRNA for a metabotropic glutamate receptor (mGluR1) in the central nervous system: an in situ hybridization study in adult and developing rat. J Comp Neurol 322:121–135

Shigemoto R, Nomura S, Ohishi H, Sugihara H, Nakanishi S, Mizuno N (1993) Immunohistochemical localization of a metabotropic glutamate receptor, mGluR5, in the rat brain. Neurosci Lett 163:53–57

Sibley DR, Lefkowitz RJ (1985) Molecular mechanisms of receptor desensitization using the beta-adrenergic receptor-coupled adenylate cyclase system as a model. Nature 317:124–129

Sidhpura N, Weiss F, Martin-Fardon R (2010) Effects of the mGlu2/3 agonist LY379268 and the mGlu5 antagonist MTEP on ethanol seeking and reinforcement are differentially altered in rats with a history of ethanol dependence. Biol Psychiatry 67:804–811. https://doi.org/10.1016/j.biopsych.2010.01.005

Sinclair CM, Cleva RM, Hood LE, Olive MF, Gass JT (2012) mGluR5 receptors in the basolateral amygdala and nucleus accumbens regulate cue-induced reinstatement of ethanol-seeking behavior. Pharmacol Biochem Behav 101:329–335. https://doi.org/10.1016/j.pbb.2012.01.014

Slack BE, Nitsch RM, Livneh E, Kunz GM Jr, Breu J, Eldar H, Wurtman RJ (1993) Regulation by phorbol esters of amyloid precursor protein release from Swiss 3T3 fibroblasts overexpressing protein kinase C alpha. J Biol Chem 268:21097–21101

Sorensen SD, Conn PJ (2003) G protein-coupled receptor kinases regulate metabotropic glutamate receptor 5 function and expression. Neuropharmacology 44:699–706

Szumlinski KK, Ary AW, Lominac KD (2008) Homers regulate drug-induced neuroplasticity: implications for addiction. Biochem Pharmacol 75:112–133. https://doi.org/10.1016/j.bcp.2007.07.031

Szumlinski KK et al (2005) Homer2 is necessary for EtOH-induced neuroplasticity. J Neurosci: Off J Soc Neurosci 25:7054–7061. https://doi.org/10.1523/JNEUROSCI.1529-05.2005

Terrell J, Shih S, Dunn R, Hicke L (1998) A function for monoubiquitination in the internalization of a G protein-coupled receptor. Mol Cell 1:193–202

Trivedi RR, Bhattacharyya S (2012) Constitutive internalization and recycling of metabotropic glutamate receptor 5 (mGluR5). Biochem Biophys Res Commun 427:185–190

Tu JC et al (1998) Homer binds a novel proline-rich motif and links group 1 metabotropic glutamate receptors with IP3 receptors. Neuron 21:717–726

Urizar NL, Yang Z, Edenberg HJ, Davis RL (2007) Drosophila homer is required in a small set of neurons including the ellipsoid body for normal ethanol sensitivity and tolerance. J Neurosci:Off J Soc Neurosci 27:4541–4551. https://doi.org/10.1523/JNEUROSCI.0305-07.2007

Wang H, Zhuo M (2012) Group I metabotropic glutamate receptor-mediated gene transcription and implications for synaptic plasticity and diseases. Front Pharmacol 3:189. https://doi.org/10.3389/fphar.2012.00189

Winder DG, Conn PJ (1996) Roles of metabotropic glutamate receptors in glial function and glial-neuronal communication. J Neurosci Res 46:131–137

Wisniewski KE, Wisniewski HM, Wen GY (1985) Occurrence of neuropathological changes and dementia of Alzheimer's disease in Down's syndrome. Ann Neurol 17:278–282. https://doi.org/10.1002/ana.410170310

Yao HH, Ding JH, Zhou F, Wang F, Hu LF, Sun T, Hu G (2005) Enhancement of glutamate uptake mediates the neuroprotection exerted by activating group II or III metabotropic glutamate receptors on astrocytes. J Neurochem 92:948–961

Yu SS, Lefkowitz RJ, Hausdorff WP (1993) Beta-adrenergic receptor sequestration. A potential mechanism of receptor resensitization. J Biol Chem 268:337–341

Zhang J, Ferguson SS, Barak LS, Menard L, Caron MG (1996) Dynamin and beta-arrestin reveal distinct mechanisms for G protein-coupled receptor internalization. J Biol Chem 271:18302–18305

Soluble Amyloid Precursor Protein α: Friend or Foe?

13

Nicola J. Corbett and Nigel M. Hooper

Abstract

The "amyloidogenic" proteolytic processing of the cell surface amyloid precursor protein (APP) produces amyloid β, which causes a range of detrimental effects in the neuron, such as synaptic loss, and plays a key role in Alzheimer's disease. In contrast, "non-amyloidogenic" proteolytic processing, which involves the cleavage of APP by α-secretase, produces soluble amyloid precursor protein α (sAPPα) and is the most predominant proteolytic processing of APP in the healthy brain. Current research suggests that sAPPα plays a role in synaptic growth and plasticity, but whether this role is protective or detrimental is age-dependent. This review looks at the effects of increasing sAPPα during three time-points in life (in development, young adult, ageing/neurodegeneration) when synaptic plasticity plays an important role.

N. J. Corbett · N. M. Hooper (✉)
Division of Neuroscience & Experimental Psychology, School of Biological Sciences, Faculty of Biology, Medicine and Health, University of Manchester, Manchester, UK
e-mail: Nicola.corbett@manchester.ac.uk; nigel.hooper@manchester.ac.uk

Keywords

Soluble amyloid precursor protein α (sAPPα) · Amyloid precursor protein (APP) · A disintegrin and metalloproteinase domain-containing protein 10 (ADAM10) · Synaptic plasticity · Development · Ageing · Neurodegeneration · Neuroprotection · Dendritic spines

13.1 Introduction

Soluble amyloid precursor protein α (sAPPα) is a product of the proteolytic cleavage of the cell surface amyloid precursor protein (APP), through the "non-amyloidogenic" processing pathway. APP is a member of the type 1 transmembrane glycoprotein family and has a large glycosylated N-terminal domain, a hydrophobic transmembrane domain, and a short C-terminal domain (Andrew et al. 2016). The main isoforms that are expressed in the brain are APP_{695}, APP_{751}, and APP_{770}, with APP_{695} being the most abundant (Habib et al. 2017). The most well-known proteolytic processing of APP occurs when the β-secretase (BACE1) and the γ-secretase complex sequentially cleave APP to produce amyloid-β (Aβ) peptides (Andrew et al. 2016). This pathway is known as the "amyloidogenic" processing of APP (Fig. 13.1). The products of this proteolytic processing cause a range of detrimental effects in neurons and the brain and play a key role in

K. Chattopadhyay, S. C. Basu (eds.), *Biochemical and Biophysical Roles of Cell Surface Molecules*, Advances in Experimental Medicine and Biology 1112, https://doi.org/10.1007/978-981-13-3065-0_13

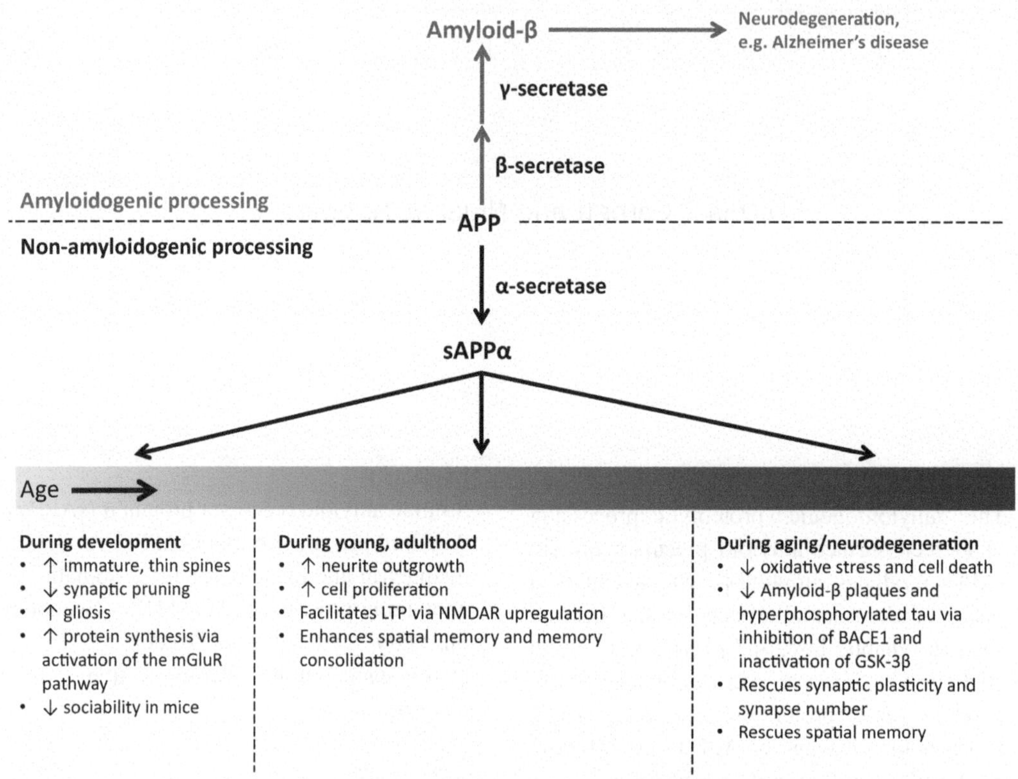

Fig. 13.1 Functions of sAPPα, a key proteolytic product of the cell surface amyloid precursor protein
In the amyloidogenic pathway, the cell surface APP is proteolytically cleaved by the β- and γ-secretases to generate the neurotoxic amyloid-β. In the non-amyloidogenic pathway, APP is cleaved by α-secretase to generate sAPPα which has either neurotoxic or neuroprotective properties depending on the stage of development

Alzheimer's disease (AD). However, the non-amyloidogenic processing pathway is the principal route for APP processing (Andrew et al. 2016). This pathway involves the cleavage of APP by α-secretase, predominantly the disintegrin and metalloproteinase domain-containing protein 10 (ADAM10), producing sAPPα and C83, an 83 amino acid C-terminal fragment (Andrew et al. 2016) (Fig. 13.1). The properties of sAPPα within the central nervous system are not thoroughly understood. However, it has been linked to synaptic plasticity in a range of models, and whether an increase in sAPPα is protective or detrimental is dependent on the stage during an organism's lifespan.

13.2 Role of sAPPα in the Developing Brain

APP expression levels are at their highest during the early stages of synaptogenesis (Priller et al. 2006), and varying expression levels appear to be important during normal development. For example, Lahiri et al. (2002) have shown that APP expression is higher postnatally than prenatally. Due to the presence of APP, and it being predominantly cleaved by α-secretase throughout the "healthy" life of a cell, sAPPα is present throughout lifespan, including during early brain development. Several studies have suggested that the expression level of sAPPα may also be important in development, particularly with regard to synaptic growth and pruning. Of particular interest is the recent study by Pasciuto et al. (2015), who highlighted the importance of maintaining balanced levels of sAPPα during a critical period in development that is important for synaptic stabilisation and elimination via protein synthesis. Using a mouse model of Fragile X syndrome (FXS; FMR1 knockout), a neurodevelopmental disorder, Pasciuto et al. (2015) paid particular attention to the Fragile X mental retardation protein (FMRP), which is absent, or mutated, in FXS. Westmark and Malter (2007) found that FMRP regulates dendritic APP expression by binding to APP mRNA, leading to APP synthesis. The authors showed that in FMR1 knockout mice, between postnatal days 21 and 28, there was an increased expression of APP and ADAM10, compared to their wild-type littermates. This led to an increase in the cleavage of APP through the non-amyloidogenic pathway, which could be seen when looking at the lack of increased cell surface APP in the knockout mice, and hence, there was an excess of sAPPα. Interestingly, this increase in APP, ADAM10, and sAPPα was also seen in fibroblasts taken from FXS patients (Pasciuto et al. 2015). The excess in sAPPα in the mouse model led to an increase in the number of thin, immature dendritic spines, which is a key pathological feature of the disorder. The authors suggest that this may be via increased protein synthesis, which has occurred due to the ability of sAPPα to activate the mGluR5 pathway (Pasciuto et al. 2015).

Other developmental disorders, such as autism, a specific form of the autism spectrum disorder (ASD), have also been associated with altered dendritic morphology and brain enlargement during this critical period. Overproduction of sAPPα may also contribute to this pathology, as elevated levels of sAPPα have been seen in the plasma and the insular cortex of autistic individuals (Sokol et al. 2006; Bailey et al. 2013; Ray et al. 2011). Autism is characterised by impaired sociability, which was seen when mice overexpressing human sAPPα were placed in a three-chambered apparatus with a stranger mouse (Bailey et al. 2013). These transgenic mice also had increased gliosis, which is correlated with elevations in interleukin-6, gp130, and Notch1 that are involved in glial differentiation. This gliosis increase may help explain, along with increased neuronal anatomy, the increase in brain volume seen in autistic individuals (Courchesne et al. 2001, 2003; Sparks et al. 2002).

13.3 Role of sAPPα in the Healthy, Young Adult Brain

In the young adult brain, expression of sAPPα does not cause detrimental effects at the synapse, as seen during development, indicating that the timing of sAPPα production is critical. Ring et al. (2007) found that APP knockout mice had decreased brain weight, grip strength, locomotor activity, and impaired learning and memory, which were not seen in mice with sAPPα knock-in on the APP null background, suggesting that it is the sAPPα fragment of APP that is involved in these neurological features. Other studies also found that sAPPα was able to enhance spatial memory and promote memory consolidation in healthy adult rodents, either via overexpressing sAPPα in the hippocampus by viral gene transfer (Fol et al. 2016) or by administering recombinant sAPPα (Taylor et al. 2008). This was further substantiated, and linked to its synaptic role, by the addition of exogenous sAPPα rescuing long-term potentiation (LTP) in the APP knockout mice (Ring et al. 2007) and when endogenous sAPPα was blocked using antibodies (Fol et al. 2016). However, it is important to note that Taylor et al. (2008) also found detrimental effects on LTP when incubating hippocampal slices with high concentrations of sAPPα. This highlights that not only the timing but also the amount of sAPPα is critical at the synaptic level.

Aiding neurite outgrowth (Gakhar-Koppole et al. 2008; Chasseigneaux et al. 2011; Hasebe et al. 2013) and cell proliferation (Caille et al. 2004; Demars et al. 2011) are also well-documented properties for sAPPα in the adult brain. However, the mechanisms of action of sAPPα in these synaptic and neuronal roles are still to be fully elucidated. The amino acid residues 319–335 of sAPPα, which have been shown to increase synaptic density and memory retention (Roch et al. 1994), have been linked to neurite outgrowth, along with residues 28–123 known as the growth factor-like domain and the C-terminal sequence residues 591–612 (Taylor et al. 2008; Furukawa et al. 1998). Hasebe et al. (2013) suggested that the role of sAPPα in neurite outgrowth is due to it binding to a cell surface receptor. The authors suggest that one potential receptor for this is p75 neurotrophic receptor (p75 ^{NTR}). Exogenous sAPPα activates membrane-associated guanylate cyclase (Barger and Mattson 1995), important in formation and regulation of synaptic scaffolding, leading to decreased N-methyl-D-aspartate (NMDA) receptor currents (Furukawa and Mattson 1998). However, Taylor et al. (2008) found that sAPPα facilitated LTP via upregulation of NMDA receptor transmission, which was further supported by Gakhar-Koppole et al. (2008), who activated NMDA receptors in the presence of sAPPα, promoting neurite outgrowth. Whatever the link between sAPPα and NMDA receptors, it is possible that the extracellular signalling-regulated kinase (ERK)/mitogen-activated protein kinase (MAPK) pathway is involved (Klatt et al. 2013; Bailey et al. 2012; Rohe et al. 2008; Claasen et al. 2009).

13.4 Role of sAPPα in the Aged Brain and in Neurodegeneration

As mentioned previously in this review, the most well-documented processing of APP is via the amyloidogenic pathway to produce Aβ that is involved in neurodegeneration (Fig. 13.1). An increase in the amyloidogenic pathway occurs in the aged and amyloid-related diseased brain; therefore, sAPPα can be reduced in these conditions. For example, sAPPα levels are decreased in mild cognitive impairment and AD patients (Lannfelt et al. 1995; Dobrowolska et al. 2014), which have been correlated with memory impairment (Almkvist et al. 1997). In aged rats, Anderson et al. (1999) found a 50% reduction in sAPPα levels in the cerebrospinal fluid (CSF) that could be correlated with poor memory performance.

The majority of research into the beneficial properties of sAPPα has been performed in different models of AD, in vitro to in vivo, all suggesting that sAPPα may be beneficial in attenuating key pathological features of the disease. Using Aβ-treated dissociated hippocampal cells, Goodman and Mattson (1994) found that exposing the cells to sAPPα attenuated oxidative injury and cell death. This was also found by Stein et al. (2004), when looking at Aβ-treated, organotypic hippocampal slices. The authors showed that not only did sAPPα prevent apoptotic cell death; it also prevented nuclear pyknosis and paired helical filaments, which are made up of hyperphosphorylated tau and are a major component of neurofibrillary tangles. Currently, the link between sAPPα and tau has only been investigated with respect to Alzheimer's disease, with Deng et al. (2015) showing the effects of sAPPα on the phosphorylation state of tau using a range of AD models: SH-SY5Y/BACE1 cells, HeLa/tau cells, and sAPPα/APPswedish/PS1 mice. They found that sAPPα was able to decrease phosphorylated tau via BACE1 and the phosphorylation of glycogen synthase kinase 3β (GSK-3β). sAPPα also had an effect on Aβ via the inhibition of BACE1 in vitro (CHO/APPswe/PS1wt, CHO/APPwt, N2a/sAPPα cells) and in vivo (sAPPα x APPswe/PS1ΔE9 mice), and this caused decreased levels of Aβ, with a decrease in Aβ plaques, fibrils, and oligomers in the cortex and hippocampus in the transgenic mice (Obregon et al. 2012). From this it is clear that sAPPα must be internalised, and Gustafsen et al. (2013) suggested that sortilin and SorLA may be potential endocytic receptors for sAPPα, as they have shown sortilin to mediate non-amyloidogenic processing of APP and both proteins to mediate internalisation of sAPPα, particularly in neurons. This was supported by Hartl et al. (2013), who found that sAPPα was unable to exert its neurotrophic properties on SORL1-deficient cortical neurons. The authors suggest that sAPPα binds to SorLA, which regulates CDK5 expression. Further to this, ERG1 was shown to increase with sAPPα, leading to an increase in CDK5 (Chasseigneaux et al. 2011), supporting the involvement of the CDK5 pathway, which reduces phosphorylation of tau and BACE1.

Again these studies showed the role of sAPPα at the synapse, if indirectly, via pathological protein regulation, whilst other studies highlighted above showed its directly related synaptic role, both showing that under these conditions sAPPα protects synapses (Fig. 13.1). The overexpression of sAPPα by viral gene transfer that showed LTP facilitation in young adult mice also found that in aged APP/PS1ΔE9 mice with plaques, this overexpression improved synaptic plasticity via LTP induction, partially rescued dendritic spines, and decreased Aβ levels and plaque load, which all may have been involved in the improvement of the animal's spatial reference memory (Fol et al. 2016). Interestingly, the authors also showed that sAPPα increased activated microglia around plaques and upregulated the insulin-degrading enzyme (IDE) and the triggering receptor expressed on myeloid cells (TREM2) expression, all of which may contribute to increased Aβ clearance. However, inflammation is a key feature of AD; therefore, the ability of sAPPα to increase activated microglia may only be beneficial at very early stages of neurodegeneration prior to major inflammatory responses. Collectively these studies have shown that sAPPα has a range of effects that are neuroprotective, particularly with regard to the synapse, in the aged, neurodegenerating brain, in complete contrast to the situation during development, in which the very same properties of sAPPα appear to be detrimental (summarised in Fig. 13.1).

13.5 Conclusion

The proteolytic processing of the cell surface APP has been well documented, particularly the amyloidogenic pathway, and the toxic effects of its products are well known. However, the non-amyloidogenic pathway produces a protein that has detrimental effects in the developing brain but neuroprotective effects in the aged, diseased brain. The exact mechanisms of action of sAPPα are still unknown, yet it, and its signalling pathway, may be a potential therapeutic target in disease.

References

Almkvist O, Basun H, Wagner SL, Rowe BA, Wahlund LO, Lannfelt L (1997) Cerebrospinal fluid levels of alpha-secretase-cleaved soluble amyloid precursor protein mirror cognition in a Swedish family with Alzheimer disease and a gene mutation. Arch Neurol 54(5):641–644

Anderson JJ, Holtz G, Baskin PP, Wang R, Mazzarelli L, Wagner SL, Menzaghi F (1999) Reduced cerebrospinal fluid levels of alpha-secretase-cleaved amyloid precursor protein in aged rats: correlation with spatial memory deficits. Neuroscience 93(4):1409–1420

Andrew RJ, Kellett KA, Thinakaran G, Hooper NM (2016) A Greek tragedy: the growing complexity of Alzheimer amyloid precursor protein proteolysis. J Biol Chem 291(37):19235–19244. https://doi.org/10.1074/jbc.R116.746032

Bailey AR, Hou H, Obregon DF, Tian J, Zhu Y, Zou Q, Nikolic WV, Bengtson M, Mori T, Murphy T, Tan J (2012) Aberrant T-lymphocyte development and function in mice overexpressing human soluble amyloid precursor protein-alpha: implications for autism. FASEB J 26(3):1040–1051. https://doi.org/10.1096/fj.11-195438

Bailey AR, Hou H, Song M, Obregon DF, Portis S, Barger S, Shytle D, Stock S, Mori T, Sanberg PG, Murphy T, Tan J (2013) GFAP expression and social deficits in transgenic mice overexpressing human sAPPalpha. Glia 61(9):1556–1569. https://doi.org/10.1002/glia.22544

Barger SW, Mattson MP (1995) The secreted form of the Alzheimer's beta-amyloid precursor protein stimulates a membrane-associated guanylate cyclase. Biochem J 311(Pt 1):45–47

Caille I, Allinquant B, Dupont E, Bouillot C, Langer A, Muller U, Prochiantz A (2004) Soluble form of amyloid precursor protein regulates proliferation of progenitors in the adult subventricular zone.

Development 131(9):2173–2181. https://doi.org/10.1242/dev.01103

Chasseigneaux S, Dinc L, Rose C, Chabret C, Coulpier F, Topilko P, Mauger G, Allinquant B (2011) Secreted amyloid precursor protein beta and secreted amyloid precursor protein alpha induce axon outgrowth in vitro through Egr1 signaling pathway. PLoS One 6(1): e16301. https://doi.org/10.1371/journal.pone.0016301

Claasen AM, Guevremont D, Mason-Parker SE, Bourne K, Tate WP, Abraham WC, Williams JM (2009) Secreted amyloid precursor protein-alpha upregulates synaptic protein synthesis by a protein kinase G-dependent mechanism. Neurosci Lett 460 (1):92–96. https://doi.org/10.1016/j.neulet.2009.05.040

Courchesne E, Karns CM, Davis HR, Ziccardi R, Carper RA, Tigue ZD, Chisum HJ, Moses P, Pierce K, Lord C, Lincoln AJ, Pizzo S, Schreibman L, Haas RH, Akshoomoff NA, Courchesne RY (2001) Unusual brain growth patterns in early life in patients with autistic disorder: an MRI study. Neurology 57 (2):245–254

Courchesne E, Carper R, Akshoomoff N (2003) Evidence of brain overgrowth in the first year of life in autism. JAMA 290(3):337–344. https://doi.org/10.1001/jama.290.3.337

Demars MP, Bartholomew A, Strakova Z, Lazarov O (2011) Soluble amyloid precursor protein: a novel proliferation factor of adult progenitor cells of ectodermal and mesodermal origin. Stem Cell Res Ther 2 (4):36. https://doi.org/10.1186/scrt77

Deng J, Habib A, Obregon DF, Barger SW, Giunta B, Wang YJ, Hou H, Sawmiller D, Tan J (2015) Soluble amyloid precursor protein alpha inhibits tau phosphorylation through modulation of GSK3beta signaling pathway. J Neurochem 135(3):630–637. https://doi.org/10.1111/jnc.13351

Dobrowolska JA, Kasten T, Huang Y, Benzinger TL, Sigurdson W, Ovod V, Morris JC, Bateman RJ (2014) Diurnal patterns of soluble amyloid precursor protein metabolites in the human central nervous system. PLoS One 9(3):e89998. https://doi.org/10.1371/journal.pone.0089998

Fol R, Braudeau J, Ludewig S, Abel T, Weyer SW, Roederer JP, Brod F, Audrain M, Bemelmans AP, Buchholz CJ, Korte M, Cartier N, Muller UC (2016) Viral gene transfer of APPsalpha rescues synaptic failure in an Alzheimer's disease mouse model. Acta Neuropathol 131(2):247–266. https://doi.org/10.1007/s00401-015-1498-9

Furukawa K, Mattson MP (1998) The transcription factor NF-kappaB mediates increases in calcium currents and decreases in NMDA- and AMPA/kainate-induced currents induced by tumor necrosis factor-alpha in hippocampal neurons. J Neurochem 70(5):1876–1886

Furukawa N, Hatano M, Fukuda H, Koga T (1998) Non-N-methyl-D-aspartate receptors may mediate the transmission of emetic signals between visceral vagal afferents and the solitary nucleus in dogs. Neurosci Lett 258(1):53–56

Gakhar-Koppole N, Hundeshagen P, Mandl C, Weyer SW, Allinquant B, Muller U, Ciccolini F (2008) Activity requires soluble amyloid precursor protein alpha to promote neurite outgrowth in neural stem cell-derived neurons via activation of the MAPK pathway. Eur J Neurosci 28(5):871–882. https://doi.org/10.1111/j.1460-9568.2008.06398.x

Goodman Y, Mattson MP (1994) Secreted forms of beta-amyloid precursor protein protect hippocampal neurons against amyloid beta-peptide-induced oxidative injury. Exp Neurol 128(1):1–12. https://doi.org/10.1006/exnr.1994.1107

Gustafsen C, Glerup S, Pallesen LT, Olsen D, Andersen OM, Nykjaer A, Madsen P, Petersen CM (2013) Sortilin and SorLA display distinct roles in processing and trafficking of amyloid precursor protein. J Neurosci 33(1):64–71. https://doi.org/10.1523/JNEUROSCI.2371-12.2013

Habib A, Sawmiller D, Tan J (2017) Restoring soluble amyloid precursor protein alpha functions as a potential treatment for Alzheimer's disease. J Neurosci Res 95(4):973–991. https://doi.org/10.1002/jnr.23823

Hartl D, Klatt S, Roch M, Konthur Z, Klose J, Willnow TE, Rohe M (2013) Soluble alpha-APP (sAPPalpha) regulates CDK5 expression and activity in neurons. PLoS One 8(6):e65920. https://doi.org/10.1371/journal.pone.0065920

Hasebe N, Fujita Y, Ueno M, Yoshimura K, Fujino Y, Yamashita T (2013) Soluble beta-amyloid precursor protein alpha binds to p75 neurotrophin receptor to promote neurite outgrowth. PLoS One 8(12):e82321. https://doi.org/10.1371/journal.pone.0082321

Klatt S, Rohe M, Alagesan K, Kolarich D, Konthur Z, Hartl D (2013) Production of glycosylated soluble amyloid precursor protein alpha (sAPPalpha) in Leishmania tarentolae. J Proteome Res 12(1):396–403. https://doi.org/10.1021/pr300693f

Lahiri DK, Nall C, Chen D, Zaphiriou M, Morgan C, Nurnberger JI Sr (2002) Developmental expression of the beta-amyloid precursor protein and heat-shock protein 70 in the cerebral hemisphere region of the rat brain. Ann N Y Acad Sci 965:324–333

Lannfelt L, Basun H, Wahlund LO, Rowe BA, Wagner SL (1995) Decreased alpha-secretase-cleaved amyloid precursor protein as a diagnostic marker for Alzheimer's disease. Nat Med 1(8):829–832

Obregon D, Hou H, Deng J, Giunta B, Tian J, Darlington D, Shahaduzzaman M, Zhu Y, Mori T, Mattson MP, Tan J (2012) Soluble amyloid precursor protein-alpha modulates beta-secretase activity and amyloid-beta generation. Nat Commun 3:777. https://doi.org/10.1038/ncomms1781

Pasciuto E, Ahmed T, Wahle T, Gardoni F, D'Andrea L, Pacini L, Jacquemont S, Tassone F, Balschun D, Dotti CG, Callaerts-Vegh Z, D'Hooge R, Muller UC, Di Luca M, De Strooper B, Bagni C (2015) Dysregulated ADAM10-mediated processing of APP during a critical time window leads to synaptic deficits in fragile X syndrome. Neuron 87(2):382–398. https://doi.org/10.1016/j.neuron.2015.06.032

Priller C, Bauer T, Mitteregger G, Krebs B, Kretzschmar HA, Herms J (2006) Synapse formation and function is modulated by the amyloid precursor protein. J Neurosci 26(27):7212–7221. https://doi.org/10.1523/JNEUROSCI.1450-06.2006

Ray B, Long JM, Sokol DK, Lahiri DK (2011) Increased secreted amyloid precursor protein-alpha (sAPPalpha) in severe autism: proposal of a specific, anabolic pathway and putative biomarker. PLoS One 6(6):e20405. https://doi.org/10.1371/journal.pone.0020405

Ring S, Weyer SW, Kilian SB, Waldron E, Pietrzik CU, Filippov MA, Herms J, Buchholz C, Eckman CB, Korte M, Wolfer DP, Muller UC (2007) The secreted beta-amyloid precursor protein ectodomain APPs alpha is sufficient to rescue the anatomical, behavioral, and electrophysiological abnormalities of APP-deficient mice. J Neurosci 27(29):7817–7826. https://doi.org/10.1523/JNEUROSCI.1026-07.2007

Roch JM, Masliah E, Roch-Levecq AC, Sundsmo MP, Otero DA, Veinbergs I, Saitoh T (1994) Increase of synaptic density and memory retention by a peptide representing the trophic domain of the amyloid beta/A4 protein precursor. Proc Natl Acad Sci U S A 91(16):7450–7454

Rohe M, Carlo AS, Breyhan H, Sporbert A, Militz D, Schmidt V, Wozny C, Harmeier A, Erdmann B, Bales KR, Wolf S, Kempermann G, Paul SM, Schmitz D, Bayer TA, Willnow TE, Andersen OM (2008) Sortilin-related receptor with A-type repeats (SORLA) affects the amyloid precursor protein-dependent stimulation of ERK signaling and adult neurogenesis. J Biol Chem 283(21):14826–14834. https://doi.org/10.1074/jbc.M710574200

Sokol DK, Chen D, Farlow MR, Dunn DW, Maloney B, Zimmer JA, Lahiri DK (2006) High levels of Alzheimer beta-amyloid precursor protein (APP) in children with severely autistic behavior and aggression. J Child Neurol 21(6):444–449

Sparks BF, Friedman SD, Shaw DW, Aylward EH, Echelard D, Artru AA, Maravilla KR, Giedd JN, Munson J, Dawson G, Dager SR (2002) Brain structural abnormalities in young children with autism spectrum disorder. Neurology 59(2):184–192

Stein TD, Anders NJ, DeCarli C, Chan SL, Mattson MP, Johnson JA (2004) Neutralization of transthyretin reverses the neuroprotective effects of secreted amyloid precursor protein (APP) in APPSW mice resulting in tau phosphorylation and loss of hippocampal neurons: support for the amyloid hypothesis. J Neurosci 24(35):7707–7717. https://doi.org/10.1523/JNEUROSCI.2211-04.2004

Taylor CJ, Ireland DR, Ballagh I, Bourne K, Marechal NM, Turner PR, Bilkey DK, Tate WP, Abraham WC (2008) Endogenous secreted amyloid precursor protein-alpha regulates hippocampal NMDA receptor function, long-term potentiation and spatial memory. Neurobiol Dis 31(2):250–260. https://doi.org/10.1016/j.nbd.2008.04.011

Westmark CJ, Malter JS (2007) FMRP mediates mGluR5-dependent translation of amyloid precursor protein. PLoS Biol 5(3):e52. https://doi.org/10.1371/journal.pbio.0050052

C. elegans Locomotion: Finding Balance in Imbalance

14

Shruti Thapliyal and Kavita Babu

Abstract

The excitation-inhibition (E-I) imbalance in neural circuits represents a hallmark of several neuropsychiatric disorders. The tiny nematode *Caenorhabditis elegans* has emerged as an excellent system to study the molecular mechanisms underlying this imbalance in neuronal circuits. The *C. elegans* body wall muscles receive inputs from both excitatory cholinergic and inhibitory GABAergic motor neurons at neuromuscular junctions (NMJ), making it an excellent model for studying the genetic and molecular mechanisms required for maintaining E-I balance at the NMJ. The cholinergic neurons form dyadic synapses wherein they synapse onto ipsilateral body wall muscles allowing for muscle contraction as well as onto GABAergic motor neurons that in turn synapse on the contralateral body wall muscles causing muscle relaxation. An alternating wave of contraction and relaxation mediated by excitatory and inhibitory signals maintains locomotion in *C. elegans*. This locomotory behavior requires an intricate balance between the excitatory cholinergic signaling and the inhibitory GABAergic signaling mechanisms.

Studies on the *C. elegans* NMJ have provided insights into several molecular mechanisms that could regulate this balance in neural circuits. This review provides a discussion on multiple genetic factors including neuropeptides and their receptors, cell adhesion molecules, and other molecular pathways that have been associated with maintaining E-I balance in *C. elegans* motor circuits. Further, it also discusses the implications of these studies that could help us in understanding the role of E-I balance in mammalian neural circuits and how changes in this balance could give rise to brain disorders.

Keywords

Excitation · Inhibition · Acetylcholine · GABA · NMJ · *C. elegans*

14.1 Introduction

Normal brain development and function requires a tight coordination between excitation and inhibition during synaptic transmission in neural circuits. Hence, excitation and inhibition are required for the synchronized activity of neural circuits regulating diverse behaviors (Isaacson and Scanziani 2011). Excitatory signals from the presynaptic neuron to the postsynaptic neuron make the latter more likely to fire, while an inhibitory signal prevents firing of the postsynaptic

S. Thapliyal (✉) · K. Babu (✉)
Indian Institute of Science Education and Research (IISER), Mohali, Punjab, India
e-mail: shruti.thapliyal@babulab.org; kavita.babu@babulab.org

K. Chattopadhyay, S. C. Basu (eds.), *Biochemical and Biophysical Roles of Cell Surface Molecules*, Advances in Experimental Medicine and Biology 1112, https://doi.org/10.1007/978-981-13-3065-0_14

neuron. Impaired excitation-inhibition (E-I) balance has been reported as a hallmark for several neuropsychiatric disorders including autism, epilepsy, schizophrenia, and Alzheimer's disease (Eichler and Meier 2008; Lei et al. 2016; Nelson and Valakh 2015; Yizhar et al. 2011). Thus understanding genetic factors and molecular mechanisms coordinating this balance in neural circuits mandates greater attention.

Locomotion is a prominent behavioral output in *C. elegans*. Neural circuits that generate coordinated dorsoventral sinusoidal bends maintain normal locomotion in *C. elegans* (illustrated in Fig. 14.1a). Locomotion in *C. elegans* is synchronized at multiple levels of the neural circuit and involves diverse sensory cues that are integrated by sensory neurons and further processed by the interneurons ultimately resulting in adjustments at the neuromuscular junction (NMJ) (Bargmann 2012; de Bono and Maricq 2005). The neural circuit that regulates locomotion in *C. elegans* consists of a wide array of sensory neurons that can detect diverse environmental stimuli. These environmental cues include volatile and soluble chemicals (chemotaxis), temperature (thermotaxis), pH, oxygen (aerotaxis), as well as various forms of mechanical stimuli (Bargmann et al. 1993; Bretscher et al. 2011; Chalfie et al. 1985; Culotti and Russell 1978; Hedgecock and Russell 1975; Kaplan and Horvitz 1993; Sassa et al. 2013; Ward 1973). Sensory neurons in *C. elegans* express distinct sensory receptors that are involved in diverse functions such as in taste reception (Bargmann and Horvitz 1991), olfaction (Bargmann et al. 1993), mechanosensation (Kaplan and Horvitz 1993), thermal perception (Mori and Ohshima 1995), and pheromone detection (McGrath et al. 2011) among others. The information collected by these sensory neurons is transferred onto several layers of interneurons. Interneurons mediate information flow from sensory to motor neurons and are the site of sensory information integration and processing. Processed information from several layers of interneurons ultimately reaches five pairs of distinct premotor interneurons (AVA, AVB, AVD, AVE, and PVC) that direct locomotory behaviors and are called the command interneurons. The processed information from the interneurons is finally transferred onto the neuromuscular junction (NMJ). The *C. elegans* NMJ has three main components: the excitatory cholinergic motor neuron, the inhibitory GABAergic motor neuron, and the muscle (illustrated in Fig. 14.1b). Information from the command interneurons is passed onto the cholinergic motor neurons that then form dyadic synapses. The cholinergic motor neurons synapse both onto the ipsilateral body wall muscles causing contraction and onto inhibitory GABAergic motor neurons that then synapse onto the contralateral body wall muscles causing muscle relaxation (White et al. 1976, 1986). Hence, at any given point of time on one side, muscle is contracted, and on the other side, the body wall muscle is relaxed. This alternating wave of muscle contraction and relaxation mediated by excitatory and inhibitory signals maintains sinusoidal locomotion in *C. elegans* (illustrated in Fig. 14.1c). The overall magnitude of the dorsal and ventral contraction must be equal for normal coordinated locomotion. Any alteration in this E-I balance modifies locomotory behavior in *C. elegans* (Isaacson and Scanziani 2011; Stawicki et al. 2013). Thus, the *C. elegans* NMJ provides a unique and sophisticated model to understand key genetic players and molecular mechanisms maintaining E-I balance in neural circuits.

This review is an attempt to address the significance of the *C. elegans* NMJ as an attractive model to conduct such studies. Several previous reports have identified distinct genetic factors and molecular mechanisms that operate at various levels of the locomotory circuit to regulate E-I balance at the NMJ. These molecular players have been discussed below and summarized in Table 14.1. However, despite providing a simple genetically modifiable and tractable system, the *C. elegans* NMJ has not been exploited to its full potential. Future studies in this area could shed light onto some crucial molecular factors and their regulatory control in maintaining balance in neural circuits.

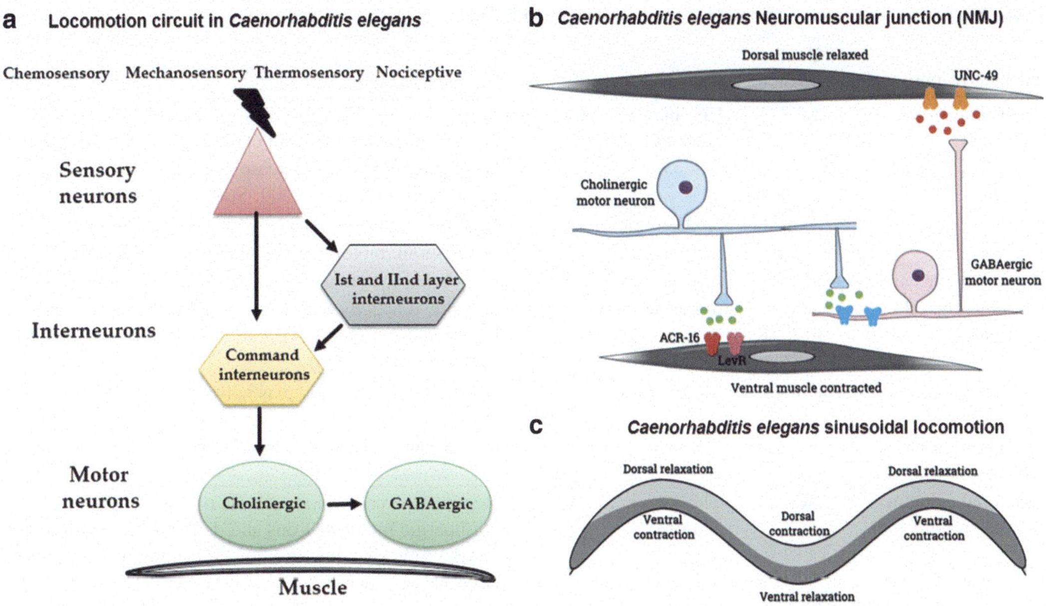

Fig. 14.1 Schematic representing the *C. elegans* locomotory circuit and neuromuscular junction (NMJ). (**a**) Diverse environmental cues sensed by the sensory neurons are processed and integrated at interneurons. The processed information reaches the command interneurons, the premotor neurons, that decide the direction of locomotion and manifest the outcome at the neuromuscular junction by coordinated action of cholinergic and GABAergic motor neurons synapsing onto the body wall muscles. (**b**) At the NMJ, this circuit works through a "cross-inhibition model" where the excitatory cholinergic motor neurons form dyadic synapses both onto the ipsilateral muscle causing contraction and onto inhibitory GABAergic motor neurons that then synapse onto the contralateral muscle causing relaxation. (**c**) The alternating waves of contraction and relaxation along the body of the animals drive locomotion in *C. elegans*

14.2 Genetic Factors Regulating Excitation-Inhibition (E-I) Balance at the *C. elegans* Neuromuscular Junction

14.2.1 Neurotransmitter Receptors

Diversity in neurotransmitter receptors provides functional heterogeneity resulting in variability in neural circuits and behaviors. Acetylcholine is the major excitatory neurotransmitter at the nematode neuromuscular junction (NMJ), and more than one third of the *C. elegans* neurons release acetylcholine (Rand 2007). *C. elegans* expresses at least 29 genes that encode for acetylcholine receptor (AChR) subunits (Rand 2007). These AChR subunits were identified in genetic screens for resistance to AChR agonists or through gene homology studies (Fleming et al. 1997; Jones and Sattelle 2004; Lewis et al. 1980). The best characterized *C. elegans* receptors are present on the muscle and include L-AChRs that respond to the levamisole and N-AChRs that respond to nicotine but not to levamisole (Richmond and Jorgensen 1999). The nicotine-sensitive receptors are homopentameric, ligand-gated ion channels that contain only the ACR-16 α-subunit (Francis et al. 2005; Touroutine et al. 2005; Unwin 2005). The levamisole-sensitive receptors on the body wall muscle are heteromeric, containing three essential subunits UNC-29, UNC-38, and UNC-63 along with two nonessential LEV-1 and LEV-8 subunits (Culetto et al. 2004; Fleming et al. 1997; Richmond and Jorgensen 1999). Genome-wide transgene expression studies have revealed the presence of several different acetylcholine receptor subunits that are expressed in neurons as well as on muscles (Jones and Sattelle 2004).

Table 14.1 Genetic factors regulating excitation-inhibition balance at the *C. elegans* NMJ

Genetic factors	Genes involved	Site of action	Proposed mechanism
Neurotransmitter receptors	ACR-2	Cholinergic motor neurons	*Gain-of-function* mutant of *acr-2* exhibited increased acetylcholine release from cholinergic motor neurons and decreased GABA release from GABAergic motor neurons (Jospin et al. 2009; Stawicki et al. 2013)
	ACR-12	GABAergic motor neurons	Maintains synaptic coupling between cholinergic and GABAergic motor neurons (Petrash et al. 2013)
Neuropeptides and G-protein-coupled receptors	FLP-1 and FLP-18	Acts on GABAergic motor neurons and muscles	FLP-1 and flp-18 neuropeptides maintain homeostatic synaptic plasticity by binding to NPR-1 and NPR-5 receptors present in GABA motor neurons and muscle (Stawicki et al. 2013)
Endoplasmic reticulum (ER) chaperone protein	RIC-3	Motor neurons	Maintains homeostatic plasticity by regulating the functional expression of inhibitory GABAA receptors on the muscle (Safdie et al. 2016)
Neuronal calcium sensor protein	NCS-2	Cholinergic motor neurons	Regulates asynchronous release of acetylcholine from cholinergic motor neurons and dampens the inhibitory GABAergic signaling in a non-cell autonomous manner (Zhou et al. 2017)
Co-chaperones	DNJ-17	Motor neurons	Increased cholinergic signaling that intensify the convulsive phenotype of *acr-2 (gf)* worms (Takayanagi-Kiya and Jin 2016)
Cell adhesion molecules	CASY-1	Sensory neurons and GABAergic motor neurons	Specific isoform functions in sensory neurons to regulate glutamate release, thus maintaining excitatory signaling. Shorter isoforms function in GABAergic motor neurons to maintain GABA release by regulating trafficking of GABA synaptic vesicle precursors (Thapliyal et al. 2018a, b)

Recently, a neuronal acetylcholine receptor, ACR-2 comprising five subunits – UNC-38, UNC-63, ACR-12 (α subunits), ACR-2, and ACR-3 (non-α subunits) – has been reported to maintain E-I balance at the NMJ (Jospin et al. 2009; Stawicki et al. 2013). A *gain-of-function* mutant of *acr-2* exhibited increased acetylcholine release from cholinergic motor neurons. Interestingly, there was a simultaneous decrease in GABA release from GABAergic motor neurons. This imbalance in excitation and inhibition resulted in loss of coordinated muscle activity (Jospin et al. 2009). This study revealed that the *gain-of-function* in *acr-2* was caused by a valine to methionine change at the 13 amino acid position of the pore-forming transmembrane domain of the ACR-2 subunit. This residue in acetylcholine receptor subunits is highly conserved, and a V13'M mutation in the β1-subunit of human muscle acetylcholine receptor has been reported to cause myasthenia gravis (Engel et al. 1996). Functional characterization revealed that *acr-2 (gf)* resulted in a hyperactive channel in vivo, giving rise to spontaneous convulsions in animals (Jospin et al. 2009). The convulsive phenotype of *acr-2(gf)* mutants shares some neurological features characteristic of several forms of epilepsy (Mann and Mody 2008; Stitzel 2008). The molecular mechanisms of how ACR-2-mediated neurotransmission regulates motor circuit activity, however, have not been fully worked out.

More recently, using genetic, behavioral, and electrophysiological approaches, the α-subunit ACR-12 of the nicotinic acetylcholine receptor ACR-2 has been shown to regulate motor circuit activity (Petrash et al. 2013). This study revealed that the ionotropic AChR subunit, *acr-12*, was required to maintain synaptic coupling between excitatory cholinergic and inhibitory GABAergic motor neurons through its function in GABA motor neurons. ACR-12 subunit-mediated signaling in the motor circuit limits the hyperactivity of cholinergic motor neurons via activation of GABA motor neurons.

Having a large variety of acetylcholine receptor subunits suggests diverse neuronal functions for these subunits. Understanding the role of other uncharacterized subunits could shed some light on the role of these neurotransmitter receptors in regulating E-I balance in the motor circuits.

14.2.2 Neuropeptides and Their G-Protein-Coupled Receptors

Neuropeptides are short stretches of amino acids that modulate synaptic activity either directly or indirectly. They can act as short-range signals, released close to the site of action or as hormones, acting on tissues outside the nervous system, thereby enhancing the intricate complexity of the activity of neural circuits. A large repertoire of neuropeptides in *C. elegans* functions to fine-tune the neural networks involved in diverse functions and also increases the complexity of behavioral outcomes in *C. elegans*.

Neuropeptide signaling has been demonstrated to modulate several different behaviors in *C. elegans*, including dauer formation, locomotion, egg-laying, sleep, learning, and mechano- and chemosensation among others (Artan et al. 2016; Banerjee et al. 2017; Bhardwaj et al. 2018; Braeckman and Vanfleteren 2007; Chalasani et al. 2010; Chang et al. 2015; Chen et al. 2016; Cheong et al. 2015; Cornils et al. 2011; Delaney et al. 2017; Harris et al. 2010; Hu et al. 2011, 2015; Hung et al. 2014; Kao et al. 2007; Laurent et al. 2015; Leinwand and Chalasani 2013, 2014; Li and Kim 2008; Li et al. 2013; Nagy et al. 2014; Nath et al. 2016; Nathoo et al. 2001; Nelson et al. 2013; Pierce et al. 2001; Stawicki et al. 2013).

Recently, neuropeptides released from the cholinergic motor neurons have been shown to modulate E-I balance in *acr-2(gf)* mutant. Using functional genetic and electrophysiological approaches, neuropeptides belonging to the FMRFamide family, *flp-1* and *flp-18*, have been shown to reduce E-I balance in the locomotory circuit. The upregulation of FLP-18 expression in *acr-2(gf)* has been proposed to be a homeostatic mechanism to readjust the balance in motor circuit activity. The study proposes that FLP-18 neuropeptide functions by acting on GABAergic motor neurons primarily using the G-protein-coupled receptors NPR-1 and NPR-5 (Stawicki et al. 2013).

Further studies highlighting the role of other neuropeptides, receptors, and their regulatory mechanisms could provide a clearer picture on how neuropeptides and their receptors could affect the E-I balance at the NMJ.

14.2.3 Endoplasmic Reticulum-Resident Chaperone Protein RIC-3

RIC-3 is an evolutionarily conserved endoplasmic reticulum (ER) chaperone protein that regulates maturation and expression of multiple classes of nicotinic acetylcholine receptors (nAChRs) (Cheng et al. 2005; Halevi et al. 2002, 2003; Lansdell et al. 2005). Previously, RIC-3 has been shown to affect motor coordination by regulating the expression and function of nAChRs in motor neurons (Jospin et al. 2009). Recently, phosphorylation of RIC-3 at serine-164 has been shown to regulate muscle excitability, thus altering the E-I balance at the NMJ (Safdie et al. 2016). Reciprocal action of the phosphatase TAX-6 (calcineurin A homologue) and kinase KIN-10 (casein kinase II homologue) has been shown to regulate the phosphorylation state of RIC-3. KIN-10-mediated RIC-3 phosphorylation at S-164 increases muscle excitability by inhibiting the functional expression of inhibitory GABAA receptors on the muscle. This decrease in inhibitory signal modulates the motor coordination mediated by excitatory and inhibitory inputs. This study stresses upon the role of TAX-6 as a novel sensor and regulator of muscle excitability, acting as a homeostatic balancer. Under conditions of increased muscle excitability, TAX-6 results in dephosphorylation of downstream targets like RIC-3, releasing inhibitory GABAA receptors from suppression by phosphorylated RIC-3, thus mediating homeostatic decrease in muscle excitability (Safdie et al. 2016).

14.2.4 Neuronal Calcium Sensor Protein (NCS-2)

Neuronal activation results in a calcium influx through voltage-gated calcium channels and changes cytoplasmic calcium levels. These alterations in calcium levels are detected by multiple calcium sensors that can modulate diverse signaling pathways. The neuronal calcium sensor protein (NCS) is a conserved molecule that can bind to numerous proteins including ion channels, synaptic proteins, membrane receptors, and multiple kinases depending upon the calcium load in the cytoplasm of the neuron (Braunewell and Gundelfinger 1999; Burgoyne 2007; Burgoyne and Haynes 2012; Flaherty et al. 1993). Despite numerous reports supporting the role of these calcium sensors in the pathogenesis of diseases like Alzheimer's disease and schizophrenia (Braunewell 2005; Seaton et al. 2011), the molecular mechanisms for how NCS proteins regulate E-I balance in neural circuits are not clear.

Recently, the duration of excitatory cholinergic motor neuron activity has been shown to result in two major outcomes. A short-duration activation results in reversible strengthening of the presynaptic synapses, maintaining E-I balance, and a long-duration or persistent activation results in asynchronous presynaptic release resulting in reduced cholinergic signaling. This decreased cholinergic signaling further depresses the inhibitory GABAergic signaling resulting in an E-I imbalance at the NMJ (Zhou et al. 2017). This study reports the role of NCS-2 in regulating the asynchronous release of acetylcholine from cholinergic motor neurons that in turn dampen the inhibitory GABAergic signaling in a non-cell autonomous manner. This role of NCS-2 requires Ca^{2+} binding as well as its membrane association domains (Zhou et al. 2017).

Influx of Ca^{2+} ions is important for neuronal cell activation. Several different calcium-binding proteins are present at the site of influx. Further studies characterizing the role of calcium-binding proteins in maintaining E-I balance at NMJ could uncover novel genetic factors regulating this balance.

14.2.5 Co-chaperone DnaJ/Hsp40 Family Proteins

The DnaJ/Hsp40 family of proteins are an evolutionarily conserved class of proteins that act as co-chaperones that interact and activate ATPase activity of Hsp70 chaperone proteins (Ohtsuka and Suzuki 2000; Qiu et al. 2006). These proteins protect the cells from stress generated by misfolded or aggregated proteins. Several different neurological disorders including Huntington's and Parkinson's disease have been attributed to misregulated cellular protein homeostasis (Muchowski and Wacker 2005; Sherman and Goldberg 2001). Thus understanding the key genetic factors regulating cellular protein homeostasis is essential.

Recently, the *C. elegans dnj-17* gain-of-function mutant has been reported to intensify the convulsive phenotype of *acr-2(gf)* animals, suggesting a role for increased cholinergic signaling in *dnj-17(gf)* mutants in disturbing the E-I balance at the NMJ (Takayanagi-Kiya and Jin 2016).

14.2.6 Cell Adhesion Molecules (CAMs)

The precise pattern of connectivity between neurons is a key factor that determines the flow of information in neural circuits. The nervous system employs multiple mechanisms to accomplish this challenging task (Sanes and Yamagata 2009). Many cell adhesion molecules are assembled at synaptic sites. Current literature advocates the role of neuronal CAMs important not only for adhesion but for various aspects of synapse development and function (Cavallaro and Dejana 2011; Dalva et al. 2007; Missler et al. 2012; Shapiro et al. 2007; Sheng et al. 2013). Adhesion molecules include the cadherin family, neurexin/neuroligin family, integrins, SynCAM, NCAM, SALMs (synaptic cell adhesion-like molecules) (Togashi et al. 2009), and the newly discovered LRRTMs (leucine-rich repeat transmembrane molecules) (Linhoff et al. 2009). During brain development, CAMs play crucial roles in neuronal migration, axon fasciculation, and neurite outgrowth (Chih et al. 2005; Cremer et al. 1997; Graf et al. 2004; Hirano and Takeichi 2012; Ko et al. 2009; Pettem et al. 2013; Siddiqui et al. 2010; Sudhof 2008; Togashi et al. 2002, 2006; Walsh and Doherty 1997; Yasuda et al. 2007). In the adult brain, they play essential roles in the regulation of synaptic plasticity as well as in axon regeneration in an injured nervous system (Blackmore and Letourneau 2006; Hansen et al. 2008; Nacher et al. 2013; Tessier-Lavigne and Goodman 1996; Walsh and Doherty 1997).

The *C. elegans casy-1* is an ortholog of mammalian Calsyntenin genes. Calsyntenins are type-I transmembrane proteins belonging to the cadherin superfamily of cell adhesion molecules. They are characterized by the presence of two cadherin-like tandem repeats, an LG/LNS domain in the extracellular region and an intracellular region that carries two kinesin light-chain binding domains (Hintsch et al. 2002; Konecna et al. 2006). All these regions are entirely conserved in the three mammalian Calsyntenin genes, *clstn1*, *clstn2*, and *clstn3*, as well as in the sole *C. elegans* Calsyntenin ortholog, *casy-1* (Hintsch et al. 2002; Ikeda et al. 2008).

CASY-1 has recently been shown to regulate the E-I balance at the NMJ in a unique way. The *casy-1* locus in *C. elegans* encodes three isoforms expressed under alternative promoters based on EST evidence (Reboul et al. 2001). CASY-1A, a 984 residue full-length protein, contains all the conserved domains of mammalian Calsyntenins. CASY-1B and CASY-1C are truncated proteins encoding 167 and 160 residues, respectively, and lack most of the conserved N-terminal of the Calsyntenin gene. The shorter isoforms of CASY-1, CASY-1B, and CASY-1C have been reported to function specifically in GABA motor neurons to regulate GABA release. The conserved C-terminal in the shorter isoforms carries a putative kinesin motor binding domain that mediates fast axonal transport of GABA synaptic vesicle precursors, thus modulating the release kinetics of GABA at the NMJ (Thapliyal et al. 2018b). The longer CASY-1A isoform has been reported to regulate excitatory acetylcholine release via functioning in higher levels of neural circuit regulating locomotion in *C. elegans*. This study proposes the role of CASY-1A in modulating the activity of sensory neurons and allows for regulating glutamate release from these neurons. In the absence of *casy-1*, excessive glutamate is released from the sensory neurons resulting in an overall increase in motor circuit activity (Thapliyal et al. 2018a).

The *C. elegans* CASY-1 represents an interesting model where three isoforms of the same gene are expressed spatially to regulate entirely different synaptic signaling processes (illustrated in Fig. 14.2). The full-length CASY-1A functions in sensory and interneurons to regulate glutamate release, which could then modulate acetylcholine release at the NMJ. On the other hand, the shorter isoforms, CASY-1B and CASY-1C, function in GABAergic motor neurons to regulate GABA release dynamics via mediating trafficking of GABA-specific synaptic vesicle precursors. This highlights interesting isoform-specific functions of a gene in *C. elegans* that ultimately regulates the overall E-I balance at the NMJ.

Deregulation of the mammalian Calsyntenins is thought to be coupled with age-related conditions like Alzheimer's and Parkinson's disease; future investigations in this area based on our work could provide a deeper understanding of the pathology of these disorders.

14.3 Conclusions

An imbalance in excitatory-inhibitory (E-I) signaling has been reported to be a major factor in the pathogenesis of several neurological disorders including Alzheimer's disease, Parkinson's disease, autism spectrum disorders, epilepsy, etc. Despite such wide acceptance for E-I imbalance as a cause for these age-related disorders, our

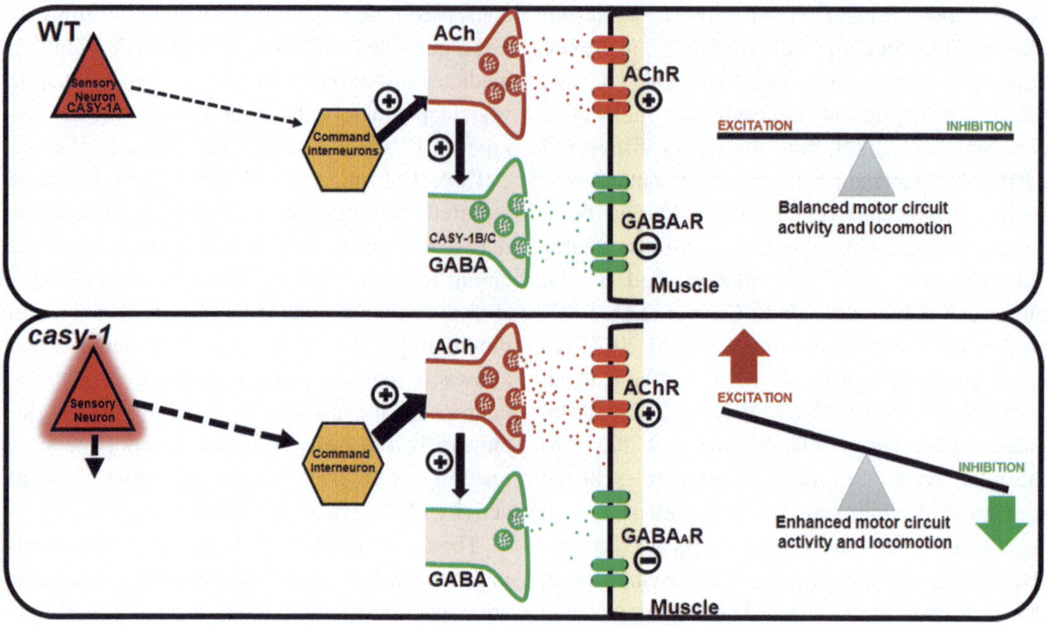

Fig. 14.2 Proposed model for CASY-1 function in regulating excitation-inhibition (E-I) balance at *C. elegans* NMJ. Mutants in *casy-1* have an E-I imbalance at the NMJ. Cholinergic transmission is enhanced, while the GABAergic transmission is reduced resulting in an overall increase in the motor circuit activity. The CASY-1A isoform functions in sensory neurons to maintain excitatory cholinergic transmission, while the shorter isoforms, CASY-1B and CASY-1C, function in GABAergic motor neurons to maintain normal GABA release

understanding for the key genetic factors and molecular mechanisms regulating this balance has just started to evolve. This review is an effort to establish the *C. elegans* NMJ as a viable model to conduct such studies. The *C. elegans* NMJ provides an efficient system amenable to a wide variety of behavioral, genetic, and pharmacological assays that allows well-defined readouts to decipher mechanisms coordinating this balance at the NMJ.

Future investigations aiming to identify other genetic factors regulating this balance using *C. elegans* NMJ as a model might provide novel insights into mechanisms maintaining this balance in neural circuits.

Funding ST was funded by the Council of Scientific and Industrial Research (CSIR) for a graduate fellowship. KB is an Intermediate Fellow of the Wellcome Trust- DBT India Alliance (Grant no. IA/I/12/1/500516) and thanks the Alliance for funding support.

References

Artan M, Jeong DE, Lee D, Kim YI, Son HG, Husain Z, Kim J, Altintas O, Kim K, Alcedo J, Lee SJ (2016) Food-derived sensory cues modulate longevity via distinct neuroendocrine insulin-like peptides. Genes Dev 30:1047–1057

Banerjee N, Bhattacharya R, Gorczyca M, Collins KM, Francis MM (2017) Local neuropeptide signaling modulates serotonergic transmission to shape the temporal organization of *C. elegans* egg-laying behavior. PLoS Genet 13:e1006697

Bargmann CI (2012) Beyond the connectome: how neuromodulators shape neural circuits. Bioessays 34:458–465

Bargmann CI, Horvitz HR (1991) Chemosensory neurons with overlapping functions direct chemotaxis to multiple chemicals in *C. elegans*. Neuron 7:729–742

Bargmann CI, Hartwieg E, Horvitz HR (1993) Odorant-selective genes and neurons mediate olfaction in *C. elegans*. Cell 74:515–527

Bhardwaj A, Thapliyal S, Dahiya Y, Babu K (2018) FLP-18 functions through the G-protein-coupled receptors NPR-1 and NPR-4 to modulate reversal length in *Caenorhabditis elegans*. J Neurosci Off J Soc Neurosci 38:4641–4654

Blackmore M, Letourneau PC (2006) L1, beta1 integrin, and cadherins mediate axonal regeneration in the embryonic spinal cord. J Neurobiol 66:1564–1583

Braeckman BP, Vanfleteren JR (2007) Genetic control of longevity in C. elegans. Exp Gerontol 42:90–98

Braunewell KH (2005) The darker side of Ca2+ signaling by neuronal Ca2+-sensor proteins: from Alzheimer's disease to cancer. Trends Pharmacol Sci 26:345–351

Braunewell KH, Gundelfinger ED (1999) Intracellular neuronal calcium sensor proteins: a family of EF-hand calcium-binding proteins in search of a function. Cell Tissue Res 295:1–12

Bretscher AJ, Kodama-Namba E, Busch KE, Murphy RJ, Soltesz Z, Laurent P, de Bono M (2011) Temperature, oxygen, and salt-sensing neurons in *C. elegans* are carbon dioxide sensors that control avoidance behavior. Neuron 69:1099–1113

Burgoyne RD (2007) Neuronal calcium sensor proteins: generating diversity in neuronal Ca2+ signalling. Nat Rev Neurosci 8:182–193

Burgoyne RD, Haynes LP (2012) Understanding the physiological roles of the neuronal calcium sensor proteins. Mol Brain 5:2

Cavallaro U, Dejana E (2011) Adhesion molecule signalling: not always a sticky business. Nat Rev Mol Cell Biol 12:189–197

Chalasani SH, Kato S, Albrecht DR, Nakagawa T, Abbott LF, Bargmann CI (2010) Neuropeptide feedback modifies odor-evoked dynamics in *Caenorhabditis elegans* olfactory neurons. Nat Neurosci 13:615–621

Chalfie M, Sulston JE, White JG, Southgate E, Thomson JN, Brenner S (1985) The neural circuit for touch sensitivity in *Caenorhabditis elegans*. J Neurosci Off J Soc Neurosci 5:956–964

Chang YJ, Burton T, Ha L, Huang Z, Olajubelo A, Li C (2015) Modulation of locomotion and reproduction by FLP neuropeptides in the nematode *Caenorhabditis elegans*. PLoS One 10:e0135164

Chen YC, Chen HJ, Tseng WC, Hsu JM, Huang TT, Chen CH, Pan CL (2016) A *C. elegans* thermosensory circuit regulates longevity through crh-1/CREB-dependent flp-6 neuropeptide signaling. Dev Cell 39(2):209–223

Cheng A, McDonald NA, Connolly CN (2005) Cell surface expression of 5-hydroxytryptamine type 3 receptors is promoted by RIC-3. J Biol Chem 280:22502–22507

Cheong MC, Artyukhin AB, You YJ, Avery L (2015) An opioid-like system regulating feeding behavior in *C. elegans*. elife 4:e06683

Chih B, Engelman H, Scheiffele P (2005) Control of excitatory and inhibitory synapse formation by neuroligins. Science 307:1324–1328

Cornils A, Gloeck M, Chen Z, Zhang Y, Alcedo J (2011) Specific insulin-like peptides encode sensory information to regulate distinct developmental processes. Development 138:1183–1193

Cremer H, Chazal G, Goridis C, Represa A (1997) NCAM is essential for axonal growth and fasciculation in the hippocampus. Mol Cell Neurosci 8:323–335

Culetto E, Baylis HA, Richmond JE, Jones AK, Fleming JT, Squire MD, Lewis JA, Sattelle DB (2004) The *Caenorhabditis elegans* unc-63 gene encodes a levamisole-sensitive nicotinic acetylcholine receptor alpha subunit. J Biol Chem 279:42476–42483

Culotti JG, Russell RL (1978) Osmotic avoidance defective mutants of the nematode *Caenorhabditis elegans*. Genetics 90:243–256

Dalva MB, McClelland AC, Kayser MS (2007) Cell adhesion molecules: signalling functions at the synapse. Nat Rev Neurosci 8:206–220

de Bono M, Maricq AV (2005) Neuronal substrates of complex behaviors in *C. elegans*. Annu Rev Neurosci 28:451–501

Delaney CE, Chen AT, Graniel JV, Dumas KJ, Hu PJ (2017) A histone H4 lysine 20 methyltransferase couples environmental cues to sensory neuron control of developmental plasticity. Development 144:1273–1282

Eichler SA, Meier JC (2008) E-I balance and human diseases – from molecules to networking. Front Mol Neurosci 1:2

Engel AG, Ohno K, Milone M, Wang HL, Nakano S, Bouzat C, Pruitt JN II, Hutchinson DO, Brengman JM, Bren N, Sieb JP, Sine SM (1996) New mutations in acetylcholine receptor subunit genes reveal heterogeneity in the slow-channel congenital myasthenic syndrome. Hum Mol Genet 5:1217–1227

Flaherty KM, Zozulya S, Stryer L, McKay DB (1993) Three-dimensional structure of recoverin, a calcium sensor in vision. Cell 75:709–716

Fleming JT, Squire MD, Barnes TM, Tornoe C, Matsuda K, Ahnn J, Fire A, Sulston JE, Barnard EA, Sattelle DB, Lewis JA (1997) *Caenorhabditis elegans* levamisole resistance genes lev-1, unc-29, and unc-38 encode functional nicotinic acetylcholine receptor subunits. J Neurosci Off J Soc Neurosci 17:5843–5857

Francis MM, Evans SP, Jensen M, Madsen DM, Mancuso J, Norman KR, Maricq AV (2005) The Ror receptor tyrosine kinase CAM-1 is required for ACR-16-mediated synaptic transmission at the *C. elegans* neuromuscular junction. Neuron 46:581–594

Graf ER, Zhang X, Jin SX, Linhoff MW, Craig AM (2004) Neurexins induce differentiation of GABA and glutamate postsynaptic specializations via neuroligins. Cell 119:1013–1026

Halevi S, McKay J, Palfreyman M, Yassin L, Eshel M, Jorgensen E, Treinin M (2002) The *C. elegans* ric-3 gene is required for maturation of nicotinic acetylcholine receptors. EMBO J 21:1012–1020

Halevi S, Yassin L, Eshel M, Sala F, Sala S, Criado M, Treinin M (2003) Conservation within the RIC-3 gene family. Effectors of mammalian nicotinic acetylcholine receptor expression. J Biol Chem 278:34411–34417

Hansen SM, Berezin V, Bock E (2008) Signaling mechanisms of neurite outgrowth induced by the cell adhesion molecules NCAM and N-cadherin. Cell Mol Life Sci 65:3809–3821

Harris G, Mills H, Wragg R, Hapiak V, Castelletto M, Korchnak A, Komuniecki RW (2010) The monoaminergic modulation of sensory-mediated aversive responses in *Caenorhabditis elegans* requires glutamatergic/peptidergic cotransmission. J Neurosci Off J Soc Neurosci 30:7889–7899

Hedgecock EM, Russell RL (1975) Normal and mutant thermotaxis in the nematode *Caenorhabditis elegans*. Proc Natl Acad Sci U S A 72:4061–4065

Hintsch G, Zurlinden A, Meskenaite V, Steuble M, Fink-Widmer K, Kinter J, Sonderegger P (2002) The calsyntenins – a family of postsynaptic membrane proteins with distinct neuronal expression patterns. Mol Cell Neurosci 21:393–409

Hirano S, Takeichi M (2012) Cadherins in brain morphogenesis and wiring. Physiol Rev 92:597–634

Hu Z, Pym EC, Babu K, Vashlishan Murray AB, Kaplan JM (2011) A neuropeptide-mediated stretch response links muscle contraction to changes in neurotransmitter release. Neuron 71:92–102

Hu Z, Vashlishan-Murray AB, Kaplan JM (2015) NLP-12 engages different UNC-13 proteins to potentiate tonic and evoked release. J Neurosci Off J Soc Neurosci 35:1038–1042

Hung WL, Wang Y, Chitturi J, Zhen M (2014) A *Caenorhabditis elegans* developmental decision requires insulin signaling-mediated neuron-intestine communication. Development 141:1767–1779

Ikeda DD, Duan Y, Matsuki M, Kunitomo H, Hutter H, Hedgecock EM, Iino Y (2008) CASY-1, an ortholog of calsyntenins/alcadeins, is essential for learning in *Caenorhabditis elegans*. Proc Natl Acad Sci U S A 105:5260–5265

Isaacson JS, Scanziani M (2011) How inhibition shapes cortical activity. Neuron 72:231–243

Jones AK, Sattelle DB (2004) Functional genomics of the nicotinic acetylcholine receptor gene family of the nematode, *Caenorhabditis elegans*. Bioessays 26:39–49

Jospin M, Qi YB, Stawicki TM, Boulin T, Schuske KR, Horvitz HR, Bessereau JL, Jorgensen EM, Jin Y (2009) A neuronal acetylcholine receptor regulates the balance of muscle excitation and inhibition in *Caenorhabditis elegans*. PLoS Biol 7:e1000265

Kao G, Nordenson C, Still M, Ronnlund A, Tuck S, Naredi P (2007) ASNA-1 positively regulates insulin secretion in *C. elegans* and mammalian cells. Cell 128:577–587

Kaplan JM, Horvitz HR (1993) A dual mechanosensory and chemosensory neuron in *Caenorhabditis elegans*. Proc Natl Acad Sci U S A 90:2227–2231

Ko J, Fuccillo MV, Malenka RC, Sudhof TC (2009) LRRTM2 functions as a neurexin ligand in promoting excitatory synapse formation. Neuron 64:791–798

Konecna A, Frischknecht R, Kinter J, Ludwig A, Steuble M, Meskenaite V, Indermuhle M, Engel M, Cen C, Mateos JM, Streit P, Sonderegger P (2006) Calsyntenin-1 docks vesicular cargo to kinesin-1. Mol Biol Cell 17:3651–3663

Lansdell SJ, Gee VJ, Harkness PC, Doward AI, Baker ER, Gibb AJ, Millar NS (2005) RIC-3 enhances functional expression of multiple nicotinic acetylcholine receptor subtypes in mammalian cells. Mol Pharmacol 68:1431–1438

Laurent P, Soltesz Z, Nelson GM, Chen C, Arellano-Carbajal F, Levy E, de Bono M (2015) Decoding a neural circuit controlling global animal state in *C. elegans*. elife 4:4

Lei M, Xu H, Li Z, Wang Z, O'Malley TT, Zhang D, Walsh DM, Xu P, Selkoe DJ, Li S (2016) Soluble Abeta oligomers impair hippocampal LTP by disrupting glutamatergic/GABAergic balance. Neurobiol Dis 85:111–121

Leinwand SG, Chalasani SH (2013) Neuropeptide signaling remodels chemosensory circuit composition in *Caenorhabditis elegans*. Nat Neurosci 16:1461–1467

Leinwand SG, Chalasani SH (2014) From genes to circuits and behaviors: neuropeptides expand the coding potential of the nervous system. WormBook 3:e27730

Lewis JA, Wu CH, Berg H, Levine JH (1980) The genetics of levamisole resistance in the nematode *Caenorhabditis elegans*. Genetics 95:905–928

Li C, Kim K (2008) Neuropeptides. WormBook 25:1–36

Li C, Timbers TA, Rose JK, Bozorgmehr T, McEwan A, Rankin CH (2013) The FMRFamide-related neuropeptide FLP-20 is required in the mechanosensory neurons during memory for massed training in *C. elegans*. Learn Mem 20:103–108

Linhoff MW, Lauren J, Cassidy RM, Dobie FA, Takahashi H, Nygaard HB, Airaksinen MS, Strittmatter SM, Craig AM (2009) An unbiased expression screen for synaptogenic proteins identifies the LRRTM protein family as synaptic organizers. Neuron 61:734–749

Mann EO, Mody I (2008) The multifaceted role of inhibition in epilepsy: seizure-genesis through excessive GABAergic inhibition in autosomal dominant nocturnal frontal lobe epilepsy. Curr Opin Neurol 21:155–160

McGrath PT, Xu Y, Ailion M, Garrison JL, Butcher RA, Bargmann CI (2011) Parallel evolution of domesticated Caenorhabditis species targets pheromone receptor genes. Nature 477:321–325

Missler M, Sudhof TC, Biederer T (2012) Synaptic cell adhesion. Cold Spring Harb Perspect Biol 4:a005694

Mori I, Ohshima Y (1995) Neural regulation of thermotaxis in *Caenorhabditis elegans*. Nature 376:344–348

Muchowski PJ, Wacker JL (2005) Modulation of neurodegeneration by molecular chaperones. Nat Rev Neurosci 6:11–22

Nacher J, Guirado R, Castillo-Gomez E (2013) Structural plasticity of interneurons in the adult brain: role of PSA-NCAM and implications for psychiatric disorders. Neurochem Res 38:1122–1133

Nagy S, Tramm N, Sanders J, Iwanir S, Shirley IA, Levine E, Biron D (2014) Homeostasis in *C. elegans* sleep is characterized by two behaviorally and genetically distinct mechanisms. elife 3:e04380

Nath RD, Chow ES, Wang H, Schwarz EM, Sternberg PW (2016) *C. elegans* stress-induced sleep emerges from the collective action of multiple neuropeptides. Curr Biol 26:2446–2455

Nathoo AN, Moeller RA, Westlund BA, Hart AC (2001) Identification of neuropeptide-like protein gene families in *Caenorhabditis elegans* and other species. Proc Natl Acad Sci U S A 98:14000–14005

Nelson SB, Valakh V (2015) Excitatory/inhibitory balance and circuit homeostasis in autism spectrum disorders. Neuron 87:684–698

Nelson MD, Trojanowski NF, George-Raizen JB, Smith CJ, Yu CC, Fang-Yen C, Raizen DM (2013) The neuropeptide NLP-22 regulates a sleep-like state in *Caenorhabditis elegans*. Nat Commun 4:2846

Ohtsuka K, Suzuki T (2000) Roles of molecular chaperones in the nervous system. Brain Res Bull 53:141–146

Petrash HA, Philbrook A, Haburcak M, Barbagallo B, Francis MM (2013) ACR-12 ionotropic acetylcholine receptor complexes regulate inhibitory motor neuron activity in *Caenorhabditis elegans*. J Neurosci Off J Soc Neurosci 33:5524–5532

Pettem KL, Yokomaku D, Luo L, Linhoff MW, Prasad T, Connor SA, Siddiqui TJ, Kawabe H, Chen F, Zhang L, Rudenko G, Wang YT, Brose N, Craig AM (2013) The specific alpha-neurexin interactor calsyntenin-3 promotes excitatory and inhibitory synapse development. Neuron 80:113–128

Pierce SB, Costa M, Wisotzkey R, Devadhar S, Homburger SA, Buchman AR, Ferguson KC, Heller J, Platt DM, Pasquinelli AA, Liu LX, Doberstein SK, Ruvkun G (2001) Regulation of DAF-2 receptor signaling by human insulin and ins-1, a member of the unusually large and diverse *C. elegans* insulin gene family. Genes Dev 15:672–686

Qiu XB, Shao YM, Miao S, Wang L (2006) The diversity of the DnaJ/Hsp40 family, the crucial partners for Hsp70 chaperones. Cell Mol Life Sci 63:2560–2570

Rand JB (2007) Acetylcholine. WormBook 30:1–21

Reboul J, Vaglio P, Tzellas N, Thierry-Mieg N, Moore T, Jackson C, Shin-i T, Kohara Y, Thierry-Mieg D, Thierry-Mieg J, Lee H, Hitti J, Doucette-Stamm L, Hartley JL, Temple GF, Brasch MA, Vandenhaute J, Lamesch PE, Hill DE, Vidal M (2001) Open-reading-frame sequence tags (OSTs) support the existence of at least 17,300 genes in *C. elegans*. Nat Genet 27:332–336

Richmond JE, Jorgensen EM (1999) One GABA and two acetylcholine receptors function at the *C. elegans* neuromuscular junction. Nat Neurosci 2:791–797

Safdie G, Liewald JF, Kagan S, Battat E, Gottschalk A, Treinin M (2016) RIC-3 phosphorylation enables dual regulation of excitation and inhibition of *Caenorhabditis elegans* muscle. Mol Biol Cell 27:2994–3003

Sanes JR, Yamagata M (2009) Many paths to synaptic specificity. Annu Rev Cell Dev Biol 25:161–195

Sassa T, Murayama T, Maruyama IN (2013) Strongly alkaline pH avoidance mediated by ASH sensory neurons in *C. elegans*. Neurosci Lett 555:248–252

Seaton G, Hogg EL, Jo J, Whitcomb DJ, Cho K (2011) Sensing change: the emerging role of calcium sensors in neuronal disease. Semin Cell Dev Biol 22:530–535

Shapiro L, Love J, Colman DR (2007) Adhesion molecules in the nervous system: structural insights into function and diversity. Annu Rev Neurosci 30:451–474

Sheng L, Leshchyns'ka I, Sytnyk V (2013) Cell adhesion and intracellular calcium signaling in neurons. Cell Commun Signal 11:94

Sherman MY, Goldberg AL (2001) Cellular defenses against unfolded proteins: a cell biologist thinks about neurodegenerative diseases. Neuron 29:15–32

Siddiqui TJ, Pancaroglu R, Kang Y, Rooyakkers A, Craig AM (2010) LRRTMs and neuroligins bind neurexins with a differential code to cooperate in glutamate synapse development. J Neurosci Off J Soc Neurosci 30:7495–7506

Stawicki TM, Takayanagi-Kiya S, Zhou K, Jin Y (2013) Neuropeptides function in a homeostatic manner to modulate excitation-inhibition imbalance in *C. elegans*. PLoS Genet 9:e1003472

Stitzel JA (2008) Naturally occurring genetic variability in the nicotinic acetylcholine receptor alpha4 and alpha7 subunit genes and phenotypic diversity in humans and mice. Front Biosci 13:477–491

Sudhof TC (2008) Neuroligins and neurexins link synaptic function to cognitive disease. Nature 455:903–911

Takayanagi-Kiya S, Jin Y (2016) Altered function of the DnaJ family cochaperone DNJ-17 modulates locomotor circuit activity in a *Caenorhabditis elegans* seizure model. G3 (Bethesda) 6:2165–2171

Tessier-Lavigne M, Goodman CS (1996) The molecular biology of axon guidance. Science 274:1123–1133

Thapliyal S, Ravindranath S, Babu K (2018a) Regulation of glutamate signaling in the sensorimotor circuit by CASY-1A/calsyntenin in *Caenorhabditis elegans*. Genetics 208:1553–1564

Thapliyal S, Vasudevan A, Dong Y, Bai J, Koushika SP, Babu K (2018b) The C-terminal of CASY-1/Calsyntenin regulates GABAergic synaptic transmission at the *Caenorhabditis elegans* neuromuscular junction. PLoS Genet 14:e1007263

Togashi H, Abe K, Mizoguchi A, Takaoka K, Chisaka O, Takeichi M (2002) Cadherin regulates dendritic spine morphogenesis. Neuron 35:77–89

Togashi H, Miyoshi J, Honda T, Sakisaka T, Takai Y, Takeichi M (2006) Interneurite affinity is regulated by heterophilic nectin interactions in concert with the cadherin machinery. J Cell Biol 174:141–151

Togashi H, Sakisaka T, Takai Y (2009) Cell adhesion molecules in the central nervous system. Cell Adhes Migr 3:29–35

Touroutine D, Fox RM, Von Stetina SE, Burdina A, Miller DM III, Richmond JE (2005) acr-16 encodes an essential subunit of the levamisole-resistant nicotinic receptor at the *Caenorhabditis elegans* neuromuscular junction. J Biol Chem 280:27013–27021

Unwin N (2005) Refined structure of the nicotinic acetylcholine receptor at 4A resolution. J Mol Biol 346:967–989

Walsh FS, Doherty P (1997) Neural cell adhesion molecules of the immunoglobulin superfamily: role in axon growth and guidance. Annu Rev Cell Dev Biol 13:425–456

Ward S (1973) Chemotaxis by the nematode *Caenorhabditis elegans*: identification of attractants and analysis of the response by use of mutants. Proc Natl Acad Sci U S A 70:817–821

White JG, Southgate E, Thomson JN, Brenner S (1976) The structure of the ventral nerve cord of *Caenorhabditis elegans*. Philos Trans R Soc Lond Ser B Biol Sci 275:327–348

White JG, Southgate E, Thomson JN, Brenner S (1986) The structure of the nervous system of the nematode *Caenorhabditis elegans*. Philos Trans R Soc Lond Ser B Biol Sci 314:1–340

Yasuda S, Tanaka H, Sugiura H, Okamura K, Sakaguchi T, Tran U, Takemiya T, Mizoguchi A, Yagita Y, Sakurai T, De Robertis EM, Yamagata K (2007) Activity-induced protocadherin arcadlin regulates dendritic spine number by triggering N-cadherin endocytosis via TAO2beta and p38 MAP kinases. Neuron 56:456–471

Yizhar O, Fenno LE, Prigge M, Schneider F, Davidson TJ, O'Shea DJ, Sohal VS, Goshen I, Finkelstein J, Paz JT, Stehfest K, Fudim R, Ramakrishnan C, Huguenard JR, Hegemann P, Deisseroth K (2011) Neocortical excitation/inhibition balance in information processing and social dysfunction. Nature 477:171–178

Zhou K, Cherra SJ III, Goncharov A, Jin Y (2017) Asynchronous cholinergic drive correlates with excitation-inhibition imbalance via a neuronal Ca2+ sensor protein. Cell Rep 19:1117–1129

Part V

Cell Surface Macromolecules: Infection, Immunity and Disease

Induction of Apoptosis in Metastatic Breast Cancer Cells: XV. Downregulation of DNA Polymerase-α – Helicase Complex (Replisomes) and Glyco-Genes

15

Subhash C. Basu, Patrick Boyle, Rui Ma, Arun Agarwal, Manju Basu, Joseph R. Moskal, Sipra Banerjee, and Narendra Tuteja

Abstract

In normal and cancer cells, successful cell division requires accurate duplication of chromosomal DNA. All cells require a multiprotein DNA duplication system (replisomes) for their existence. However, death of normal cells in our body occurs through the apoptotic process. During apoptotic process several crucial genes are downregulated with the upregulation of caspase pathways, leading to ultimate degradation of genomic DNA. In metastatic cancer cells (SKBR-3, MCF -7, and MDA-462), this process is inhibited to achieve immortality as well as overexpression of the enzymes for the synthesis of marker molecules. It is believed that the GSL of the lacto family such as Le^X, SA-Le^X, Le^Y, Le^a, and Le^b are markers on the human colon and breast cancer cells. Recently, we have characterized that a few apoptotic chemicals (*cis*-platin, L-PPMP, D-PDMP, GD3 ganglioside, GD1b ganglioside, betulinic acid, tamoxifen, and melphalan) in low doses kill metastatic breast cancer cells. The apoptosis-inducing agent (e.g., *cis*-platin) showed inhibition of DNA polymerase/helicase (part of the replisomes) and also modulated (positively) a few glycolipid-glycosyltransferase (GSL-GLTs) transcriptions in the early stages (within 2 h after treatment) of apoptosis. These Lc-family GSLs are also present on the surfaces of human breast and colon carcinoma cells. It is advantageous to deliver these apoptotic chemicals through the metastatic cell surfaces containing high concentration of marker glycolipids (Lc-GSLs). Targeted application of apoptotic chemicals (in micro scale) to kill the cancer cells would be an ideal way to inhibit the metastatic growth of both breast and colon cancer cells. It was observed in three different breast cancer lines (SKBR-3, MDA-468, and MCF-7) that in 2 h very little apoptotic process had started, but predominant biochemical changes (including inactivation of replisomes) started between 6 and 24 h of the

[Supported by NIH-NINDS (NS-18005; Jacob Javits Research Grant awards) and NCI (RO1-CA-14764)] to SB and a grant-in-aid from Siemens Corporation to MB.

S. C. Basu (✉) · P. Boyle · A. Agarwal · M. Basu
University of Notre Dame, Notre Dame, IN, USA
e-mail: sbasu@nd.edu; subhash.c.basu.2@nd.edu

R. Ma
Diagnostic Division, Siemens Corporation, Shanghai, People's Republic of China

J. R. Moskal
The Falk Center for Molecular Therapeutics, Northwestern University, Evanston, IL, USA

S. Banerjee
Department of Cancer Biology, Cleveland Clinic Foundation, Cleveland, OH, USA

N. Tuteja
ICGEB, New Delhi, India

K. Chattopadhyay, S. C. Basu (eds.), *Biochemical and Biophysical Roles of Cell Surface Molecules*, Advances in Experimental Medicine and Biology 1112, https://doi.org/10.1007/978-981-13-3065-0_15

drug treatments. The contents of replisomes (replisomal complexes) during induction of apoptosis are not known. It is known that DNA helicase activities (major proteins catalyze the melting of dsDNA strands) change during apoptotic induction process. Previously DNA Helicase-III was characterized as a component of the replication complexes isolated from carcinoma cells and normal rapid growing embryonic chicken brain cells. Helicase activities were assayed by a novel method (combined immunoprecipitation-ROME assay), and DNA polymerase-alpha activities were determined by regular chain extension of nicked "ACT-DNA," by determining values obtained from +/− aphidicolin added to the incubation mixtures. Very little is known about the stability of the "replication complexes" (or replisomes) during the apoptotic process. DNA helicases are motor proteins that catalyze the melting of genomic DNA during replication, repair, and recombination processes. In all three breast carcinoma cell lines (SKBR-3, MCF-7, and MDA-468), a common trend, decrease of activities of DNA polymerase-alpha and Helicase-III (estimated and detected with a polyclonal antibody), was observed, after *cis*-platin- and L-PPMP-induced apoptosis. Previously our laboratory has documented downregulation (within 24–48 h) of several GSL-GLTs with these apoptotic reagents in breast and colon cancer cells also. Perhaps induced apoptosis would improve the prognosis in metastatic breast and colon cancer patients.

Keywords

Apoptosis · Anticancer drugs · Carcinoma cells · *Cis*-platin · Caspase-3 · Caspase-9 · DNA polymerase-α · Helicase-III · GD3 ganglioside · GD1a ganglioside · D-PDMP · L-PPMP · Replication complex · Replisomes · SAT-2 · SAT-3 · SAT-4

15.1 Introduction

A cellular phenomenon, apoptosis (also known as programmed cell death), was recognized over a century ago (Wyllie 1987; Strasser et al. 2000). In normal tissues, apoptosis is a regulated process, which plays an important role in embryonic development and cell turnover. It is also involved in immune regulation and hormone-inducing atrophy (Arends and Wyllie 1991). Apoptosis maintains homeostasis by inducing cell death in senescent cells (Fig. 15.1). On the other hand, cancer cells die in a necrotic process. Precise mechanism of auto-induction of the necrosis or externally induced death in cancer cells is not known. Cleavage of genomic DNA during onset of apoptosis was recognized first in mouse thymocytes treated with glucocorticoids (Wyllie 1980). The DNA laddering analyses after apoptotic induction by *cis*-platin were associated with induction of Caspase-8 and Caspase-3 and have been established in several metastatic carcinoma cells of human colon (Colo-205) and breast (SKBR-3, MCF-7, and MDA-468) (Basu et al. 2004a) origin. Cytotoxicity and induction of apoptosis via activation of caspases and with mitochondrial membrane potential disruption of human breast cancer cells (MCF-7 and MDA-MB-231) by two novel platinum (II) complexes Pt-10 [Pt_2 (3-ethylpyridine)$_4$(berenil)$_2$] and Pt-11 [[Pt_2 (3-butylpyridine)$_4$(berenil)$_2$] have been reported (Bielawski et al. 2013). The same berenil-Pt (II) compounds affected the cellular metabolism of estrogen-positive breast cancer cells with degradation of genomic DNA and induction of Caspase-3, Caspase-8, and Caspase-9 (Agnieszka et al. 2014). Galagin (a flavonol found in *Alpinia officinarum* galangal root) induces colon cancer cell death via the mitochondrial dysfunction and caspase-dependent DNA degradation pathway (Ha et al. 2013).

In addition to breakdown of genomic DNA, the inhibition of replication of genomic DNA

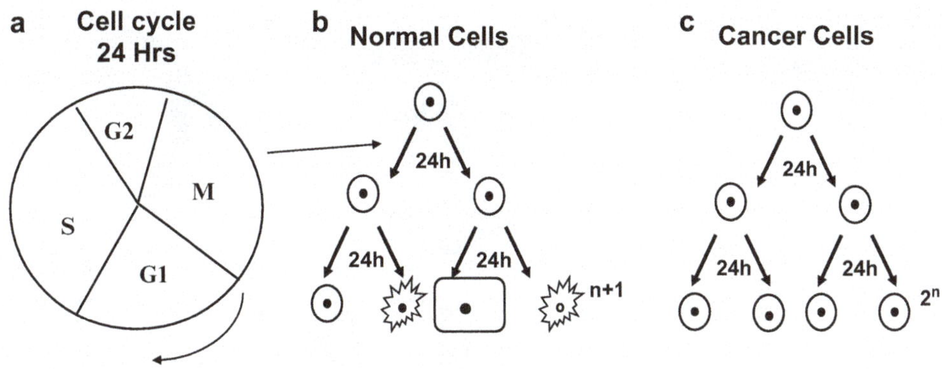

Fig. 15.1 Normal and cancer cell growth regulation. (**a**) DNA Replication-S phase. (**b**) Apoptotic Death. (**c**) Nacrotic Death. Induction of apoptosis by potential anti-cancer drugs

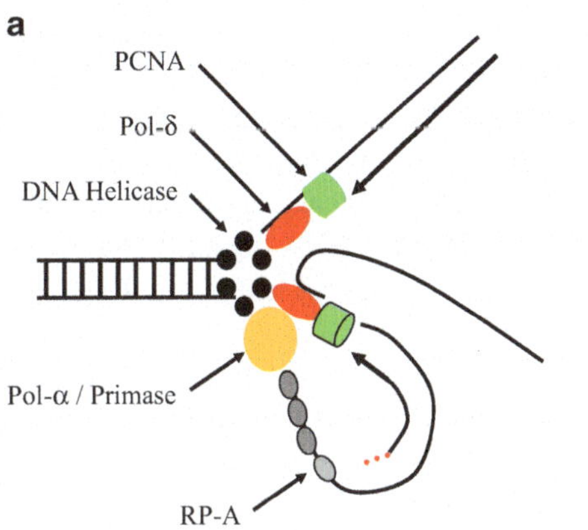

Fig. 15.2a Model for a functional replication complex

during apoptotic induction in carcinoma cells has been recognized only recently (Boyle 2005; Boyle et al. 2006; Basu et al. 2012b, 2014). DNA replication is an intricate process requiring coordinated activity of a number of enzymes and accessory proteins (Figs. 15.2a and 15.2b). These proteins must rapidly and accurately coordinate the discontinuous synthesis of the genomic DNA once per cell cycle. The idea that an elaborate multiprotein complex, the replisome, contains and coordinates all the enzymatic activities of DNA replication is highly plausible. After the formation of the initiation complex, Cdc45 along with other proteins begins the recruitment of replication-related proteins (Mimura et al. 2000; Aparicio et al. 1999; Zou and Stillman 2000; Walter and Newport 2000; Boyle 2005).

The eukaryotic genome DNA is replicated by a multienzyme complex (Boyle 2005; Figs. 15.2a and 15.2b), replisomes (Kornberg and Baker 1992; Benkovic et al. 2001; Chang et al. 1984; Kelley 1993; Kroes et al. 2006; Kunkel and Burgers 2008; Yao and O'Donnell 2009, 2010, 2016, 2017), which is expected to be associated with over 30 different proteins. These complexes contain a large number of replication-related proteins including pol-alpha, pol-delta, primase, PCNA, DNA ligase I, RF-C, RP-A, DNA topoisomerases I and II, and DNA helicase activities (or mini-chromosome maintenance factor (MCM)). Of the proteins included in the replisomes, the proteins involved in elongation of a DNA chain (pol-alpha, pol-delta, RF-C, DNA helicases, MCM, DNA ligase I, and topoisomerase II) (Fig. 15.2a) are tightly associated with the replisome complexes, while the proteins involved in the initiation (RP-A, PCNA) have a loose association with the complexes (Fig. 15.2a). The observation that replisome complexes are bound to the nuclear matrix suggests that replisomes remain immobile while the DNA is fed through these fixed complexes. A later study in HeLa cells identified similar complexes of 620 kDa and 500 kDa along with a complex of approximately 1 MDa containing proliferating cell nuclear protein (PCNA) (Tom et al. 2002).

Fig. 15.2b A model for DNA-fork propagation. (Taken from Patrick John Boyle's Ph.D. Thesis, 2005)

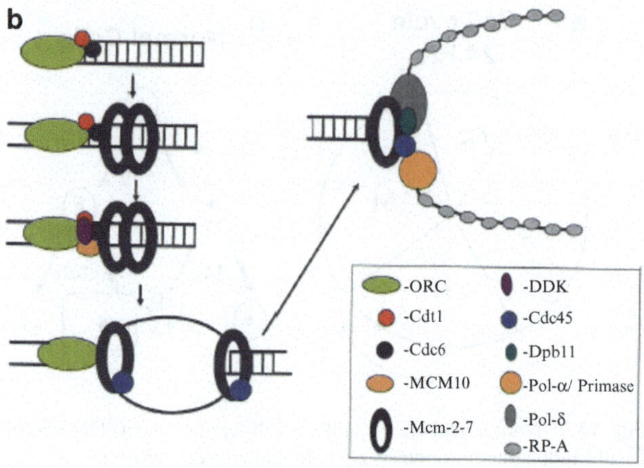

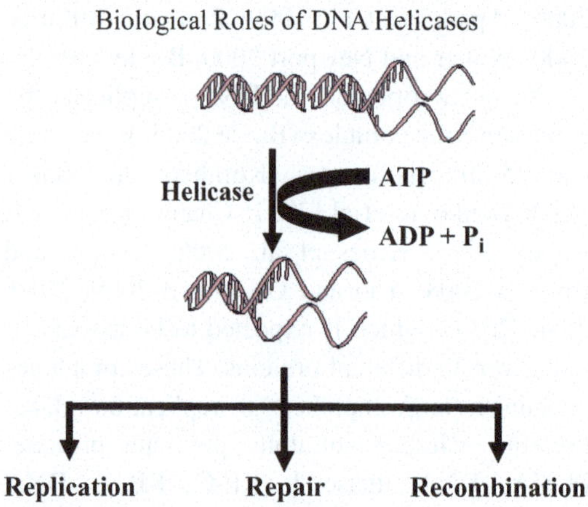

Fig. 15.3 Biological roles of DNA helicases

A 900 kDa complex containing replication protein was isolated from embryonic fibroblasts (Maga and Hubscher 1996). An 18S complex was identified from MCF-7 breast cancer cells that was competent in SV-40 replication (Jiang et al. 2002). SDS-PAGE analysis of the complex showed the presence of at least 25 proteins ranging from 20 kDa to 240 kDa. The isolation of replication-competent eukaryotic replisomes has been reported from several eukaryotic sources. An early report showed a putative replisome, isolated from human HeLa cells, was able to catalyze SV-40 replication (Malkas et al. 1990). Later reports showed the presence of a similar complex exists in murine carcinoma (FMA-3) cells (Wu et al. 1994), human breast carcinoma cells (MCF-7) (Jiang et al. 2002), and human leukemia (HL-60) cells (Lin et al. 1997). These complexes contain a large number of replication-related proteins including pol-alpha, pol-delta, primase, PCNA, DNA ligase I, RF-C, RP-A, DNA topoisomerases I and II, and DNA helicase activities (Figs. 15.2a, 15.2b and 15.3; Boyle 2005) and are able to catalyze SV-40 replication.

DNA helicases (Table 15.1) catalyze one of the most fundamental reactions in DNA metabolism. The ability of DNA to transmit its genetic information requires the transient formation of ssDNA. Processes such as DNA replication, recombination, and repair all require helicase activity to proceed. Helicases, having these fundamental roles, are being characterized from viruses (Goetz et al. 1988; Seki et al. 1990; Seo et al. 1980; Costa et al. 1999; Lohman and Bjornson 1996; Seo et al. 1993; Crute et al. 1991; Boehmer et al. 1993; Im and Muzyczka 1992; Wilson et al. 1991), prokaryotic cells (Zyskind and Smith 1977; Shlomai and Kornberg 1980; Matson et al. 1983), and eukaryotic cells (Holmes et al. 1986; Labib et al. 2000; Budd and Campbell 1995; Yan et al. 1998). DNA helicases are a class of enzymes (Table 15.1) that catalyze the energetically unfavorable separation of double-stranded DNA (dsDNA) to form single-stranded DNA (ssDNA) using the hydrolysis of nucleoside triphosphates, commonly ATP

Table 15.1 Different DNA helicases isolated from human cells

Helicase	Source	References
I	HeLa	Tuteja et al. (1990)
II (Ku)	HeLa	Tuteja et al. (1994)
III	HeLa	Tuteja et al. (1992)
IV (nucleolin)	HeLa	Tuteja et al. (1991)
V (FBP)	HeLa	Tuteja et al. (1993)
VI	HeLa	Tuteja et al. (1995)
VIII (G3BP)	HeLa	Costa et al. (1999)

Patric John Boyle, Ph.D. thesis; 2005

(Fig. 15.3), as the source of energy for the reaction (Lohman 1993; Lohman and Bjornson 1996; Tuteja and Tuteja 2004). DNA helicases generally bind to ssDNA or ssDNA/dsDNA junctions and translocate in a characteristic direction into the adjacent dsDNA area (Fig. 15.2b). Movement of the helicase into the dsDNA region disrupts the hydrogen bonds linking and the subsequent disruption of the dsDNA containing ssDNA. The movement and subsequent disruption of hydrogen bonding between the DNA strands require a source of energy. As such, all known helicases contain an intrinsic ssDNA-dependent NTPase/dNTPase (commonly ATPase) activity which hydrolyzes the gamma phosphate from a ribonucleoside or deoxyribonucleoside-5′-triphosphate and uses the resultant energy to catalyze the helicase activity. Although detailed mechanism of helicase activity is presently unknown (Tuteja and Tuteja 2004), it has been suggested that two general mechanisms may exist. The mechanism used by a particular helicase may be dependent on the oligomeric nature of the enzyme (Fig. 15.2b). Monomeric helicases are thought to simply bind to ssDNA and translocate unidirectionally (Lohman and Bjornson 1996). The level of protein assembly for most helicases investigated has been oligomeric (often hexameric) (Hotta and Stern 1978; McCulloch and Kunkel 2008). The consequence of the oligomeric structure of helicase is that the helicase will often contain multiple nucleic acid binding sites. Perhaps oligomeric helicases unwind dsDNA by binding dsDNA at several sites to all along the dsDNA, thereby separating the dsDNA (Tuteja and Tuteja 2004). Helicases have been shown to have a specific polarity of translocation along the DNA strand. Helicases can be classified as 3′- > 5′ or 5′- > 3′; after binding to ssDNA, they move according to the directions. In addition to polarity, helicases can be classified according to processivity. Processivity is defined as the number of base pairs unwound by the enzyme per binding event. Highly processive helicases are often involved in DNA replication, while the helicases with lower processivity are often involved in DNA repair (Enomoto et al. 1985).

Cis-platin is the first in a series of platinum-based compounds that have demonstrated significant and clinically relevant anticancer activity. It was first synthesized and characterized almost seven decades ago (Peyrone 1844). The antitumor activity was recognized by Rosenberg and his associates when they observed that *E. coli* grown in an electric field exhibited filamentous DNA growth as a DNA-damaging agents and inhibition of bacterial cell division (Rosenberg et al. 1965). It was reported that the cytostatic effect was caused by electrochemical reactions taking place at the platinum electrode forming *cis*-platin compound, which was further proven to have potent antitumor activity (Rosenberg et al. 1969). *Cis*-platin as a messenger of cell death by apoptosis induction was recognized by Rossi and Gaidano (2003) and our group as well, independently, in colon (Colo-205) and breast carcinoma cells (SKBR-3) (Boyle et al. 2003, 2006; Basu et al. 2004b; Boyle 2005).

It is commonly believed that the *cis*-platin binds to DNA and causes DNA damage by inter- and intra-strand lesions by binding to the N7 atoms of adjacent guanines. *Cis*-platin has been shown to induce apoptosis in breast cancer cells (SKBR-3, MCF-7, and MDA-468 (Basu et al. 2004b) through activation of Caspase-3, Caspase-8, and Caspase-9).

In addition to *cis*-platin, several other inducers of apoptosis (tamoxifen, betulinic acid, L-PPMP,

D-PDMP, GD3, and GD1b) (Fig. 15.4; Structures) appeared to induce apoptosis in three different breast carcinoma cells (SKBR3, MCF-7, and MDA-468) (Basu et al. 2012b, 2004b, 2014, 2017a; Rui et al. 2004, 2009, 2011; Boyle et al. 2006) and colon carcinoma Colo-205 cells (Basu et al. 2004a).

Our initial investigations (Kelley 1993; Kelley et al. 1993b) with prostate cancer cells (PA-2 and PA-3) established that *c*is-platin inhibited immune-purified DNA polymerase-alpha by binding to four –SH groups on the C-terminal end of the DNA polymerase-alpha chain (Fig. 15.5), displacing Zn^{+2} at the active site (Figs. 15.6 and 15.7; Kelley et al. 1993b; Bose et al. 1995, 1999). It is shown that Pt (II) binds to thiol groups (in the protein chains) much stronger than the N-7 of a guanine in free GMP or guanine in the DNA chain (Bose et al. 1995). This led to the hypothesis that *c*is-platin inhibits DNA polymerase-alpha binding to the zinc-binding domain, displacing the zinc and perturbing the structure of the domain (Fig. 15.7). This hypothesis was investigated further through a series of NMR structural studies (Bose et al. 1995, 1999) investigating the ability of the enzyme to bind *cis*-platin. It was shown first that *cis*-platin is able to bind near the zinc-binding domain—a synthetic nonapeptide from DNA polymerase-alpha (Fig. 15.6)—through the two cysteines separated by a proline (Fig. 15.7) almost 5 Å apart. It was later predicted that cis-platin binds to a 38-mer from the Zn^{+2}-binding domain of DNA polymerase-alpha and causes severe structural changes in the protein as determined by spectroscopic studies (Bose 2002). The results mentioned above regarding *c*is-platin and DNA polymerase-alpha inactivation had been determined from the in vitro enzyme studies. In the current studies, we sought to determine whether *cis*-platin inhibition of DNA polymerase-alpha occurs in the in vivo experiment with cultured cells, in some marked biological effect, apoptosis.

Cis-platin appears to be an apoptosis-inducer ultimately killing the breast cancer cells (Figs. 15.8, 15.9, 15.10, and 15.11). Our recent studies (Basu et al. 2004a, b, c; Boyle et al. 2006) indicated that *c*is-platin inactivation occurs in both DNA polymerase-alpha and Helicase-III perhaps by binding through Zn^{+2}-binding domains (Figs. 15.5, 15.6, and 15.7). Several proteins in the replisomes are predicted to contain Zn^{+2}-binding domain. Further proof is necessary to establish this hypothesis. Our recent observation showed inactivation of Helicase-III occurred in the presence of *cis*-platin at the apoptosis induction level (added 10 to 80 micromolar concentration outside the breast cancer cells), which will be discussed in the latter part of this article.

15.2 Methods

15.2.1 Cell-Culture Conditions

Human breast carcinoma cell lines (SKBR-3, MDA-468, and MCF-7) were grown in DMEM medium, supplemented with 10% fetal bovine serum, 100 unit/ml penicillin, 100 microgram/ml streptomycin, and 50 mM L-glutamine. When cells were 90% confluent, they were used for passage or harvested for biochemical investigations. Cell synchronization was performed by treating the semi-confluent cells twice each time with 0.5 mM hydroxyurea for 24 h under the same culturing conditions. Hydroxyurea was then removed, and the cells were treated with the apoptotic reagents (Fig. 15.4) in the presence or absence of serum (Basu et al. 2004a, b; Rui et al. 2004).

Breast cancer cells (MCF-7, MDA-468, and SKBR-3) were treated with cis-platin under indicated conditions (Figs. 15.8, 15.10, and 15.11). After treatment, the cells were harvested and were suspended in four volumes of 50 mM Tris-HCl (pH 8.0) buffer containing 1 mM

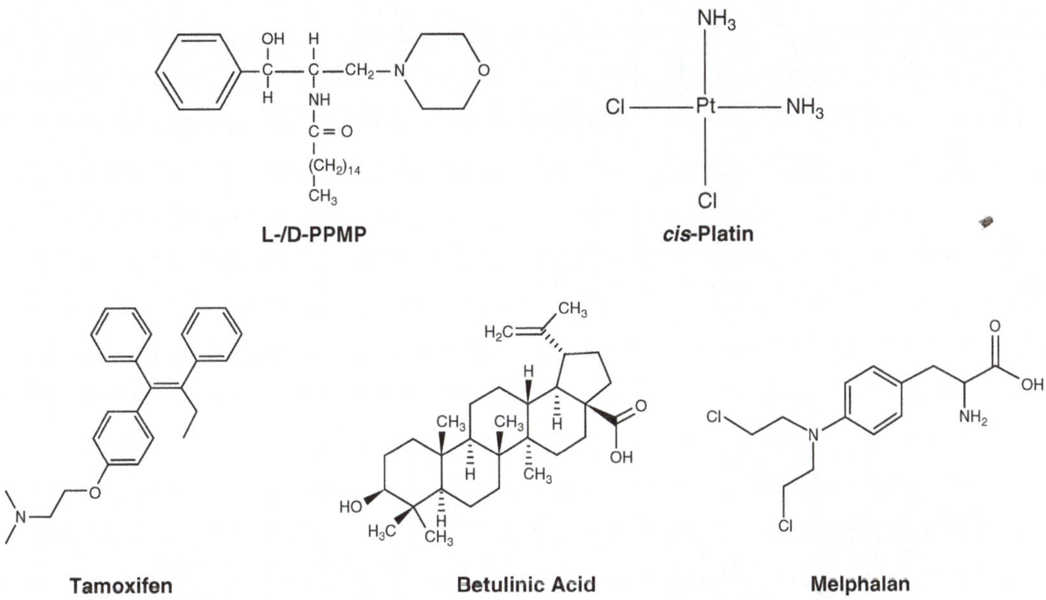

Fig. 15.4 Structures of anticancer apoptotic agents. (Basu et al. 2004a, b, c, 2012a)

HUMAN- LTDEEKYRDCERFKCPCPTCGTEN IYDNVFDGSGTDMEPSLYRCSN IDCKASPLTF
YEAST- I TDVERFKDTVT LELS CPSCDKRFPFGGIVSS NYYRVSYNGLQCKHCEQLFT PLQL

HUMAN- TVQLSNKLIMDIRRFIKKYYDGWL ICEEPT CRNRTRHL PLQFSRTGPL CPACMKATL
YEAST- TSQI EH- - - -S IRAHISLYYAGWLQCDDSTCGI V TRQVSVFGKRCLNDGCT- - - -GVM

HUMAN- QPEYSDKSLYTQ
YEAST- RYKYSDKQLYNQ

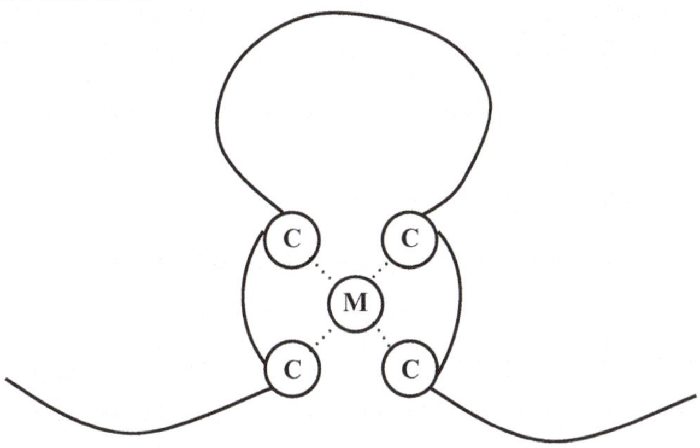

Fig. 15.5 Comparison of human DNA polymerase-α and yeast pol-I zinc-binding domain sequences

EDTA, 2 mM $MgCl_2$, 5 mM KCl, 0.1% PEG (MW. 8000), and 5 mM aminoacetonitrile bisulfate (TEMPAK). Aminoacetonitrile bisulfate was added to the buffer just prior to the homogenization. When the total DNA polymerase-alpha or Helicase-III activities were determined, the homogenate was prepared by three 15-s pulses by a Sonicator (Ultrasonics,

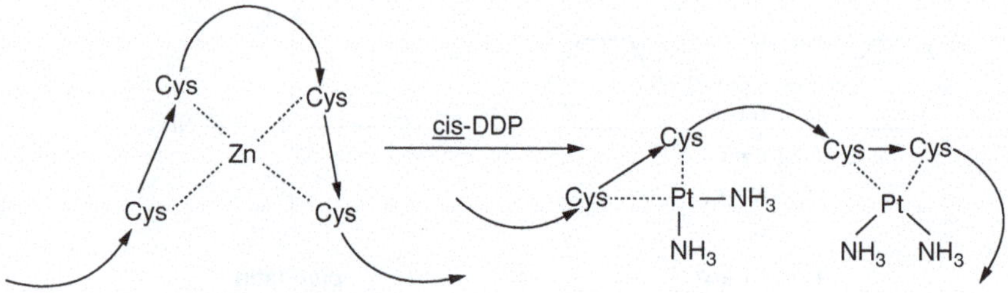

Fig. 15.6 A model for *cis*-platin inhibition of DNA polymerase-α

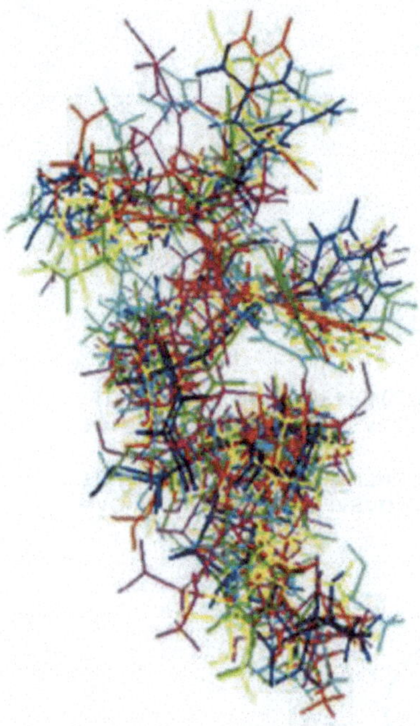

Fig. 15.7 Ensemble of simulated annealing structures including side chains of the nonapeptide (ERGFKCPCPT) of the Zn + 2-binding domain of DNA pol-α. (Bose et al. 1999)

Inc. Model W370 Sonicator; at setting 5). The homogenate was maintained in the ice water temperature (4 °C) during homogenization and ultrasonic treatment. The sonicated homogenate was then centrifuged at 16,000×g for 1 h at 4 °C. The resulting supernatant was divided (50–-200 microliter portions) in the glass baby tubes, covered with parafilm and stored at −18 °C for further experiments. These fractions were assayed for DNA polymerase-alpha (in the presence and absence of aphidicolin) as described before (Bhattacharya and Basu 1978; Ray et al. 1994; Kelley 1993; Kelley et al. 1993b; Boyle 2005) and DNA helicase (Boyle et al. 2003, 2006; Boyle 2005) activities.

15.2.2 Isolation of the DNA Polymerase-α/Helicase-III Complex by Glycerol Density Gradient Centrifugation

Three to 4 mg of soluble protein from breast cancer cell-free extract in *TEMPAK* buffer without sonic treatment were layered on the top of a 4.5-ml linear gradient of 10 to 30% glycerol in 20 mM HEPES buffer pH 8.0, 5 mM DTT, and 100 mM KCl. Centrifugation was carried out in a SW 50/1-type rotor (Beckman) at 105,000×g for 16 h at 4 °C. Fractions of 0.5 ml were assayed for immuno-detection (Fig. 15.12) of DNA polymerase-alpha and helicase-III using specific antibodies (Kelley et al. 1993a, b; Ray 1988; Ray et al. 1994; Boyle et al. 2003; Boyle 2005) as described below.

15.2.3 Dot-Blot Spot Tests for Detection of DNA Pol-α and Helicase-III Proteins in Glycerol Gradient Fractions

Drops from the glycerol gradients were subjected to Western dot-blot spot tests on nitrocellulose papers (membranes) using a modified Western blot detection procedure used to detect activation of Caspase-3 as described below. Nitrocellulose

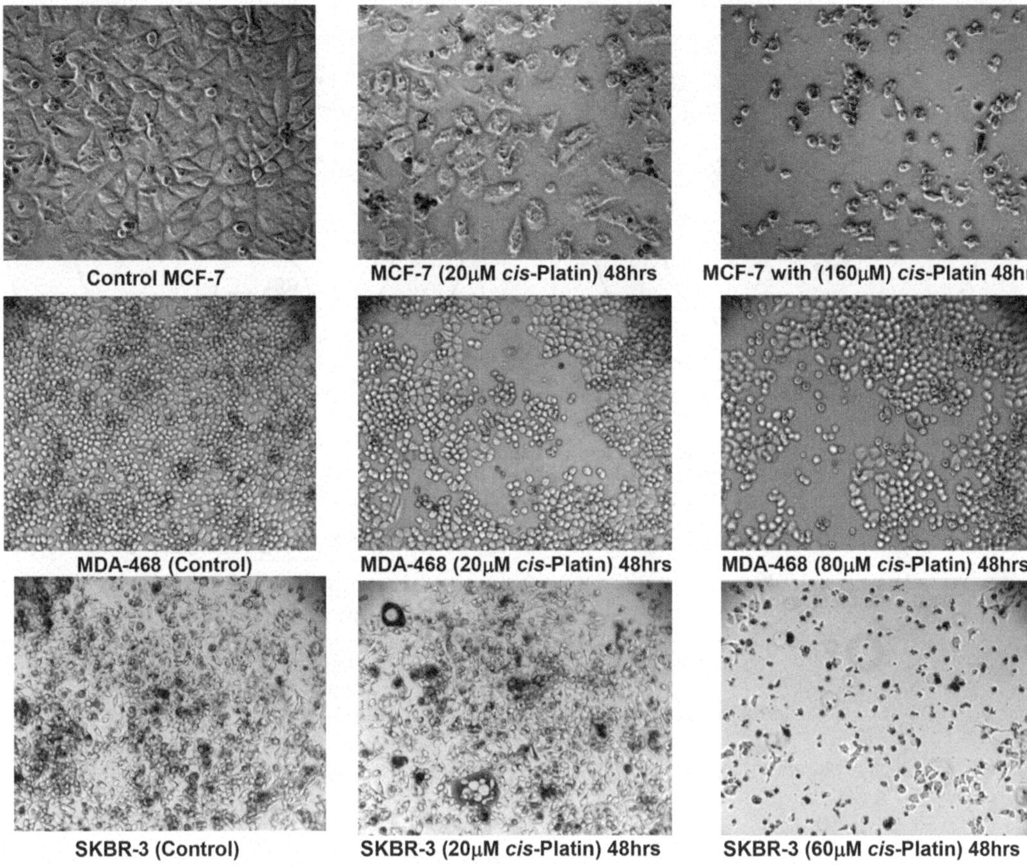

Fig. 15.8 Morphological changes of apoptotic breast cancer cells treated with *cis*-platin

papers were washed with Tris-buffered saline containing 8% bovine serum albumin for 1 h at room temperature. The membrane was then incubated for 12 h in blocking buffer containing rabbit polyclonal anti-Helicase-III antibodies (diluted 1:1000). The polyclonal antibody was raised against purified helicase from HeLa cells by Tuteja and his coworkers (Tuteja et al. 1992). Similar nitrocellulose strips were treated with mouse monoclonal anti-DNA pol-alpha SJK 132–20 (from ATCC) antibodies (diluted 1:500) prepared in our laboratory (Ray 1988; Kelley 1993; Ray et al. 1994). The membranes were then washed and incubated with anti-rabbit/or anti-mouse IgG-alkaline phosphatase conjugate (diluted 1:3000). Spots (antibody-alkaline phosphatase activity) were visualized (Fig. 15.12) using NBT-BCIP reagent in AP buffer (0.1 M Tris-HCl pH 9.5, 0.1 mM NaCl, 5 mM $MnCl_2$).

15.2.4 Morphological Changes of the Apoptotic Breast Carcinoma Cells

Three breast carcinoma cell lines of human origin (MCF-7, MDA-468, and SKBR-3) were synchronized in 0.5% FBS-containing RPMI 1640 medium for 24 h twice. When the cells were treated with *cis*-platin (Fig. 15.8) at the indicated concentration (20–160 micromolar for 48 h), loss of cell numbers was observed. Minimum loss was visible in MCF-7 cells; this may be due to the fact that MCF-7 is a multidrug-resistant

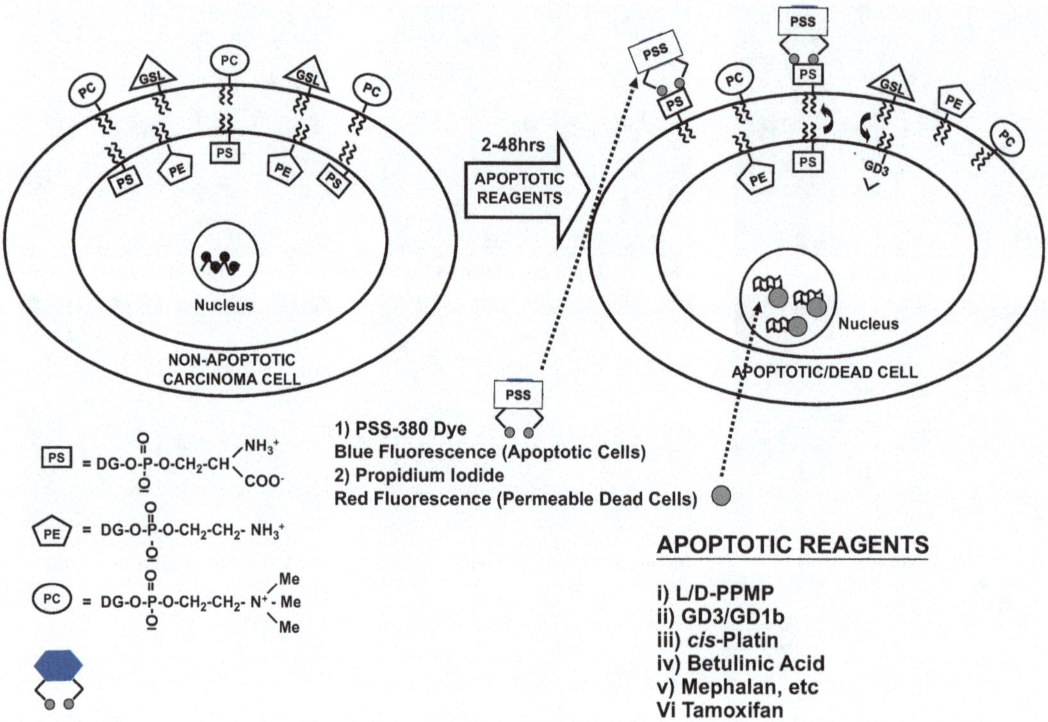

Fig. 15.9 Detection of phosphatidylserine flopping out with PSS-380 [a newly synthesized fluorescent dye. (Basu et al. 2004b, Koulov et al. 2003)]

(MDR) cell line. Nucleus condensation with all three cell lines was observed. Enlargement of volumes of cell lines suggested a marked apoptosis occurred in 48 h at 20 to 60–160 micromolar concentrations. Under similar conditions in all three cell lines, apoptosis was observed at 2-–8 micromolar L-PPMP concentrations (data has been published before; Rui et al. 2004; Rui 2008). In order to confirm the apoptosis process, the following fluorescence studies were performed.

15.2.5 Fluorescent Staining of Migration of Phosphatidylserine to the Upper Leaflet (FSMPS-UL) of the Apoptotic Cells

In a schematic diagram (Fig. 15.9), the principle of the fluorescence experiment (FSMPS-UL) is described. Breast carcinoma cells (MCF-7, MDA-468, and SKBR-3) cultured on Falcon Microslide System were synchronized two times for 24 h by treatment with medium containing 0.5 mM hydroxyurea (as given above) before treatment with apoptotic agents (e.g., *cis*-platin/L-PPMP). The microplate (attached with cells) was then washed two times with 5 mM TES buffer (*N*-tris[hydroxymethyl]-2-aminoethanesulfonic acid) containing 150 mM NaCl. The cells were then incubated in TES buffer containing 25 μM PSS-380 and 0.25 microgram/ml propidium iodide for 10 min at 37 °C. The cell-staining buffer was removed; the wells were washed with 0.2 mL TES buffer mixture and soaked in 0.1 mL of fresh TES buffer mixture for fluorescence observation (Fig. 15.10 after 6 h and Fig. 15.11 after 24 h of treatment with *cis*-platin). This is an easier and cheaper method (Hanshaw and Smith 2005) than the Ca^{2+}–mediated annexin V binding of the phosphatidylserine on the cell surfaces (Gerke et al. 2005).

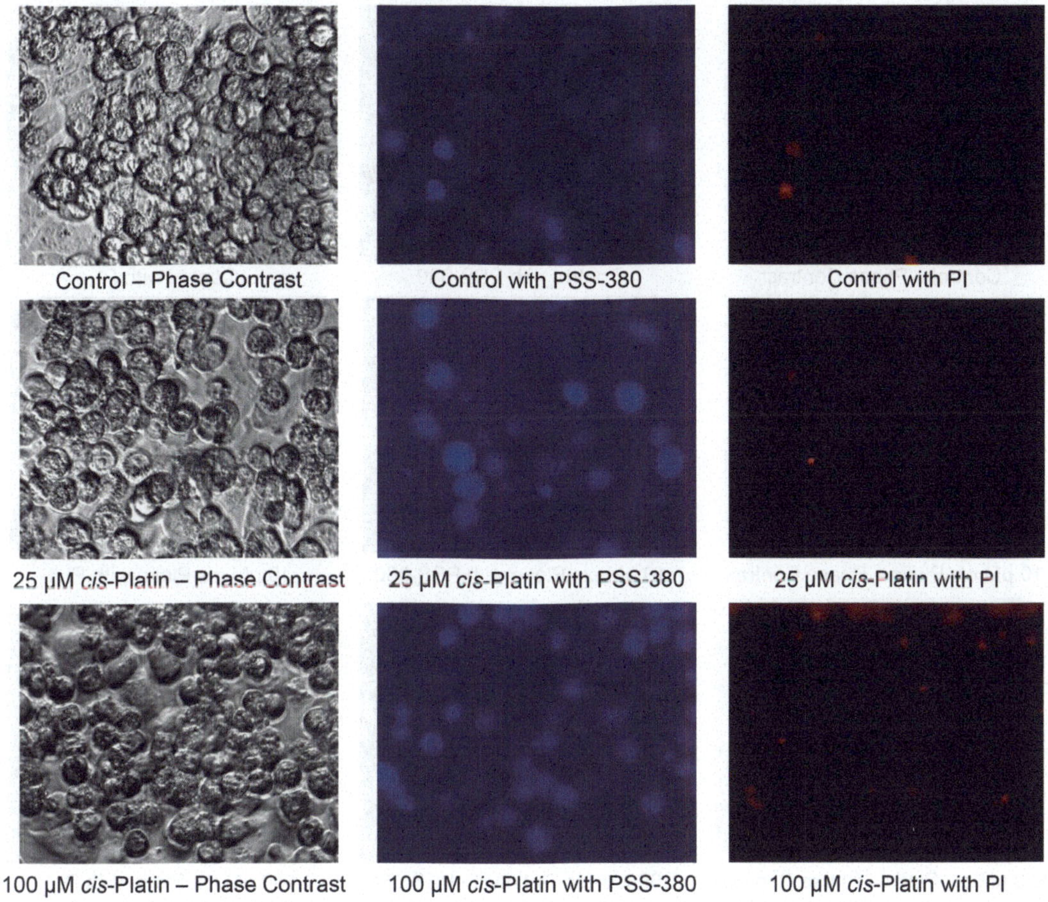

Fig. 15.10 Staining of apoptotic SKBR-3 cells with PSS-380 and propidium iodide after treatment with *cis*-platin for 6 h

15.2.6 ROME Assay (Radioactive Oligonucleotides in Membrane Filtration Effluent Assay)

Helicase activities were routinely measured by the extent of [3H]-oligonucleotides (>60 nt) released from a [3H] calf thymus DNA substrate prepared in the laboratory (Bhattacharya and Basu 1978; Bhattacharya et al. 1979). Helicase activity was determined by incubating [^{3}H]-calf thymus DNA (100,000 cpm/microgram CT DNA) in 20 mM Tris-HCl pH 8.0, 1 mM $MgCl_2$, 4 mM ATP, 150 mM KCl, 8 mM DTT, 4% sucrose, and 100 microgram/ml BSA for 2 h at 37 °C. The assay mixture was then diluted to 0.15 ml with water and filtered through a Microcon-30 centrifugal filtration device (Fig. 15.13).

An aliquot of the supernatant was spotted on Whatman 1 MM chromatography paper and subjected to descending chromatography using 0.1 M K-PO4 buffer pH 6.8 (containing 60 g/100 ml $(NH_4)_2SO_4$ plus 2% *n*-propanol) (Fig. 15.13). The radioactivity at the origin and at the next 7 inches was quantitated (Boyle et al. 2003, 2006) by liquid scintillation-toluene system (Moskal et al. 1974).

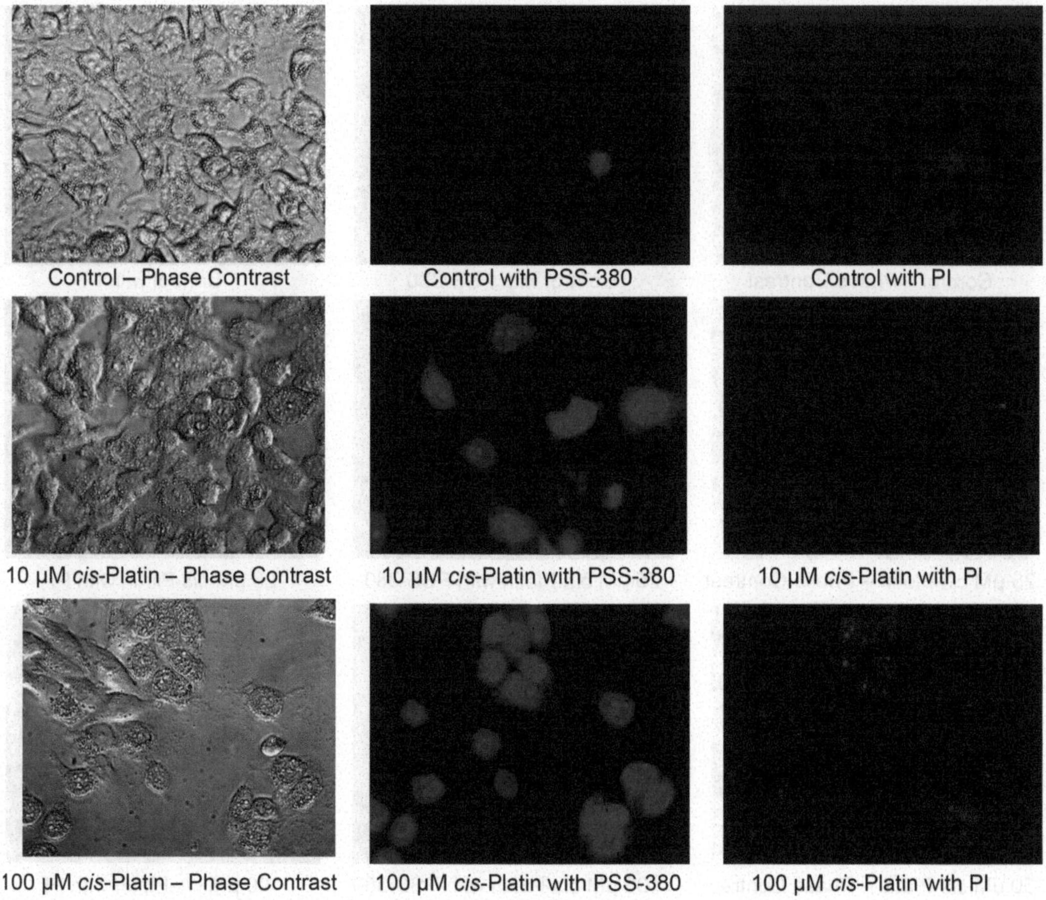

Fig. 15.11 Staining of apoptotic SKBR-3 cells with PSS-380 and PI after treatment with *cis*-platin for 24 h

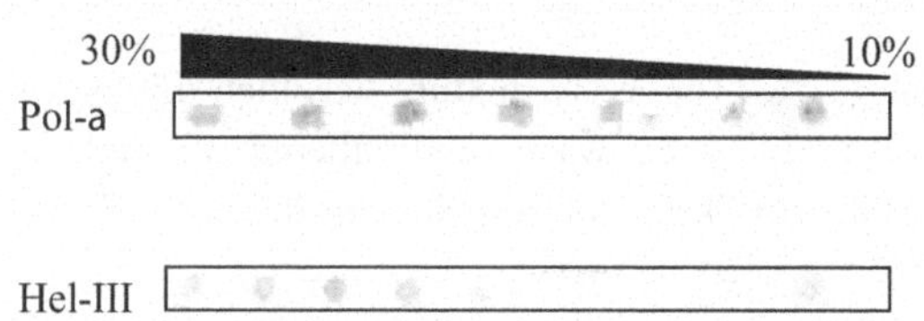

Fig. 15.12 Detection of DNA polymerase-alpha and Helicase-III proteins by specific antibodies. (DNA pol-a/SJK 237–70 MAb and helicase-III/PAb raised by Tuteja et al. 1992)

15.2.7 Detection of Activation of Caspase-3 in *Cis*-Platin-Induced Breast Cancer Cells

Breast carcinoma cells (SKBR-3 or MDA-468 cells, 0.5×10^6) were harvested after treatment with *cis*-platin (50–150 micromolar) for 48 h and were pelleted and resuspended in 0.1 ml lysis buffer (62.5 mM Tris-HCl pH 6.8, 2.0% (w/v) SDS, 10% glycerol, 50 mM DTT) followed by breaking open the cells (homogenization) for 3 × 10 s sonication at 4 °C. The homogenates were incubated for 1 h at 37 °C followed by denaturation for 5 min at 95 °C. The samples (containing 20–30 microgram proteins) were loaded onto gels and subjected to SDS-PAGE and transferred to a nitrocellulose membrane. Nonspecific binding was blocked by incubation in Tris-buffered bovine serum albumin as described above. The membrane was then incubated for 12 h in blocking buffer containing rabbit polyclonal anti Caspase-3 antibodies (diluted 1:1000; raised against full-length Caspase-3). The N/C membranes were then washed and incubated with anti-IgG-AP

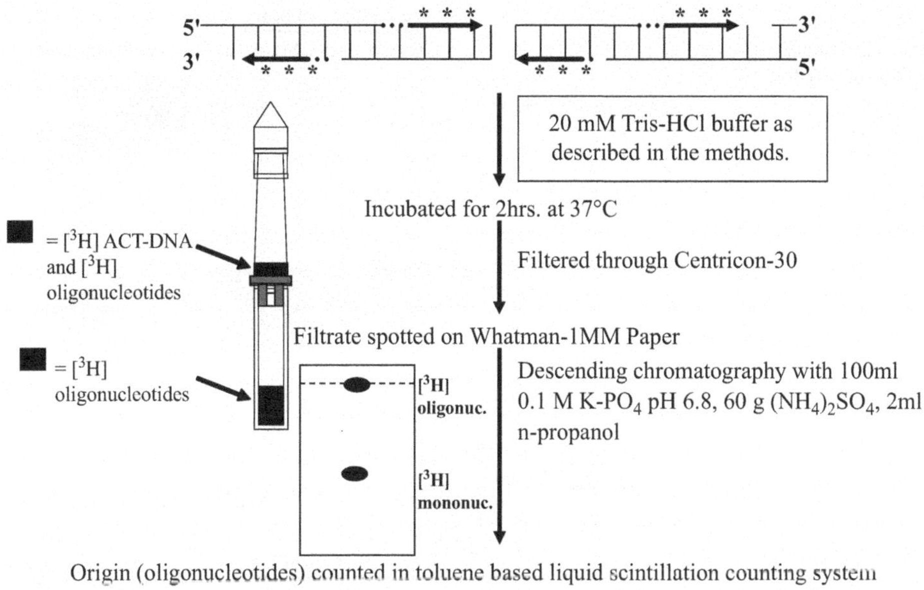

Fig. 15.13 ROME (radioactive oligonucleotide in membrane filtration effluent) assay of total helicase activities. (Boyle 2005; Boyle et al. 2003)

conjugate (diluted (1:3000) and visualized using NBI-BCIP (Fig. 15.14). Similar results were obtained with MDA-468 cells (results reported elsewhere; Rui 2008). Since MCF-7 has the defective Caspase-3 gene, similar results were not obtained.

15.2.8 Determination of Posttranslational Glycolipid: Glycosyltransferase (GLTs) Activities in Apoptotic Carcinoma Cells (SKBR-3, MDA-468, MCF-7, and Colo-205)

Activities of at least 10 GLTs (GalT-2, GalT-4, GalT-5, GalNAcT-1, GlcNAcT-1, SAT-1, SAT-2, SAT-3, SAT-4, and FucT-3, Fig. 15.15) were determined before and after treatment with the apoptosis inducers. The carcinoma cells treated with *cis*-platin (20–80 micromolar) were harvested in 0.2 M HEPES buffer (pH 7.2) containing 0.32 M sucrose, 0.025 M EDTA, and 2 mM 2-mercaptoethanol or DTT. Cells were homogenized in the Swiss Polytron homogenizer (attached with a mini generator) for 3× 15-s pulse (at 4 °C) for the apoptotic cells produced by *cis*-platin inducers. The homogenates were pelleted at 10,000×g, and aliquots of small portions were stored at −18 °C in micro-baby tubes or in 0.5-ml Eppendorf snap-cap tubes after being properly covered.

These GLTs (glycosyltransferases) catalyze the intermediate steps for GD1a, GD3, GD1b (in the Basu-Roseman pathway for ganglioside biosynthesis; Fig. 15.15, left side), and SA-LE^X or H blood group active glycolipid biosynthesis (in the Basu-Hakomori pathway, Fig. 15.15, right side). Detailed procedures for the assay methods for individual GLT have been published previously (Basu and Basu 1972, 1984; Basu et al. 1982, 1987, 1991, 1999, 2017a).

15.2.9 DNA Microarray Assay for Transcriptional Expression of mRNAs of GLTs

The total RNA from a cell sample was prepared with Qiagen RNeasy Mini Kit. After the inducer treatment (20–80 micromolar *cis*-platin for 2–24 h) in a T-25 cm flask (containing 5 × 10^6

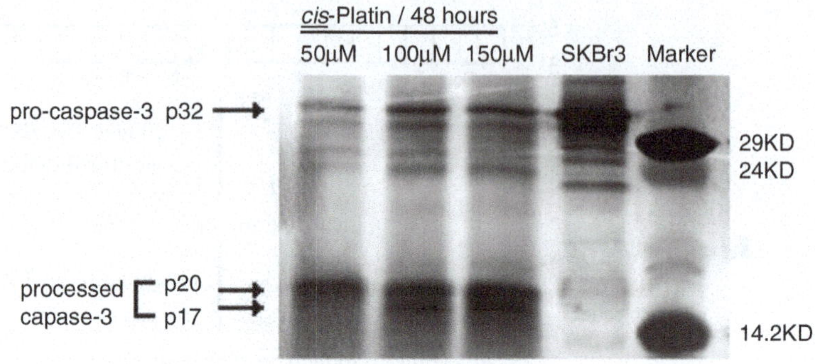

Fig. 15.14 Identification of Caspase-3 activations by Western blot (in SKBR-3 cells). (Basu et al. 2004a, b, 2012a; Rui et al. 2004, 2009)

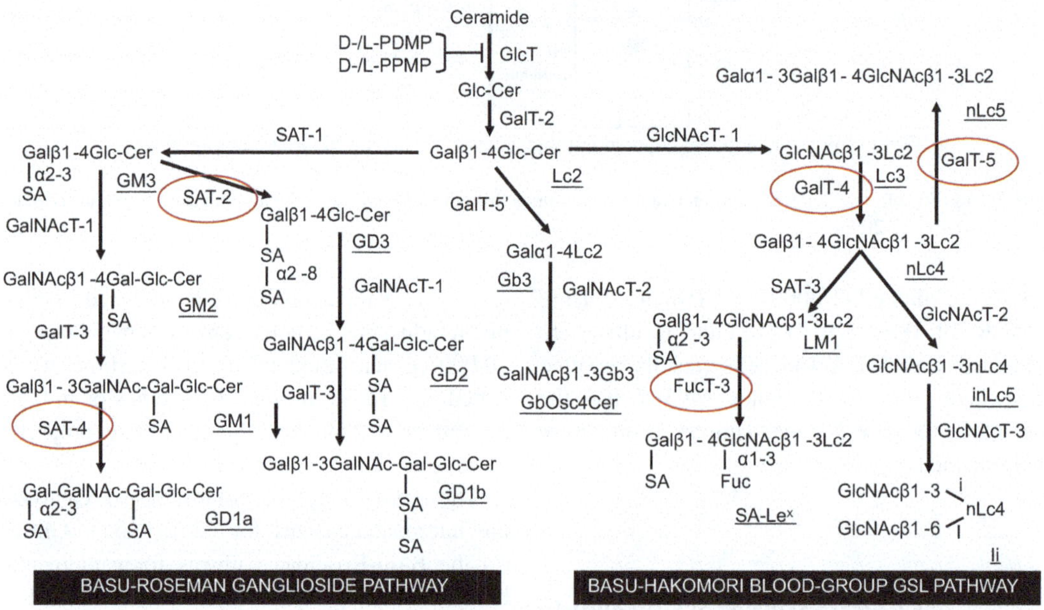

Fig. 15.15 Biosynthesis of mono- and di-sialosyl glycosphingolipids via Basu-Roseman ganglioside and Basu-Hakomori blood group GSL pathway. (Basu, S. and collaborators (Rui et al. 2009); Basu et al. (2012a, b)

cells), the treated cells were collected with trypsin digestion and spun in 15-ml conical graduated sterile tubes. The cells were washed with sterile 1X PBS once, and the pelleted cells were treated with 350 microliter of BLT buffer, and the lysate was passed through a blunt 21-gauge needle fitted to an RNase-free sterile syringe. Then 350 microliter of 95% ethanol was added and mixed well. The sample was transferred to an RNeasy spin column placed in a 2-ml collection tubes, and the column was washed with 700 microliter buffers, followed by centrifuging at 10,000×g for 15 s. The flow through and two more washes with RPE buffer wash were also discarded. RNeasy spin columns were filled with 25-microliter RNase-free water and placed in new 1.5-ml collection tubes. The columns were centrifuged for 2 min at 10,000×g to elute the RNA. This step was then repeated once with 25-microliter RNase-free water, and the total RNA samples were then stored at −70 °C for further analysis. The following steps (Rui 2008; Kroes et al. 2006) were performed using our previously published methods until each chip was analyzed:

1. *RNA integrity examination*: (Kroes et al. 2006; Rui 2008; Rui et al. 2011).
2. *Preparation of aminoallyl: RNA (amplified aRNA)* (Van Gelder et al. 1990; Kroes et al. 2006).
3. *First-strand cDNA synthesis*: (Rui 2008).
4. *Second-strand cDNA synthesis and its purification*: (Kroes et al. 2006).
5. *In vitro transcription of antisense RNA (aRNA)*: (Kroes et al. 2006; Rui 2008).
6. *Purification of aRNA*: (Kroes et al. 2006; Rui 2008; Rui et al. 2011).
7. *Dye coupling of aRNA*: (Kroes et al. 2006; Rui 2008).
8. *Chip hybridization*: (Kroes et al. 2006; Rui 2008; Rui et al. 2011).

15.3 Results and Discussion

15.3.1 Detection of Apoptosis by Phase-Contrast and Fluorescence Microscopy

Synchronized breast carcinoma cells (SKBR-3, MCF-7, and MDA-468) were treated with varying concentrations of *cis*-platin (20 to 60--80 microM) for 48 h and at first examined under phase-contrast microscopy. MCF-7 was most resistant and SKBR-3 was most sensitive for cell death (Fig. 15.8).

To characterize the phosphatidylserine externalization (Fig. 15.9) on cell membrane, the newly synthesized phosphatidylserine-binding fluorescent dye (PSS-380) (Hanshaw and Smith 2005; Koulov et al. 2003) and DNA staining dye propidium iodide (PI) were used with SKBR-3 carcinoma cells after treatment with *cis*-platin (25–100 micromolar for 6 h (Fig. 15.10) and 10--100 micromolar for 24 h (Fig. 15.11). During our studies in last one and half decades (results are published elsewhere), these cells were treated with different apoptotic agents (structures; Fig. 15.4) for 6–24 h to obtain the minimum time required for phosphatidylserine externalization during apoptosis (Rui et al. 2004, 2009, 2011; Rui 2008; Basu et al. 2014, 2017a). The treatment conditions are given above. Our detailed study (Rui 2008; Rui et al. 2009, 2011; Basu et al. 2012a, b, 2014, 2017a) showed that PE externalization happened as early as 6 h of drug treatment. At that time most of the cells (>80) were still intact and not permeable to PI (propidium iodide) since not many nuclei were stained red. As apoptosis proceeded and the cells were near the stage of lysis, they (cell membranes) allowed the propidium iodide to permeate all the way through the plasma membrane and nuclear membrane up to the binding of the dye to the fragmented DNA. We saw the dark red spots in the center of the cells (Fig. 15.11). At 6 h the PSS-380 was staining PE migrated outside the cell leaflet (Fig. 15.10), whereas at 24 h the PSS-380 stained the nuclear membranes also (Fig. 15.11) (detailed results were shown elsewhere; Rui 2008; Rui et al. 2011).

15.3.2 Characterization of Replisome Complex and Its Inhibition During Apoptosis Induced by *Cis*-Platin

After treatment with inducer, the same batch of cells (with which fluorescence was observed) was harvested without the fluorescence dyes and processed up to cell-free extract as described above and subjected to DE-23 column chromatography (Ray 1988; Ray et al. 1994) or on glycerol (10–30%) density-gradient ultracentrifugation (Fig. 15.12) for separation of DNA polymerase-alpha and helicases. Fractions were assayed for DNA polymerase-alpha and Helicase-III (assayed using ROME assay combined with specific antibody-binding assay) (Figs. 15.12, and 15.13); the effect of *cis*-platin concentrations added to three different cell lines (MCF-7, MDA-468, and SKBR-3) revealed that DNA pol-alpha in all these cell lines were inhibited (75–95%) by *cis*-Platin concentration of 80 micromolar (Fig. 15.16). When the same replisome fractions were tested for DNA helicase activities (Fig. 15.17) at 80 micromolar concentration of *cis*-platin, almost 70–95% inhibition

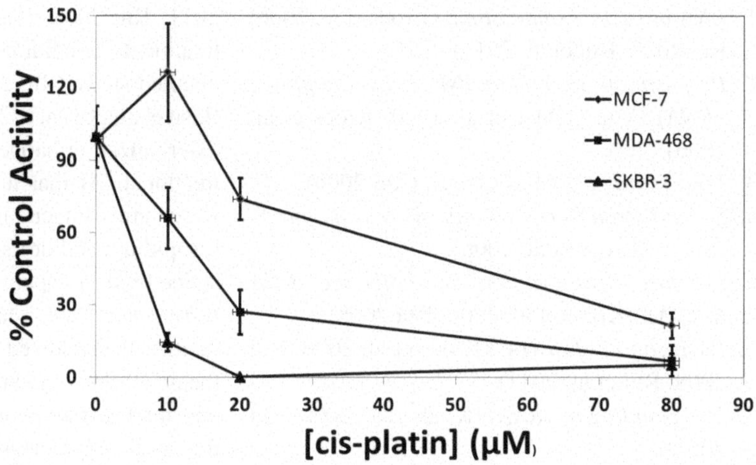

Fig. 15.16 Effect of *cis*-platin treatment on DNA polymerase-α activity present in different cell lines

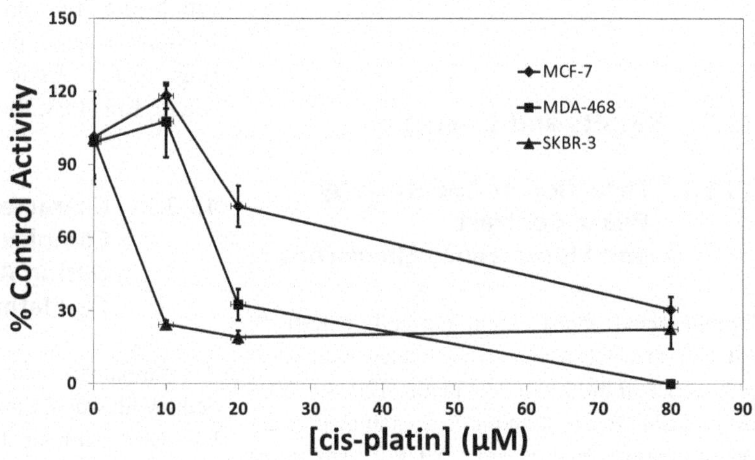

Fig. 15.17 Effect of *cis*-platin treatment on helicase activity present in different cell lines

occurred in MCF-7 and SKBR-3 cells, and almost 95% inhibition occurred in MDA-468 cells. We previously established (Kelly 1993; Kelly et al. 1993b; Bose et al. 1995, 1999; Bose 2002) that in the in vitro system, *cis*-platin with immunopurified DNA pol-alpha from prostate tumor (PA-2 and PA-3 cells) was inhibited by binding in the Zn^{+2}-binding domain and changed the conformation of the vicinal nonapeptide (Bose et al. 1999) and 35-mer peptide (Bose 2002). Whether the DNA helicase is also Zn^{+2}-bound protein is not established yet. Our present report suggests that perhaps *cis*-platin passively inhibit interaction of DNA helicase and DNA polymerase-alpha during the induction of apoptosis.

15.3.3 Activation of Caspases During *Cis*-Platin Induction of Apoptosis and Degradation of DNA

A schematic diagram is drawn for the apoptotic activation of caspases (Caspase-3, Caspase-8, and Caspase-9 (Fig. 15.18). *Cis*-platin appeared to activate Caspase-8 (Boyle et al. 2006), which in turn activates Caspase-3 and finally causes DNA

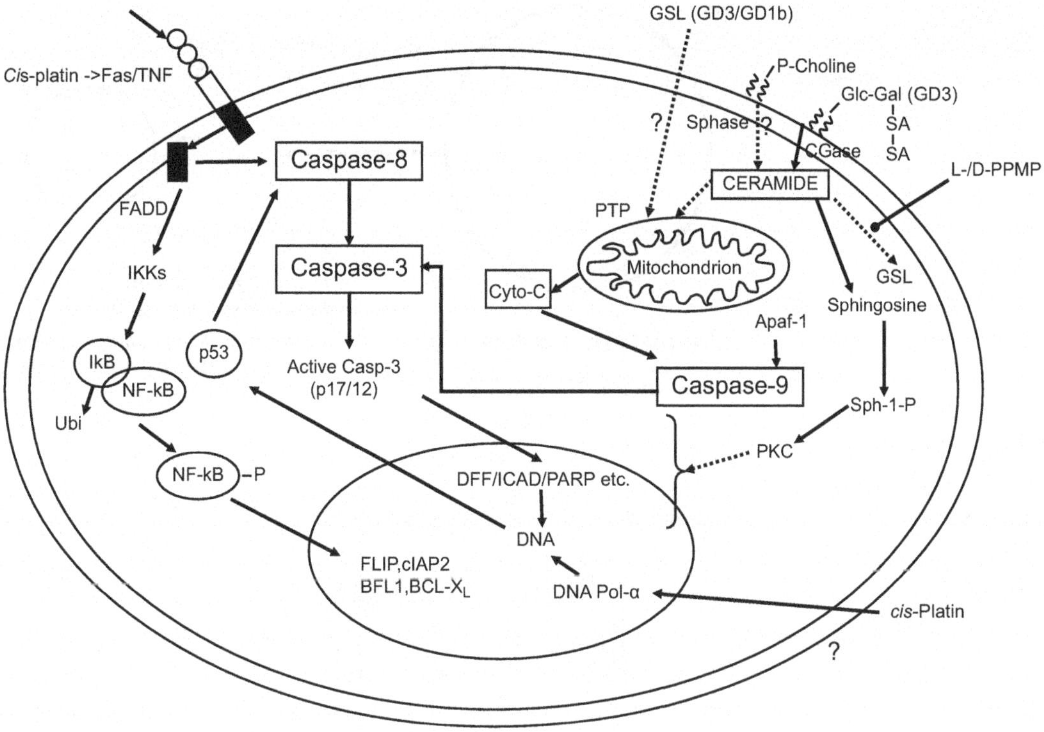

Fig. 15.18 Apoptosis signal pathways activated by different anticancer agents. (Rui et al. 2011)

degradation. It is expected that *cis*-platin probably acts (Fig. 15.18) on the cancer cells (SKBR-3 or MDA-468) through the EFSAP (external FADD stimulated apoptosis pathway). Since MCF-7 cells have the defective gene for Caspase-3, the final apoptosis in these cells may not occur by FESAP or IMSAP (intrinsic mitochondrial stimulated apoptosis pathway). Bypassing the Caspase-3 activation step, how Caspase-8 is causing apoptosis in MCF-7 cells is not known yet. Recently, with two dinuclear platinum (II) complexes Pt-10 and Pt-11, strong antitumor agents (capable of high apoptosis) were shown in human breast cancer cells (MCF-7 and MDA-MB-231) (Bielawski et al. 2013). During the apoptotic effect, FADD and Caspase-8 were activated in these human breast cancer cells more efficiently than *cis*-platin at 50 micromolar concentration. Considering the toxic effect of *cis*-platin as an anticancer drug, perhaps these new generations of Pt-II compounds would be more suitable to treat breast cancer patients provided suitable drug delivery systems are discovered (Fig. 15.19). The future feasibility of these new Pt-II polynucleated drugs could be used in picomolar quantities instead of toxic high micromolar concentrations (Table 15.2) needed for *cis*-platin (in use) for treatment at the present time.

15.3.4 Modulation of Acidic Glycolipid Biosyntheses During Apoptosis

The acidic glycolipids (GSLs) containing N-acetylglucosamine, fucose, and sialic acids are ubiquitous on the surfaces of breast and colon carcinoma cells. At least eight GSL-GLTs have been studied in the pathway of Le^X and SA-Le^X (Fig. 15.15) in normal tissues (Basu et al. 1982, 1987, 1999, 2000, 2017a, b; Basu and Basu 1984; Higashi et al. 1985) as well as in cancer cells in culture (Basu et al. 1991, 2012a,

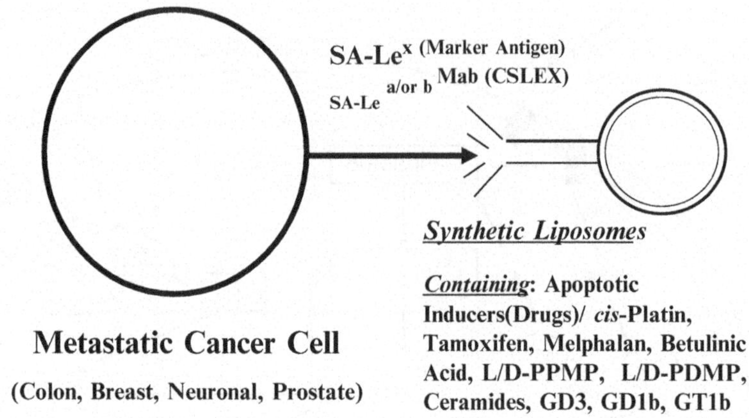

Fig. 15.19 Proposed targeted delivery of apoptotic agents containing liposomes coated with antibodies

Table 15.2 Expected lowering of apoptotic chemical concentration for therapeutic use in cancer treatment without toxicity

Conc. used	MDA-468	MCF-7	SK-BR-3
L-PPMP (1–20 μM)	Do	Do	Do
Cis-platin (10–150 μM)	Do	Do	Do
Betulinic acid (40–160 μM)	Do	Do	Do
Doses to be lowered-	μM to pM	μM to pM	μM to pM

2014; Rui et al. 2011). Each step for in vitro biosynthesis for Le^X and $SA\text{-}Le^X$ is well established through their product characterization (Basu 1991). All these GSL-GLTs are proven to have different gene sequences as well. These GSL-GLTs are modulated during induction of apoptosis by *cis*-platin and other inducers (e.g., L-PPMP, D-PDMP, tamoxifen, betulinic acid, and melphalan; Fig. 15.4) (Rui et al. 2011; Basu et al. 2012a, b, 2014, 2017a). Apoptosis induction was visible by fluorescence studies between 6 h and 24 h. (Fig. 15.9, 15.10 and 15.11). The breast carcinoma cells incubated between 24 h and 48 h with *cis*-platin or other inducers (Fig. 15.4) showed marked inhibition (80–90%) of posttranslational expression (activities) of those GSL-GLT activities as summarized in Table 15.3 when checked by in vitro assay systems described previously (Basu et al. 1987, 1999, 2017b). However, it was evident by DNA microarray studies that during the first 2 h of induction of apoptosis, mRNA expression of some of the crucial GSL-GLTs involved in the marker antigen biosynthesis is stimulated (1.3- to 1.5-fold) (Table 15.4). They all are downregulated after 48 h as it was evident from GLT activity studies (Table 15.3).

The cause of these glyco-gene regulations during apoptotic induction in metastatic breast and colon cancer cells is unknown and needs future investigations for further explanations. These apoptotic agents reported here could be employed as a new generation of anticancer drugs after proper delivery to patients at a nontoxic level (Table 15.2). Our suggested targeted delivery of nontoxic level apoptotic drugs (Table 15.4) could be tested through the use of proper liposomes coated with antibodies (polyclonal or monoclonal, raised against cancer cell surface-specific marker antigens, e.g.,GD3, GD1b, Le^X, $SA\text{-}Le^X$, Le^a, Le^b, etc.) (Fig. 15.15).

The cancer chemotherapy through apoptotic induction depends on the differences of the apoptotic pathways between normal cells and cancer cells (Letai 2017). Our present observation on killing in vitro the three different human

Table 15.3 Overall conclusion of posttranslational activities of glycosphingolipid: GLTs (cell/agents/time treated)

GSL-GLT	Catalyzed reaction	Enzymatic activity
GalT-4	Lc3 (GlcNAc-Gal-Glc-Cer) → Galβ-Lc3 (nLc4)	Decrease (MCF-7/L-PPMP-2 h,6 h; SKBR-3//***cis*Platin**, L-PPMP and MDA-468/L-PPMP, 48 h)
GalT-5	nLc4) (Gal-GlcNAc-Gal-Glc-Cer) → Galα-Lc4 (nLc5)	Decrease (SKBR-3/L-PPMP-2 h,6 h; MCF-7,MDA-468/L-PPMP-6 h; SKBR-3//***cis*Platin** and MDA-468/L-PPMP-48 h)
SAT-2	GM3 → GD3	Decrease (MCF-7/*cis*-Platin; L-PPMP-48 h)
SAT-4	GM1 → GD1a	Decrease (MCF-7/L-PPMP, SKBR-3//***cis*Platin** and MDA-468/L-PPMP-48 h)
SAT-4′	Gg4 (Gal-GalNAc-Gal-Glc-Cer) → GM1b	Decrease (SKBR-3//***cis*Platin** and MDA-468/L-PPMP-48 h)
FucT-3	LM1 → SA-Le^X	Decrease (SKBR-3/L-PPMP-48 h)

Table 15.4 Changes in GLT-RNA expressions in apoptotic breast cancer cells (DNA microarray assay after 2-h treatment with 2 μM L-PPMP)

Cell line	GLT gene name	Linkage formed	Fold change
MCF-7	B3GALT5	Galβ1-3GlcNAc-R1	1.19–1.33
	B3GNT3	GlcNAcβ1-3 Gal-R2	1.20
MDA-468	UGCGL2	Glcβ1-1 Cer	1.28
	B3GNT4	GlcNAcβ1-3 Gal-R2	1.34
SKBR-3	B3GNT4	GlcNAcβ1-3 Gal-R2	1.35
	ST6GAL1	NeuAcα2-6Galβ1-3GlcNAc-R3	−1.55

Core Blood Type: Galβ1-3/4GlcNAcβ1-3Galβ1-4-R1/R2
R1 = Galβ1-4Glcβ1-1 Cer/R2 = OligoN-Protein

carcinoma cell lines (SKBR-3, MCF-7, and MDA-468) after induction of apoptosis with different apoptotic agents (*cis*-platin, tamoxifen, L-PPMP, and betulinic acid) supports the hypothesis that the use of mixture of these agents would be helpful for breast cancer treatment. In recent years anti-apoptotic proteins of BCL-2 family along with apoptotic inducing agents have gained therapeutic importance in different cancer treatments (Letai 2017).

Acknowledgment We thank Mrs. Dorisanne Nielsen and Mr. Eric Kuehner for their help during the preparation of this manuscript. We thank coworkers of Dr. Narendra Tuteja for supplying us mono- and polyclonal antibodies against DNA Helicase-III, coworkers of Dr. Sipra Banerjee for supplying us all three (SKBR-3, MCF-7, and MDA-468) human breast carcinoma cells, and coworkers of Professor Joseph R. Moskal for teaching our students in detail about the DNA microarray experiments.
The Jacob Javits Research Award from NIH-NINDS NS-18005, Coleman Cancer Foundation, and NCI grant-CA-14764 to S. Basu and a grant-in-aid from Siemens Corporation to M. Basu supported this work.

References

Agnieszka G, Ewa A, Anna B, Krzyszlof B, Monika C, Elzbieta S (2014) Dinuclear berenil-platinum (II) complexes as modulator of apoptosis in human MCF-7 and MDA-MB231 breast cancer cells. Anti Cancer Agents Med Chem 14(8):1179–1186
Aparicio O, Stout A, Bell S (1999) Differential assembly of Cdc45p and DNA Polymerases at early and late origins of DNA replication. Proc Natl Acad Sci U S A 06(16):9130–9135
Arends MJ, Wyllie AH (1991) Apoptosis: mechanisms and roles in pathology. Int Rev Exp Pathol 32:223–254
Basu S (1991) The serendipity of ganglioside biosynthesis: pathway to CARS and HY-CARS glycosyltransferases. Glycobiology 1:469–475
Basu M, Basu S (1972) Enzymatic synthesis of a tetraglycosylceramide by a galactosyltransferase from rabbit bone marrow. J Biol Chem 247:1489–1495
Basu M, Basu S (1984) Biosynthesis *in vitro* of Ii-core glycolipids from neolactotetraosylceramide by β1-3 and β1-6 N-acetylglucosaminyltransferases from mouse T-lymphoma. J Biol Chem 259:12557–12562
Basu M, Basu S, Stoffyn A, Stoffyn P (1982) Biosynthesis *in vitro* of sialyl(α2-3)neolactotetraosylceramide by a sialyltransferase from embryonic chicken brain. J Biol Chem 257:12765–12769

Basu M, De T, Das K, Kyle JW, Chon HC, Schaeper RJ, Basu S (1987) Glycosyltransferases involved in glycolipid biosynthesis. In: Ginsburg V (ed) Methods in enzymology, vol 138. Academic, New York, pp 575–607

Basu M, Hawes JW, Li Z, Ghosh S, Khan FA, Zhang B, Basu S (1991) Biosynthesis *in vitro* of SA-Le^x and SA-diLe^x by α1-3 fucosyltransferases from colon carcinoma cells and embryonic brain tissues. Glycobiology 1:527–535

Basu S, Basu M, Dastgheib S, Hawes JW (1999) Biosynthesis and regulation of glycosphingolipids. In: Barton D, Nakanishi K, Meth-Cohen O, Pinto BM (eds) Comprehensive natural products chemistry, vol 3. Pergamon Press, New York, pp 107–128

Basu S, Basu M, Das K (2000) Glycosyltransferases in glycosphingolipid biosynthesis. In: Ernst B, Sinay P, Hart G (eds) Oligosaccharides in chemistry and biology – a comprehensive handbook. Wiley-VCH Verlag GmbH, Weinheim

Basu S, Rui M, Mikulla B, Bradley M, Moulton C, Basu M, Banerjee S, Inokuchi JI (2004a) Apoptosis of human carcinoma cells in the presence of inhibitors of Glycosphingolipid biosynthesis: I. Treatment of Colo-205 and SKBR3 cells with isomers of PDMP and PPMP. Glycoconj J 20(3):157–168

Basu S, Rui M, Boyle PJ, Mikulla B, Bradley M, Smith B, Basu M, Banerjee S (2004b) Apoptosis of human carcinoma cells in the presence of potential anti-cancer drugs. III. Treatment of Colo-205 and SKBR-3 cells with cis-platin, Tamoxifen, Melphalan, Betulinic acid, L-PDMP, L-PPMP, and GD3 ganglioside. Glycoconj J 20:563–577

Basu S, Rui M, Basu M, Goodson H, Smith B, Banerjee SV (2004c) Glycosphingolipid metabolism and signaling in apoptotic cancer cells, lipids. In: Haldar D, Das SK (eds) Sphingolipid metabolizing enzymes. Research Signpost, Trivandrum, pp 81–100

Basu S, Rui M, Moskal JR, Basu M, Banerjee S (2012a) Apoptosis of breast cancer cells: XI. Modulation of genes of glycoconjugate biosynthesis and targeted drug delivery. In: Sudhakaran PR, Surolia A (eds) Proceedings of 9th international symposium, biochemical roles of eukaryotic cell surface macromolecules, Advances in experimental medicine and biology, vol 749. Springer, Trivandrum, pp 233–255

Basu S, Rui M, Moskal JR, Basu M (2012b) Ganglioside biosynthesis in developing brains and apoptotic cancer cells. X. Regulation of glyco-genes involved in GD3 and Sialyl-LeX/a syntheses. Neurochem Res 27:1245–1255

Basu S, Rui M, Moskal JR, Basu M (2014) Regulation of glycolipids: XII. Glycosyltransferase genes involved in SA-LeX and related GSLs biosynthesis in carcinoma cells by biosimilar apoptotic agents: potential anticancer drugs. In: Chakrabarti A, Surolia A (eds) Proceedings of 10th international symposium, biochemical roles of eukaryotic cell surface macromolecules, Advances in experimental medicine and biology, vol 842. Springer, Trivandrum, pp 329–354

Basu S, Rui M, Moskal JR, Basu M (2017a) Chapter 4: Biosynthesis of LeX family glycosphingolipids and ifs gene regulation in horizons in cancer research, vol 64 (ed. Watanabe HS). Nova Science Publisher (E-Book), pp 85–100

Basu S, Agarwal A, Basu M, Rui M, Moskal JR (2017b) Protocols of glycosyltransferase assays: gangliosides globoside and Lewis-X intermediate-lactosylceramide biosyntheses in eukaryotic systems. In: Sonnino S, Alessandro (eds) Protocols of gangliosides, Advances in experimental medicine and biology. Springer, New York

Benkovic SJ, Valentine AM, Salinas F (2001) Replisome-mediated DNA replication. Ann Rev Biochem 70:181–208

Bhattacharya P, Basu S (1978) DNA polymerase activities in differentiating mouse neuroblastoma N-18 cells. Proc Natl Acad Sci U S A 75(3):1289–1293

Bhattacharya P, Simet I, Basu S (1979) Inhibition of human neuroblastoma DNA polymerase activities by plant lectins and toxins. Proc Natl Acad Sci U S A 76 (5):2218–2221

Bielawski K, Czarnomysy R, Bielawska A, Poplawska B (2013) Cytotoxicity and induction of apoptosis of human breast cancer cells by novel platinum (II) complexes. Environ Toxicol Pharmacol 25 (2):254–264

Boehmer PE, Dodson M, Lehman I (1993) The herpes simplex virus type-1 origin binding protein. DNA helicase activity. J Biol Chem 268(2):1220–1225

Bose RN (2002) Biomolecular targets for platinum antitumor drugs. Mini Rev Med Chem 2(2):103–111

Bose RN, Li D, Kennedy M, Basu S (1995) Facile formation of cis-platin Nonapeptide complex of human DNA polymerase-alpha origin. J Chem Soc Commun R Soc Chem 1731–1732

Bose RN, Li D, Yang WW, Basu S (1999) NMR structures of a Nonapeptide from DNA binding domain of human DNA polymerase-alpha determined by iterative complete-relaxation-matrix approach. J Biomol Struct Dyn 16(5):1075–1085

Boyle PJ (2005) Characterization of DNA helicase-III in replication complexes isolated from embryonic chicken brains and breast carcinoma cells. PhD. thesis, University of Notre Dame

Boyle PJ, Campbell BA, Ma R, Moulton C, Tuteja N, Basu S (2003) Differential association of DNA helicase-III in the replisomal complex isolated from developing embryonic chicken brains: as assayed by a new method. FASEB J 17(4):A600

Boyle PJ, Ma R, Tuteja N, Banerjee S, Basu S (2006) Apoptosis of human breast carcinoma cells in the presence of cis-platin and L–/D-PPMP: IV. Modulation of replication complexes and glycolipid: Glycosyltransferases. Glycoconj J 23(3–4):175–187

Budd M, Campbell J (1995) A yeast gene required for DNA replication encodes a protein with homology to

DNA helicases. Proc Natl Acad Sci U S A 92 (17):7612–7616

Chang LM, Rafter E, Augl C, Bollum FJ (1984) Purification of DNA polymerase-DNA primase complex from calf thymus glands. J Biol Chem 259 (23):14679–14687

Costa M, Ochem A, Staub A, Falaschi A (1999) Human DNA helicase VII: a DNA and RNA helicase corresponding to the G3BP protein, an element of the ras transduction pathw2ay. Nucleic Acids Res 27 (3):817–821

Crute J, Bruckner R, Dodson M, Lehman I (1991) Herpes simplex-1 helicase-primase. Identification of two nucleoside triphosphatase sites that promotes DNA-helicase action. J Biol Chem 286 (31):21252–21256

Enomoto T, Suzuki M, Takahashi M, Kawasaki K, Watanabe Y, Nagata K, Hanaoka F, Yamada M (1985) Purification and characterization of two forms of DNA polymerase-alpha from mouse FM3A cells: DNA polymerase-alpha-primase complex and a free DNA polymerase-alpha. Cell Struct Funct 10:101–107

Gerke V, Creutz CE, Moss SE (2005) Annexin-linking Ca2+ signalling to membrane dynamics. Nat Rev Mel Cell Biol 6(6):449–461

Goetz G, Dean F, Hurwitz J, Matson S (1988) Unwinding of duplex region in DNA by the simian virus 40 large tumor-associated DNA helicase activity. J Biol Chem 263(1):383–392

Ha TK, Kim ME, Yoon JH, Bae SJ, Yoon J, Lee JS (2013) Galagin induces human colon cancer death via the mitochondrial dysfunction and caspase-dependent pathway. Exp Biol Med 238(9):1047–1054

Hanshaw RG, Smith BD (2005) New Reagents for phosphatidylserine recognition and detection of apoptosis. Bioorg Med Chem 13(17):5035–5042

Higashi H, Basu M, Basu S (1985) Biosynthesis *in vitro* of diasialosyl-neolactotetraosylceramide by a solubilized sialyltransferase from embryonic chicken brain. J Biol Chem 260:824–828

Holmes AM, Cheriathundam E, Bollum FJ, Chang LM (1986) Immunological analysis of the polypeptide structure of calf thymus DNA-polymerase-primase complex. J Biol Chem 261(25):11824–11830

Hotta Y, Stern H (1978) DNA unwinding protein from meiotic cells of Lilium. Biochemistry 17 (10):1872–1880

Im DS, Muzyczka N (1992) Partial purification of adeno-associated virus Rep78, Rep52, and Rep40 and their biochemical characterization. J Virol 66(2):1119–1128

Jiang H, Hickey R, Abdel-Aziz W, Tom T, Wills P, Liu J, Malakas L (2002) Human cell DNA replication is mediated by a discrete multiprotein complex. J Cell Biochem 85(4):762–774

Kelley TJ (1993). Subunit structure of DNA polymerase-alpha and its inhibition by antitumor drugs. PhD thesis, Department of Chemistry and Biochemistry, The University of Notre Dame

Kelley TJ, Groll TM, Basu S (1993a) Purification of a galactose binding 56kDa protein from embryonic chicken brain. Its probable role in the initiation of replication. In: Proceedings of Cold Spring Harbor symposium on Euk. DNA Replication September 8–12, 1993, p 75

Kelley TJ, Moghaddas S, Bose R, Basu S (1993b) Inhibition of immunopurified DNA polymerase-alpha from PA-3 prostate tumor cells by platinum (II) antitumor drugs. Cancer Biochem Biophys 13(3):135–146

Kornberg A, Baker TA (1992) DNA replication-2. W.H. Freeman, New York

Koulov AV, Stucker KA, Lakshmi C, Robinson JP, Smith BD (2003) Detection of apoptotic cells using synthetic fluorescent sensor for membrane surfaces that contain phosphatidylserine. Cell Death Differ 10 (12):1357–1359

Kroes RA, Panksepp J, Burgdorf J, Otto NJ, Moskal JR (2006) Modeling depression: social-dominance-submission gene expression patterns in rat neocortex. Neuroscience 137(1):47–49

Kunkel TA, Burgers PM (2008) Dividing the workload at a eukaryotic replication fork. Trends Cell Biol 18 (11):521–527

Labib K, Torcero J, Diffley J (2000) Uninterrupted MCM2-7 function required for DNA –replication fork progression. Science 286(5471):1643–1647

Letai A (2017) Apoptosis and cancer. Ann Rev Cancer Biol 1:275–294

Lin S, Hickey R, Malakas L (1997) The isolation of DNA synthesome from human leukemia cells. Leuk Res 21:501–512

Lohman TM (1993) Helicase-catalyzed DNA unwinding. J Biol Chem 268(4):2269–2272

Lohman TM, Bjornson KP (1996) Mechanisms of helicase-catalyzed DNA unwinding. Annu Rev Biochem 65:169–214

Maga G, Hubscher U (1996) DNA replication machinery: functional characterization of a complex containing DNA polymerase-alpha, DNA polymerase-delta, and replication factor C suggests an asymmetric DNA polymerase dimer. Biochemistry 35(18):5764–5777

Malkas L, Hickey R, Li C, Pederson N, Baril EA (1990) 21S enzyme complex from HeLa cells that functions in simian virus 40 DNA replication in vitro. Biochemistry 29(27):8362–8374

Matson S, Tabor S, Richardson C (1983) The gene 4 protein of bacteriophage 17. Characterization of helicase activity. J Biol Chem 258(22):14017–14024

McCulloch SD, Kunkel TA (2008) The fidelity of DNA synthesis by eukaryotic replicative and translational synthesis polymerases. Cell Res 18(1):148–161

Mimura S, Masuda T, Matsui T, Takisawa H (2000) Central role for cdc45 in establishing an initiation complex of DNA replication in Xenopus egg extract. Gene Cells 5(6):439–452

Moskal JR, Gardner DA, Basu S (1974) Changes in glycolipid Glycosyl transferases and glutamate decarboxylase and their relationship to differentiation in neuroblastoma cells. Biochem Biophys Res Commun 61:751–758

Peyrone M (1844) Ueber die einwirkung des ammoniaks and platinchlorur. Ann Chem Pharm 51:1–29

Ray S (1988) Studies on DNA replication in vitro by DNA polymerase-alpha/primase complex from embryonic chicken brains. PhD thesis, Department of Chemistry and Biochemistry, University of Notre Dame, Indiana

Ray S, Kelley TJ, Fan L, Basu S (1994) Characterization of DNA polymerase-alpha/primase complex from developing embryonic chicken brains. Indian J Biochem Biophys 3(4):226–235

Rosenberg B, VanCamp L, Krigas T (1965) Inhibition of cell division in Escherichia coli by electrolysis products from a platinum electrode. Nature 205:698–699

Rosenberg B, VanCamp L, Trosko JE, Mansour VH (1969) Platinum compounds: a new class of potent antitumor agents. Nature 222(191):385–386

Rossi D, Gaidano G (2003) Messenger of cell death: apoptotic signalling in health and disease. Haematologica 88(2):212–218

Rui M (2008) Apoptosis of breast and colon cancer cells by inhibitors of glycolipid and DNA biosynthesis. PhD thesis, University of Notre Dame, Notre Dame, IN, pp 1–271

Rui M, Koulov A, Moulton C, Basu M, Banerjee S, Goodson H, Basu S (2004) Apoptosis of human breast carcinoma cells in the presence of Disialosyl gangliosides: II. Treatment of SKBR3 cells with GD3 and GD1b gangliosides. Glycoconj J 20(5):319–330

Rui M, Decker NM, Anilus V, Moskal JR, Bergdorf J, Johnson J, Basu M, Banerjee S, Basu S (2009) Post-translational and transcriptional regulation of glycolipid glycosyltransferase genes in apoptotic breast carcinoma cells: VII. After treatment with L-PPMP. Glycoconj J 26:647–661

Rui M, Hopp EA, Decker NM, Loucks A, Johnson JP, Moskal JR, Basu M, Banerjee S, Basu S (2011) VIII. Regulation of glycosyltranferase genes in apoptotic breast cancer cells by inhibitors of glycolipid and DNA biosynthesis in immunology of complex carbohydrates. Adv Exp Med Biol 705:621–642 (editor: Albert Wu)

Seki M, Enomoto T, Eki T, Miyajima A, Murakami Y, Hanaoka F, Ui M (1990) DNA helicase and nucleoside-5′-triphosphatase activities of polyoma virus large antigen. Biochemistry 29(4):1003–1009

Seo YS, Muller F, Lusky M, Gibbs E, Kim H, Phillips B, Hurwitz J, Shlomai J, Kornberg A (1980) A prep riming DNA replication enzyme of *Escherichia coli.* I. Actions of protein n': a sequence-specific, DNA dependent ATPase. J Biol Chem 255(14):6794–6798

Seo YS, Mollar F, Lusky M, Gibbs F, Kim H, Phillips B, Hurwitz J (1993) Bovine Papilloma virus (BVPV)-encoded E2 protein enhances binding of E1 protein to the BPV replication origin. Proc Natl Acad Sci U S A 90(7):2865–2869

Shlomai J, Kornberg A (1980) A prepriming DNA replication enzyme of *Escherichia coli* II. Actions of protein n: a sequence-specific, DNA dependent ATPase. J Biol Chem 255(14):6794–6798

Strasser A, O'Connor L, Dixit VM (2000) Apoptosis signaling. Annu Rev Biochem 69:217–245

Tom T, Malakas L, Hickey R (2002) Identification of multiprotein complexes containing DNA replication factors by native immunoblotting 0f HeLa cell protein preparation with T-antigen-dependent SV-40 replication activity. J Cell Biochem 85(4):762–774

Tuteja N, Tuteja R (2004) Unraveling DNA helicases. Motif structure, mechanism and function. Eur J Biochem 271(10):1849–1863

Tuteja N, Tuteja R, Rehman K, Kang LY, Falaschi A (1990) A DNA helicase from human cells. Nucleic Acids Res 18(23):6785–6802

Tuteja N, Rahman K, Tuteja R, Falaschi A (1991) DNA helicase IV from HeLa cells. Nucleic Acids Res 19 (3):3613–3618

Tuteja N, Rahman K, Tuteja R, Ochem A, Slopac D, Falaschi A (1992) DNA helicase-III from HeLa cells: an enzyme that acts preferentially on partially unwound duplexed. Nucleic Acids Res 20 (20):5329–5337

Tuteja N, Rahman K, Tuteja R, Falaschi A (1993) Human DNA helicase V. a novel DNA unwinding enzyme from HeLa cells. Nucleic Acid Res 21(10):2323–2329

Tuteja N, Tuteja R, Ochem A, Taneja P, Huang NW, Simonies A, Susic S, Rahman K, Marusic L, Chen J (1994) Human DNA ligase-II: a novel DNA unwinding enzyme identified as the Ku autoantigen. Eur J 13(20):4991–5001

Tuteja N, Cohen A, Tuteja R, Skopac D, Falaschi A (1995) Purification and properties of human DNA helicase-VI. Nucleic Acids Res 23(3):2457–2463

Van Gelder RN, von Zastrow ME, Yool A, Dement WC, Barchas JD, Eherwine JH (1990) Amplified RNA synthesizes from limited quantities of hetero generous cDNA. Proc Natl Acad Sci U S A 87(5):1963–1967

Walter J, Newport J (2000) Initiation of eukaryotic DNA replication: origin unwinding and sequential chromatin association of Cdc45, RPA, and DNA polymerase-alpha. Mol Cell 5(4):617–627

Wilson G, Jindal H, Yeung D, Chen W, Ashwell G (1991) Expression of minute virus of mice major nonstructural protein in insect cells: purification and identification of AYP and helicase activities. Virology 185(1):90–98

Wu Y, Hickey R, Lawlor K, Wills P, Yu F, Ozer H, Starr R, Quant J, Lee M, Malakas LA (1994) A 17S multiprotein form of murine cell DNA polymerase mediates polyoma virus DNAs replication in vitro. J Cell Biochem 54(1):32–46

Wyllie A (1980) Glucocorticoid-induced thymocyte apoptosis is associated with endogenous endonuclease activation. Nature 284:555–556

Wyllie AH (1987) Apoptosis: cell death in tissue regulation. J Pathol 153:313–316

Yan H, Chen C, Kobayashi R, Newport J (1998) Replication focus-forming activity 1 and the Werner syndrome gene product. Nat Genet 19(4):375–378

Yao NY, O'Donnell M (2009) Replisome structure and conformational dynamics underline fork progression past obstacles. Curr Opin Cell Biol 21:336–343

Yao NY, O'Donnell M (2010) The replisome (a review). Cell 141(6):1088–1092

Yao NY, O'Donnell M (2016) DNA replication machines. Clin Rev Biochem Mol Biol 51:135–149

Yao NY, O'Donnell ME (2017) DNA replication: how does a sliding clamp slide? Curr Biol 27(4):R174–R176

Zou J, Stillman B (2000) Assembly of a complex containing Cdc45p, replication protein A, and Mcm2p of replication origins controlled by S-phase cyclin dependent kinases and Cdc7p-Dbf4p kinase. Mol Cell Biol 20(9):3086–3096

Zyskind L, Smith D (1977) E coli dnaB mutant: direct involvement of the dnaB252 gene product in the synthesis of an origin-ribonucleic acid species during initiation of a round of deoxyribonucleic replication. J Bacteriol 129(3):1476–1486

Dynamic Function of DPMS Is Essential for Angiogenesis and Cancer Progression

16

Zhenbo Zhang, Jesús E. Serrano-Negrón, Juan A. Martínez, Krishna Baksi, and Dipak K. Banerjee

Abstract

Dolichol phosphate mannose synthase (DPMS) is an inverting GT-A-folded enzyme and classified as GT2 by CAZy. DPMS sequence carries a metal-binding DXD motif, a PKA motif, and a variable number of hydrophobic domains. Human and bovine DPMS possess a single transmembrane domain, whereas that from *S. cerevisiae* and *A. thaliana* carry multiple transmembrane domains and are superimposable. The catalytic activity of DPMS is documented in all spheres of life, and the 32kDa protein is uniquely regulated by protein phosphorylation. Intracellular activation of DPMS by cAMP signaling is truly due to the activation of the enzyme and not due to increased Dol-*P* level. The sequence of DPMS in some species also carries a protein *N*-glycosylation motif (Asn-X-Ser/Thr). Apart from participating in *N*-glycan biosynthesis, DPMS is essential for the synthesis of GPI anchor as well as for O- and C-mannosylation of proteins. Because of the dynamic nature, DPMS actively participates in cellular proliferation enhancing angiogenesis and breast tumor progression. In fact, overexpression of DPMS in capillary endothelial cells supports increased *N*-glycosylation, cellular proliferation, and enhanced chemotactic activity. These are expected to be completely absent in congenital disorders of glycosylation (CDGs) due to the silence of DPMS catalytic activity. DPMS has also been found to be involved in the cross talk with N-acetylglucosaminyl 1-phosphate trans-

Z. Zhang
Department of Biochemistry, School of Medicine, University of Puerto Rico, San Juan, PR, USA

Toronto Western Research Institute, University Health Network, Toronto, ON, Canada

J. E. Serrano-Negrón
Department of Biochemistry, School of Medicine, University of Puerto Rico, San Juan, PR, USA

Department of Natural Sciences & Mathematics, Interamerican University of Puerto Rico, Bayamón, PR, USA

J. A. Martínez
Department of Biochemistry, School of Medicine, University of Puerto Rico, San Juan, PR, USA

Ology Bioservices, Alachua, FL, USA

K. Baksi
Universidad Central del Caribe, Department of Anatomy and Cell Biology, School of Medicine, Universidad Central del Caribe, Bayamón, PR, USA

D. K. Banerjee (✉)
Department of Biochemistry, School of Medicine, University of Puerto Rico, San Juan, PR, USA

Institute of Functional Nanomaterials, University of Puerto Rico, San Juan, PR, USA
e-mail: dipak.banerjee@upr.edu

K. Chattopadhyay, S. C. Basu (eds.), *Biochemical and Biophysical Roles of Cell Surface Molecules*, Advances in Experimental Medicine and Biology 1112, https://doi.org/10.1007/978-981-13-3065-0_16

ferase (GPT). Inhibition of GPT with tunicamycin downregulates the DPMS catalytic activity quantitatively. The result is impairment of surface *N*-glycan expression, inhibition of angiogenesis, proliferation of human breast cancer cells, and induction of apoptosis. Interestingly, nano-formulated tunicamycin is three times more potent in inhibiting the cell cycle progression than the native tunicamycin and is supported by downregulation of the ratio of phospho-p53 to total-p53 as well as phospho-Rb to total Rb. DPMS expression is also reduced significantly. However, nano-formulated tunicamycin does not induce apoptosis. We, therefore, conclude that DPMS could become a novel target for developing glycotherapy treating breast tumor in the clinic.

Keywords

Angiogenesis · Asparagine-linked glycoprotein · Breast cancer · Carbohydrate-active enzyme · Dolichol phosphate mannose synthase · Endoplasmic reticulum · Glycosyltransferase · Phosphoprotein · Protein N-glycosylation · Unfolded protein response

16.1 Introduction

Asparagine-linked (*N*-linked) glycoproteins are evolutionarily conserved (Banerjee et al. 2017; Calo et al. 2010; Szymanski and Wren 2005; Sinhoara and Maruyama 1973). The *N*-glycan provides structural and functional stability while adding flexibility to the glycoprotein. Glycan biosynthesis is elaborative and multi-compartmental, requiring a stepwise addition of sugars either from sugar nucleotides or from lipid intermediates by specific glycosyltransferases. Inadequate assembly of *N*-glycan leads to phenotypic changes developing infection, cancer, and congenital disorders of glycosylation (CDGs), among others. The biosynthetic process begins at the endoplasmic reticulum (ER) with the assembly of dolichol-linked tetra-decasaccharide (i.e., $Glc_3Man_9GlcNAc_2$-PP-Dol) and is completed in Golgi prior to departing to their final destinations (Kornfeld and Kornfeld 1985; Helenius and Aebi (2004)). Dolichol phosphate mannose synthase (dolichyl-phosphate β-D mannosyltransferase, DPMS; EC 2.4.1.83) plays a critical role in this process, especially for the elongation of $Man_5GlcNAc_2$-PP-Dol to $Man_9GlcNAc_2$-PP-Dol. DPMS is an inverting GT-A-folded enzyme and classified as GT2 by CAZy (carbohydrate active enZyme; http://www.cazy.org). In addition to participation in *N*-glycan biosynthesis, DPMS is also essential for GPI anchor biosynthesis as well as for O- and C-mannosylation of proteins in yeast and in mammalian cells (Banerjee et al. 2017; Banerjee 2012; Fig. 16.1). The catalytic activity of DPMS is present in vertebrates, non-vertebrates, fruit fly, plants, fungi, slime molds, protozoan parasites, as well as archaea (Banerjee et al. 2017). Human chromosome 20q13.13 harbors the *DPMS* gene.

16.2 DPMS Exemplifies Unity Among Structural Diversities

DPMS gene has been cloned from 39 different species (archaea to human). Diversities are present in their amino acid sequences, but similarities have been observed between the *B. Taurus* (bDPMS) and *H. sapiens* (hDPMS) homologs, in one hand, and the DPMS between *S. cerevisiae* and *A. thaliana*, on the other (Fig. 16.2). The sequence alignment has also detected a metal-binding signature (i.e., a DAD motif) in all 39 species, a cAMP-dependent protein phosphorylation motif (i.e., PKA motif) in 38 species, and a potential *N*-glycosylation site only in 20 species (Fig. 16.3). Computing the hydropathy index (hydrophobicity) under ExPASy Bioinformatics Resource Portal (http://web.expasy.org/protscale/) (Gasteiger et al. 2005) using the algorithm of Kyte and Doolittle (1982) allowed examination of the hydrophobic interior of the membrane-spanning region of DPMS (Fig. 16.4). The positive indices around the

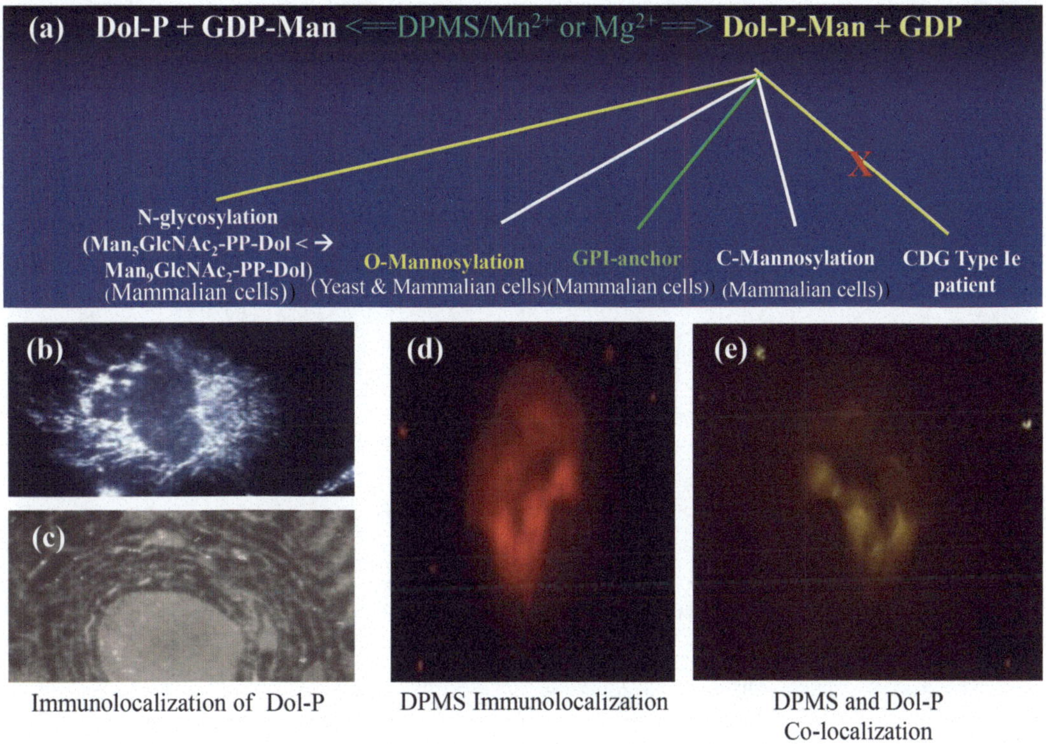

Fig. 16.1 DPMS-catalyzed reaction and the ER localization of Dol-P and DPMS. (**a**) Schematic of the DPMS-catalyzed reaction and the donor participation of dol-*P*-man. (**b**) Detection of Dol-*P* by immunofluorescence microscopy: Fixed capillary endothelial cells are incubated with Ca^{2+}-amphomycin (1 mg/mL) followed by anti-amphomycin mouse monoclonal antibody and stained with FITC-conjugated rabbit anti-mouse IgG (Gastl et al. 1997). (**c**) Detection of Dol-*P* by electron microscopy: Thin sections from capillary endothelial cells on copper grids are treated with Ca^{2+}-amphomycin (1 mg/mL) followed by anti-amphomycin mouse monoclonal antibody and stained with 20 nm gold-conjugated rabbit anti-mouse IgG (x20,000 magnification). (**d**) Detection of DPMS by immunofluorescence microscopy: Fixed capillary endothelial cells are incubated with anti-DPMS rabbit polyclonal antibody (Kyte and Doolittle 1982) and stained with Rhodamine-conjugated goat anti-rabbit IgG. (**e**) Detection of co-localized Dol-*P* and DPMS by immunofluorescence microscopy: Fixed cells are incubated with Ca^{2+}-amphomycin (1 mg/mL) followed by anti-amphomycin antibody (mouse monoclonal) and anti-DPMS antibody (rabbit polyclonal) before staining with FITC-conjugated rabbit anti-mouse IgG and Rhodamine-conjugated goat anti-rabbit IgG and collecting the images

amino acid residues aa104 to 124 and aa166 to 167 are separated by a hydrophilic region of aa125 to 165 supported bDPMS with one membrane-spanning region. The hydropathy plot of bDPMS (Fig. 16.4a) is similar, if not identical to that of the DPMS from human and *A. thaliana* (Fig. 16.4b, c). The hydropathy plots from *S. cerevisiae* and archaea however are almost superimposable (Figs. 16.4d–f). The PKA motif (solid lines in Fig. 16.4) is present in between the two hydrophobic domains (one preceding and the other succeeding the motif) (Fig. 16.4a–d). The hydropathy plots of DPMS from *H. sapiens*, *B. Taurus*, and *A. thaliana* (Fig. 16.4a–c) show four major hydrophobic regions, whereas that for *S. cerevisiae* DPMS displays a possibility for more than one membrane-spanning domain (Fig. 16.4d). Given the polarity index window of 19, any of these regions could function as an ER membrane-

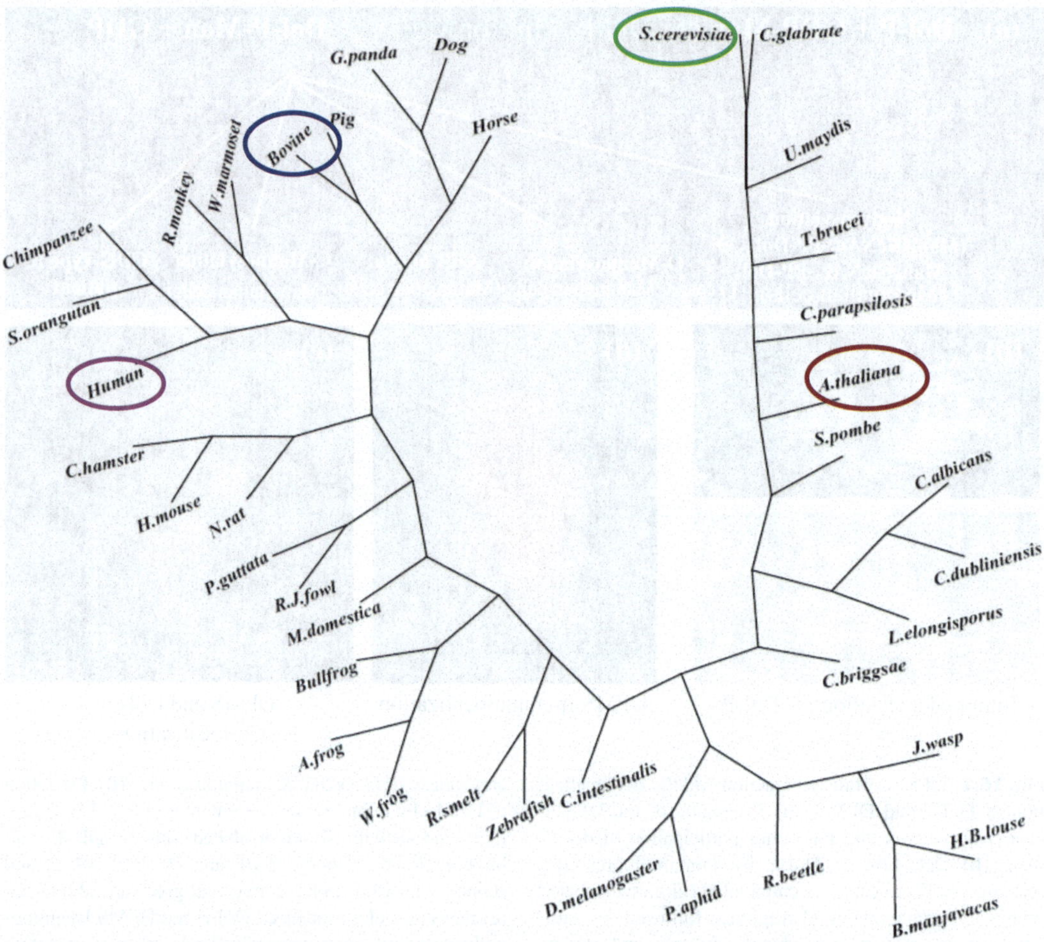

Fig. 16.2 Unrooted phylogenetic tree with branch lengths for DPMS sequences from 39 species. The sequences of DPMS from different species are aligned, and the dendrogram is generated using the CLUSTAL W program (Kyte and Doolittle 1982 on the GenomeNet server (http://www.genome.jp/tools/clustalw/))

spanning domain and the binding site for Dol-*P* (Fig. 16.5a). Sensitive fold recognition analysis (http://bioinfo.pl/Meta) indicates they are similar to one another at the primary, secondary, and tertiary structure levels (Bujnicki et al. 2001) and resemble the structure of the spore coat polysaccharide biosynthesis protein A (SpsA) glycosyltransferase (GT) from *Bacillus subtilis* (Tarbouriech et al. 2001). A 3-D model of DPMS agrees nearly perfectly with the GT and allows the identification of the active site cleft, its properties, and a putative functional mechanism (Lamani et al. 2006). Recent crystallographic analysis of archaea DPMS strongly supports Dol-*P* docks to the Ser135 in the PKA motif (RKLISK) (Gandini et al. 2017).

16.3 Dynamicity in DPMS Catalytic Function

Because of an apparent biosensing ability, the catalytic activity of DPMS undergoes regulation by cellular microenvironment as well as by extracellular signaling (Banerjee et al. 1987; Banerjee 1988). For example, capillary endothelial cells when cultured in the absence of 5% CO_2 increase the DPMS catalytic activity by lowering the K_m for GDP-mannose by 33%. The V_{max} however remains unchanged (Table 16.1). The *N*-glycosylation of cellular proteins increases ~4.3 fold, but the cells exhibit poor adherence to the matrix. The result is inhibition of cellular proliferation (Banerjee 1988).

Insulin, a vital hormone for metabolism and a growth factor, downregulates angiogenesis in vitro irrespective of (i) the presence of high (4.0×10^{-11} M)- and low-affinity (4.7×10^{-9} M) insulin receptors; (ii) 210,000 insulin receptors per cell; (iii) ~30% increase in glucose transport; (iv) increased glycosylation of a cellular marker *N*-linked glycoprotein Factor VIIIc; (v) increased rate of $Glc_3Man_9GlcNAc_2$-PP-Dol (lipid-linked oligosaccharide; LLO) biosynthesis and turnover; and (vi) 1.5–2-fold increase in the DPMS activity. The conclusion of increased LLO synthesis in insulin-treated cells may be independent of elevating either the intracellular Dol-*P* level or the *DPMS* gene transcription. Increased secretion of Factor VIIIc, however, suggests the possibility of existence of additional LLO pool (s) and argues favorably that insulin retards the growth of capillary endothelial cells but turning a proliferative cell into a highly secretory cell (Banerjee et al. 2015).

Epinephrine, on the other hand, a "fright, flight, or fight" hormone prepares the cells to either combat or escape upon binding and activating the β-adrenoreceptors. Competitive ligand ([^{3}H]-DHA; dihydroalprenol) binding indicates the

```
S-cerevisiae      ------------------------------------------------------------ 0
C-glabrata        ------------------------------------------------------------ 0
U-maydis          ----------------------------------------------------MSIALDMD 8
T-brucei          ------------------------------------------------------------ 0
Bovine            -------------------------------------------------------MAAEE 5
Pig               -------------------------------------------------------MASEE 5
G-panda           -------------------------------------------------------MASEE 5
Dog               -------------------------------------------------------MASEE 5
Horse             -------------------------------------------------------MASEE 5
Human             -------------------------------------------------------MASLE 5
S-orangutan       MYKSRIFLIEPRNGPRRTSDQSHRSSPASEFGTVQPVLFRASLRNSASGIWLSSAMASLE 60
Chimpanzee        -------------------------------------------------------MASLE 5
R-monkey          MHKSCIFVIEPHNSSRRTSAQSHRSSPASEFGTVRPVLFRAALRNSASGVWLSSAMASLE 60
W-marmoset        -------------------------------------------------------MASLE 5
N-rat             -------------------------------------------------MASTGASR--- 8
H-mouse           -------------------------------------------------MASTGASR--- 8
C-hamster         -------------------------------------------------MASPGASRGAS 11
RJ-fowl           ------------------------------------------------------------ 0
P-guttata         ------------------------------------------------------------ 0
G-opossum         ---------------------------------------------------MASQSDIPN 9
W-frog            ------------------------------------------------------------ 0
A-frog            ------------------------------------------------------------ 0
Bullfrog          ------------------------------------------------------------ 0
Zebrafish         ------------------------------------------------------------ 0
R-smelt           ------------------------------------------------------------ 0
C-intestinalis    ------------------------------------------------------------ 0
J-wasp            ------------------------------------------------------------ 0
HB-louse          ------------------------------------------------------------ 0
R-beetle          ------------------------------------------------------------ 0
P-aphid           ------------------------------------------------------------ 0
D-melanogaster    ------------------------------------------------------------ 0
B-manjavacas      ------------------------------------------------------------ 0
C-Albicans        ------------------------------------------------------------ 0
C-dubliniensis    ------------------------------------------------------------ 0
L-elongisporus    ------------------------------------------------------------ 0
S-pombe           ------------------------------------------------------------ 0
C-briggsae        ------------------------------------------------------------ 0
A-thaliana        ------------------------------------------------------------ 0
C-parapsilosis    ---------------MYVIYVSATIAVLCVLVYVTLILFSHKPRPPTSSESYYKTNDNAN 45
```

Fig. 16.3 Alignment of amino acid sequence of DPMS. The alignment of DPMS sequences from all 39 species used the CLUSTAL W program (Kyte and Doolittle 1982) (http://www.genome.jp/tools/claustalw/). The PKA motif is highlighted in blue (3d). The metal (Mn^{2+} or Mg^{2+})-binding DAD motif is present in DPMS sequence from all 39 species and highlighted in green (3d). DPMS sequence from some species has *N*-glycosylation site (Asn-X-Ser/Thr) and highlighted in magenta (3e)

```
S-cerevisiae       -----------------MSIEYSVIVPAYHEKLNIKPLTTRLFAGMSPEMAK---KTELI 40
C-glabrata         -----------------MSIENSVIVPAYHEKLNIKPLTTRLFAALGNEGSK---ITELI 40
U-maydis           ASAKMRKQPGSSGWSTSSTPSCSVIVPAFRENLNLRPLVTRLSSAFASQSSSELANTEII 68
T-brucei           -----------------MAVKYSIIVPAYKECGNLEPLTKQVFDALADDGFSK-NEVEMV 42
Bovine             ASRSSPRFRREPKGRVSRQDKYSVLLPTYNERENLP----FIVWLLVKSFSESGFNYEII 61
Pig                ASRNS-RSRREPEGRFPRQDKYSVLLPTYNERENLP----LIVWLLVKSFSESGINYEII 60
G-panda            ASRSPRRSRREPEGRAPRQDKYSVLLPTYNERENLP----LIVWLLVKSFSESGINYEII 61
Dog                ASGSPRRSRREPEGRAPRQDKYSVLLPTYNERENLP----LIVWLLVKSFSESGINYEII 61
Horse              ASRSPPRSRREPGRRAPRQDKYSVLLPTYNERENLP----LIVWLLVKSFSQSGINYEII 61
Human              VSRSPRRSRRELEVRSPRQNKYSVLLPTYNERENLP----LIVWLLVKSFSESGINYEII 61
S-orangutan        VSRSPRRSRRELEVRSPRQNKYSVLLPTYNERENLP----LIVWLLVKSFSESGINYEII 116
Chimpanzee         VSRSPRRSRRELEVRSPRQNKYSVLLPTYNERENLP----LIVWLLVKSFCESGINYEII 61
R-monkey           VRRSPPMSRRQLEVRSPRRDKYSVLLPTYNERENLP----LIVWLLVKSFSESGINYEII 116
W-marmoset         VSGSPPRSRRELEVRSTLQDKYSVLLPTYNERENLP----LIVWLLVKSFSESGINYEII 61
N-rat              ---SLAASQRPPQGRSSRQDKYSVLLPTYNERENLP----LIVWLLVKSFSESAINYEII 61
H-mouse            ---SLAASPRPPQGRSSRQDKYSVLLPTYNERENLP----LIVWLLVKSFSESAINYEII 61
C-hamster          AATAAAASPRPPQGRSSRRDKYSVLLPTYNERENLP----LIVWLLVKSFSESSINYEII 67
RJ-fowl            -------------MAARGGNKFSVLLPTYNERENLP----LVVWLLVRTFRESGTDFEVI 43
P-guttata          -------------MAARGPGRVSVLLPTYNERDNLP----LVVWLLVRTFRHSGIDFEII 43
G-opossum          SPGSQRRGVKKATTEGTGVDKYSVLLPTYNERENLP----LIVWLLVKSFGESGNNYEII 65
W-frog             --------MAASGNKRKSGDKYSVLLPTYNERENLP----LIVWLLVRCFRDSGYNYEII 48
A-frog             --------MATSGNKRKSGDKYSVLLPTYNERENLP----IIVWLLVRCFRDSGYNYEII 48
Bullfrog           -------MAAVRERRERAGDVYSVLLPTYNERENLP----IIVWLLVKYFGESGYKYEII 49
Zebrafish          -----MASRRSNAKSRDKPDKYSVLLPTYNERENLP----LIVWLLVKYFGESGYNYEII 51
R-smelt            -----MASRRGSQKSRGDGNKYSVLLPTYNERENLP----LIVWLLVKYFSESGYDYEII 51
C-intestinalis     --------------MSKSENKYSVLLPTYNERENLP----IITWLLVKYFTESNYNFEII 42
J-wasp             ----MSTEAKVASASDIGDNKYSILLPTYNEVENLP----IIVWLIVKYMNEAKVSYEII 52
HB-louse           --MVTLTSGRVSQNEKINKNKYSILLPTYNEKENLP----IIVWLIVKYMNEGNYDYEII 54
R-beetle           ----------MGKESKSSEDKYSILLPTYNEVENLP----IIVWLIVKYMEKSGYNYEVI 46
P-aphid            -----------------MPPKYSILLPTYNEKENLP----VIVYLIHKYLEISEIEYEII 39
D-melanogaster     -------RFLRTITMPTNGHKYSILMPTYNEKDNLP----IIIWLIVKYMKASGLEYEVI 49
B-manjavacas       ----------------MNELKYSIILPTYNESENLP----LIIWLIIQHMENE--NFEII 38
C-Albicans         ----------------MTQNKYSVILPTYNEKRNLP----ILIYLLNKTFTANKLDWEVI 40
C-dubliniensis     ----------------MTQNKYSVILPTYNEKRNLP----ILIYLLNKTFTANKLDWEVI 40
L-elongisporus     ----------------MTANKYSVILPTYNEKRNLP----ILVYLLNKTFTEHKLEWEVI 40
S-pombe            ------------------MSKYSVILPTYNERKNLP----IITYLIAKTFDQEKLDWEIV 38
C-briggsae         ------------MIAVLSTPKYSIILPTYNEKENLP----ICIWLIEK--YLKGISYEII 42
A-thaliana         -------MADEMETKGEKKYKYSIIIPTYNERLNIA----IIVYLIFK--HLRDVDFEII 47
C-parapsilosis     ASYELPTRIDSTSLERKPQVEISVVIPCYNETKRLS---KMLSEAAGYLEQHYQGNYEII 102
                                         *:::* :.*  .:                      *::

S-cerevisiae       FVDDNSQDGSVEEVDALAHQGYNV--RIIVRTNER--GLSSAVLKGFYEAKGQYLVC 93
C-glabrata         FVDDNSQDGSVEEVEALHKQGYNV--RIIVRTDER--GLSSAVLKGFHEAKGEYLIC 93
U-maydis           IVDDNSRDGSVETVSALQSEGYNV--RIIVATSER--GLSSAVVRGFREARGQRMIC 121
T-brucei           IVDDNSRDGSVEVVEKVRNEGYGV--RIEVRTNDR--GLSSAVIHGISVSKGSFILV 95
Bovine             IIDDGSPDGTRDIAEQLEKIYGSD--RILLRPREKKLGLGTAYIHGMKHATGNYIII 116
Pig                IIDDGSPDGTRDIAEQLVKIYGSD--KILLRPREKKLGLGTAYIHGMKHATGNYIII 115
G-panda            IIDDGSPDGTRDVAEQLEKIYGSD--KILLRPREKKLGLGTAYIHGMKHATGNYIII 116
Dog                IIDDGSPDGTRDVAEQLEKIYGSD--KILLRPREKKLGLGTAYIHGMKHATGNYIII 116
Horse              IIDDGSPDGTRDVAEQLEKIYGSN--KILLRPREKKLGLGTAYIHGLKYATGNYIII 116
Human              IIDDGSPDGTRDVAEQLEKIYGSD--RILLRPREKKLGLGTAYIHGMKHATGNYIII 116
S-orangutan        IIDDGSPDGTRDVAEQLEKIYGSD--RILLRPREKKLGLGTAYIHGMKHATGNYIII 171
Chimpanzee         IIDDGSPDGTRDVAEQLEKIYGSD--RILLRPREKKLGLGTAYIHGMKHATGNYIII 116
R-monkey           IIDDGSPDGTRDVAEQLEKIYGSD--RILLKPREKKLGLGTAYIHGMKHATGNYIII 171
W-marmoset         IIDDGSPDGTRDVAEQLEKIYGSD--RILLRPREKKLGLGTAYIHGMKHATGNYIII 116
N-rat              IIDDGSPDGTREVAEQLEKIYGPD--RILLRPREKKLGLGTAYIHGIKHATGNYVII 116
H-mouse            IIDDGSPDGTREVAEQLAEIYGPD--RILLRPREKKLGLGTAYIHGIKHATGNYVII 116
C-hamster          IIDDGSPDGTREVAEQLEKIYGPD--RILLRPREKKLGLGTAYIHGIKHATGNYVII 122
RJ-fowl            IIDDGSPDGTQQVAEQLEKIYGSD--KILLRPRPKKLGLGTAYIHGMKHATGNFIVI 108
P-guttata          IIDDGSPDGTKEVAQQLEKIYGSD--KILLRPRARKLGLGTAYIHGMKYATGNFIVI 108
G-opossum          IIDDGSPDGTLEVAEQLKKIYGSD--KILLRPRAKKLGLGTAYIHGMQHATGNFIII 120
W-frog             VIDDGSPDGTLEVAQQLQKIYGSD--KILLRPRAKKLGLGTAYVHGMQHATGNFIII 103
A-frog             VIDDGSPDGTLEVAQQLQKIYGSD--KILLRPRAKKLGLGTAYVHGMQHATGNFIII 103
Bullfrog           IIDDGSPDGTLEVAKQLQKIYGGD--KILLRPREKKLGLGTAYVHGMQHASGNFIII 104
Zebrafish          VIDDGSPDGTLQIAEQLQKIYGAD--KILLRPRAEKLGLGTAYIHGIKHATGNFVII 106
R-smelt            VIDDGSPDGTLEVAQQLQQIYGED--KILLRPRKSKLGLGTAYIHGIKHATGNFVFI 106
C-intestinalis     VIDDGSPDGTLEVAKQLQTIYGED--KIVLRPRAKKLGLGTAYVHGVKHATGNFVII 97
J-wasp             VIDDGSPDGTLDMARQLQNLYGDD--KIVLNPRERKLGLGTAYIHGIKHATGNYIII 107
HB-louse           IIDDGSPDGTLEVAKQLQKIFGSD--KLILRPREKKLGLGTAYIHGLKHSSGDYIVI 109
R-beetle           IIDDGSPDGTQDAAKQLQKIYGEN--KILLKPREKKLGLGTAYIHGMKHASGNFILI 101
P-aphid            IIDDGSPDGTLEVAKQLETIYGKD--KILLRPRGKKLGLGTAYIHGMKHATGDFIII 94
D-melanogaster     VIDDGSPDGTLDVAKDLQKIYGED--KIVLRPRGSKLGLGTAYIHGIKHATGDFIVI 104
B-manjavacas       IVDDNSPDGTTQVAKDLQKIYGDE--KIVVTGREKKLGLGSAYIHGIKFAKGDFVVI 93
C-Albicans         IVDDNSPDGTQEIAKKLIDIFGPE--HIQLRPRAGKLGLGTAYVHGLQFVTGNFVII 95
C-dubliniensis     IVDDNSPDGTQEIAKKLIDIFGPE--HIQLRPRAGKLGLGTAYVHGLQFVTGNFVII 95
L-elongisporus     IVDDNSPDGTQEVAKKLIDIYGSK--HIQLRPRAGKLGLGTAYVHGLQYVTGNFVII 95
S-pombe            IIDDASPDGTQEVAKELQKIYGED--KILLKPRSGKLGLGTAYIHGLKFATGDFVII 93
C-briggsae         IVDDASPDGTQDVAKLLEKEYGEN--KILIKPRAGKLGLGTAYSHGLSFARGEFIIL 97
A-thaliana         VVDDGSPDGTQEIVKQLQQLYGED--RILLRARAKKLGLGTAYIHGLKHATGDFVVI 103
C-parapsilosis     IVDDGSSDGTAEYAIQLAKEYKLEPHTMKVVQLSKNRGKGGAVTHGLLHTQGKYALF 159
                   .:** * **: : .  :          : :       * . *  :*.    *.  .
```

Fig. 16.3 (continued)

```
S-cerevisiae      MDAD----LQHPPETVPKLFESLH--DHAFTLGTRYAPGVGIDKDWPMYRRVISSTARMMARP 150
C-glabrata        MDAD----LQHPPESVPSLFDSLR--KHPFVLGTRYAPGVGIDKDWPLHRRVISSTARMMARP 150
U-maydis          MDAD----LQHPPEAVPSLLLALNG-QKSFVLGTRYGVGVSMDKDWPLHRRIISSGARMLARP 179
T-brucei          MDAD----LQHPPKTVPCLLRALEKPGVEFVCGTRYGAGVEIDKDWPLHRRFISWGARLLARP 154
Bovine            MDAD----LSHHPKFIPEFIRKQKEGNFDIVSGTRYKGNGGVYG-WDLKRKIISRVANFITQI 174
Pig               MDAD----LSHHPKFIPEFIRKQKEGNFDIVSGTRYKGNGGVYG-WDLKRKIISRGANFITQI 173
G-panda           MDAD----LSHHPKFIPEFIRKQKEGNFDIVSGTRYKGNGGVYG-WDLKRKIISRGANFITQI 174
Dog               MDAD----LSHHPKFIPEFIRKQKEGNFDIVSGTRYKGNGGVYG-WDLKRKIISRGANFLTQI 174
Horse             MDAD----LSHHPKFIPEFIRKQKEGNFDIVSGTRYKGNGGVYG-WDLKRKIISRGANFITQI 174
Human             MDAD----LSHHPKFIPEFIRKQKEGNFDIVSGTRYKGNGGVYG-WDLKRKIISRGANFLTQI 174
S-orangutan       MDAD----LSHHPKFIPEFIRKQKEGNFDIVSGTRYKGNGGVYG-WDLKRKIISRGANFLTQI 229
Chimpanzee        MDAD----LSHHPKFIPEFIRKQKEGNFDIVSGTRYKGNGGVYG-WDLKRKIISRGANFLTQI 174
R-monkey          MDAD----LSHHPKFIPEFIRKQKEGNFDIVSGTRYKGNGGVYG-WDLKRKIISRGANFLTQI 229
W-marmoset        MDAD----LSHHPKFIPEFIRKQKEGNFDIVSGTRYKGNGGVYG-WDLKRKIISRGANFLTQI 174
N-rat             MDAD----LSHHPKFIPEFIRKQKEGNFDIVSGTRYKGNGGVYG-WDLKRKIISRGANFITQI 174
H-mouse           MDAD----LSHHPKFIPEFIRKQKEGNFDIVSGTRYKGNGGVYG-WDLKRKIISRGANFITQI 174
C-hamster         MDAD----LSHHPKFIPEFIRKQKEGNFDIVSGTRYKGNGGVYG-WDLKRKIISRGANFITQI 180
RJ-fowl           MDAD----LSHHPKFIPEFIRKQKEGNFDIVSGTRYKGDGGVYG-WDLKRKLISRGANFITQV 156
P-guttata         MDAD----LSHHPKFIPEFIRKQQEGNFDIVSGTRYKGNGGVYG-WDLKRKLISRGANFLTQV 156
G-opossum         MDAD----LSHHPKFIPEFIRKQKEGNFDIVSGTRYKGNGGVYG-WDLKRKIISRGANFLTQI 178
W-frog            MDAD----LSHHPKFIPEFIRKQKEGSYDIVSGTRYSGNGGVYG-WDLKRKLISRGANFVTQV 161
A-frog            MDAD----LSHHPKFIPEFIRKQKEGSYDIVSGTRYAGNGGVYG-WDLKRKLISRGANFITQV 161
Bullfrog          MDAD----LSHHPKFIPEFINKQKEGDFDVVSGTRYIGNGGVYG-WDLKRKLISRGANYLSQV 162
Zebrafish         MDAD----LSHHPKFIPQFIEKQKEGGYDLVSGTRYRGDGGVYG-WDLRRKLISRGANFVTQV 164
R-smelt           MDAD----LSHHPKFIPEFIEKQRQGGYDVVAGTRYRGDGGVYG-WDLRRKLISRGANFVTQV 164
C-intestinalis    MDAD----LSHHPKFIPQYIKKQEELDCDIVTGTRYEGEGGVYG-WDWKRKLVSRTANFVTQV 155
J-wasp            MDAD----LSHHPKFIPEMIKEQVKNNFDVVTGTRYAGNGGVYG-WDFKRKLISRGANFITQI 165
HB-louse          MDAD----LSHHPKFIPQFIEFQKTQDFDVVTGTRYKGNGGVYG-WDFKRKLVSRGANFLTQF 167
R-beetle          MDAD----LSHHPKFIPQFIEKQKSKDYDVVSGTRYVGSGGVYG-WDFRRKLISRGANFLTQL 159
P-aphid           MDAD----LSHHPKFILDFLKKQAEQDYDVVTGSRYIGNGGVSG-WDFKRKLVSRGANFVSQL 152
D-melanogaster    IDAD----LSHHPKFIPEFIKLQQEGNYDIVSGTRYAGNGGVFG-WDFKRKLISRGANFLSQV 162
B-manjavacas      MDAD----LSHHPKFIKQFIKHQKEKDYDIVTGTRYAGDGGVHG-WNLKRKAISRTANYITQV 151
C-Albicans        MDAD----FSHHPEAIPEFIAKQKSQDYDIVTGTRYAGDGGVFG-WDFKRKLISRGANFLASV 153
C-dubliniensis    MDAD----FSHHPEAIPEFIAKQKSQDYDIVTGTRYAGDGGVYG-WDFKRKLISRGANFLASV 153
L-elongisporus    MDAD----FSHHPEAIPEFIAKQKTEDYDIVTGTRYAGNGGVYG-WDFKRKLISRGANFLATV 153
S-pombe           MDAD----FSHHPKYLPEFIKLQKEHNYDIVLGTRYAKDGGVYG-WNLKRKFISRGANLLAST 151
C-briggsae        MDAD----LSHHPKFIPEMIALQQKYKLDIVTGTRYKNGGGYSG-WDLKRKTISKGANFLAQF 155
A-thaliana        MDAD----LSHHPKYLPSFIKKQLETNASIVTGTRYVKGGGVHG-WNLMRKLTSRGANVLAQT 160
C-parapsilosis    ADADGATQFSDVAKLIDYLGSYPNEPAIAIGSRAHMVNTDAVVK-RSLIRNFLMYGLHTLVFI 221
                   ***    :.. .: :             .    ::             *.      . :

S-cerevisiae      LT-IASDPMSGFFGLQKKYLENCNPRDINSQGFKIALELLAKLPLPRDPR----VAIGEV 205
C-glabrata        LT-TASDPMSGFFGLQKKYLTKANPQDINSQGFKIALELLAKLPLPAN------EPIGEI 203
U-maydis          LT-SASDPMSGFFGITKHSFHTAD-HHINAQGFKIALDLLVKSGVHST-------DIAEV 230
T-brucei          LT-PLSDPMSGFFGLRVDVFQRGR-EVVNPIGYKIALELFVKCAVRKY---------EEV 203
Bovine            LLRPGASDLTGSFRLYRKEVLQKLIGKCISKGYVFQMEMIVRARQLNYT-------IGEV 227
Pig               LLRPGASDLTGSFRLYRKEVLQKLIEKCVSKGYVFQMEMIVRARQLNYT-------IGEV 226
G-panda           LLRPGASDLTGSFRLYRKEVLQKLIEKCVSKGYVFQMEMIVRARQLNYT-------IGEV 227
Dog               LLRPGASDLTGSFRLYRKEVLQKLIEKCVSKGYVFQMEMIVRARQLNYT-------IGEV 227
Horse             LLRPGASDLTGSFRLYRKEVLQKLIEKCVSKGYVFQMEMIIRARQLNYT-------IGEV 227
Human             LLRPGASDLTGSFRLYRKEVLEKLIEKCVSKGYVFQMEMIVRARQLNYT-------IGEV 227
S-orangutan       LLRPGASDLTGSFRLYRKEVLEKLIEKCVSKGYVFQMEMIVRARQLNYT-------IGEV 282
Chimpanzee        LLRPGASDLTGSFRLYRKEVLEKLIEKCVSKGYVFQMEMIVRARQLNYT-------IGEV 227
R-monkey          LLRPGASDLTGSFRLYRKEVLEKLIEKCVSKGYVFQMEMIVRARQLNYT-------IGEV 282
W-marmoset        LLRPGASDLTGSFRLYRKEVLEKLIEKCVSKGYVFQMEMIVRARQLNYT-------IGEV 227
N-rat             LLRPGASDLTGSFRLYRKEVLQKLIEKCVSKGYVFQMEMIVRARQMDYT-------VGEV 227
H-mouse           LLRPGASDLTGSFRLYRKEVLQKLIEKCVSKGYVFQMEMIVRARQMNYT-------IGEV 227
C-hamster         LLRPGASDLTGSFRLYRKEVLQKLIEKCVSKGYVFQMEMIVRARQLNYT-------IGEV 233
RJ-fowl           LLRPGASDLTGSFRLYRKEVLEKLMEKCVSKGYVFQMEMIVRARQLGYT-------IGEV 209
P-guttata         LLRPGASDLTGSFRLYRKEVLEKLMEKCVSKGYVFQMEMIVRARQLGFT-------VGEV 209
G-opossum         LLRPGVSDLTGSFRLYRKEVLQKLMEKCVSKGYVFQMEMIVRARQLNYT-------IGEV 231
W-frog            LLRPGASDLTGSFRLYRKDVLQKLVERCVSKGYVFQMEMIVRARQLNYT-------IGEV 214
A-frog            LLRPGASDLTGSFRLYRKDVLQKLVERCVSKGYVFQMEMIVRARQLNYT-------IGEV 214
Bullfrog          LLRPGASDLTGSFRLYRRSVLQKLVENCVSKGYVFQMEMIVRARQFNYT-------IGEV 215
Zebrafish         LLRPGASDLTGSFRLYKKEVLEKLVEQCVSKGYVFQMEMIVRARQLGYT-------IGEV 217
R-smelt           LLRPGASDLTGSFRLYKKEVLESLVSCCVSKGYVFQMEMIVRARQLKYT-------IGEV 217
C-intestinalis    LLRPGVSDLTGSFRLYRKDVLKQLVGLCVSKGYVFQMEMMVRARQLNLS-------IAEV 208
J-wasp            MLRPGVSDLTGSFRLYKKDVLEKLTKSCVSKGYVFQMEMIVRARQFSFS-------IGEV 218
HB-louse          LLRPGVSDLTGSFRLYKKPVLEKLVSSCVSKGFVFQMEMIIRAKENLFT-------VGEV 220
R-beetle          LLRPGASDLTGSFRLYKKDVLEKLIKSCVSKGYVFQMEMIVRARQFNYT-------VGEV 212
P-aphid           LLRPKVSDLTGSFRLYKKPVLEKLIESCISKGYVFQMEMMVRARQLNYT-------IGEV 205
D-melanogaster    LLRPNASDLTGSFRLYKKDVLEKCIASCVSKGYVFQMEMLVRARQHGYT-------IAEV 215
B-manjavacas      LLRPGVSDLTGSFRLYKKQVLEKLVQSCISKGYVFQMEMMVRANQFGFS-------VGEV 204
C-Albicans        VLRPHVSDLTGSFRLYKTDVLRKIIDVTQSKGYVFQMEMMVRAKAMGFT-------VGEV 206
C-dubliniensis    VLRPHVSDLTGSFRLYKTEVLRKIIDVTQSKGYVFQMEMMVRAKAMGFT-------VGEV 206
L-elongisporus    VLRPNVSDLTGSFRLYKTDVLKKIIDVTQSKGYVFQMEMMVRARSLGFT-------VGEV 206
S-pombe           VLGTGVSDVTGSFRLYKKPVLETLMSEVTSKGYVFQMEIIARAREHNYT-------IGEV 204
C-briggsae        LLNPGVSDLTGSFRLYKKEVLATLIAESVSKGYVFQMEMMFRAKKSGYR-------IGEV 208
A-thaliana        LLWPGVSDLTGSFRLYKKSALEDVISSCVSKGYVFQMEMIVRATRKGYH-------IEEV 213
C-parapsilosis    FGIRKIHDTQCGFKMFNMDAVKQIFPHMHTERWIFDVEVLLLGQMQGMKMKEIAVNWQEI 281
                  .           * :              .  : : ::::                  *:
```

Fig. 16.3 (continued)

```
S-cerevisiae        PFTFGVRTEGESKLSGKVIIQYLQQLKELYVFKFGANNLILFITFWSILFFYVCYQLYHL 265
C-glabrata          PFSFGVRVEGESKLSGKVIIQYLEQLKELYVFKYGANTLIAVFAIAVLLALYILLKLRSL 263
U-maydis            PFSFGLRQEGESKLDGKVMFKYLQQLVELYRFRFGTVPIVFVLIVLLVLALYIWSHVLAP 290
T-brucei            GFNFAARTVGESKLTGKVIINYLEHLKLLYFVYGTALTVLLVLLPLIFYCFYILL----- 259
Bovine              PISFVDRVYGESKLGGNEIVSFLKGLLTLFATT--------------------------- 260
Pig                 PISFVDRVYGESKLGGNEIVSFLKGLLTLFATT--------------------------- 259
G-panda             PISFVDRVYGESKLGGNEIVSFLKGLLTLFATT--------------------------- 260
Dog                 PISFVDRVYGESKLGGNEIVSFLKGLLTLFATT--------------------------- 260
Horse               PISFVDRVYGESKLGGNEIVSFLKGLLTLFATT--------------------------- 260
Human               PISFVDRVYGESKLGGNEIVSFLKGLLTLFATT--------------------------- 260
S-orangutan         PISFVDRVYGESKLGGNEIVSFLKGLLTLFATT--------------------------- 315
Chimpanzee          PISFVDRVYGESKLGGNEIVSFLKGLLTLFATT--------------------------- 260
R-monkey            PISFVDRVYGESKLGGNEIVSFLKGLLTLFATT--------------------------- 315
W-marmoset          PISFVDRVYGESKLGGNEIVSFLRGLLTLFATT--------------------------- 260
N-rat               PISFVDRVYGESKLGGNEIVSFLKGLLTLFATT--------------------------- 260
H-mouse             PISFVDRVYGESKLGGNEIVSFLKGLLTLFATT--------------------------- 260
C-hamster           PISFVDRVYGESKLGGNEIVSFLKGLLTLFATT--------------------------- 266
RJ-fowl             PISFVDRVYGESKLGGNEIVSFLKGLLTLFATT--------------------------- 242
P-guttata           PISFVDRVYGESKLGGNEIVSFLKGLLTLFATT--------------------------- 242
G-opossum           PISFVDRVYGESKLGGNEIVSFLKGLLTLFATT--------------------------- 264
W-frog              PISFVDRVYGESKLGGNEIVSFLKGLLTLFATT--------------------------- 247
A-frog              PISFVDRVYGESKLGGNEIVSFLKGLLTLFATT--------------------------- 247
Bullfrog            PISFVDRVYGESKLGGNEIVSFLKGLLTLFATT--------------------------- 248
Zebrafish           PISFVDRVYGESKLGGNEIVSFLKGLLTLFATT--------------------------- 250
R-smelt             PITFVDRVYGESKLGGNEIVSFLKGLLTLFATT--------------------------- 250
C-intestinalis      PISFVDRFYGESKLGGNEIFRFLSGLLYLFATT--------------------------- 241
J-wasp              PISFVDRVYGESKLGGSEIVQFVKGLLYLFATT--------------------------- 251
HB-louse            PISFVDRVYGESKLGQTEIFQFIKALLYLFFTT--------------------------- 253
R-beetle            PITFVDRVYGESKLGGSEIFQFAKALLYLFATT--------------------------- 245
P-aphid             PITFVDRVYGESKLGGSEIVQFVKSLLYLFATT--------------------------- 238
D-melanogaster      PITFVDRIYGTSKLGGTEIIQFAKNLLYLFATT--------------------------- 248
B-manjavacas        PISFVDRQFGESKLGGNEIFQFLKGLLYFFAPT--------------------------- 237
C-Albicans          PISFVDRLYGESKLGGDEIVQYAKGVWTLFTSV--------------------------- 239
C-dubliniensis      PISFVDRLYGESKLGGDEIVQYAKGVWTLFTNV--------------------------- 239
L-elongisporus      PISFVDRLYGESKLGGDEIVLYAKGVWALFTSV--------------------------- 239
S-pombe             PIAFVDRLYGESKLGMDDILGYLKGVFSLLFI---------------------------- 236
C-briggsae          PISFVDRFFGESKLGSQEIVDYAKGLLYLFAIVW-------------------------- 242
A-thaliana          PITFVDRVFGTSKLGGSEIVEYLKGLVYLLLTT--------------------------- 246
C-parapsilosis      DGSKIDLARDSIEMAIDLVVTRLAYLLGIYELDECGRAGKKNQ----------------- 324
                             .  ::    :.     :  :

S-cerevisiae        VF-- 267
C-glabrata          LF-- 265
U-maydis            MLGA 294
T-brucei            ---- 259
Bovine              ---- 260
Pig                 ---- 259
G-panda             ---- 260
Dog                 ---- 260
Horse               ---- 260
Human               ---- 260
S-orangutan         ---- 315
Chimpanzee          ---- 260
R-monkey            ---- 315
W-marmoset          ---- 260
N-rat               ---- 260
H-mouse             ---- 260
C-hamster           ---- 266
RJ-fowl             ---- 242
P-guttata           ---- 242
G-opossum           ---- 264
W-frog              ---- 247
A-frog              ---- 247
Bullfrog            ---- 248
Zebrafish           ---- 250
R-smelt             ---- 250
C-intestinalis      ---- 241
J-wasp              ---- 251
HB-louse            ---- 253
R-beetle            ---- 245
P-aphid             ---- 238
D-melanogaster      ---- 248
B-manjavacas        ---- 237
C-Albicans          ---- 239
C-dubliniensis      ---- 239
L-elongisporus      ---- 239
S-pombe             ---- 236
C-briggsae          ---- 242
A-thaliana          ---- 246
C-parapsilosis      ---- 324
```

Fig. 16.3 (continued)

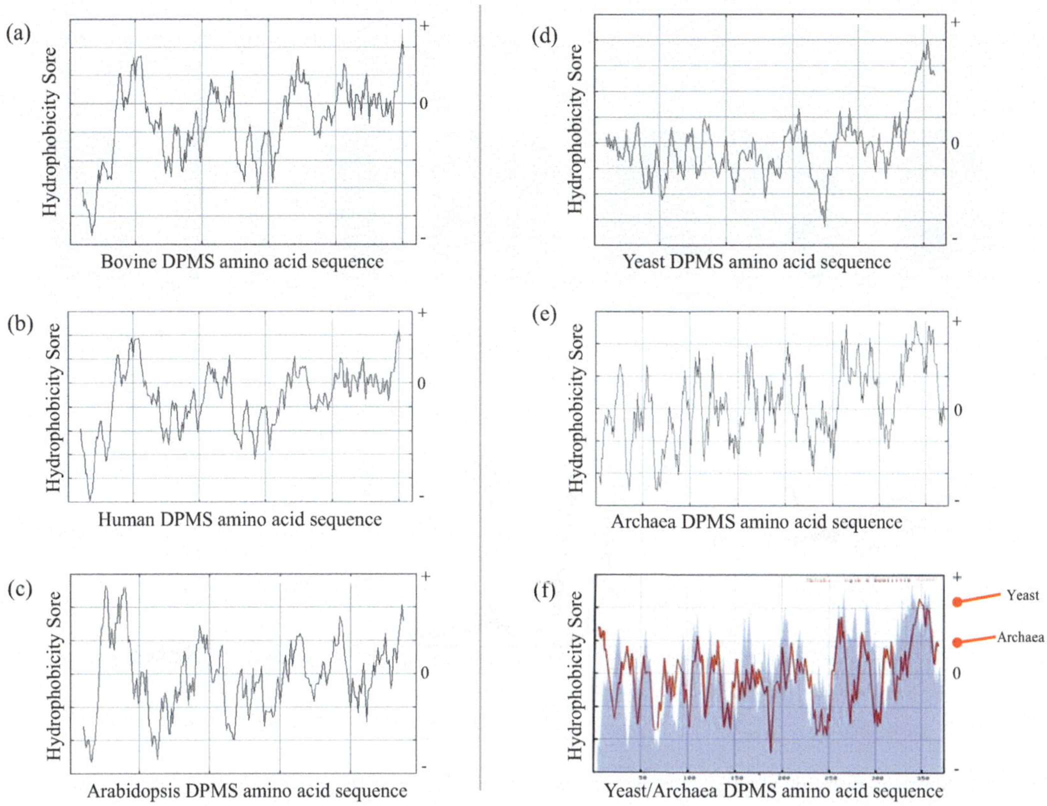

Fig. 16.4 Hydropathy profile of DPMS from five different species. For scanning the polypeptide sequence of DPMS, 19 amino acids are faulted as one window and hydropathy index calculated based on one window. (**a**) Bovine capillary endothelial cells; (**b**) human; (**c**) *A. thaliana* (Arabidopsis plant); (**d**) *S. cerevisiae* (yeast); (**e**) Archaea; and (**f**) superimposed yeast and Archaea hydropathy profile. The hydropathic character of DPMS was computed under ExPASy Bioinformatics Resource Portal (http://web.expasy.org/portscale/) on the logarithm established (Gasteiger et al. 2005) with a window of 19

presence of high ($K_d = 0.27 \pm 0.9 \times 10^{-9}$ M)- and low ($K_d = 2.96 \pm 0.31 \times 10^{-9}$ M)-affinity β-adrenoreceptors (i.e., β_1 and β_2) on the capillary endothelial cell plasma membrane. β_1 adrenoreceptors, however, is three times more potent than the β_2 adrenoreceptors. 1.5-fold increase in intracellular cAMP following isoproterenol (a nondiscriminatory β-agonist) treatment indicates both receptors are coupled to adenylate cyclase. Cholera toxin (100 ng/mL), forskolin (1×10^{-6} M), as well as prostaglandin E_1 (1×10^{-3} M) mimic the β-agonist isoproterenol in enhancing the intracellular cAMP level in capillary endothelial cells.

A time-dependent (2 h–32 h) increase in DPMS activity is observed in microsomal membranes from capillary endothelial cells treated with 2 mM 8Br-cAMP (Fig. 16.5b). High DPMS activity in isoproterenol-treated cells is not due to an increased gene expression because *actinomycin D* failed to block the upregulation of DPMS activity (Baksi et al. 2008). Phosphorylation in vitro of microsomal membranes by PKA also increases the DPMS catalytic activity. Furthermore, purification of phosphorylated DPMS by antibody affinity chromatography detects a 32kDa phosphoprotein co-migrating with the DPMS activity (Banerjee et al. 1999) (Fig. 16.5c, d), thus establishing DPMS as a phosphoprotein and corroborating with the presence of the PKA motif in DPMS protein sequence.

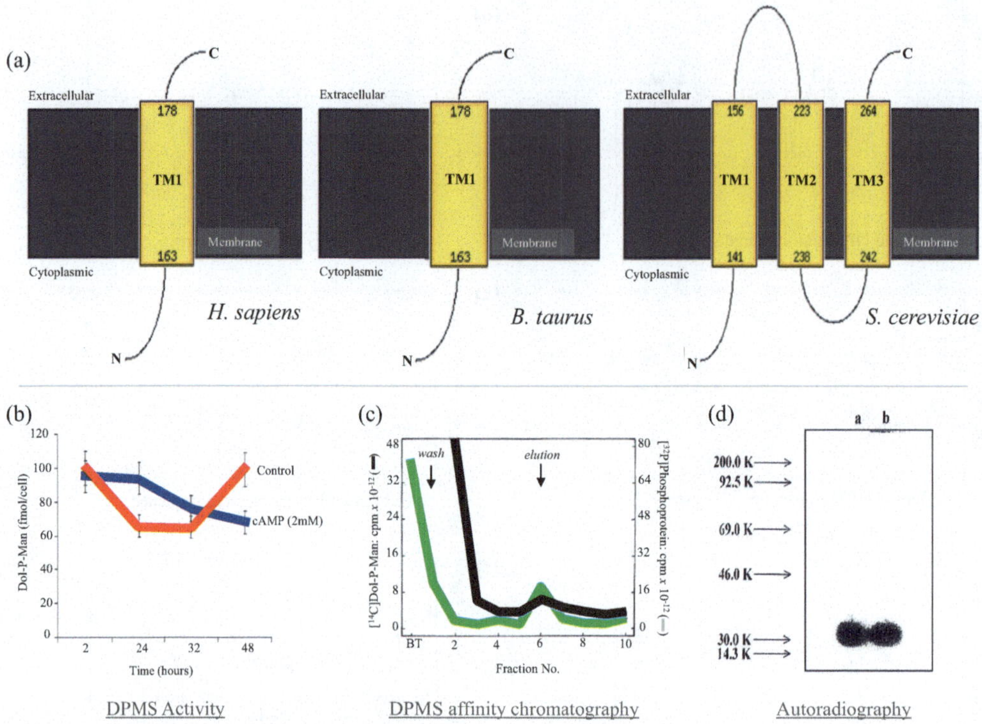

Fig. 16.5 Structure prediction of DPMS and the localization of the phosphorylation target. (**a**) 99% of residues modelled at >90% confidence. The model is generated with Phyre2, a web-based services for protein structure prediction. The phosphorylation target for human and bovine DPMS is the amino acid 163, whereas it is the amino acid 141 for the yeast DPMS. (**b**) DPMS activity in ER membrane from capillary endothelial cells treated with 2mM 8Br-cAMP for 0 h–48 h., control (blue); 8Br-cAMP (red). (**c**) Antibody-affinity column chromatography detects co-migration of the DPMS catalytic activity and the phosphorylated DPMS. DPMS catalytic activity (green); ^{32}P-protein (black). (**d**)^{32} P-labeled DPMS is immunoprecipitated with anti-DPMS antibody, separated by a 7.5% SDS-polyacrylamide gel electrophoresis, and detected by autoradiography. (*a*) 1:1000 antibody dilution and (b) 1:2000 antibody dilution

Table 16.1 Effect of CO_2 deprivation on DPMS activity in capillary endothelial cells

	Dol-P-man synthase activity	
Culture condition	K_m (μM)	V_{max} (pmol/mg protein/min)
Normal	0.09 ($r = 0.98$)	18.4
Cultured in the absence of CO_2	0.6 ($r = 0.9$)	8.8

The cells are cultured for 24 h at 37 °C in the presence or absence of 5% CO_2. The DPMS activity is assayed in cell homogenate (Banerjee 1988)

16.4 Consequences of Increased DPMS Level on Lipid-Linked Oligosaccharide (LLO) and Asparagine-Linked (*N*-Linked) Glycoprotein

$Glc_3Man_9GlcNAc_2$-PP-Dol in the ER lumen is a prerequisite for *N*-linked protein glycosylation. Dol-*P*-man, the product of DPMS-catalyzed reaction, is a key intermediate in the elongation of $Man_5GlcNAc_2$-PP-Dol to $Man_9GlcNAc_2$-PP-Dol for the synthesis of $Glc_3Man_9GlcNAc_2$-PP-Dol (LLO). Therefore, the *N*-glycosylation process is expected to be highly regulated. To evaluate the cellular consequences when dol-*P*-man and/or DPMS catalytic activity is upregulated or downregulated, we have considered the following:

First, we have analyzed when the dol-*P*-man level is higher compared to the control. These studies are conducted in situ. A time-dependent increase of 2-^{3}H-mannose incorporation is shown in dol-*P*-man, in LLO, and in proteins in capillary endothelial cells treated with the β-agonist isoproterenol (Fig. 16.6a–c). The synthesis of LLO is almost identical when the process is repeated with intracellular cAMP-generating agents such as 8Br-cAMP, forskolin, cholera toxin, or PGE_1 (Fig. 16.6d). Selective β_1- and β_2-antagonist atenolol or ICI 118,551 and the nonselective β-antagonist propranolol block the stimulatory effect of isoproterenol, quantitatively (Fig. 16.6e). Importantly, the LLO turns over at a faster rate (i.e., 5–7 min) in the presence of isoproterenol compared to the control (~12 min) (Fig. 16.6f). To verify that the increase in dol-*P*-man in β-adrenoreceptor stimulates cells is not the result of an increased level of intracellular Dol-*P*; the experiments are conducted with exogenously added Dol-*P*. The addition of exogenous Dol-*P* enhances the LLO biosynthesis, but the effect is additive when both isoproterenol and Dol-*P* are present together (Fig. 16.6g).

Second, we have analyzed the influence of β-adrenoreceptor activation on *N*-glycosylation status. When the capillary endothelial cells are pulsed labeled with [2-^{3}H]-mannose for 0–4 h, the mannose incorporation in glycoproteins is higher in isoproterenol-treated cells (Fig. 16.6c). To clarify, if the effect is related to enhanced glycosylation and not due to increased protein synthesis, the cells are labeled with [2-^{3}H]-mannose (10 μCi/mL) and [^{14}C]-leucine (1.25 μCi/mL) in a continuous pulse for 1 h at 37 °C in the presence of a β-agonist isoproterenol (1×10^{-9} M) with or without a β-antagonist atenolol or ICI 118,551 (5×10^{-8} M). The results indicate the ratio of [^{3}H]-mannose to [^{14}C]-leucine incorporation (a protein glycosylation index) into protein is reduced by ~45% in cells in the presence of β-antagonist. To analyze a specific *N*-linked glycoprotein, we select the capillary endothelial cell marker FVIIIc (Fig. 16.7a, b) (Banerjee et al. 1992). Factor VIIIc, in general, is a plasma glycoprotein with 2332 amino acid residues and acts as a regulatory cofactor for blood coagulation (Fay 1999; Kane and Davie 1988; Lenting et al. 1998).

Factor VIIIc (FVIIIc) is collected from cell extracts and from the conditioned media by immunoprecipitation after dually labeling the capillary endothelial cells with [2-^{3}H]-mannose (10 μCi/mL) and [^{35}S]-methionine (40 μC/mL) for 1 h in a serum-free/methionine-free low-glucose DMEM (Bio-Fluids, Inc.) in the presence of isoproterenol (1×10^{-7} M). The immunoprecipitates are separated on 10% SDS-PAGE, and the radioactivity in excised bands is quantified in a liquid scintillation spectrometer. A 73% increase in the ratio of [2-^{3}H]-mannose to [^{35}S]-methionine is observed in cellular Factor VIIIc, whereas the increase is 45% for the secretory FVIIIc.

16.5 Reflection of Increased Protein *N*-Glycosylation on Angiogenesis

The hallmark of cancer follows the core principles of sustainability to proliferative signaling, ability to evade growth suppressors, ability to resist cell death, and ability to enable replicative

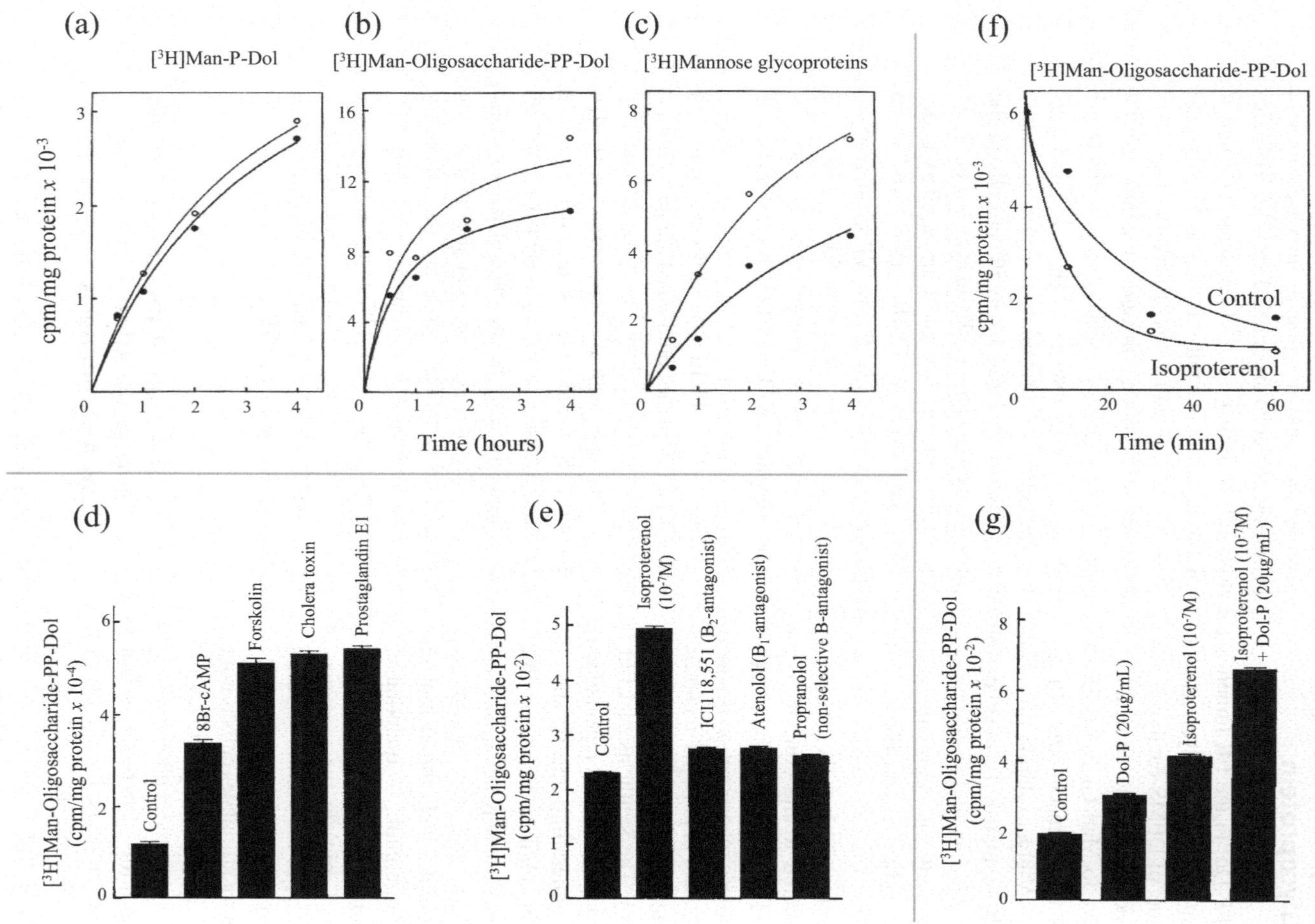

Fig. 16.6 Effect of β-agonist isoproterenol and other cAMP enhancers on dol-*P*-man, LLO, and glycoprotein synthesis in capillary endothelial cells. The cells are labeled with D-[2-^{3}H]-mannose (50 μCi/mL) for 1 h–4 h in the presence (o--o) and absence (●--●) of 1×10^{-7} M isoproterenol. (**a**) Dol-*P*-man, (**b**) LLO, and (**c**) glycoproteins are extracted and quantified in a liquid scintillation spectrometer. (**d**) Evaluation of other cAMP enhancers mimic the isoproterenol effect. The cells are labeled with radioactive mannose for 1 h but in the presence of none or 2 mM 8Br-cAMP, 1 μM forskolin, 1 ng/mL of cholera toxin, or 10 μM PEG1. The LLO is extracted and quantified. (**e**) Determining the specificity of the β-adrenergic stimulation. The cells are labeled with the radioactive mannose for 1 h but in the presence of none or 1×10^{-7} M isoproterenol alone or together with fivefold excess of ICI 181551 (β_1 antagonist) or atenolol (β_2 antagonist) or propranolol (nonselective β-antagonist). The LLO is extracted and quantified. (**f**) Evaluation that the LLO pool turns over faster in the presence of isoproterenol. The cells are pulse-labeled with for 1 h with radioactive mannose, washed, resuspended in a media containing 20 mM unlabeled mannose, and chased in the absence (o—o) or in the presence of 1×10^{-7} M isoproterenol (●--●) for 0–60 min. The LLO is extracted and quantified. (**g**) Evaluation that increased LLO synthesis in isoproterenol-treated cells is not due to increased *dol-P* level. The cells are labeled with 2-^{3}H mannose for 1 h in the absence or in the presence of Dol-*P* (20 μg/mL) or in the presence of isoproterenol (1×10^{-7}M) plus *Dol-P* (20 μg/mL). The LLO is extracted and quantified

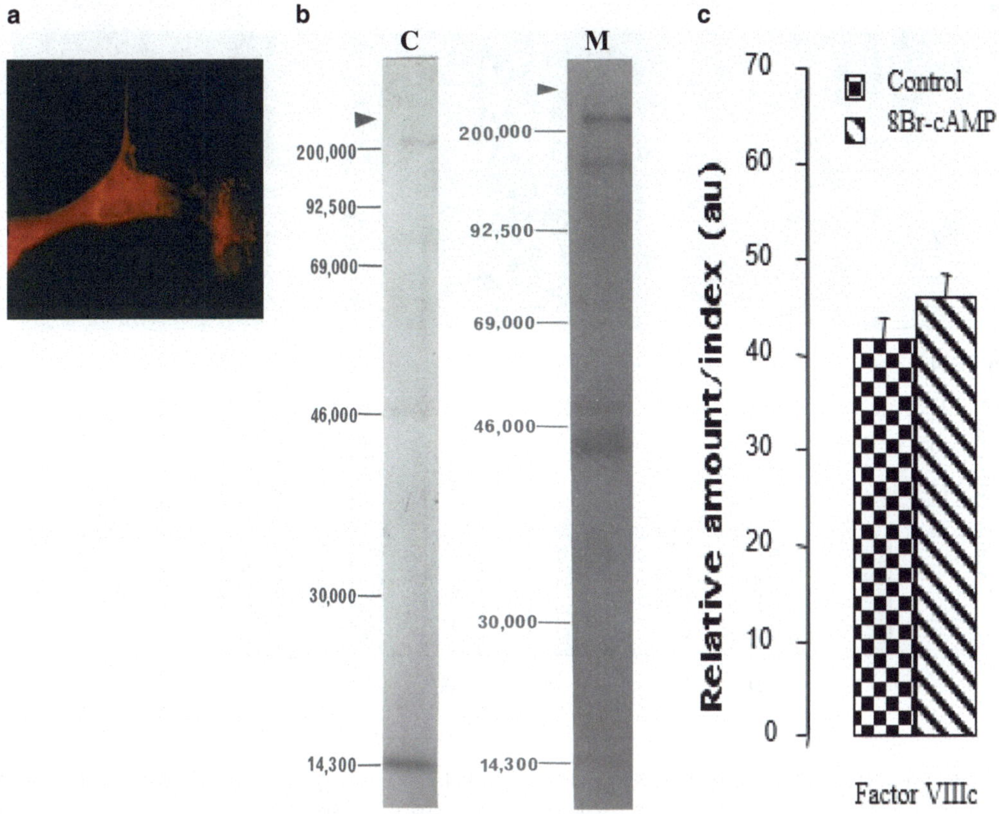

Fig. 16.7 Effect of cAMP enhancer on Factor VIIIc expression. Detection of Factor VIIIc: (**a**) by fluorescence microscopy. Cells cultured in chamber slides detect Factor VIIIc in the perinuclear region of the cell after fixation and processing for immunofluorescence microscopy with anti-Factor VIIIc mouse monoclonal antibody. (**b**) By autoradiography. Cells labeled with [^{35}S]-methionine (40 μCi/mL) for 1 h and lysed after separating from the media. Factor VIIIc from both cell lysate and conditioned media is immunoprecipitated with a mouse monoclonal anti-Factor VIIIc antibody, separated on 10% SDS-PAGE and exposed to X-ray film for detection. (**c**) Factor VIIIc expression in cells treated with 8Br-cAMP. Cells cultured with 2 mM 8Br-cAMP for 32 h harvested and processed for two-color flow cytometric analysis

immortality, angiogenesis induction, and activating invasion and metastasis (Hanahan and Weinberg 2011). Angiogenesis, i.e., neovascularization, is also a key to breast tumor progression (Schneider and Miller 2005; Uhr et al. 1997; Gastl et al. 1997). Significant components of angiogenesis are endothelial cell migration, capillary budding, establishment of capillary loops, and neovascular remodeling (Folkman 2007).

The *N*-linked glycoproteins have been found to play an important role in capillary endothelial cell proliferation and differentiation (Banerjee et al. 2015). Inhibition of "hybrid"- and "complex"-type *N*-glycan synthesis, however, has been suggested in the inhibition of capillary tube formation. In contrast, inhibition of only "complex"-type *N*-glycan, but not "hybrid"-type *N*-glycan, does not inhibit the tube formation (Nguyen et al. 1992, 1993; Pili et al. 1995).

Mentioned earlier, cells treated with a β-agonist isoproterenol and/or various cAMP-producing pharmacological agents exhibit increased DPMS activity and the levels of dol-*P*-man, the LLO, as well as the *N*-linked glycoproteins. To evaluate if

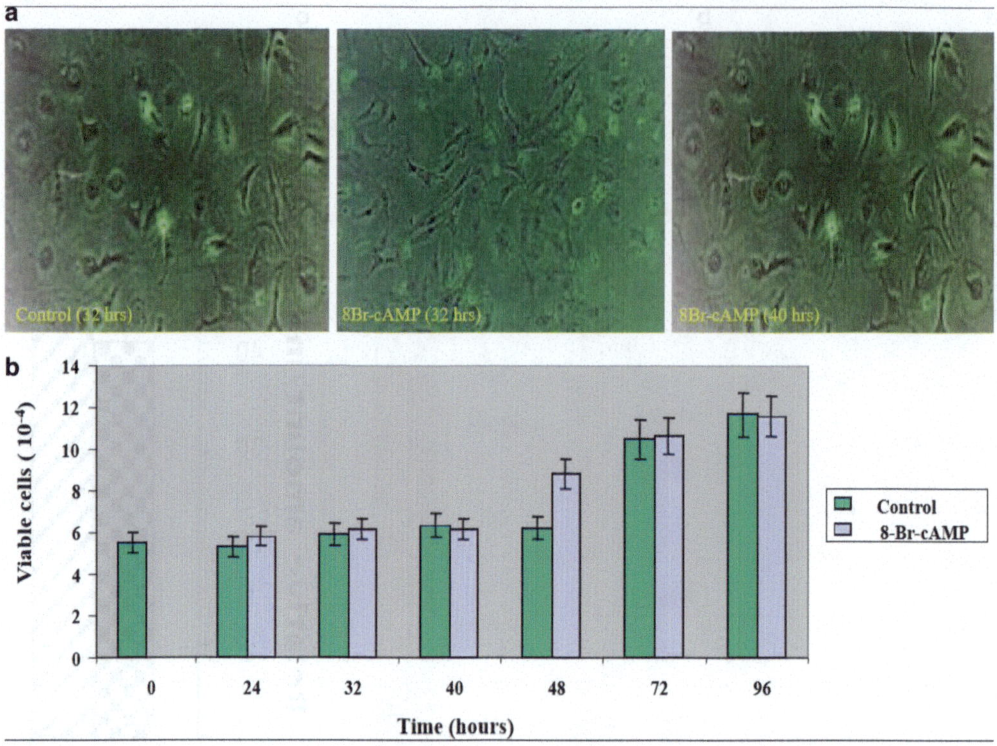

Fig. 16.8 Effect of cAMP enhancer on capillary endothelial cell proliferation. (**a**) Cellular morphology following 8Br-cAMP treatment. Cells cultured in the presence or absence of 2 mM 8Br-cAMP are examined microscopically after 32 h and 40 h. The images explain distinctive cellular morphology following 8Br-cAMP treatment. (**b**) Accelerated cellular proliferation in the presence of 8Br-cAMP. Cells are cultured in the absence or presence of 2 mM 8Br-cAMP for 0 h–96 h. At desired times the cells are removed and counted

these agents would also increase the capillary endothelial cell marker *N*-linked glycoprotein Factor VIIIc and enhance angiogenesis, the level of Factor VIIIc and the proliferation of capillary endothelial cells are monitored after treating with 8Br-cAMP. The results indicate that cells treated with 2mM 8-Br-cAMP increase Factor VIIIc level (Fig. 16.7c), exhibit morphological changes (Fig. 16.8a), and increase cellular proliferation in a time-dependent manner (Fig. 16.8b).

Since Factor VIIIc is a cofactor in Factor IXa-dependent activation of Factor Xa during blood coagulation, we propose Factor VIIIc would analogously activate matrix metalloproteinases (MMPs) for capillary invasion to support tumor growth. To evaluate if this could be true, the capillary endothelial cells are cultured on plates coated with growth factor-reduced Matrigel™ in the presence or absence of anti-Factor VIIIc monoclonal antibody. The control inserts contained no Matrigel™ coating. The chemoattractant used is the conditioned media from MCF-7 human breast cancer cells. The results indicate that anti-Factor VIIIc monoclonal antibody blocks the Matrigel™ invasion of capillary endothelial cells (Table 16.2) (Banerjee et al. 2011a, b).

16.6 DPMS Overexpression and Angiogenesis

Activation of G-protein-coupled receptor (GPCR) and enhancing the intracellular cAMP level by pharmacological agents have claimed to impact

Table 16.2 Anti-factor VIIIc antibody prevents Matrigel™ chemoinvasion of capillary endothelial cells

Sample	Control insert (cell number)	Growth factor reduced (cell number)	% Invasion
Control	5668 ± 788	2078 ± 0	37
Anti-Factor VIIIc antibody (100 ng)	2372 ± 612	80 ± 0	2.4

The cells are cultured on Biocoat plates with growth factor-reduced Matrigel™ in the presence or absence of anti-Factor VIIIc mouse monoclonal antibody for 24 h at 37 °C. The chemoattractant used is conditioned media from human breast cancer cells MCF7. The inserts are processed, and the number of cells passed through the membrane pores is counted under a microscope

angiogenesis by enhancing the protein *N*-glycosylation through activation of DPMS. This interesting observation however has also raised a question about the sustainability of the process. The reasons are simple. These biological processes are short-lived and need a steady supply of the signal enhancer(s). It is therefore expected that the tumors may circumvent these processes to maintain the higher DPMS activity through an alternative pathway. We then ask if cells constitutively overexpressing DPMS could mimic the above observation and promote angiogenesis. To answer the question, we developed a stable cell line of capillary endothelial cell overexpressing DPMS (Zhang et al. 2010). These cells express (i) high level of DPMS mRNA as judged by qPCR (Fig. 16.9a); (ii) four times higher DPMS protein as judged by the Western blotting and immunofluorescence microscopy (Fig. 16.9b, c); and (iii) most importantly ~108% higher DPMS activity (Table 16.3). Immunofluorescence microscopy with Texas Red-conjugated WGA indicates a high level of GlcNAc-β-(1,4)-GlcNAc-(1,4)-GlcNAc-NeuAc glycans (*N*-glycans) on the plasma membrane of the cell (Fig. 16.9d). When examined, the DPMS overexpressing cells outperform the cellular proliferation and their chemotactic activity of the cells expressing scrambled gene (i.e., control) or the wild-type capillary endothelial cells (Fig. 16.9e, f). Thus, supporting unequivocally DPMS is indeed an intracellular regulator of angiogenesis.

16.7 Inhibition of DPMS Catalytic Activity Is Anti-angiogenic and Anti-tumorigenic

(a) *Targeting DPMS directly*: One lingering question is, what happens when the DPMS or its catalytic product dol-*P*-man is in short supply? To answer the question, we have used a naturally occurring lipopeptide amphomycin (M_r1290 Da). Amphomycin forms a complex with Dol-*P* in the presence of Ca^{2+}. The interaction is highly specific and is lost if Ca^{2+} is either chelated, or replaced with other divalent cation, or the N-terminal fatty-acylated aspartic acid is removed by mild acid hydrolysis (Banerjee 1989, 1994). The apparent K_m for GDP-mannose for dol-*P*-man synthesis in the presence or absence of amphomycin was 1.08 μM and 1.37 μM, respectively, whereas the V_{max} was 0.17 pmol/mg protein/min in the presence of amphomycin as compared to 1.86 pmol/mg protein/min in its absence. On the other hand, when the DPMS activity is measured as a function of Dol-*P*, the presence of amphomycin changes the substrate-velocity curve from a rectangular hyperbola to a sigmoid. The Hill coefficient (n) for each of this reaction is found to be 2.02 in the presence and 1.22 in the absence of amphomycin. The corresponding K_m values however for Dol-*P* are 333 μM and 47.3 μM, respectively. The inhibitory effect

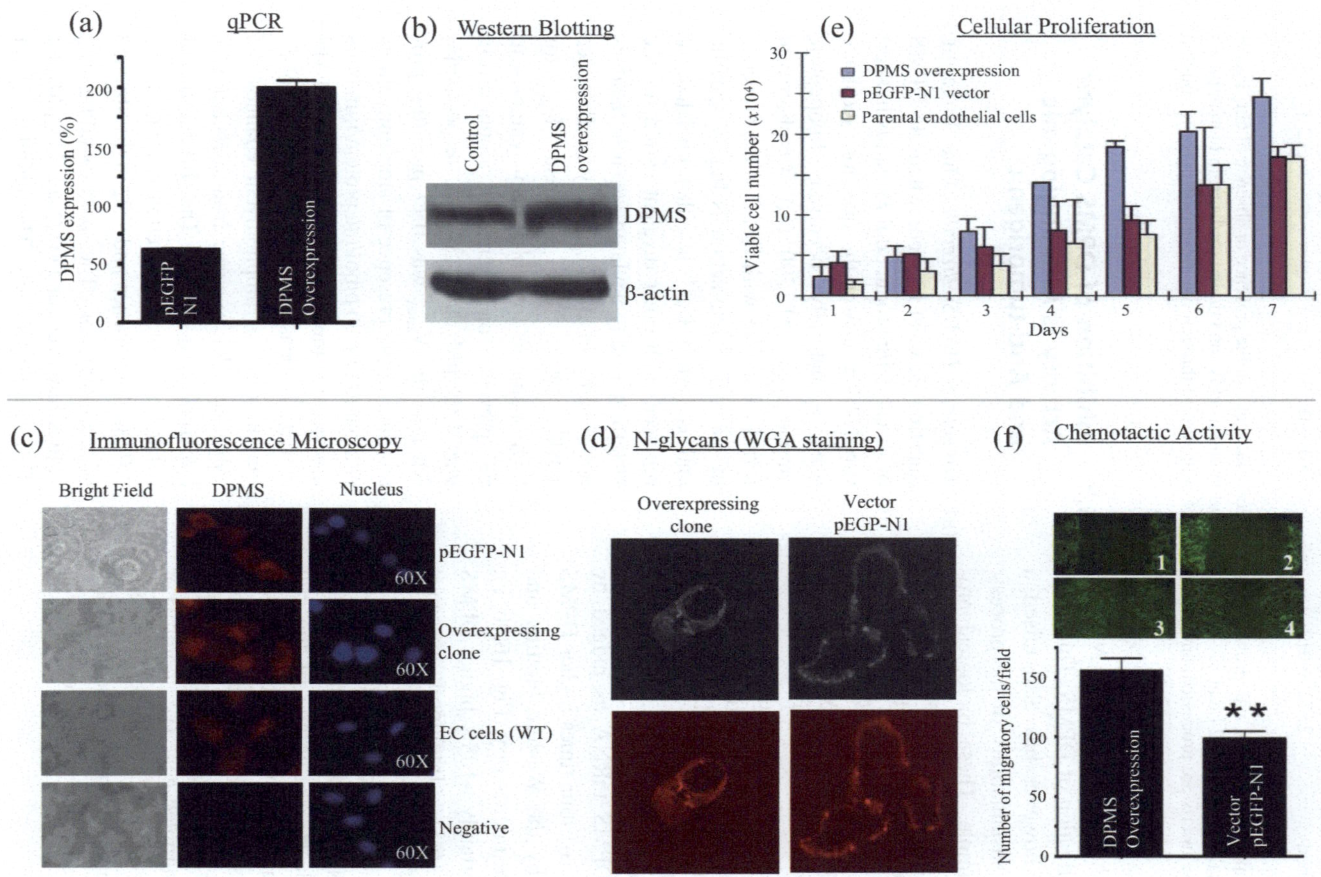

Fig. 16.9 DPMS overexpression increases N-Glycan levels and activates "angiogenic switch". qPCR, Western blot developed with anti-DPMS antibody and the immunofluorescence microscopy confirm increased DPMS expression in capillary endothelial cells overexpressing the *DPMS* gene (**a, b, and c**). The primers used for DPMS gene are forward primer 5′-GCTGAGCAGTTGGAGAAG-3′ and reverse primer 5-′-TGGATGGTGTGAGAGGTC-3′ (PCR product size: 153 bp). (**d**) *N*-glycan expression is detected in control and DPMS overexpressing cells by immunofluorescence microscopy with Texas Red-conjugated WGA. (**e**) Cellular proliferation of DPMS overexpressing cells. The cell numbers were plotted against time and compared with wild-type capillary endothelial cells. (**f**) Chemotactic behavior due to DPMS overexpression is conducted in a scratch assay. 1-mm-wide scratch is made across (one and two), and the plates are photographed after 6 h. Three = DPMS overexpression clone; Four = pEGFP-N1 vector with scrambled RNA sequence (control). Bottom panel is quantification of the cell migration

Table 16.3 DPMS activity in DPMS-overexpressing capillary endothelial cells

Sample	Dol-P-Man (pmol/mg protein/5 min mean ± SD	*p*-value[a]
pEGFP-N1 vector	44.06 ± 1.2	
DPMS overexpressing cells	47.73 ± 2.1	≈0.04

Total cell lysates from control and DPMS overexpressing cells are used for the enzyme assay (Zhang et al. 2010)
[a]Analyzed by student *t*-test

of amphomycin could only be overcome with increasing amount of Dol-*P* (Banerjee et al. 1981). When added to a culture of capillary endothelial cells, amphomycin inhibited cellular proliferation in a dose-dependent manner, thus supporting further that angiogenesis indeed depends on DPMS.

(b) *Inhibiting the cross talk between DPMS and N-acetylglucosaminyl 1-phosphate transferase (GPT)*: GPT is essential for *N*-glycan biosynthesis. Based on the studies we have performed earlier with microsomes from β-agonist isoproterenol-treated rat, parotid acinar cells exhibit enhanced catalytic activity of both DPMS and GPT (Banerjee et al. 1985). The subsequent studies further confirm exogenous addition of dol-*P*-man activates the GPT activity in vitro (Kean 1982). To evaluate if there is a cross talk between DPMS and GPT, we use a specific inhibitor of GPT tunicamycin and examine the DPMS activity in capillary endothelial cells. Excitingly, the tunicamycin-treated capillary endothelial cells lose their ability to synthesize dol-*P*-man (Fig. 16.10a) (Martinez 2002). The result is inhibition of LLO synthesis and reduction/or no protein *N*-glycosylation (Fig. 16.10b). The consequences are cell cycle arrest and induction of apoptosis (Fig. 16.9c, d). Tunicamycin equally inhibits the human breast cancer cell MDA-MB-231 (triple negative; $ER^-/PR^-/Her2^-$) proliferation (Fig. 16.10e), and the progression of a double-negative ($ER^-/PR^-/Her2^+$) breast tumor progression in athymic nude mice (Fig. 16.11) (Banerjee et al. 2011a, b).

16.8 Nanoformulation Enhances Anti-angiogenic Efficacy of Tunicamycin

Nanoparticle (<100 nm) evades the immune system. To evaluate if nano-formulated tunicamycin would have a better efficacy for treating breast cancer, we have synthesized tunicamycin nanoparticles. When tested, tunicamycin gold nanoparticles (Au NPs) inhibit DPMS expression (Fig. 16.12a) and inhibit capillary endothelial cell proliferation, i.e., angiogenesis ~50% within 1 h, whereas the native tunicamycin has no effect (Fig. 16.12b). The nano-formulated tunicamycin also blocks the cell cycle progression, but the effect is much sooner than with native tunicamycin (Figs. 16.12c and 16.13a, b). Mechanistically, tunicamycin downregulates the cell cycle progression by inhibiting the phosphorylation of p53 and Rb proteins but does not induce apoptosis (Fig. 16.13c). The observation is novel and has opened the door for further investigation on the noncanonical pathway of apoptosis (Banerjee et al. 2013).

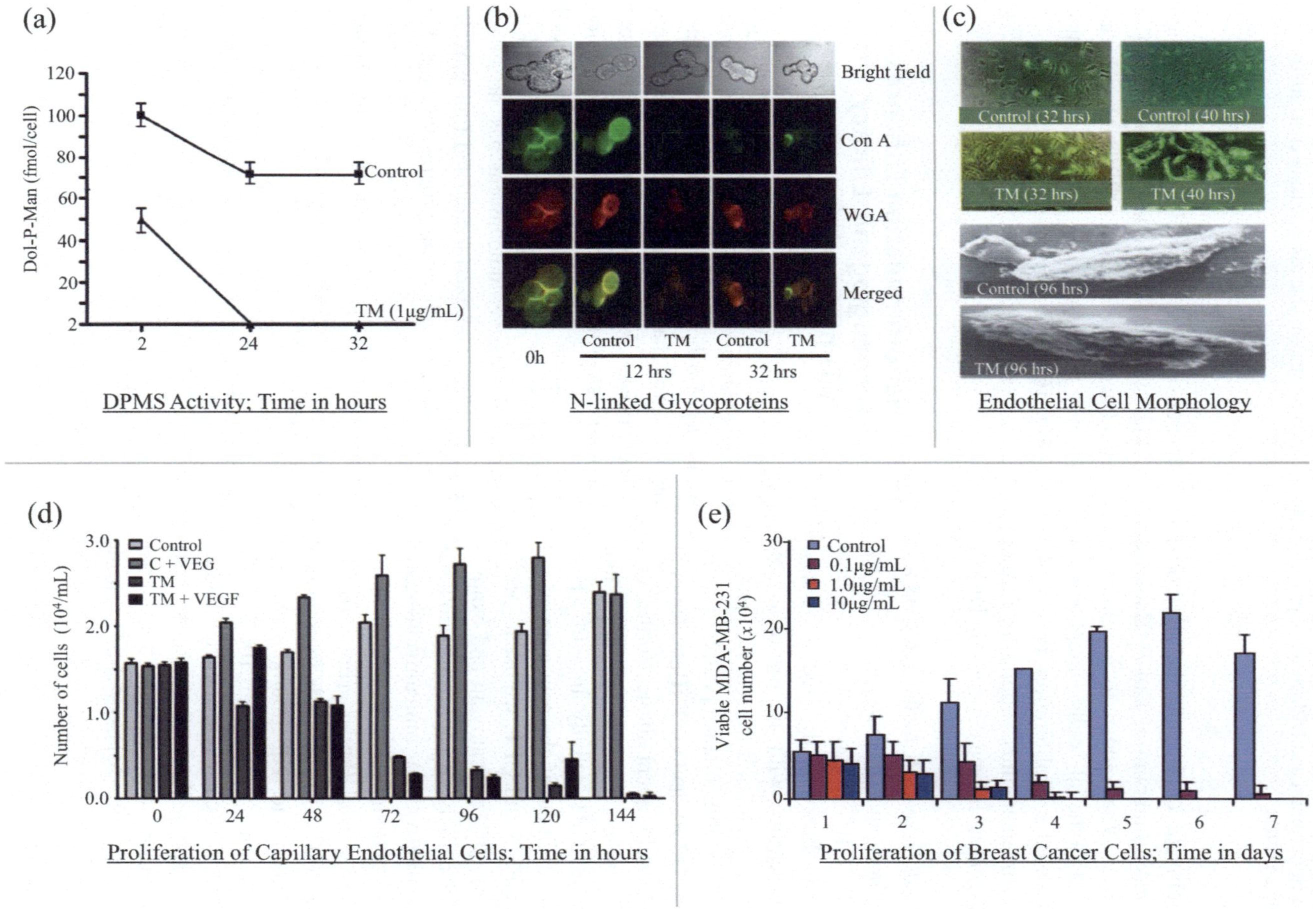

Fig. 16.10 Loss of DPMS activity reduces angiogenesis and inhibits proliferation of human breast cancer cells. (**a**) Tunicamycin (1 μg/mL) treatment makes capillary endothelial cells to lose DPMS activity. (**b**) Cells express decreased "high" mannose and "complex" type N-glycans. (**c**) Light and scanning electron microscopy images project surface blebbing, membrane fragmentation, nuclear condensation, and pyknotic appearance reminiscence of apoptotic cell death. (**d** and **e**) Proliferation of capillary endothelial cells and triple-negative ($ER^-/PR^-/Her2^-$) human breast cancer cells (MDA-MB-231) inhibited with time

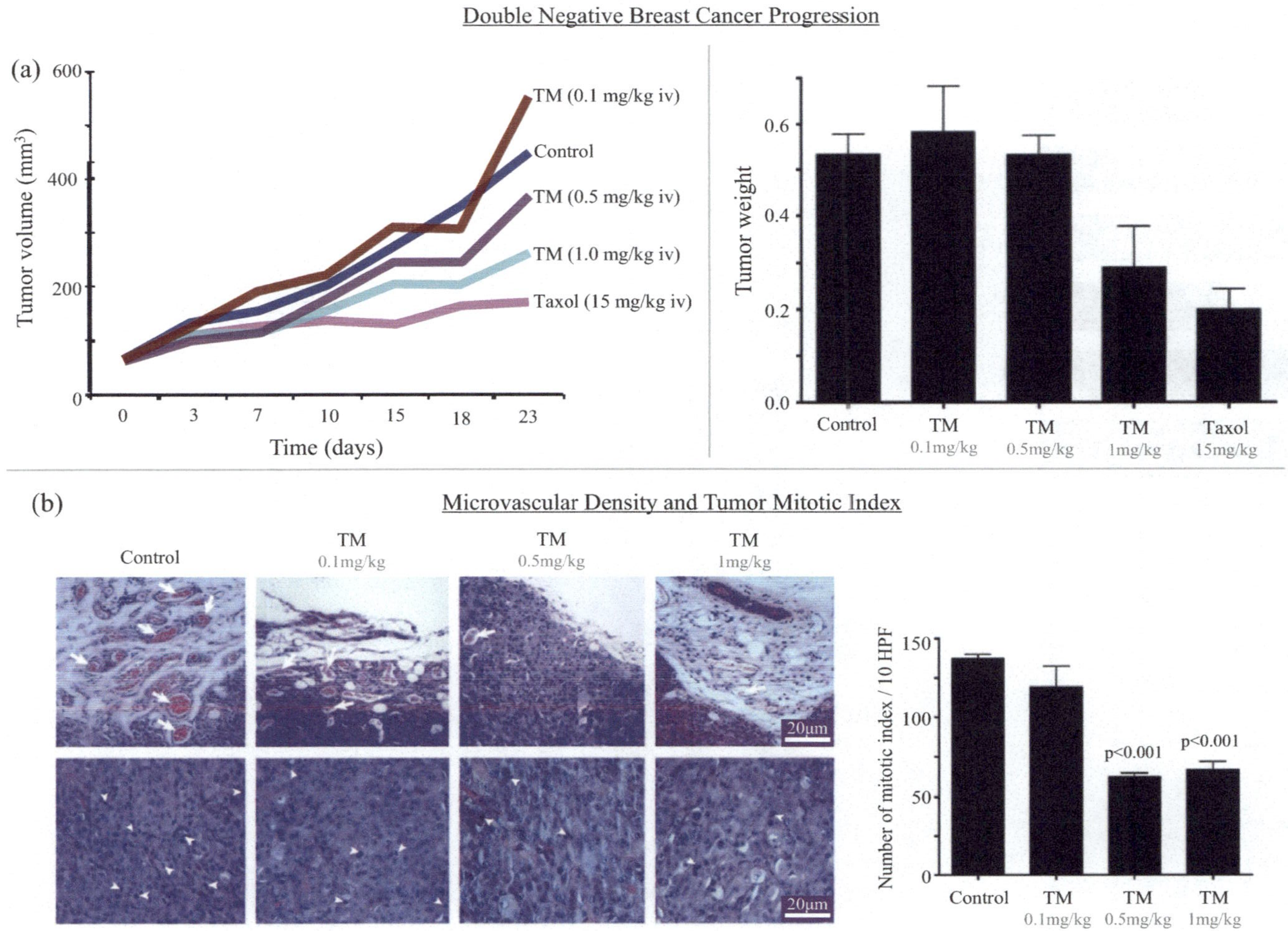

Fig. 16.11 Loss of DPMS activity reduces breast tumor progression in athymic nude mice. (**a**) The progression of double-negative ($ER^-/PR^-/EGFR^+$) orthotopic breast tumor in athymic nude mice is reduced ~55% in 3 weeks when treated with tunicamycin (1 mg/Kg) intravenously once a week. Taxol, a FDA-approved breast cancer drug requires 15 times more drug (i.e., 15 mg/Kg) to match the effect of tunicamycin (1 mg/kg) to see the same effect. (**b**) *Upper panel* is the H&E-stained tumor sections identifying the microvascular density (arrows) in control and after tunicamycin treatment. *Bottom panel* is H&E staining identifying the mitotic index (arrow heads) in control and tunicamycin-treated breast tumor section. The histogram on the right is the quantification of the mitotic index averaged from ten representative areas of each tumor

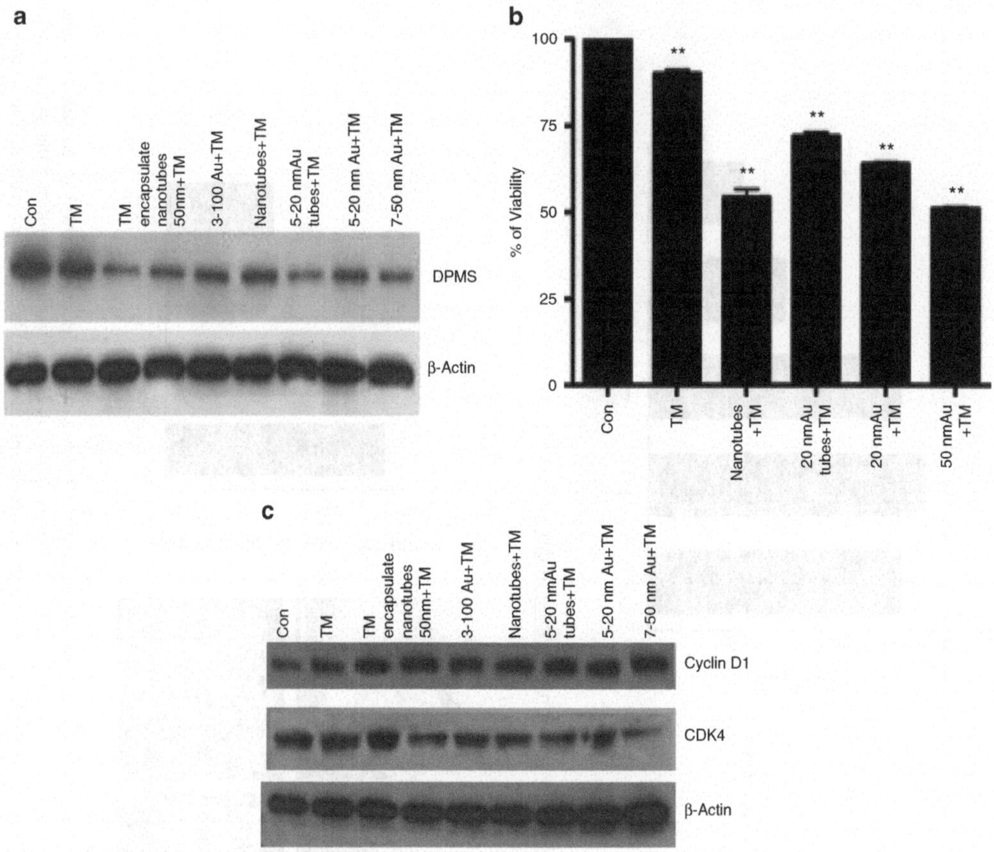

Fig. 16.12 Nano-formulated tunicamycin is three times more potent than the native formulation. (**a**) Nano-formulated tunicamycin (1 μg/mL) reduces the DPMS activity by ~75% in 1 h, whereas the native formulation reduces the activity by only 15%. (**b**) The viability of capillary endothelial cells also reduces to almost 50%. (**c**) The level of cyclin D1, CDK4

16.9 Conclusion

The presence of an active DPMS either in tumor microvascular endothelial cells or in tumor tissue is essential for tumor progression. By developing an appropriate biochemical as well as cell/molecular biological tools and/or taking genetic approaches, we have concluded that dynamic functioning of DPMS is the key to upregulation of angiogenesis and breast tumor progression. Therefore, DPMS is expected to be a new target for developing next-generation anti-angiogenic/anti-tumorigenic glycotherapy for treating breast cancer.

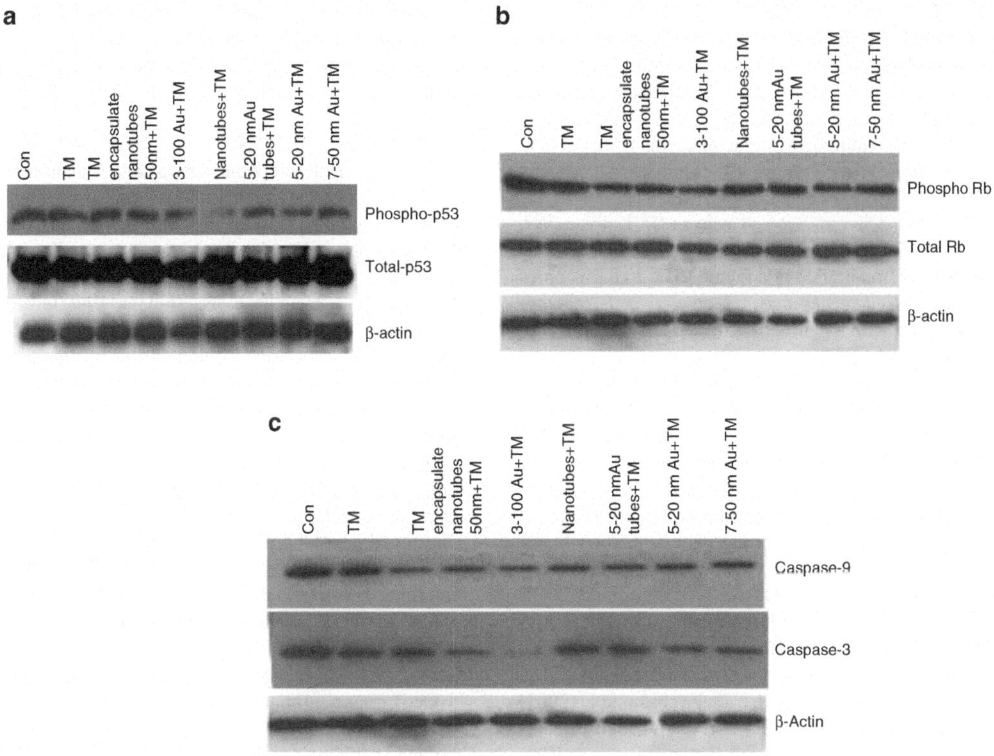

Fig. 16.13 Nano-formulated tunicamycin downregulates cell cycle progression but does not activate apoptosis in capillary endothelial cells. P53 is a transcription factors and often considered as the gatekeeper for the cell cycle. (**a**) The expression of total and phosphorylated (serine-392) is measured by Western blotting. The ratio of phospho-p53 to total p53 is 1.0 in cells treated with native tunicamycin, but the ratio is 0.1–0.2 in cells treated with nano-formulated tunicamycin. (**b**) Rb is a cell cycle inhibitor protein. The ratio of phosphorylated Rb (ser-249/Thr252) to total Rb reduces by 34.4% and 64.2%, and it is only 20% in cells treated with native tunicamycin. Activation of pro-caspase 9 to caspase 9 in the aptosome is the initiation of the apoptotic process. (**c**) Downregulation of caspase-9 indicates that inhibition of cellular proliferation is not due to apoptosis

Acknowledgment This work is partly supported by funds from the Office of the Dean, School of Medicine, University of Puerto Rico, and grants from the Department of Defense DAMD17-03-1-0754, the National Institutes of Health NIH U54-CA096297, Susan G. Komen for the Cure BCTR0600582, the National Science Foundation NSF EPS-1002410 (DKB), and the National Institutes of Health NIH/NIMHD 2G12MD007583 (KB).

References

Baksi K, Tavárez-Pagán JJ, Martínez JA, Banerjee DK (2008) Unique structural motif supports mannosylphospho dolichol synthase: an important angiogenesis regulator. Curr Drug Targets 9:262–271

Banerjee DK (1988) Microenvironment of endothelial cell growth and regulation of protein N-glycosylation. Indian J Biochem Biophys 25:8–13

Banerjee DK (1989) Amphomycin inhibits mannosylphosphoryldolichol synthesis by forming a complex with dolichylmonophosphate. J Biol Chem 264:2024–2028

Banerjee DK (1994) A recent approach to the study of dolichyl monophosphate topology in the rough endoplasmic reticulum. Acta Biochim Pol 41:275–280

Banerjee DK (2012) N-glycans in cell survival and death: cross-talk between glycosyltransferases. Biochim Biophys Acta 1820:1338–1346

Banerjee DK, Scher MG, Waechter CJ (1981) Amphomycin: effect of the lipopeptide antibiotic on the glycosylation and extraction of dolichyl monophosphate in calf brain membranes. Biochemistry 20:1561–1568

Banerjee DK, Kousvelari EE, Baum BJ (1985) beta-Adrenergic activation of glycosyltransferases in the dolichylmonophosphate-linked pathway of protein N-glycosylation. Biochem Biophys Res Commun 126:123–129

Banerjee DK, Kousvelari EE, Baum BJ (1987) cAMP-mediated protein phosphorylation of microsomal membranes increases mannosylphospho dolichol synthase activity. Proc Natl Acad Sci (USA) 84:6389–6393

Banerjee DK, Tavárez JJ, Oliveira CM (1992) Expression of blood clotting factor VIII:C gene in capillary endothelial cells. FEBS Lett 306:33–37

Banerjee DK, DaSilva JJ, Bigio B (1999) Mannosylphosphodolichol synthase activity is associated with a 32 kDa phosphoprotein. Biosci Rep 19:169–177

Banerjee A, Lang JY, Hung MC, Sengupta K, Banerjee SK, Baksi K, Banerjee DK (2011a) Unfolded protein response is required in nu/nu mice microvasculature for treating breast tumor with tunicamycin. J Biol Chem 286:29127–29138

Banerjee DK, Oliveira CM, Tavárez JJ, Katiyar VN, Saha S, Martínez JA, Banerjee A, Sánchez A, Baksi K (2011b) Importance of a factor VIIIc-like glycoprotein expressed in capillary endothelial cells (eFactor VIIIc) in angiogenesis. Adv Exp Med Biol 705:453–464

Banerjee A, Johnson KT, Banerjee IA, Banerjee DK (2013) Nanoformulation enhances anti-angiogenic efficacy of tunicamycin. Transl Cancer Res 2:240–255

Banerjee A, Martinez JA, Longas MO, Zhang Z, Santiago J, Baksi K, Banerjee DK (2015) N-acetylglucosaminyl 1-phosphate transferase: an excellent target for developing new generation breast cancer therapeutic. Adv Exp Med Biol 842:355–374

Banerjee DK, Zhang Z, Baksi K, Serrano-Negrón JE (2017) Dolichol phosphate mannose synthase: A Glycosyltransferase with Unity and molecular diversities. Glycoconj J 34:467–479

Bujnicki JM, Elofsson A, Fischer D, Rychlewski L (2001) Structure prediction meta server. Bioinformatics 17:750–751

Calo D, Kaminski L, Eichler J (2010) Protein glycosylation in Archaea: sweet and extreme. Glycobiology 20:1065–1076

Fay PJ (1999) Regulation of factor VIIIa in the intrinsic factor Xase. Thromb Haemost 82:193–200

Folkman J (2007) Angiogenesis: an organizing principle for drug discovery? Nat Rev Drug Discov 6:273–286

Gandini R, Reichenbach T, Tan T, Divne C (2017) Structural basis for dolichylphosphate mannose biosynthesis. Nat Commun 8:1–12

Gasteiger E, Hoogland C, Gattiker A, Duvaud S, Wilkins MR, Appel RD, Bairoch A (2005) Protein identification and analysis tools on the ExPASy server. In: Walker JM (ed) The proteomics protocols handbook. Humana Press, Totowa, pp 571–607

Gastl G, Hermann T, Steurer M, Zmija J, Gunsilius E, Unger C, Kraft A (1997) Angiogenesis as a target for tumor treatment. Oncology 54:177–178

Hanahan D, Weinberg RA (2011) Hallmarks of cancer: the next generation. Cell 144:646–674

Helenius A, Aebi M (2004) Roles of N-linked glycans in the endoplasmic reticulum. Annu Rev Biochem 73:1019–1049

Kane WH, Davie EW (1988) Blood coagulation factors V and VIII: structural and functional similarities and their relationship to hemorrhagic and thrombotic disorders. Blood 71:539–555

Kean EL (1982) Activation by dolichol phosphate-mannose of the biosynthesis of N-acetylglucosaminylpyrophosphoryl polyprenols by the retina. J Biol Chem 257:7952–7954

Kornfeld R, Kornfeld S (1985) Assembly of asparagine-linked oligosaccharides. Annu Rev Biochem 54:6631–6636

Kyte J. Doolittle RF (1982) A simple method for displaying the hydropathic character of a protein. J Mol Biol 157:105–132

Lamani E, Mewbourne RB, Fletcher DS, Maltsev SD, Danilov LL, Veselovsky VV, Lozanova AV, Grigorieva NY, Pinsker OA, Xing J, Forsee WT, Cheung HC, Schutzbach JS, Shibaev VN, Jedrzejas MJ (2006) Structural studies and mechanism of Saccharomyces cerevisiae dolichyl-phosphatemannose synthase: insights into the initial step of synthesis of dolichyl-phosphate-linked oligosaccharide chains in membranes of endoplasmic reticulum. Glycobiology 16:666–678

Lenting PJ, van Mourik JA, Mertens K (1998) The life cycle of coagulation factor VIII in view of its structure and function. Blood 92:3983–3996

Martinez JA (2002) Angiogenesis and Glycosylation: interplay between dolichol cycle and cell cycle. In: Ph.D. thesis, University of Puerto Rico School of Medicine, San Juan, PR, pp 1–233

Nguyen M, Folkman J, Bischoff J (1992) 1-Deoxymannojirimycin inhibits capillary tube formation in vitro. Analysis of N-linked oligosaccharides in bovine capillary endothelial cells. J Biol Chem 267:26157–26165

Nguyen M, Strubel NA, Bischoff J (1993) A role for sialyl Lewis-X/A glycoconjugates in capillary morphogenesis. Nature 365:267–269

Pili R, Chang J, Partis RA, Mueller RA, Chrest FJ, Passaniti A (1995) The alpha-glucosidase I inhibitor castanospermine alters endothelial cell glycosylation, prevents angiogenesis, and inhibits tumor growth. Cancer Res 55:2920–2926

Schneider BP, Miller KD (2005) Angiogenesis of breast cancer. J Clin Oncol 23:1782–1790

Sinhoara H, Maruyama T (1973) Evolution of glycoproteins as judged by the frequency of occurrence of the tripeptides Asn-X-Ser and Asn-X-Thr in proteins. J Mol Evol 2:117–122

Szymanski CM, Wren BW (2005) Protein glycosylation in bacterial mucosal pathogens. Nat Rev Microbiol 3:225–237

Tarbouriech N, Charnock SJ, Davies GJ (2001) Three-dimensional structures of the Mn and Mg dTDP complexes of the family GT-2 glycosyltransferase SpsA: a comparison with related NDP-sugar glycosyltransferases. J Mol Biol 314:655–661

Uhr JW, Scheuermann RH, Street NE, Vitetta ES (1997) Cancer dormancy: opportunities for new therapeutic approaches. Nat Med 3:505–509

Zhang Z, Banerjee A, Baksi K, Banerjee DK (2010) Mannosylphosphodolichol synthase overexpression supports angiogenesis. Biocatal Biotransformation 28:90–98

Benzothiophenes as Potent Analgesics Against Neuropathic Pain

17

Saurabh Yadav, Vishnu Kumar Dwivedi, Sarika Gupta, and Avadhesha Surolia

Abstract

Neuropathic pain arises because of neuronal injury. Unlike inflammatory pain which can be managed by classical nonsteroid anti-inflammatory drugs (NSAIDs), neuropathic pain is difficult to treat. The classical NSAIDs work through inhibition of cyclooxygenase 2 (COX2) enzyme. However, COX2 inhibitors are insufficient to treat neuropathic pain. Hence, it becomes important to explore for novel molecules acting through cell surface molecules like ion channels, for the treatment of neuropathic pain. We investigated multiple bromobenzothiophene carboxamides for their efficacy against neuropathic pain. Interestingly, AS6 was found to be very effective in treating neuropathic pain through inhibition of Kv4.3 ion channel. AS6 also reduced the COX2 overexpression associated with neuropathic pain. These results as well as results from our previous study indicate that AS6 can be a potent antinociceptive agent against both inflammatory and neuropathic pain.

Keywords

Neuropathic pain · NSAIDs · COX2 · Bromobenzothiophenes · Inflammation

17.1 Introduction

Pain evolved as a warning toward potential hazardous stimuli and can be broadly classified into inflammatory and neuropathic pain. Inflammatory pain is associated with tissue injury due to physical injury or infection. It is accompanied with a heightened immune response. As a result, inflammatory pain is modulated by a plethora of inflammatory mediators majority of which are well characterized (Kidd and Urban 2001). Consequently, inflammatory pain is comparatively well managed. Nonsteroid anti-inflammatory drugs (NSAIDs), most of which are inhibitors of COX2, substantially reduce inflammatory pain (Bruno et al. 2014; Rainsford 2007). On the contrary, neuropathic pain originates as a result of nerve damage, and often not associated with heightened inflammation and immune response (Millan 1999). As a result, notion that immune modulators have little or no role to play in the pathophysiology of neuropathic pain was prevalent. However, subsequent to nerve damage, Wallerian degeneration leading to myelin breakdown, activation of Schwann cells, and influx of macrophages occurs, and pro-inflammatory mediators, predominantly from microglia and

S. Yadav · A. Surolia (✉)
Molecular Biophysics Unit, Indian Institute of Science, Bangalore, India
e-mail: surolia@iisc.ac.in; surolia@mbu.iisc.ernet.in

V. K. Dwivedi · S. Gupta
National Institute of Immunology, Aruna Asaf Ali Marg, Jawaharlal Nehru University, New Delhi, Delhi, India

K. Chattopadhyay, S. C. Basu (eds.), *Biochemical and Biophysical Roles of Cell Surface Molecules*, Advances in Experimental Medicine and Biology 1112, https://doi.org/10.1007/978-981-13-3065-0_17

infiltrating macrophage, play important role in the development of neuropathic pain (Ma and Quirion 2008; Durrenberger et al. 2004, 2006). The idea that inflammatory mediators play important role in neuropathic pain has gained interest.

Chronic neuropathic pain is a major debilitating condition and difficult to treat. About 7–10% human population suffers with neuropathic pain-like characteristics (Van Hecke et al. 2014). The mechanism of neuropathic pain is still not completely understood. Treatments against neuropathic pain are often inadequate as most of the patients either do not respond to classical NSAIDs or the response is very weak (Vo et al. 2009). Multiple drugs have been proposed based on various new pathways discovered; however so far none of the lead molecules have been found to be universally effective.

The cyclooxygenase 2 (COX2) enzyme plays an important role during inflammatory pain development. COX2-dependent prostaglandin E2 (PGE2) induces inflammatory and nociceptive effects through the EP1–4 receptors. These EP receptors are upregulated in the neurons during neuropathic pain (Ma and Eisenach 2003). Although no significant upregulation of COX2 is observed in the neurons during neuropathic pain, it is dramatically upregulated in Schwann cells and invading macrophages in injured nerves of both rats and humans (Durrenberger et al. 2006). Concurrently, PGE2 is also known to be upregulated in these tissues. Despite all these components being there, COX2 inhibitors are surprisingly ineffective in treating neuropathic pain (Vo et al. 2009). In our previous study, we showed that benzothiophene carboxamide compounds (AS compounds) were effective in relieving inflammatory pain (Pathak et al. 2014). In the present study, we show benzothiophene carboxamide AS6 as potent analgesic against neuropathic pain.

17.2 Results

17.2.1 Treatment of AS Compounds Alleviates Neuropathic Pain in Rats

Neuropathic pain models and control rats were randomly divided into groups of 8–12 animals each. To test the effect of AS compounds on neuropathic pain, we injected (IP intraperitoneal) three compounds AS4, AS6, and AS8 at the dosage of 15 mg/kg body weight of the animals (the dosage which was found to be effective in our previous study) (Pathak et al. 2014). Sham animals were injected with the vehicle solution. Injection of AS6 completely abrogated mechanical hyperalgesia and thermal hyperalgesia in the neuropathic pain rat models (Fig. 17.1a and b). Interestingly AS4 and AS8 were not effective in relieving neuropathic pain.

17.2.2 AS6 Effectively Reduces Pain through Multiple Routes of Application

Next, we tested the different routes of application of AS6 compound. AS6 was given at a dosage of 15 mg/kg body weight through oral, intraperitoneal (IP), and intravenous (IV) to the neuropathic rat models (8–12 animals per group). AS6 was effective in relieving neuropathic pain through all these routes; however, intravenous injections were found to be most effective (Fig. 17.2).

17.2.3 Systemic Injection of AS6 Is Better than NSAIDs in Alleviating Neuropathic Pain

We compared the efficacy of AS6 with multiple NSAIDs in managing neuropathic pain upon intraperitoneal injections to the rat models. Rats were randomly divided into multiple groups

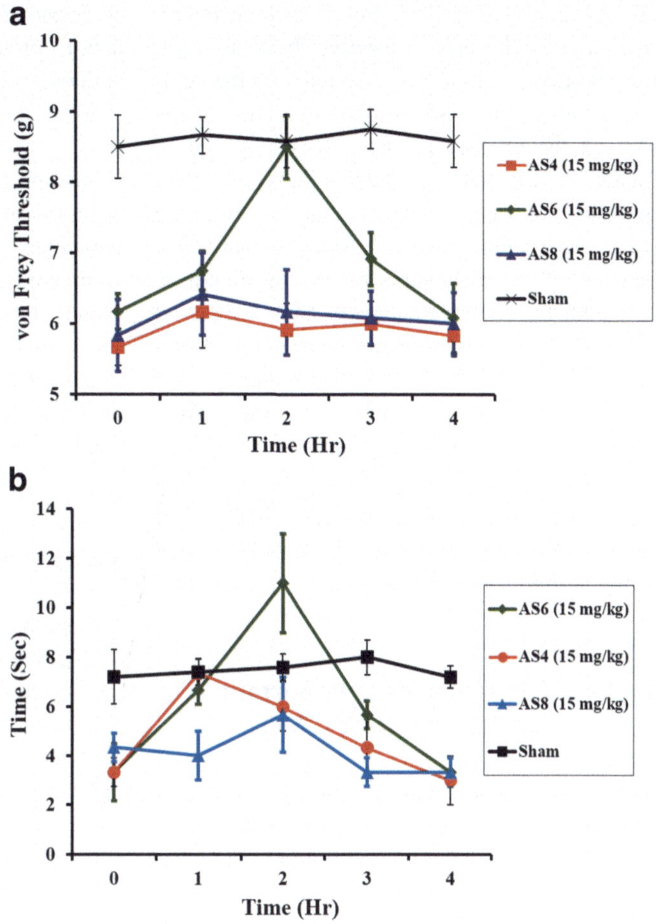

Fig. 17.1 AS6 injections relieved neuropathic pain in the rat models. Intraperitoneal (IP) injections of AS6 (15 mg/kg body weight) relieved both mechanical allodynia (**a**) and thermal hyperalgesia (**b**) in rat models. Injections of AS4 and AS8 (both at 15 mg/kg body weight) did not affect neuropathic pain in the rat models. Sham surgery rats were used as healthy (Sham) control (n = 8–12 animals per group)

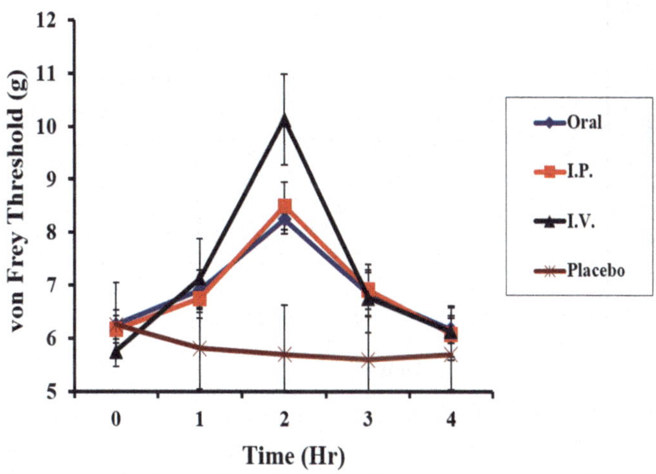

Fig. 17.2 Systemic administration of AS6 relieved neuropathic pain in the rat models. Oral administration and intraperitoneal (IP) and intravenous injections of AS6 (15 mg/kg body weight) relieved mechanical allodynia in rat models. Neuropathic pain models treated with vehicle solution (placebo) showed no pain relief (n = 8–12 animals per group)

(8–12 animals each). Each group was treated with one drug with their respective best dosages. Placebo-treated group was injected with the vehicle solution for AS6 compound. The results showed the efficacy of the AS6 compound in treating neuropathic pain. Interestingly, the effect of AS6 was comparable to that of morphine (Fig. 17.3a). Previously we showed that these benzothiophene carboxamides reverse inflammatory pain by downregulating COX2. We tested the COX2 expression in the spinal cord of rat models treated with AS6. Fascinatingly, we found that treatment of AS6 reduced the expression of COX2 in neuropathic rats as compared to placebo-treated rats (Fig. 17.3b). These results suggest that AS6 is a better analgesic than classical NSAIDs, and it exerts its effect through downregulation of COX2.

17.2.4 AS6 Molecule Is Important for the Analgesic Effect

AS6 being a carboxamide compound is prone to degradation upon injection. We tested the degradation products of AS6 compound for their analgesic effects. Both BTCA and 4FBA (degradation products of AS6) did not have any antinociceptive effect on neuropathic pain models (Fig. 17.4a). The carboxamide compounds are prone to degradation; we changed the carboxamide bonds to thioamide. The benzothiophene thioamide AST6 (AST6 thioamide AS6), along with other AST compounds, was injected intraperitoneally to a group of neuropathic pain models. The results show that like AS6, AST6 was effective in treating neuropathic pain. Other AST compounds were not found to be very effective for neuropathic pain treatment (Fig. 17.4b).

17.2.5 AS6 Inhibits $KV_{4.3}$ Ion Channel

To identify the mechanism behind regulation of COX2 by the AS6, we tested the effects of AS6 on ion channel inhibition. We tested the AS6 compound against $Kv_{4.3}$. AS6 significantly inhibited $Kv_{4.3}$ channel current at a concentration of 30 μM. This result indicates that AS6 may regulate the neuronal electrical properties by interfering with the ion channel conductance (Fig. 17.5).

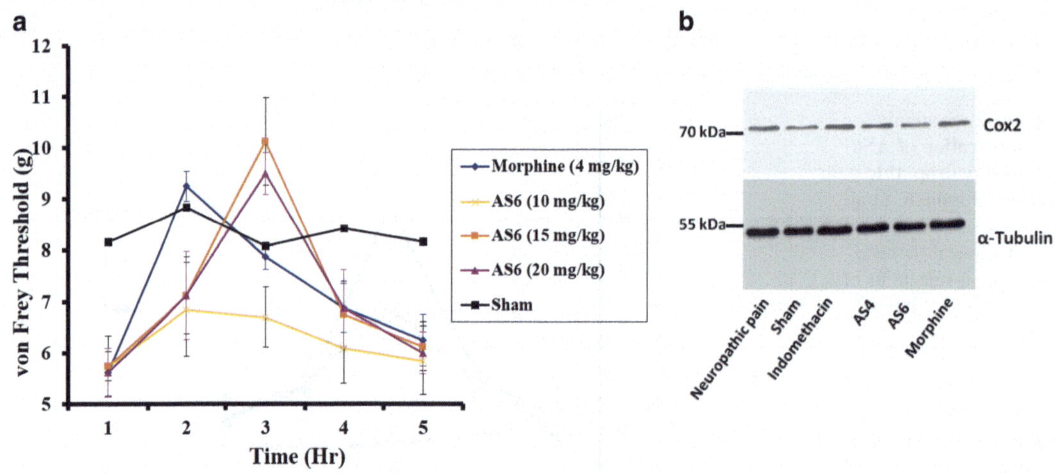

Fig. 17.3 (a) **Minimum effective dosage of AS6 for neuropathic pain relief in the rat models. Intravenous injections of AS6 (at 15 and 20 mg/kg body weight) relieved mechanical allodynia in rat models**. Neuropathic pain models treated with 10 mg/kg AS6 dose showed no significant pain relief. A group of neuropathic pain rats treated with morphine (4 mg/kg body weight) was used as positive control. Sham surgery rats were used as healthy (Sham) control. (n = 8–12 animals per group). (b) **Intravenous injections of AS6 (at 15 mg/kg body weight) COX2 expression in the spinal cord lumbar region L4-L6 of neuropathic pain rat models**. Lysates from the same region of neuropathic pain models treated with indomethacin, AS4, or morphine showed no significant difference in COX2 expression as compared to vehicle-treated neuropathic pain animals

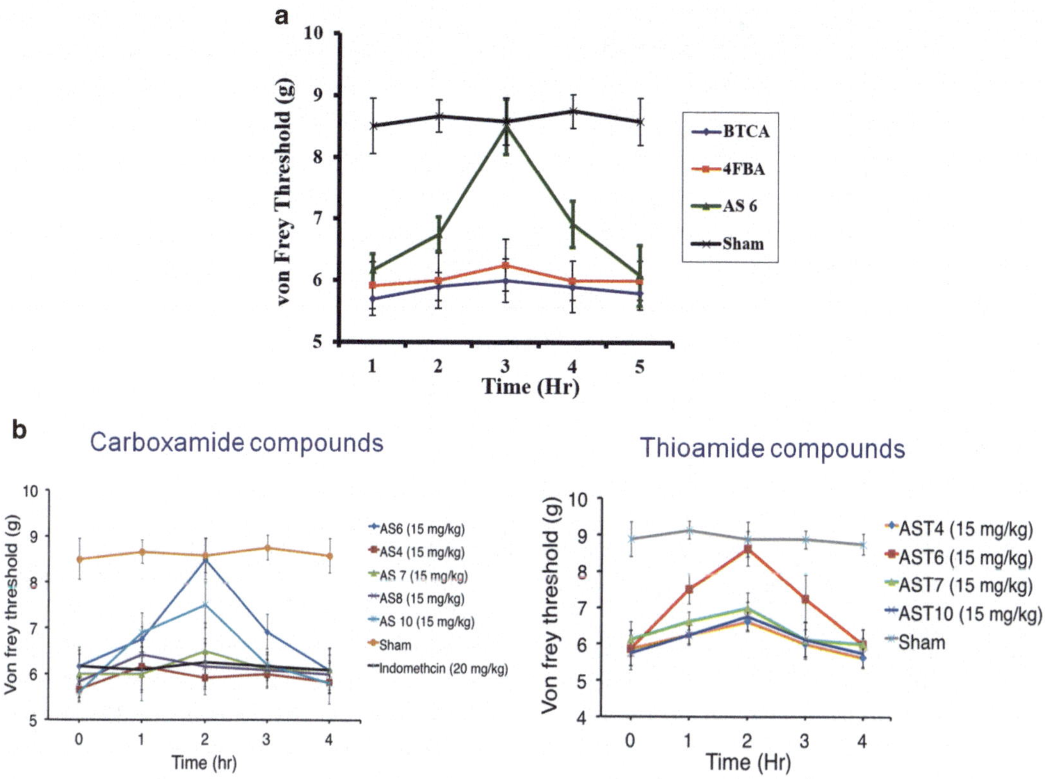

Fig. 17.4 (a) **AS6 as a molecule was effective for neuropathic pain treatment in the rat models.** Intravenous injections of AS6 (at 15 mg/kg body weight) relieved mechanical allodynia in rat models. Neuropathic pain models treated with the degradation product of AS6, BTCA, and 4FBA at the same concentration of 15 mg/kg showed no significant pain relief. Sham surgery rats were used as healthy (Sham) control (n = 8–12 animal per group). (**b**) **AS6 and AST6 molecules were most effective for neuropathic pain treatment in the rat models.** Intravenous injections of AS6 (at 15 mg/kg body weight) relieved mechanical allodynia in rat models. Neuropathic pain models treated with other AS compounds at the same concentration of 15 mg/kg showed no significant pain relief. Indomethacin treatment was ineffective in treating neuropathic pain in the rat models. Sham surgery rats were used as healthy (Sham) control. (Fig. 17.4b lower panel) Intravenous injections of AS6 thioamide analogue (AST6) at a concentration of 15 mg/kg body weight reduced neuropathic pain in the rat models. All other AS thioamide analogues (at 15 mg/kg) were ineffective in treatment of neuropathic pain. Sham surgery rats were used as healthy (Sham) control. (n = 8–12 animals per group)

17.3 Discussion

The mechanism behind initiation and sustenance of neuropathic pain were thought to be different than inflammatory pain. However, of late inflammatory mediators have been reported to play important role in neuropathic pain development. Multiple inflammatory mediators like IL-1β, TNFα, etc. have been shown to play important role in neuropathic pain development (Zelenka et al. 2005). Interestingly, the role of COX2, an important mediator of inflammatory pain, is not very clear in the pathophysiology of neuropathic pain. Upregulation of COX2 during neuropathic pain has been established in the infiltrating macrophages. Similarly, increased expression of both PGE2 and EP1–4 receptors has been shown (Ma et al. 2010; Durrenberger et al. 2006). However, the efficacy of COX2 inhibitors (most of which are classical NSAIDs) in neuropathic pain management remains debatable (Vo et al. 2009). Though enhanced expression of all the components of COX2 pathway has been shown in the spinal cord of the neuropathic pain models, why NSAIDs fail in providing pain relief during neuropathic pain is still an open question.

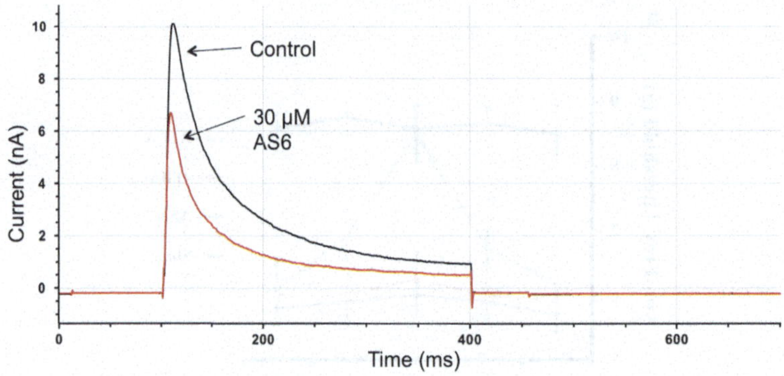

Fig. 17.5 **AS6 inhibited $Kv_{4.3}$ ion channel.** $Kv_{4.3}$ potassium channel (encoded by the human KCND3 gene) was overexpressed in HEK293 cells. The assay was performed in antagonist mode. AS6 was bath applied at a concentration of 30 μM for 3 min. AS6 bath application significantly reduced $Kv_{4.3}$ channel current indicating its inhibition by AS6 compound. (*nA* nano-Ampere, *ms* millisecond)

In our previous study, we have shown that the AS compounds act through downregulation of COX2 to relieve inflammatory pain (Pathak et al. 2014). To test the efficacy of AS compounds in neuropathic pain, we treated the neuropathic pain models with these compounds. The results showed that AS6 was exceptionally effective in neuropathic pain treatment at concentrations much less than that of classical NSAIDs. Its efficacy was equivalent to morphine. The systemic injection of AS6 reduced the pain sensitivity of the animal models, whereas the systemic injections of multiple NSAIDs even at their best reported dosages did not affect the pain threshold in the same models. AS6 was effective upon any type of systemic injection (viz., intravenous, oral, or intraperitoneal). This led us to ask the question if systemic injection of AS6 can downregulate COX2 in the spinal cord. We tested for the expression of COX2 in the spinal cord of the treated rats and compared it with the placebo-treated rats. Western blot show that AS6 treatments reduce the expression of COX2 in the spinal cord of the rats. This result confirms that systemic injections of AS6 can reduce COX2 expression in the spinal cord. The possible reasons behind inconsistent results of COX2 inhibitors could be multiple. Most of the NSAIDs are acidic moieties and hence may not cross the blood-brain barrier effectively. Their IC_{50} against COX2 varies. Although they inhibit the enzymatic activity, they have been found to upregulate the expression of COX2 upon application (Paik et al. 2000). All these variables may be responsible for the inconsistencies with the NSAID treatment. However, benzothiophene carboxamides being hydrophobic could cross the blood-brain barrier easily, and it works through downregulation of COX2 expression and hence is more effective. We tested the degradation product of AS6 to confirm if AS6 molecule or its degradation products are responsible for its analgesic effects. The data confirmed that AS6 as a molecule is responsible for the observed pain relief. We tested if changing the carboxamide bond to thioamide bond could increase the duration of pain relief by reducing the rate of degradation of carboxamide bond. However, changing carboxamide to thioamide did not have any effect on the efficacy of the compound. This led us to conclude that the time course of the effects of AS6 is because of its metabolic clearance. We also tested the AS6 for its inhibitory potential against ion channels. The results show its efficacy in inhibiting $Kv_{4.3}$ ion channel. This shows that the molecule can interfere with the ion channel conductance to inhibit pain perception.

The role of COX2 in pathophysiology of inflammatory pain is well established. During neuropathic pain, the upregulation of COX2, PGE2, and EP1–4 could induce neuronal hyperexcitability. However, the role of COX2 in neuropathic pain is still not very clear. This is mainly due to inconsistent results with the treatment with NSAIDs. There exists no clear evidence if downregulation of COX2 may abrogate neuropathic pain. In the present study, we show for the first time that downregulation of COX2 can reduce neuropathic pain. We also show here that AS6 could alleviate neuropathic pain, to an extent which rivals morphine, through downregulation of COX2 in the spinal cord of neuropathic pain rats. Our previous study confirmed that AS compounds are safer to use and get cleared easily through the body. Taken together we demonstrate AS6 as a safe and effective molecule to treat both inflammatory and neuropathic pain.

Table: List of AS compounds with code

Structure	Code
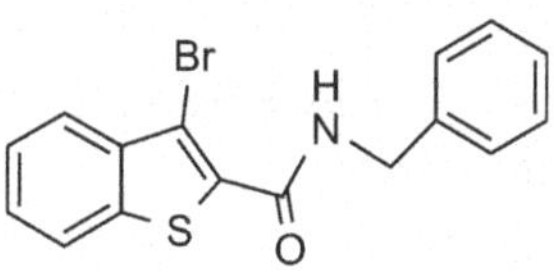	4
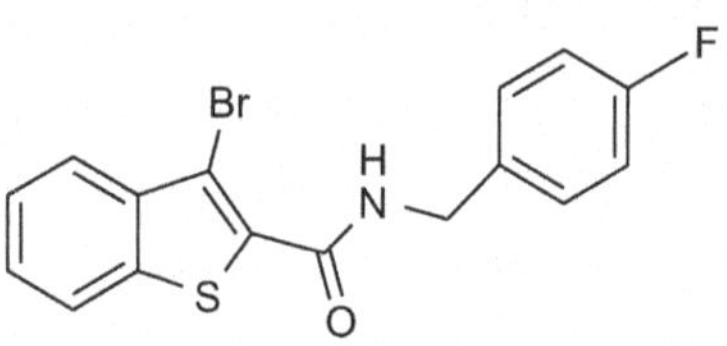	6
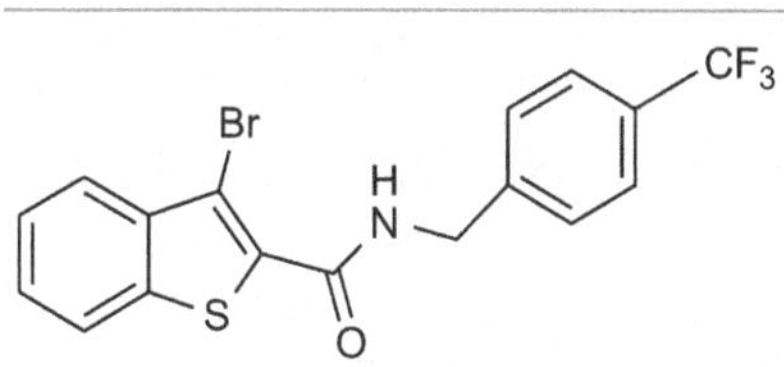	7
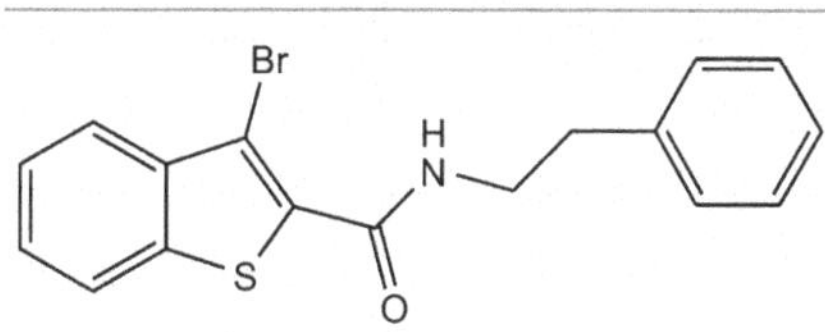	8
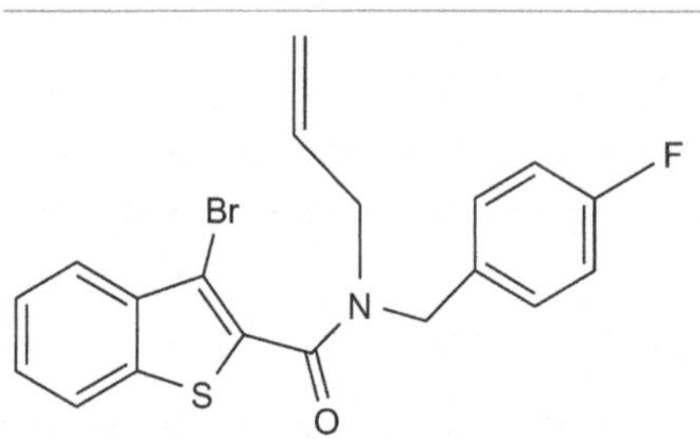	10
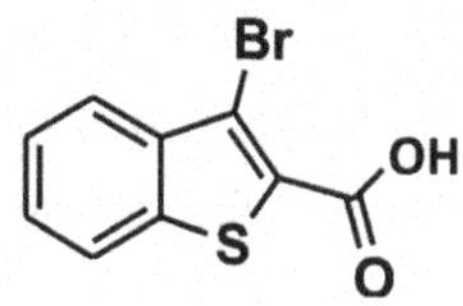	BTCA
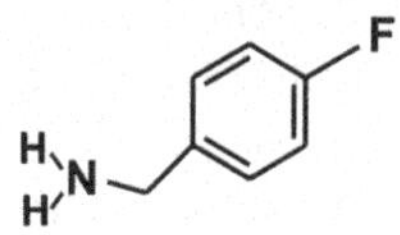	4FBA

17.4 Materials and Methods

17.4.1 General Procedure for Synthesis of Compounds 4–9

3-Bromobenzo[b]thiophene-2-carboxylic acid (02) (0.60 g, 2.33 mmol) and boric acid (0.02 mmol) were suspended in toluene (20 ml). To the stirred suspension, the appropriate arylalkylamine (2.45 mmol) was added in one lot. Dean-Stark trap mounted with a reflux condenser fitted with a nitrogen inlet was attached to the round-bottomed flask to remove water from the reaction mixture. The reaction mass was refluxed under nitrogen atmosphere. The reaction was monitored on TLC for completion of the reaction. The reaction was completed after 14 hrs (as indicated by TLC). The mixture was allowed to cool to ambient temperature and then was poured with stirring into 200 mL of hexane leading to the immediate precipitation of a white solid. Stirring was continued for an additional 30 min, and then the precipitate was filtered off with suction through a sintered glass filter funnel. The collected solid was successively washed with three 10-mL portions of hexanes and three 20-mL portions of distilled water and then was dried in vacuo at room temperature to afford the target compound as white solid. The product was recrystallized from acetone.

N-Benzyl-3-bromobenzo[b]thiophene-2-carboxamide (04)

Crystalline colorless solid (0.72 g, 90%). M.p. 115–117 °C. R_f: 0.40 Hexane/ethyl acetate (1:1)

ES-MS m/z: Cal for $C_{16}H_{12}BrNOS$: 346.24; Obs. 346.^{1}H-NMR ($CDCl_3$, 300 MHz): 4.72 (d, 2H, CH_2), 7.31–7.45 (m, 5H, ArH benzyl), 7.48 (m, 2H, ArH), 7.83 (dd, 2H, ArH).^{13}C-NMR: 44.5, 106.6, 122.9, 124.7, 125.8, 127.9, 128.0, 129.1, 137.7, 138.6, 138.7 and 161.1.

3-Bromo-N-(4-methoxybenzyl)-benzo[b]thiophene-2-carboxamide (05)

Crystalline colorless solid (0.80 g, 92%). Mp. 123–124 °C. R_f: 0.44 Hexane/ethyl acetate (1:1)

ES-MS m/z: Cal for $C_{17}H_{14}BrNO_2S$: 376.27; Obs. 376.26.^{1}H-NMR ($CDCl_3$, 300 MHz): 3.80 (s, 3H, OMe), 4.65 (d, 2H, CH_2), 6.91 (m, 2H), 7.34 (m, 2H), 7.46 (m, 2H, ArH), 7.83 (d, 2H, ArH).^{13}C-NMR: 44.0, 55.5, 106.6, 114.4, 122.9, 124.7, 125.7, 125.7, 129.4, 135.1, 138.7, 159.4 and 161.0.

3-Bromo-N-(4-fluorobenzyl)-benzo[b]thiophene-2-carboxamide (06)

Crystalline colorless solid (0.85 g, 93%) M.p. 120–125 °C. R_f: 0.41 Hexane/ethyl acetate (1:1)

ES-MS m/z: Cal for $C_{16}H_{11}BrNOSF$; 364.23; Obs 364.20.^{1}H-NMR ($CDCl_3$, 300 MHz) 4.70 (d, 2H), 7.45 (m, 2H, ArH) 7.83 (dd, 2H, ArH) 7 (m, 2H), 7.36 (m, 2H).^{13}C-NMR 44.0, 106.6, 125.7, 124.7, 123.9, 124.5, 130, 138.7, 163.5 and 161.0.

3-Bromo-N-[4-(trifluoromethyl)benzyl]-benzo[b]thiophene-2-carboxamide (07)

Crystalline colorless solid (0.85 g, 88%) Mp. 128–130 °C. R_f: 0.45 Hexane/ethyl acetate (1:1).

ES-MS m/z: Cal for $C_{17}H_{11}BrF_3NOS$ 414.24, Obs. 414.^{1}H-NMR ($CDCl_3$, 300 MHz): 4.70 (d, 2H), 7.47 (m, 2H, ArH) 7.12 (m, 2H), 7.38 (m, 2H).^{13}C-NMR 44.0, 106.6, 125.7, 124.7, 123.9, 124.5, 130, 138.7 and 161.0.

3-Bromo-N-(2-phenylethyl)-benzo[b]thiophene-2-carboxamide (08)

Crystalline colorless solid (0.76 g,91%) R_f: 0.4 Hexane/ethyl acetate (1:1)

ES-MS m/z: Cal for $C_{17}H_{14}BrNOS$ 360.27; Obs 360.

^{1}H-NMR ($CDCl_3$, 300 MHz): 3.5 (d, 2H), 3.2 (d, 2H), 7.47 (m, 2H, ArH) 7.12 (m, 2H), 7.38 (m, 2H).

^{13}C-NMR 42.9, 35.5, 106.6, 125.7, 124.7, 123.9, 124.5, 130, 138.7 and 161.0.

3-Bromo-N-(naphthalen-1-ylmethyl)-benzo[b]thiophene-2-carboxamide (09)

Crystalline colorless solid (0.85 g,93%) R_f: 0.38 Hexane/ethyl acetate (1:1)

ES-MS m/z: Cal for $C_{20}H_{14}BrNOS$ 396.30; Obs 396.

^{1}H-NMR ($CDCl_3$, 300 MHz): 4.9 (d, 2H), 7.47 (m, 2H, ArH), 7.12 (m, 2H), 7.38 (m, 2H), 7.1 (m, 2H), 7.3 (m, 2H), 7.6 (s, 1H).^{13}C-NMR (75 MHz): 44.0, 125.7, 124.7, 123.9, 124.5, 127.4, 127.5, 128.3, 128.2, 130, 138.7 and 161.0.

17.4.2 General Procedure for Synthesis of 10 and 11

Potassium carbonate (1.62 g, 12 mmol) and potassium hydroxide (0.67 g, 12 mmol) were ground in a mortar. To the resulting mixture, tetrabutylammonium bromide (0.1 g, 0.3 mmol) and the carboxamide 06 or 07 (3.0 mmol) were thoroughly mixed and placed in an open conical flask. Allyl bromide (4.5 mmol) was adsorbed dropwise to the above mixture with glass rod for a few seconds, placed in a domestic microwave oven, and irradiated for 2 min at 900 W. Upon completion of the reaction, the product was extracted into methylene chloride, solvent removed, and the residue recrystallized from ethanol to afford the product.

3-Bromo-N-(4-fluorobenzyl)-N-(prop-2-en-1-yl)-benzo[b]thiophene-2-carboxamide (10)

Colorless solid (1.04 g, 87%)

ES-MS m/z: Cal for $C_{19}H_{15}BrFNOS$ 404.30; Obs 404.0.

^{1}H-NMR ($CDCl_3$, 300 MHz): 4.70 (d, 2H), 5.18 (d, 2H), 7.45 (m, 2H, ArH), 7.83 (2H, ArH), 7 (m, 2H), 7.36 (m, 2H).^{13}C-NMR (75 MHz): 58.0, 106.6, 125.7, 124.7, 123.9, 124.5, 130, 138.7, 163.5, 136, 137.5 and 161.0.

3-bromo-N-(prop-2-en-1-yl)-N-[4-(trifluoromethyl)benzyl]-1-benzo[b]thiophene-2-carboxamide (11)

Colorless solid (1.2 g, 89%)

ES-MS m/z: Cal for $C_{20}H_{15}BrF_3NOS$ 454.30; Obs 454.4.

^{1}H-NMR ($CDCl_3$, 300 MHz): 4.70 (d, 2H), 5.18 (s, 2H), 7.47 (m, 2H, ArH) 7.12 (m, 2H), 7.38 (m, 2H).^{13}C-NMR 58.0, 106.6, 125.7, 124.7, 123.9, 124.5, 130, 136.4, 138.7, 142.2, and 161.0.

17.5 Animals

All experiments were conducted according to the guidelines of the International Association for the study of pain and approved by the "Institutional Animal Ethical Committee of the National Institute of Immunology" for animal research. The animals were housed in an animal house facility of National Institute of Immunology, India, with controlled temperature (22 ± 2 °C) under a 12/12 h light/dark cycle. They had free access to food and water ad libitum. All in vivo experiments were carried out on inbred male SD rats (6–8 weeks of age and ~ 200–300 g BW). Same age and equal body weight of animals were used according to experiments. No animals were used for more than one experiment.

17.6 Neuropathic Pain Models

Animals were anesthetized with sodium pentobarbital (50 mg/kg IP), and partial sciatic nerve ligation was performed. In brief, dorsal 1/3–1/2 of the right sciatic nerve was tightly ligated using an 8–0 nylon suture. The wounds were closed, and the animals were allowed to recover. Two animals per cage were housed postsurgical procedure. Sham surgery was done on a separate group of animals of the same age and body weight where no nerve ligation was done (Seltzer et al. 1990).

17.7 Mechanical and Thermal Hyperalgesia

Animals were placed on a wire mesh platform and were covered with transparent plastic cage. Von Frey filaments (IITC Inc., USA) were applied perpendicularly to the planter surface of hind paw (ipsilateral to the site of nerve injury). Sensitivity to the mechanical pressure was accessed by sequentially increasing and decreasing the stimulus (up-and-down method). A crisp withdrawal of hind paw was taken as positive response as described earlier (Seltzer et al. 1990). Thermal hyperalgesia was measured by hot plate (IITC Inc., USA) assay for nociception as described earlier (Seltzer et al. 1990) at 45 °C.

17.8 Western Blotting

Intact spinal cord regions were isolated immediately after decapitation, washed in PBS, and frozen. Ten percent (w/v) homogenates were prepared in Tissue Lytic M buffer (Sigma) containing protease and phosphatase inhibitor cocktail (Sigma). Protein samples were normalized for total protein content and run on SDS polyacrylamide gel (12–15%) followed by transferring to nitrocellulose membrane. The membrane was then blocked with Tris-buffered saline (TBS) containing 5% BSA and 0.1% Tween 20 for 2 h at room temperature, incubated with respective primary antibodies overnight at 4 °C. Membranes were then washed four times with TBS with 1% Tween 20 (TBST), probed with HRP-conjugated secondary antibodies (Santa Cruz, CA, USA) for 1 h at room temperature, washed again four times with TBST, and were developed using chemiluminescent substrate (Ez-ECL, Biological Industries) in gel documentation system (BioRad). Blots were stripped using stripping buffer (Sigma) and probed for α-tubulin which was used as internal loading control for the blots.

17.9 Fast Patch Assay

Kv4.3 potassium channel was expressed in HEK293 cells. The HEK293 cells were transfected with the plasmid containing the full-length cDNA (KCND3 gene) in 35-mm culture dishes. 5×10^5 cells were plated into 35-mm culture dishes with 2 ml of appropriate complete growth medium. Cells reaching approximately 50% confluence were transfected 24 h later using lipofection. The in vitro effect of AS6 was evaluated in QPatch (Sophion) and PatchXpress (Molecular Devices) automated patch clamp system, in the cells expressing $Kv_{4.3}$ channels. The assay was performed in the antagonist mode. The compound was evaluated at a concentration of 30 μM. The duration of exposure to the AS6 compound at this concentration was 3 min (Fig. 17.5).

References

Bruno A, Tacconelli S, Patrignani P (2014) Variability in the response to non-steroidal anti-inflammatory drugs: mechanisms and perspectives. Basic Clin Pharmacol Toxicol 114:56–63

Durrenberger PF, Facer P, Gray RA, Chessell IP, Naylor A, Bountra C, Banati RB, Birch R, Anand P (2004) Cyclooxygenase-2 (Cox-2) in injured human nerve and a rat model of nerve injury. J Peripher Nerv Syst 9:15–25

Durrenberger PF, Facer P, Casula MA, Yiangou Y, Gray RA, Chessell IP, Day NC, Collins SD, Bingham S, Wilson AW, Elliot D, Birch R, Anand P (2006) Prostanoid receptor EP1 and Cox-2 in injured human nerves and a rat model of nerve injury: a time-course study. BMC Neurol 6(1):1

Kidd BL, Urban LA (2001) Mechanisms of inflammatory pain. Br J Anaesth 87:3–11

Ma W, Eisenach JC (2003) Four PGE2 EP receptors are up-regulated in injured nerve following partial sciatic nerve ligation. Exp Neurol 183:581–592

Ma W, Quirion R (2008) Does COX2-dependent PGE2 play a role in neuropathic pain? Neurosci Lett 437:165–169

Ma W, Chabot JG, Vercauteren F, Quirion R (2010) Injured nerve-derived COX2/PGE2 contributes to the maintenance of neuropathic pain in aged rats. Neurobiol Aging 31:1227–1237

Millan MJ (1999) The induction of pain: an integrative review. Prog Neurobiol 57:1–164

Paik JH, Ju JH, Lee JY, Boudreau MD, Hwang DH (2000) Two opposing effects of non-steroidal anti-inflammatory drugs on the expression of the inducible cyclooxygenase. Mediation through different signaling pathways. J Biol Chem 275:28173–28179

Pathak C, Ranjan Singh R, Yadav S, Kapoor N, Raina V, Gupta S, Surolia A (2014) Evaluation of benzothiophene carboxamides as analgesics and anti-inflammatory agents. IUBMB Life 66:201

Rainsford KD (2007) Anti-inflammatory drugs in the 21st century. Subcell Biochem 42:3–27

Seltzer Z, Dubner R, Shir Y (1990) A novel behavioral model of neuropathic pain disorders produced in rats by partial sciatic nerve injury. Pain 43:205–218

Van Hecke O, Austin SK, Khan RA, Smith BH, Torrance N (2014) Neuropathic pain in the general population: a systematic review of epidemiological studies. Pain 155:654–662

Vo T, Rice AS, Dworkin RH (2009) Non-steroidal anti-inflammatory drugs for neuropathic pain: how do we explain continued widespread use? Pain 143:169–171

Zelenka M, Schafers M, Sommer C (2005) Intraneural injection of interleukin-1beta and tumor necrosis factor-alpha into rat sciatic nerve at physiological doses induces signs of neuropathic pain. Pain 116:257–263

PRR Function of Innate Immune Receptors in Recognition of Bacteria or Bacterial Ligands

18

Aakanksha Gulati, Deepinder Kaur, G. V. R. Krishna Prasad, and Arunika Mukhopadhaya

Abstract

Recognition of a bacterial attack is the first and the most important step in clearing the bacteria from the body of the host. Towards this, the host innate immune system employs pattern recognition receptors (PRRs) such as Toll-like receptors (TLRs), nucleotide-binding leucine-rich repeat-containing receptors (NLRs) and scavenger receptors (SRs) present mostly in innate immune cells. These receptors sense the presence of bacteria and help in spreading the signal to the host, which results in recruitment of other immune cells leading to the elimination of the bacteria from the system. Since their discovery, a lot has been established about these receptors. Their role has been elucidated not only in pathogen recognition but also in eradication of the dead cells from the system. This review is focussed mainly on their role in the bacterial recognition and how these receptors play a role in eliciting an immune response against bacteria in the host.

Keywords

Pattern recognition receptor · TLR · NLR · Scavenger receptor · Innate immunity

18.1 Introduction

The immune system of all the organisms is highly professional and tightly regulated. A pathogen, after entering the host, first encounters the innate immune system of the host. The innate immune system recognizes the pathogen and signals the presence of an intrusion to the whole body, thus further activating the adaptive immune system.

The cells of the innate system, such as macrophages, monocytes, dendritic cells and neutrophils, possess specialized receptors called pattern recognition receptors (PRRs) that recognize patterns unique to the pathogens. These patterns of the pathogen are called pathogen-associated molecular patterns (PAMPs), such as lipopolysaccharide in gram-negative bacteria. The idea of PRRs was proposed as the immune cell receptors which recognize microbial products and link innate and adaptive immune systems. Different kinds of PRRs are specialized to recognize different PAMPs. The PRR-PAMP interaction activates various signalling cascades which result in the production of cytokines, chemokines and interferons. These cytokines then signal other cells of the immune system about the presence of the pathogen and help in the elimination of the

A. Gulati · D. Kaur · G. V. R. Krishna Prasad · A. Mukhopadhaya (✉)
Department of Biological Sciences, Indian Institute of Science Education and Research (IISER) Mohali, Mohali, Punjab, India
e-mail: arunika@iisermohali.ac.in

K. Chattopadhyay, S. C. Basu (eds.), *Biochemical and Biophysical Roles of Cell Surface Molecules*, Advances in Experimental Medicine and Biology 1112, https://doi.org/10.1007/978-981-13-3065-0_18

infection. In addition to PAMPs, these PRRs can also recognize damage-associated molecular patterns (DAMPs) which are released by cells in response to various stimuli.

The PRRs are divided mainly into four classes based on their structure – the Toll-like receptors (TLRs), the nucleotide-binding leucine-rich repeat-containing receptors (NLRs), the RIG-1-like receptors (RLRs) and the C-type lectin receptors (CLRs). Of these, TLRs and CLRs are transmembrane receptors which are present either on the surface (both TLRs and CLRs) or in the endosomal membranes (TLRs). The NLRs and the RLRs are present in the cytoplasm of the cell. In addition to these PRRs, another family of receptors called scavenger receptors have recently been implicated in recognition of PAMPs.

In this review, we will discuss the PRRs which are involved in recognition of bacterial ligands with major emphasis on the PRR function of TLR, NLR and SR.

18.2 The Toll-like Receptors (TLRs)

Toll receptors were first discovered in *Drosophila*, where they were found to play a role in innate immunity against fungal infections (Lemaitre et al. 1996). Homologs of this receptor in vertebrates were also found to have a role in generating immune responses and named as Toll-like receptors (TLRs) (Armant and Fenton 2002; Medzhitov et al. 1997). In mice and humans, the genes encoding the TLRs are located on different chromosomes (Qureshi and Medzhitov 2003). Till date 10 functional TLRs have been found in humans and 12 in mice (Takeda and Akira 2005). TLR11, TLR12 and TLR13 are not present in humans (Baccala et al. 2007), whereas TLR10 is not functional in mice due to a retroviral insertion (Kawai and Akira 2010). Earlier, mouse TLR8 was also thought to be non-functional as it is not stimulated by its natural ligand-viral ssRNA. However, later it was observed that it is expressed during brain development in mice and can get activated by a combination of imidazoquinoline immune response modifiers and polyT oligodeoxynucleotides (Demaria et al. 2010). Apart from this, another protein RP105 is found in humans that is similar to TLR4 but lacks intracellular signalling domain (Zarember and Godowski 2002).

Signalling pathways initiated by TLRs lead to the upregulation of co-stimulatory molecules on dendritic cells which are essential for induction of adaptive immune responses (Iwasaki and Medzhitov 2004). So, it was concluded that TLRs are essential, responsible and required for an adaptive immune response (Beutler 2009). Some TLRs like TLR3, TLR4, TLR7, TLR8 and TLR9 can also induce antiviral responses by inducing type I IFN (Kawai and Akira 2006).

18.2.1 TLR Structure

Structurally, TLRs are type I integral membrane glycoproteins having cytoplasmic region similar to IL-1Rs (TIR domain) and an extracellular region having leucine-rich repeat (LRR) motif (Akira et al. 2001; Medzhitov 2001; Slack et al. 2000). The TIR domain has 20–30% conserved sequence containing five stranded parallel beta sheets surrounded by five alpha helices on each side. The LRR domain is comprised of a β-strand and an α-helix combined to form a horseshoe structure (Xu et al. 2000). The receptors present in TLR/IL-R family are one of the first signalling domains to be evolved (Kimbrell and Beutler 2001).

18.2.2 TLR Distribution

The expression of TLRs is different in different cell types. They are majorly present on cells involved in innate immunity including macrophages, neutrophils, mucosal epithelial cells, dendritic cells and dermal endothelial cells. Some TLRs like TLR2 and TLR4 are also present in B and T cells (Imler and Hoffmann 2001). TLR3 is expressed only in mature epithelial cells, whereas the expression of TLR2 and TLR4 is lost during maturation (Furrie et al. 2005). TLR3 is also present in higher levels in the colon where it acts as the first line of defence for viruses. Moreover, TLR3 and TLR5 are expressed at higher levels in spleen (Zarember and Godowski 2002). The TLRs are present both on the surface of the cells and intracellularly depending on their ligand specificities. Mostly TLR1, TLR2, TLR4, TLR5, TLR6 and TLR11 are present on the cell surface; however, TLR3, TLR7, TLR8 and TLR9 are present intracellularly (Cervantes et al. 2012). TLR9 is claimed to be expressed in the endoplasmic reticulum, which on activation is recruited to endosomal/lysosomal compartments with the help of UNC93B protein (Beutler 2009). However, in epithelial cells, TLR4 is present in Golgi apparatus due to which these cells are hypo-responsive to LPS (Hornef et al. 2002). Some reports also indicated the recruitment of the surface TLRs to the phagosomes after their stimulation (Akira and Takeda 2004).

Even microglia and astrocytes that are majorly involved in immune responses in the central nervous system have different expression of TLRs. In microglia, TLR3 is expressed intracellularly; however, astrocytes express both surface and intracellular TLR3 (Jack et al. 2005). Apart from this, TLR2 or TLR4 also help in activation of mast cells (Supajatura et al. 2002).

18.2.3 TLR Agonists

Of all the TLRs, TLR4 was the first TLR to be discovered. It was found to be a receptor for lipopolysaccharide (LPS) as C3H/HeJ mice having a mutation in *tlr4* gene failed to respond to LPS (Poltorak et al. 1998). TLR4 is capable of discriminating between LPS of different species because of a highly variable 82 amino acid region in its extracellular domain (Hajjar et al. 2002). After that, many ligands for TLR1, TLR2, TLR3, TLR5, TLR6, TLR7 and TLR9 were discovered during 1999–2003 (Akira and Takeda 2004). Some of the ligands recognized by different TLRs are shown in Table 18.1. TLR2 usually acts by forming M-shaped heterodimers with TLR1 or TLR6 to recognize different bacterial ligands. The crystal structures of these heterodimers were solved in 2007 (O'Neill et al. 2013). The TLR1/2 heterodimer recognizes the triacylated lipoproteins, and TLR2/6 heterodimers recognize diacylated lipoproteins. In TLR1/2 heterodimer, one of the three chains of triacylated lipopeptide is recognized by a hydrophobic channel of TLR1. Due to the absence of this hydrophobic channel in TLR6,

Table 18.1 Ligands for TLRs

TLR	Cell type	Adaptor proteins	Ligand (PAMPs)
Surface TLRs			
TLR1	Ubiquitous	MyD88/TIRAP	Triacyl lipopeptide, soluble factors
TLR2	DCs, PMLs and monocytes	MyD88/TIRAP	Peptidoglycan, heat shock protein 70 lipoprotein
TLR4	Macrophage, PMLs, DCs and ECs	MyD88/TIRAP, TRIF/TRAM	LPS, envelope protein, taxol
TLR5	Monocyte, immature DC, epithelial, NK and T cells	MyD88	Bacterial flagellin
TLR6	B cells, NK cells and monocytes	MyD88/TIRAP	Diacyl lipopeptide, zymosan, lipoteichoic acid
TLR10	B cells, pDCs	MyD88	Lipopeptides (probable)
TLR11	ND	MyD88	Flagellin
TLR12	ND	MyD88	Profilin-like molecule
TLR13	ND	MyD88	23S RNA
Endosomal TLRs			
TLR3	DC and NK cells, up-regulated on epithelial and endothelial cells	TRIF	dsDNA
TLR7	B cells and pDCs	MyD88	Imidazoquinoline, ssRNA
TLR8	Monocytes, low in NK and T cells	MyD88	Imidazoquinoline, ssRNA
TLR9	pDCs, B cells, macrophage, PMLs, NK cells and microglial cells	MyD88	CpG-containing DNA

TLR2/6 heterodimer cannot recognize triacylated lipopeptide. The intracellular TLRs, namely TLR7, TLR8 and TLR9, mainly recognize the nucleic acids, i.e. DNA/RNA. Despite structural similarity between human TLR7 and TLR8, they recognize different ligands. In addition to this, these ligands can target different cell types to show varied cytokine responses. TLR7 ligands are known to activate plasmacytoid dendritic cells (pDCs) and produce IFNα and IFN regulatory cytokines, whereas TLR8 ligands activate myeloid dendritic cells, monocytes and monocyte-derived dendritic cells to produce pro-inflammatory cytokines (Gorden et al. 2005). TLR9 can sometimes recognize host DNA resulting in auto-immune disorders. Also, TLR13, which is only present in mice, is shown to recognize bacterial ribosomal RNA (O'Neill et al. 2013).

There are some cases where the same ligand may be recognized by different TLRs in different cell types. For example, HMGB1, a late cytokine mediator of lethal endotoxemia and sepsis, signals through TLR4 in human whole blood and primary macrophages and through TLR2 in RAW 264.7 cells (murine macrophage cell line) and in CHO cells (Chinese hamster ovarian cell line) (Yu et al. 2006). Besides this, biglycan which is an ECM component acts as an agonist of both TLR2 and TLR4 in macrophages (Schaefer et al. 2005).

18.2.4 Signalling Pathways of TLRs

On recognition of a ligand by the LRR motif of the TLRs, they dimerize into either a homodimer or a heterodimer (Beutler 2009). Some TLRs require other proteins to form a complex, to pursue their signalling. For example, TLR4 requires MD-2, a small protein for recognition of LPS. CD14 is also involved in the signalling pathway of TLR4 and TLR2. CD36 acts as an accessory component for the TLR2/TLR6 heterodimer to sense diacylated lipoproteins (Triantafilou et al. 2006). TLR2 also utilizes Dectin-1 for recognition of certain glycans (Beutler 2009).

Recent studies have shown that the heterodimers of TLR2 are formed prior to ligand interaction. These dimers are recruited to lipid rafts on exposure to ligands and interact with CD36 and CD14 present in the raft. This complex is then internalized and moves to the Golgi apparatus. It is suggested that lipid raft formation is crucial for internalization of the TLR2 receptor, but the role of rafts in TLR2 internalization and signalling is not clear. In most of the reports, the signalling pathway is shown to be independent of internalization, but in case of *Staphylococcus aureus*, internalization is crucial for signal transduction (Triantafilou et al. 2006).

Signalling pathway by surface TLRs and not endosomal TLRs has also been shown to increase mitochondrial ROS (mROS) production by recruiting mitochondria to the macrophage phagosomes. TRAF6 translocates to the mitochondria and interacts with a protein ECSIT, leading to its ubiquitinylation. This increases the mROS production leading to bactericidal activity. Thus, mitochondria have been suggested as a hub for innate signalling that generates effector responses (West et al. 2011).

The downstream signalling of TLRs depends on the adaptor molecule binding to the cytoplasmic TIR domains of the TLR dimer upon ligand recognition. Various adaptor proteins have been implicated in TLR-mediated signalling like myeloid differentiation primary response protein 88 (MyD88), TIR-domain-containing adaptor protein (TIRAP, also known as MyD88-adaptor-like protein, MAL) and TIR-domain-containing adaptor protein inducing IFN-β (TRIF, also known as TIR-domain-containing molecule 1, TICAM1) and TRIF-related adaptor molecule (TRAM, also known as TIR-domain-containing molecule 2, TICAM2). Most of the TLRs recruit MyD88 and/or MAL; however, some also recruit TRIF and/or TRAM to elicit the response. So, the TLR signalling has been classified into either MyD88-dependent or -independent based on the involvement of the adaptor protein MyD88.

18.2.5 MyD88-Dependent Signalling

TLR mediates its signalling via different adaptor molecules, of which the canonical pathway involves the adaptor molecule MyD88. MyD88 has N-terminal death domain (DD) and C- terminal TIR domain joined by a linker sequence. It exists as a homodimer when recruited to the TLR complex. It helps in recruitment of IL-1R-associated kinases (IRAKs) by interacting with them through its death domain. Out of four family members of IRAK family, only IRAK1 and IRAK4 have intrinsic kinase activity. IRAK2 and IRAK-M lack the kinase activity, thus negatively regulate TLR signalling. IRAK-M-deficient cells show increased inflammatory cytokine production as it prevents the dissociation of IRAK4 and IRAK1 from MyD88 (Kobayashi et al. 2002a). From studies, IRAK4 having an important role in TLR signalling is shown to work upstream of IRAK1. Patients with IRAK4 deficiency failed to respond to IL-1, IL-18 or to stimulation of TLR2, TLR3, TLR4, TLR5 and TLR9 (Akira and Takeda 2004). TRAF6 acts as a signalling mediator by binding indirectly to TLR receptor complex via IRAK1/4. The TRAF6-binding domain is present in IRAK1, IRAK2 and IRAK-M but not in IRAK4. Through this domain, IRAK interacts with the conserved C-terminal region of TRAF6 (TRAF-C). TRAF6 then activates TGFβ-activated kinase (TAK1). Two adaptor proteins TAB1 and TAB2/TAB3 are involved in this process. TAB1 is known to increase kinase activity of TAK1, whereas TAB2/TAB3 help in the binding of TAK1 to TRAF6. TRAF6 functions as a ubiquitin ligase and forms a complex with ubiquitin-conjugating enzyme 13 (UBC13) and Uev1A for the synthesis of lysine 63-linked polyubiquitin chains which help in the signal transduction and hence the cellular responses (Chen 2005). Activation of TAK1 further triggers MAPK [c-Jun N-terminal kinase (JNK), p38 and extracellular signal-regulated kinase (ERK)] activation which in turn activates AP-1 transcription factor. AP-1 forms dimers that are composed of Jun, Fos and ATF (activating transcription factor) subunits. Jun family members can form homodimers, whereas Fos family cannot homodimerize. The heterodimers of Jun and Fos are more stable. These subunits bind to DNA and affect the transcription of pro-inflammatory genes.

TAK1 activation also activates IκB kinases including IKKα, IKKβ and regulatory molecule IKKγ/Nemo. These kinases then phosphorylate IκB leading to its poly-ubiquitination and thus its 26S proteasomal degradation. The degradation of IκB renders cytoplasmic subunits of NF-κB free to move to the nucleus. The NF-κB family is composed of five members – p65 (REL-A), REL-B, c-REL, p50 and p52 which can either form homodimers or heterodimers. NF-κB1 and NF-κB2 are expressed as p105 and p100, which undergo proteolytic cleavage to become p50 and p52. Although p65-p50 dimer is considered as NF-κB in most cases, many other dimers they form can act as transcription factors, but some dimers do not have DNA-binding domain due to which they can also act as repressors (Hoffmann and Baltimore 2006) like p50-p50, p52-p52 homodimers and p52-p50 heterodimers due to the absence of transcriptional activation domain (TAD) in them. The NF-κB after translocating to nucleus activates transcription of pro-inflammatory genes. Other than this, an additional adaptor TIRAP that works upstream to MyD88 is required for TLR2 and TLR4 signalling (Kawai and Akira 2006).

18.2.6 MyD88-Independent Signalling

Most TLRs mediate their signalling via the adaptor molecule MyD88; however, TLR3 doesn't require MyD88 for its signalling. Instead, TLR3 recruits TRIF upon ligand recognition to elicit the downstream signalling. TRIF is known to be involved in MyD88-independent pathway leading to the activation of IFN-regulatory factor 3 (IRF3) and thus inducing IFNβ production. This IFNβ then acts on neighbouring cells and activates the JAK-STAT pathway leading to the induction of IRF7 expression due to which there is an increase in induction of type I IFN through a positive feedback mechanism.

Also, in MyD88-deficient mice, TLRs like TLR2, TLR5, TLR7 and TLR9 on macrophages and DCs are unable to produce any pro-inflammatory cytokines when exposed to ligands, and hence they cannot work in MyD88-independent manner. However, it was observed that when these cells were exposed to LPS, NF-κB and MAPKs were activated in a delayed manner. LPS lead to the MyD88-independent induction of IFNβ in such cells (Kawai et al. 2001). Not just that, TRIF-deficient mice also show reduced inflammatory cytokine production in response to LPS when compared to wild-type mice leading to the conclusion that both MyD88-dependent and -independent pathways are required for maximal cytokine production in response to LPS via TLR4. Also, TRIF was found to be linked to the cytoplasmic portion of TLR4 through another adaptor called TRAM. It is very interesting to note that all the adaptor proteins having TIR domains are involved in TLR4 signalling. Involvement of both MyD88-dependent and -independent pathways can explain the strong immunogenic nature of LPS which leads to endotoxic shock. However, in absence of CD14, MyD88-independent pathway gets inactivated and TLR4 signals through MyD88/TIRAP pathway only (Beutler 2009).

TRIF has receptor-interacting protein (RIP)-binding motif at its C-terminal region. The interactions of C-terminal of TRIF and RIP1 are responsible for NF-κB and AP-1 activation through TAK-1, whereas RIP3 hinders these interactions. Moreover, TRIF has three TRAF6-binding domains which can recruit TRAF6 resulting in activation of NF-κB and AP-1 transcription factors leading to a pro-inflammatory response in addition to the production of IFNβ.

18.2.7 TLR-Mediated IRF Activation

IRF7 is an essential transcription factor for regulation of type I IFN induction. IRF7 acts in MyD88-dependent manner and gets phosphorylated and thus activated by IRAK1 present downstream of MyD88 and IRAK4. TRAF6 also helps in this activation. IRF5 interacts with MyD88 for the production of inflammatory cytokine rather than IFN production. TLR7, TLR8 and TLR9 may trigger IRF7-mediated type I IFN induction majorly in plasmacytoid DCs (pDC) in response to viral infections. In pDCs ligand recognition by TLR9 also involves IRF8 which results in the production of type I IFN and inflammatory cytokines.

Apart from pDC, IFN-α/β production also occurs in conventional DCs and macrophages via TLR3 and TLR4. Recognition of dsRNA by TLR3 in endosomes mediates IFN production by TRIF-dependent pathway, whereas LPS recognition by TLR4 at the cell surface induces both MyD88-dependent and TRIF-dependent pathways. The TRIF-dependent pathway leads to activation of the IRF3, AP-1 and NF-κB transcription factors that regulate IFN production. In case of the MyD88-dependent pathway, there is no IRF3 activation, but AP-1 and NF-κB get activated (Baccala et al. 2007).

18.2.8 Negative Regulators of TLR Signalling

TLR signalling is crucial for both innate and adaptive immune responses, but dis-regulations in TLR pathways are known to induce many auto-immune disorders and chronic inflammation. One such example is TLR4, which upon uncontrolled activation with LPS can cause sepsis. So, TLR signalling needs to be tightly regulated. Some of the negative regulators involved in this process include soluble TLRs that prevent a direct communication of ligand and receptor that can either inhibit signalling pathways or decrease *tlr* gene expression (Gomariz et al. 2010). Some examples of negative regulators and their target cells are given in Table 18.2.

Table 18.2 Negative regulators of TLRs

Negative regulator	Affected TLR	Mode of action
Regulators on surface		
sTLR2	TLR2	TLR2 antagonist
sTLR4	TLR4	Blocks interaction of MD2 and TLR4
Regulators affecting transmembrane proteins and adaptor molecules		
ST2L	TLR2, TLR4, TLR9	Sequesters MyD88 and MAL
RP105	TLR4	Homolog of TLR4
MyD88s	TLR4	MyD88 antagonist
SOCS1	TLR4, TLR9	Suppresses IRAK
SARM	TLR3, TLR4	Inhibits MyD88- and TRIF-dependent pathway
Fliih	TLR4	Blocks interaction of TLR4 and MyD88
Regulators affecting effector proteins		
IRAK-M	TLR4, TLR9	Inhibits IRAK1 phosphorylation
IRAK2	TLR4	Overexpression inhibits NF-κB activation
TRAF4	TLR2, TLR3, TLR4, TLR9	Interacts with TRAF6, TRIF and IRAK1
TOLLIP	TLR2, TLR4	Autophosphorylates IRAK1
A20	TLR2, TLR3, TLR4, TLR9	Deubiquitylates TRAF6
β-arrestins 1,2	TLR4	Stabilize IKBα
SHP-2	TLR3, TLR4	Suppresses TBK1 phosphorylation
PTP1B	TLR3, TLR4, TLR9	Supresses MyD88- and TRIF-dependent pathway
Regulators affecting transcription factors		
PI3K	TLR2, TLR4, TLR9	Inhibits p38, JNK and NF-κB
ATF3	TLR4	Inhibits binding of NF-κB to DNA
NOD2	TLR2	Suppresses NF-κB
Zc3h12a	TLR4	Affects mRNA stability

18.2.9 TLR and Therapeutics

The expression of TLRs is known to increase in many diseases. Atherosclerotic heart disease is due to chronic infection or high serum LPS levels. TLR4 plays an important role in causing inflammation during this disease. Apart from this, TLR knockout mice have been found to be resistant to some inflammatory diseases. But these TLRs also help in evoking a protective immune response, and thus TLR4-mutant mice are more susceptible to systemic infections by gram-negative bacteria like *Salmonella typhimurium* (O'Neill et al. 2009). These studies suggest the role of TLRs in causing disease, and thus TLRs can be used as a potential therapeutic target (Hennessy et al. 2010). To limit the inflammation mediated by TLRs, the positive feedback loop can be inhibited by preventing the reactivation of the TLRs, and this will inhibit the amplification of the response.

Another strategy can be the use of TLR agonist in the form of drugs which activate innate and consecutively the adaptive immune system, thus activating natural killer cells and cytotoxic T cells. Example of one such successful drug is imiquimod which acts through TLR7-MyD88-dependent pathway. TLR7 and TLR8 agonists induce Th1 responses to target tumours in mice, non-human primates and human cells. Despite this, an agonist of TLRs could serve as a good therapeutic target; although the problem of the weakened immune system in cancer patients can be a halt in the way, it may be resolved by combining them with anticancer agents.

Some viral proteins are also known to manipulate TLR signalling. A protein of vaccinia virus, A52R, decreases NF-κB activation by targeting IRAK2 and TRAF6, and thus peptide synthesized from it may also inhibit inflammatory responses. Clinical trials have also proved that targeting key molecules of the TLR pathway has restricted the inflammation in various diseases. So, these therapies seem to hold a great potential in the future.

18.3 The Nucleotide-Binding Leucine-Rich Repeat-Containing Receptors (NLRs)

In addition to TLRs, another class of pattern recognition receptors called NLRs or nucleotide-binding leucine-rich repeat-containing receptors aid in the recognition of bacterial ligands. NLRs are soluble, intracellular pattern recognition receptors (PRR), which recognize mostly the ligands of intracellular pathogens. These are usually present in the cytosol of immune cells, namely, lymphocytes, macrophages and dendritic cells, but can also be present in non-immune cells like epithelial cells. The earliest known NLRs, NOD-1 and NOD-2, recognize components of the peptidoglycan assembly of the bacterial outer membrane. The presence of NLR homologs in a vast variety of organisms ranging from sea urchin, zebrafish, plants and various animals attest the importance of NLRs in host defence against pathogens.

18.3.1 Structure of NLRs

NLRs consist of three domains (1) the N-terminal domain, which comprises of one of the following, a CARD (caspase recruitment) domain, a PYD (pyrin) domain, an acidic transactivation domain or a BIR (Baculovirus inhibitor repeat) domain; (2) the central domain known as the NOD domain or the NACHT domain; and the (3) the C-terminal domain – which contains the leucine-rich repeats (LRRs).

The NLRs recognize the agonist by the C-terminal LRR domain, which triggers the oligomerization of the NLRs by the central domain. NLRs further bind to the downstream molecules by their N-terminal domain leading to various signalling pathways.

NLRs are divided into four classes based on their N-terminal domains. The NLRs containing acidic transactivation domain are classified as NLRA; similarly, those with a CARD or the PYD or the BIR domain are classified as NLRC, NLRP and NLRB, respectively. In addition to these, another class of NLRs is known as NLRX whose N-terminal domain remains unclear. The following table depicts the various NLRs and their structures discovered till date.

18.3.2 NLR Agonists

A wide variety of ligands ranging from PAMPs such as bacterial membrane components like peptidoglycan and diaminopimelic acid and DAMPs (danger-associated molecular patterns) and change in the cellular ion imbalance have been known to be recognized by various NLRs. A detailed list of the ligands of various NLRs and their structure has been described in Table 18.3.

18.3.3 Signalling Pathways of NLRs

Upon activation, NLRs oligomerize and get activated. Different NLRs elicit different signalling pathways upon ligand recognition. NLR activation can lead to two kinds of signalling – they can either elicit a signalling cascade leading to the activation of transcription factors NF-κB and AP-1 similar to TLRs or can lead to the formation of a multi-protein assembly called inflammasomes (Fig. 18.1a). Mainly the NLRs – NOD-1 and NOD-2 – elicit signalling leading to NF-κB and AP-1 activation, and NLRs such as NLRP1, NLRP3 and NLRC4 have been well characterized to form inflammasomes upon their activation.

18.3.4 NOD-1- and NOD-2-Mediated Signalling

NOD-1 and NOD-2 are the most studied NLRs. They recognize the peptidoglycan (PGN) moieties or its breakdown products from both the gram-positive and gram-negative bacteria. NOD-1 mainly recognizes PGN moieties

Table 18.3 Structure of NLRs and their ligands

Class of NLR	Structure	NLRs known	Ligands
NLRA	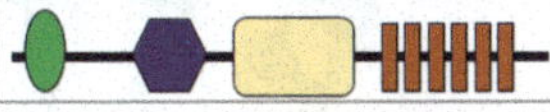	CIITA	IFN-γ
NLRB		NAIP1	T3SS rod proteins
		NAIP2	T3SS inner rod proteins
		NAIP3	Unknown
		NAIP4	Unknown
		NAIP5	Flagellin
		NAIP6	Flagellin
		NAIP7	Unknown
NLRC		NOD-1	Meso-diaminopimelic acid (mesoDAP)
		NOD-2	Muramyl dipeptide (MDP)
		NLRC3	Unknown
		NLRC4	Flagellin
		NLRC5	IFN-γ, IFN-β
NLRP	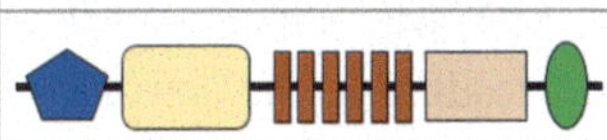	NLRP1	Muramyl dipeptide (MDP), lethal toxin
	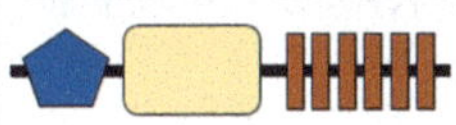	NLRP2	Unknown
		NLRP3	ATP, alum, silica, asbestos, ROS
		NLRP4	Unknown
		NLRP5	Unknown
		NLRP6	Unknown
		NLRP7	Lipopeptide
		NLRP8	Unknown
		NLRP9	Unknown
		NLRP10	Unknown
		NLRP11	Unknown
		NLRP12	*Yersinia*
		NLRP13	Unknown
		NLRP14	Unknown
NLRX	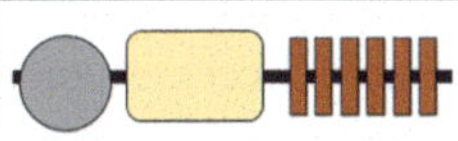	NLRX1	Unknown

containing meso-diaminopimelic acid (meso-DAP) which is mainly present in gram-negative and some gram-positive bacteria like *Bacillus* spp. NOD-2 recognizes the muramyl dipeptide (MDP) present across gram-negative and gram-positive bacteria. NOD-1 has been found to be a receptor for detecting various invasive bacteria in vitro, namely, *Listeria monocytogenes* (Hasegawa et al. 2006; Park et al. 2007), *Bacillus* spp. (Hasegawa et al. 2006), *Shigella flexneri*

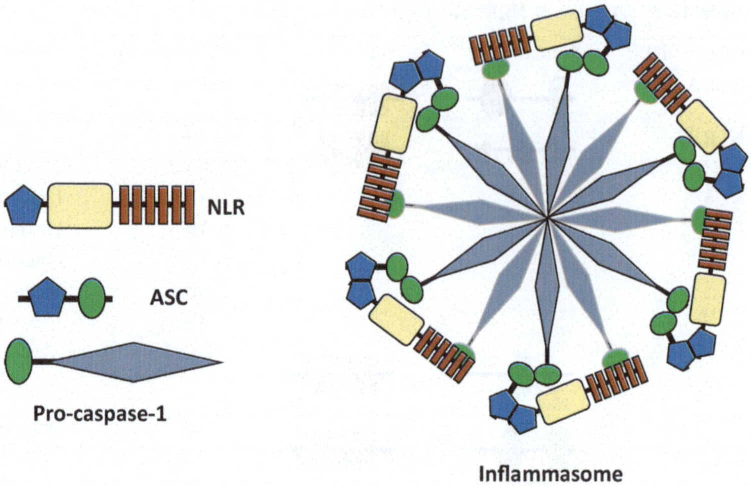

Fig. 18.1a The inflammasome assembly

(Girardin et al. 2003), *Campylobacter jejuni* (Zilbauer et al. 2007) and *Chlamydophila pneumonia* (Opitz et al. 2005). However, NOD-2 has been implicated in detecting bacteria in vitro such as *Mycobacterium tuberculosis* (Ferwerda et al. 2005), *Listeria monocytogenes* (Kobayashi et al. 2005; Opitz et al. 2004) and *Streptococcus pneumonia* (Opitz et al. 2004). These in vitro studies, however, do not directly imply a role for NOD-1 and NOD-2 alone as there is always a redundancy in the innate immune receptors. Certain in vivo studies have however established the role of NOD-1 and NOD-2 in sensing the bacteria, namely, *Listeria monocytogenes* and *Helicobacter pylori* as a PAMP. For example, when NOD-2 knockout mice (*nod2*$^{-/-}$ mice) were infected orally with *Listeria monocytogenes*, they showed lower bacterial load compared to the wild-type mice (Kobayashi et al. 2005). Also, NOD-1 knockout mice (*nod1*$^{-/-}$ mice) were more susceptible to infection with *Helicobacter pylori* than the wild-type mice. Interestingly, *Helicobacter pylori* being an extracellular pathogen secretes PGN moieties directly into the host cell by a type IV secretion system, where PGN moieties are detected by NOD-1 (Viala et al. 2004).

NOD-1 or NOD-2 upon activation by the PAMPs oligomerize and, thus, bind to a CARD-containing protein RIP2 (receptor-interacting protein 2) via a CARD-CARD homodimeric interaction (Kobayashi et al. 2002b). RIP2 is a serine/threonine kinase which further gets K63-ubiquitinylated by members of the IAPs (inhibitors of apoptosis proteins) such as cIAP and XIAP (Bertrand et al. 2009; Krieg et al. 2009). K63-ubiquitinylated RIP2 further binds to NEMO, which then phosphorylates IKKα and IKKβ, leading to the activation of the transcription factor NF-κB (similar to TLR signalling) (Fig. 18.1b).

Additionally, RIP2 can initiate a pro-inflammatory response by phosphorylating TAK-1 leading to activation of the IKK and, ultimately, NF-κB activation. However, TAK-1 can also activate MAP-kinases, namely, p38, JNK and ERK, which also lead to a pro-inflammatory response. Another protein called CARD9 has been found to be involved in binding to RIP2 and activating p38 and JNK MAP kinases (Hsu et al. 2007). However, CARD9 has not been found to have a role in NF-κB activation. Recent studies have implicated increased affinity of activated RIP2 to NOD-1 when meso-DAP is bound to the LRR domain of NOD-1 (Laroui et al. 2011). Also, NOD-2 has been shown to bind to its ligand MDP in acidic environments (Grimes et al. 2012).

Some studies have shown that NOD-1 can also sense the activation states of many Rho GTPases. Various bacterial infections like those of *Shigella* and *Salmonella* have been shown to secrete certain effector proteins, namely, GEFH1 (Fukazawa et al. 2008) and SopE (Keestra et al. 2013),

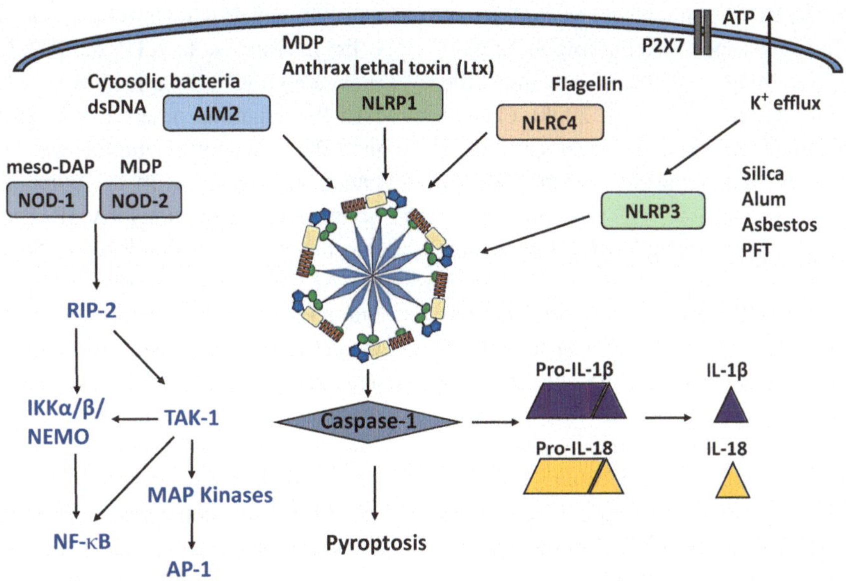

Fig. 18.1b The NLR signalling

respectively, via their type III secretion systems into the cytoplasm of the host which activate Rho GTPases. These activated Rho GTPases might be detected by NOD-1 to activate the NF-κB signalling.

18.3.5 Inflammasome Formation and Signalling

Inflammasomes are a multi-protein assembly comprising mainly of NLRs, an adaptor molecule ASC and the inactivated form of caspase-1 – pro-caspase-1. Most of the inflammasomes known till date comprise members of the NLRP or NLRC subfamilies of the NLRs. ASC, the adaptor molecule has both a pyrin domain (PYD) and a CARD domain which allows its binding to both the NLR and the pro-caspase-1 (which contains the CARD domain). This assembly is formed when NLRs recognize a ligand and oligomerize, to which the ASC adaptor protein binds via PYD-PYD interaction and recruits the pro-caspase-1 to this assembly via CARD-CARD domain interactions. After this assembly is formed, the pro-caspase-1 is cleaved, rendering it active. Active caspase-1 further activates various pro-inflammatory cytokines – mainly the activation of the pro-IL-18 and pro-IL1β to their mature forms, which are then secreted out of the cells (Fig. 18.1). In addition to this, active caspase-1 may also lead to pyroptosis ('pyro' – heat or fever and 'ptosis' – falling or death) which is a rapid, inflammatory programmed cell death and shares the features of both apoptosis and necrosis. Each inflammasome is unique in its structure, ligand recognition and effect, which has been described in the following section.

18.3.6 NLRP1

NLRP1 is the first identified NLR which forms an inflammasome in vitro (Martinon et al. 2002). NLRP1 has a property distinct from other identified NLRs – it has both the N-terminal PYD domain and the C-terminal CARD domain. In addition to these, it also has a FIIND domain (Table 18.3), whose cleavage is imperative for complete activation of NLRP1 (Finger et al. 2012; Frew et al. 2012).

Not many agonists of NLRP1 have been identified; however, MDP and anthrax lethal toxin, which are secreted by *Bacillus anthracis*, has been found to act as agonists for the activation of the NLRP1 inflammasome. A direct interaction of MDP and NLRP1 has not been reported. However, certain studies speculate the role of NOD-2 in binding to MDP and then activating the NLRP1 inflammasome (Hsu et al. 2008). Certain unknown factors seem to govern whether NOD-2 will activate the inflammasome or the NF-κB pathway.

Since NLRP1 has both the CARD domain and the PYD domain, it can bind to pro-caspase-1 directly by the CARD domain, and thus, the role of ASC seems redundant (Faustin et al. 2007). Also, it has been shown that the activation of caspase-1 directly by the NLRP1 or via ASC determines the action of activated caspase-1. If the activation of caspase-1 involves the recruitment of ASC, the activated caspase-1 causes the proteolytic processing of pro-IL-1β and pro-IL-18; however, when it is activated directly by the NLRP1, it most likely causes pyroptosis (Broz et al. 2010).

18.3.7 NLRP3

The most characterized NLR which forms the inflammasome is the NLRP3. This NLR has a PYD domain (Table 18.3); hence the role of ASC is absolutely necessary for the formation of the inflammasome. Two triggers are necessary for the formation of NLRP3 inflammasome: (1) activation of NF-κB, resulting in the transcription of various pro-forms of the cytokines and of NLRP3 itself (also known as priming) (Bauernfeind et al. 2009), and (2) the stimulus which activates the NLR. In most of the cases, lipopolysaccharide (LPS) acts as the first trigger and activates the NF-κB. However, many ligands such as extracellular ATP, monosodium urate crystals, amyloids, viral RNA, pore-forming toxins, fungal hyphae, alum, asbestos, silica, etc. may act as the second trigger. The vast varieties of ligands which stimulate the formation of NLRP3 inflammasome lead to the proposal of three models for the activation of the NLRP3 inflammasomes:

1. The first model proposed the involvement of P2X7 receptors in activation of NLRP3 inflammasomes by extracellular signals such as ATP, which is released by dead and dying cells. In response to ATP, the P2X7 receptors induce high K^+ efflux, leading to activation of NLRP3 inflammasomes. It was also thought that P2X7 receptors initiate the opening of pannexin-1 channels in the outer membrane, giving access to various PAMPs and DAMPs into the cytosol, which leads to the activation of NLRP3 inflammasomes (Pelegrin and Surprenant 2006). However, a group showed the cells deficient in pannexin-1 channels also activated the NLRP3 inflammasomes, disproving the involvement of pannexin-1 channels (Qu et al. 2011).
2. The second model proposes the involvement of ROS in activation of the NLRP3 inflammasomes since most of the known agonists of NLRP3 inflammasomes are known to generate ROS. Also, chemical inhibitors of ROS production decrease the IL-1β secreted by the cells after being stimulated with ATP (Dostert et al. 2008). It has also been observed that ROS production causes the NLRP3 to dissociate from its inhibitor and interact with TXNIP leading to inflammasome assembly (Zhou et al. 2010).
3. The third model suggests that many of the PAMPs and DAMPs which activate NLRP3 inflammasomes are known to utilize the endocytic pathways of the cell to enter it. Some of them are also known to cause damage to the lysosomes, releasing its content into the cytosol. Cathepsin B is one such protease which is thought to be involved in activating the NLRP3 inflammasome as pharmacological inhibitors of cathepsin B decrease the NLRP3 inflammasome activation. However, in macrophages deficient in cathepsin B, NLRP3 inflammasome formation is not abolished (Duncan et al. 2009). This discrepancy may be due to other cathepsins like cathepsin D present in the cells taking up the role of cathepsin B. However, this needs to be studied further in detail.

Another upcoming ideology for the activation of NLRP3 inflammasomes is disruption in the ion balance (mainly that of calcium). It is believed that all the above-mentioned agonists disrupt the ion balance of the cell which is eventually detected by the NLRP3 inflammasome leading to its activation (Lee et al. 2013; Murakami et al. 2012).

18.3.8 NLRC4

NLRC4 is the only characterized inflammasome without a PYD domain. Since, this inflammasome itself contains a CARD domain (Table 18.3), it can directly bind to pro-caspase-1 and cleave it and may cause pyroptosis. The most common agonist for NLRC4 inflammasome is flagellin and flagellar proteins which are translocated into the cytoplasm by the secretion systems of the bacteria (Miao et al. 2010). However, a direct interaction between these flagellar components and the NLRC4 has not been found indicating the involvement of a mediator for the same. Some reports also suggest the role of other NLRs, namely, Naip5 and 6, in the recognition of the flagellar components, which further activate the NLRC4 inflammasome (Lightfield et al. 2008). Also, the downstream signalling of NLRC4 is dependent on its binding to the ASC adaptor molecule, i.e. if it binds directly to the pro-caspase-1 via its CARD domain, it leads to a pro-inflammatory cell death named pyroptosis, and if it binds to pro-caspase-1 via the adaptor molecule ASC, it leads to the processing of IL-1β and IL-18, which is independent of cell death (Mariathasan et al. 2004). The involvement of NLRC4 in the recognition of pathogenic bacteria, namely, *Salmonella*, *Shigella*, *Pseudomonas* and *Legionella*, has been reported (Franchi et al. 2009). The activation of NLRC4 inflammasome after infection with *Salmonella* and *Legionella* leads to pyroptosis; however, in case of *Shigella* and *Pseudomonas* infection, NLRC4 inflammasome seems to be involved only in the IL-1β and IL-18 processing and not in causing pyroptosis (Schroder and Tschopp 2010). In case of *Legionella*, Naip5 has also been implicated in the activation of the NLRC4 inflammasome (Lightfield et al. 2011).

18.3.9 AIM2

AIM2 is not an NLR protein, but due to a high similarity with the NLR inflammasome, AIM2 inflammasome has been classified in the same group. The AIM2 protein consists of an N-terminal pyrin domain and a C-terminal HIN200 domain. The pyrin domain of AIM2 interacts with the adaptor molecule ASC, which in turn binds to the pro-caspase-1. The HIN200 domain of the AIM2 has been studied to bind to dsDNA. Pertaining to this property of the AIM2 protein, the AIM2 inflammasome gets activated in response to cytosolic bacteria such as *Francisella*. On recognition of the dsDNA, the AIM2 inflammasome gets activated which results in the activation of caspase-1 leading to cell death and processing of IL-1β and IL-18 (Fernandes-Alnemri et al. 2010).

18.3.10 NLRs and Diseases

Many of the metabolic and auto-immune diseases such as type I and type II diabetes mellitus and atherosclerosis are a result of an overwhelming immune response. Type I diabetes mellitus is a condition in which there is a reduced level of insulin in the body. This decrease in insulin production is a result of cell death of beta cells of the pancreas which are known to produce insulin. This cell death of beta cells has been found as a result of an increased production of IL-1β (Mandrup-Poulsen et al. 2010). Also, in the case of type 2 diabetes mellitus, wherein the cells develop resistance to insulin, IL-1β has been found to be the culprit. IL-1β has been found to impair insulin signalling, hence leading to the disease (Lagathu et al. 2006). Pertaining to this, some drugs have been designed to inhibit the NLRP3 inflammasome, thus reducing the production of IL-1β and hence leading to their antidiabetic effect (Lamkanfi et al. 2009).

Atherosclerosis is an auto-immune disease in which there is an accumulation of cholesterol in the blood vessels. This deposition of cholesterol results in the recruitment of immune cells to them.

An overwhelming action of the immune cells on these deposits results in atherosclerotic lesions. It has been reported that cholesterol crystals formed in this condition are recognized by the NLRP3 inflammasome, which leads to the release of IL-1β (Duewell et al. 2010). Also, the atherosclerotic lesions were observed to reduce upon the diminution of IL-1β in mice model which develop atherosclerosis in response to high cholesterol conditions (Kirii et al. 2003).

Interestingly, apart from their harmful role, inflammasomes also play a protective role in some conditions. In the case of colorectal cancer and inflammatory bowel disease, both of which result from excessive inflammatory cytokine production such as TNFα, IL-1α/β and IL-6, inflammasomes have been reported to reduce the inflammation. In a drug-induced colitis and colorectal cancer models of mice, a higher inflammatory response was observed in the absence of NLRP3, ASC or caspase-1 (Allen et al. 2010; Zaki et al. 2010). Another group confirmed an increase in NF-κB and MAPK activation in caspase-1 knockout mice (Dupaul-Chicoine et al. 2010).

18.4 The Scavenger Receptors

Scavenger receptors (SRs) are a group of membrane-bound receptors mainly known for their role in endocytosis and degradation of modified low-density lipoproteins (LDL), e.g., OxLDL (oxidative low-density lipoproteins) or AcLDL (acetylated LDL) and also in the differentiation of macrophages into foam cells leading to a chronic inflammatory condition called atherosclerosis.

In view of the growing list of SR ligands, which now includes exogenous PAMPs such as lipopolysaccharide (LPS), lipoteichoic acid (LTA), bacterial CpG DNA and yeast zymosan/β-glucans and microbial surface proteins in addition to altered-self targets (DAMPs), the SRs can be now termed as PRRs. Currently, SRs are categorized into ten classes designated from A to J (Prabhudas et al. 2014). These include structurally heterogenous proteins which share primary sequence similarity but have very less or no homology among classes. However, this disparate collection of proteins is grouped based on their ability to recognize a broad range of ligands and shared functional properties. Besides acting as PRRs, SRs can also co-operate with other types of receptors, for example, TLR2 or TLR4, thereby increasing the variability of the macrophage response to multiple extracellular ligands.

18.5 Structure, Function and Signalling of SRs Known to Act as PRR for Bacteria or Bacterial Ligands

Similar to other PRRs, scavenger receptors play a central role in innate immunity. Here we are going to discuss the structure, function and signalling mechanism of some of the scavenger receptors in response to whole bacteria or bacterial ligands (Fig. 18.2).

18.5.1 SR-A1 and SR-A6 (Macrophage Receptor with Collagenous Structure (MARCO))

Both SR-A1 and MARCO belong to class A SR and are widely known to recognize various bacterial ligands.

18.5.1.1 Distribution

These receptors are abundantly expressed on immune cells such as macrophages, dendritic cells and monocytes (Canton et al. 2013). Their expression also gets upregulated in response to a bacterial ligand such as LPS (Fitzgerald et al. 2000; Jing et al. 2013).

18.5.1.2 Structure

Class A SR are type II membrane glycoproteins with the N-terminal containing a short cytoplasmic tail, the transmembrane domain, the extracellular α-helical coiled coil, a long collagenous domain and cysteine-rich C-terminal. The collagenous domain has been identified as the ligand binding site of these receptors (Resnick et al. 1996; Zani et al. 2015).

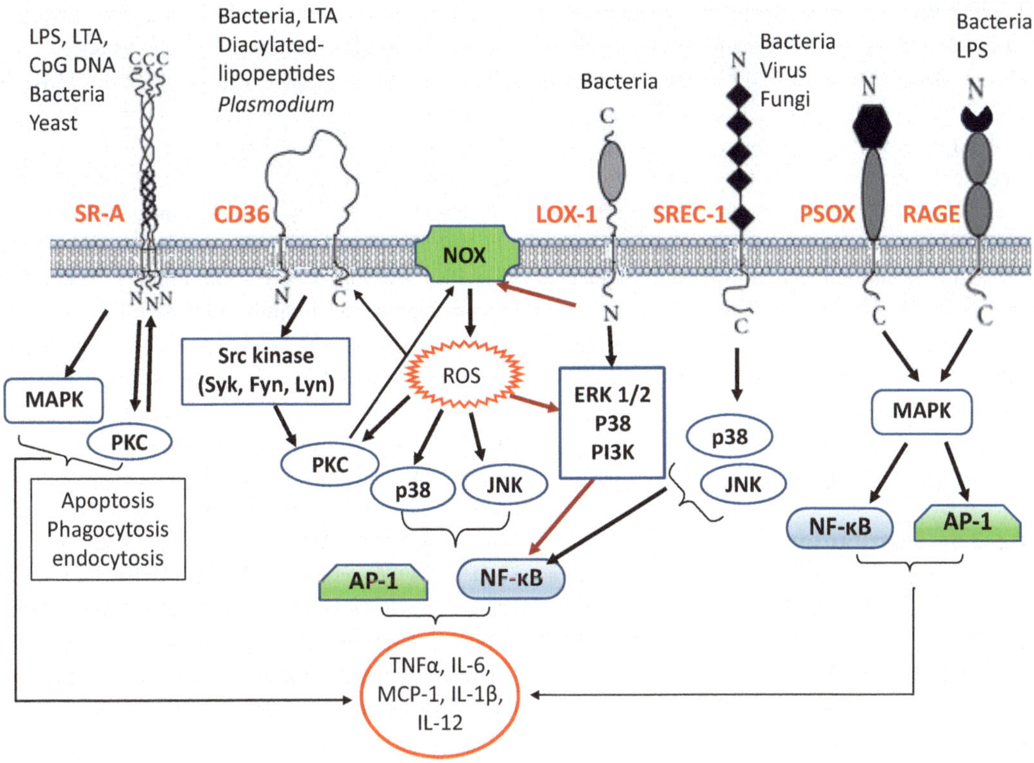

Fig. 18.2 The SR-mediated signalling

18.5.1.3 Agonists

Ligands of SR-A1 and MARCO that have been identified till date include gram-negative and gram-positive bacteria, LPS, LTA, CpG, and surface proteins of *Neisseria meningitidis*.

18.5.1.4 Function and Signalling Pathways

Both SR-A1 and MARCO regulate various macrophage activities. For example, SR-A1 binding to bacterial ligands such as LPS, LTA or bacterial surface proteins results in the non-opsonic phagocytosis and killing of the bacteria (Ben et al. 2015; Kim et al. 2004; Pluddemann et al. 2009). SR-A1 modulates the endocytosis of various exogenous and endogenous ligands in association with clathrin or caveolin-1. Also, activation of SRA1 triggers signalling pathways involving MAPKs, protein kinase C, PI3K and NF-κB resulting in the pro-inflammatory cytokine production (Coller and Paulnock 2001; Hsu et al. 2001; Hsu et al. 1998; Kim et al. 2003; Nakamura et al. 2006; Zhu et al. 2011). During infection, expression of MARCO in macrophages is highly upregulated by TLR-mediated signalling on recognition of microbial ligands such as LPS and LTA (Doyle et al. 2004; Mukhopadhyay et al. 2006; Mukhopadhyay et al. 2004). Elevated levels of MARCO result in efficient scavenging and delivery of microbial ligands to TLRs, thus enhancing their function (Mukhopadhyay et al. 2011). Moreover, MARCO binding to microbial ligands activates key signalling events involving PKA, NF-κB, PI3K and FCγR-mediated phagocytosis, etc. (Arredouani 2014). Upregulation of MARCO in dendritic cells induces phenotypic changes such as the formation of plasma membrane-derived dendrites and lamellipodia which may enhance the phagocytic ability of the cells (Granucci et al. 2003; Pikkarainen et al. 1999). Besides modulating immune cell functions, class A SRs often initiate intracellular signalling events in association with other membrane receptors, therefore acting as co-receptors. For example, SR-A receptors (such as SR-A1 and MARCO) in some cases physically associates with TLR4/

TLR2/TLR3 or NLRs resulting in enhanced pro-inflammatory responses and strong immune defence (Bowdish et al. 2009; Mukhopadhyay et al. 2011). Furthermore, class A SRs are also involved in eliciting pathogen-specific immune response as evident from the studies involving SR-A1/MARCO knockout mice that display altered spleen morphology and lower circulating IgM and IgG antibodies against bacterial surface proteins.

18.5.2 CD36/SR-B2

It is a member of class B SR.

18.5.2.1 Distribution

CD36 is present in various cell types such as monocytes, macrophages, hepatocytes, adipocytes, platelets, B-lymphocytes, epithelial and endothelial cells.

18.5.2.2 Structure

It is a type III membrane protein containing two transmembrane domains, an extracellular loop and short intracellular N- and C-terminals. SR-B1 and SR-B3 are the other members of this class. The extracellular domain is the ligand binding site of these receptors. The short intracellular tails are involved in signal transduction and trafficking. These proteins can form dimers or multimers and are highly glycosylated or acylated. Glycosylation is essential for the proper folding, trafficking and for the functionality of these receptors (Pepino et al. 2014; Silverstein and Febbraio 2009).

18.5.2.3 Agonists

Bacterial ligands that are recognized by the class B SRs include LPS, LTA, diacylated lipoproteins, gram-negative and gram-positive bacteria.

18.5.2.4 Function and Signalling Pathways Activated

CD36 is enriched in the cholesterol-rich membrane microdomains called lipid rafts, where it interacts with other receptors such as TLRs or integrins and triggers signalling events involving cellular kinases (Zani et al. 2015). For example, upon stimulation with lipoproteins or diacylglycerides, CD36 associates with TLR2/6 heterodimers within the lipid rafts. This interaction is crucial for triggering inflammatory responses (Triantafilou et al. 2006). Additionally, as co-receptors they can activate TLR2/6- or TLR4/6-mediated NF-κB activation involving MAPKs leading to pro-inflammatory responses against microbial ligands such as LTA, MALP-2 and modified self-ligands (Hoebe et al. 2005; Stewart et al. 2010). Also, it was found that on recognition of LPS, CD36 in association with TLR4 is involved in *E. coli* endocytosis by mammary epithelial cells. Further it induces inflammatory response involving NF-κB- and JNK-mediated AP-1 activation (Cao et al. 2016). CD36, when expressed in HEK 293 cells, can bind to several gram-negative and gram-positive bacteria resulting in their phagocytosis or production of inflammatory responses involving JNK-mediated signalling (Baranova et al. 2008).

18.5.3 LOX-1 and Dectin-1

These are the members of class E SR.

18.5.3.1 Distribution

These receptors are found in different cell types such as endothelial and epithelial cells, macrophages, platelets, DCs, adipocytes and neutrophils. In resting cells, the level of LOX-1 is relatively low but gets upregulated by inflammatory stimuli such as TNFα, oxidative stress, shear stress, OxLDL, phorbol esters, etc. Overexpression of LOX-1 in CHO cells resulted in bacterial recognition and uptake (Mehta et al. 2006).

18.5.3.2 Structure

These are type II transmembrane proteins with an N-terminal cytoplasmic domain, a single transmembrane region, and extracellular coiled-coil domain followed by a C-type lectin-like domain. The members of this class structurally belong to a subfamily of C-type-lectin (CLEC)-like NK cell receptor family and include LOX-1 and Dectin-1 (Ohki et al. 2005; Thakkar et al. 2015).

18.5.3.3 Agonists

LOX-1 can recognize few gram-positive and gram-negative bacteria, peptidoglycans and phosphate groups within the repeating units of LTA present on the surface of gram-positive bacteria (Park et al. 2005; Shimaoka et al. 2001). Dectin-1 recognizes β-glucan-rich ligands present in fungal cell walls and some bacteria such as *Alcaligenes faecalis* and *Mycobacterium* species (Huysamen and Brown 2009; Palma et al. 2006; Schorey and Lawrence 2008).

18.5.3.4 Function and Signalling Pathways Activated

In DCs, LOX-1 mediates uptake of antigens and antigen cross-presentation (Delneste et al. 2002). LOX-1-induced inflammatory response involves ERK1/2 or p38 MAPK, NF-κB activation and NADPH oxidase (NOX)-mediated ROS generation (Lee et al. 2008; Lubrano and Balzan 2014; Zhao et al. 2014). During bacterial infection, LOX-1 signalling stimulates inflammation, cytokine production and increased neutrophil recruitment (Wu et al. 2011). LOX-1 can bind to surface membrane proteins of *Enterobacteriaceae* and can trigger the activation of TLR2 (Jeannin et al. 2005). Dectin-1 along with TLR2 mediates inflammatory cytokine production in response to β-glucans and induces phagocytosis involving a Syk-independent pathway (Huysamen and Brown 2009; Schorey and Lawrence 2008).

18.5.4 SR-F1/SREC-1

It is a member of class F SR.

18.5.4.1 Distribution

Present in endothelial cells and in immune cells such as macrophages, DCs, etc. It was reported that the SR-F1 expression is upregulated by LPS; however, pro-inflammatory cytokines such as TNF-α and IL-1 suppress its expression.

18.5.4.2 Structure

These are type I membrane receptors and contains a long extracellular domain with multiple epidermal growth factor (EGF)-like repeats, transmembrane region, large cytoplasmic tail rich in serine/proline (SR-F1) or arginine/lysine (SR-F2) (Tamura et al. 2004). This class includes three members SR-F1, F2 and F3.

18.5.4.3 Agonists

Ligands of these receptors include gram-negative and gram-positive bacteria, LPS, fungal pathogens, zymogen granule proteins, etc.

18.5.4.4 Function and Signalling Pathways

SR-F1 mediates its signalling by localizing in cholesterol and sphingolipid rich-membrane microdomains called lipid rafts. Within the lipid rafts, it is found to interact with TLRs and induce innate immune signalling involving MAPK leading to inflammatory responses against microbial ligands such as LPS. Activation of inflammatory signalling by SR-F1/TLR also involves recruitment of other molecules such as Cdc42 and non-receptor tyrosine kinase c-Src into the lipid rafts (Murshid et al. 2015a; Murshid et al. 2015b). It was also reported that recognition of outer membrane protein A (OmpA) from *K. pneumoniae* by SR-F1 and LOX-1 present in DCs and macrophages results in TLR2-dependent IL-6 and IL-8 production (Jeannin et al. 2005). In THP-1 cells, poly I:C-induced SR-F1 and TLR3 interaction led to phosphorylation and activation of MAPKs p38, JNK and NF-κB resulting in IL-8 and IL-6 production (Murshid et al. 2016). SR-F2 in some cases forms heterodimers with SR-F1. Role of SR-F2 and SR-F3 during microbial infection is yet to be elucidated.

18.5.5 SR-PSOX or CXCL16

This is the only known member of class G SR.

18.5.5.1 Distribution

This receptor is selectively expressed in APCs such as DCs and macrophages. It exists in both membrane and soluble forms. It is highly upregulated during bacterial infection and inflammation (Wittel et al. 2015; Xu et al. 2005).

18.5.5.2 Structure

SR-PSOX is a type I transmembrane receptor with an extracellular N-terminal domain consisting of a CXC chemokine motif fused to a glycosylated mucin-like stalk, a transmembrane region and a short cytoplasmic tail. It does not share structural homology with other scavenger receptors (Shimaoka et al. 2003).

18.5.5.3 Agonists

SR-PSOX was found to recognize gram-negative and gram-positive bacteria. Since it is known to recognize negatively-charged molecules, LTA present on the gram-positive bacteria could be an agonist of SR-PSOX, while ligands of gram-negative bacteria remain to be identified (Shimaoka et al. 2003).

18.5.5.4 Function and Signalling Pathways

The soluble form acts as a ligand for CXCR6 present on naïve or type I polarized $CD4^+$ or $CD8^+$ T cells and NK cells and results in their migration to the inflammatory sites (Matloubian et al. 2000; Slauenwhite and Johnston 2015). The membrane-bound form facilitates the phagocytosis of bacteria by APCs (Shimaoka et al. 2003). SR-PSOX was found to induce immune responses on recognition of CpG-rich DNA found in bacterial pathogens by interacting with TLR9 leading to the activation of Myd88-dependent NF-κB pathway (Gursel et al. 2006).

18.5.6 SR-I1 or CD163

This is a member of class I SR.

18.5.6.1 Distribution

It is present in immune cells such as monocytes and macrophages.

18.5.6.2 Structure

These are type I transmembrane proteins with an extracellular domain containing nine scavenger receptor cysteine-rich (SRCR) domain, a transmembrane region and short intracellular cytoplasmic tail and include two members SR-I1 and SR-I2 (Etzerodt and Moestrup 2013).

18.5.6.3 Agonists

CD163 recognizes gram-positive and gram-negative bacteria.

18.5.6.4 Function and Signalling Pathways

SR-I1 functions as an innate immune sensor for gram-positive and gram-negative bacteria and can induce cytokine production against the bacteria by the cells expressing this receptor, thus preventing the growth of bacterial pathogens (Etzerodt and Moestrup 2013; Fabriek et al. 2009; Kneidl et al. 2012).

18.5.7 RAGE: Member of Class J SR

18.5.7.1 Distribution

This receptor is present in neurons, alveolar epithelial cells and endothelial cells.

18.5.7.2 Structure

These are type I transmembrane proteins containing an extracellular N-terminus with three Ig-like regions, a single transmembrane domain and a short C-terminal cytoplasmic domain. RAGE is the only member of this class, and it belongs to immunoglobulin (Ig) superfamily of cell surface molecules (Koch et al. 2010).

18.5.7.3 Agonists

Some of the known ligands of these receptors include advanced glycation end products (AGE), β-amyloids, S100 proteins, LPS, gram-negative and gram-positive bacteria (Bierhaus et al. 2001; Chuong et al. 2009; Kierdorf and Fritz 2013).

18.5.7.4 Function and Signalling Pathways Activated

RAGE act as a PRR and can recognize harmful microbial and endogenous ligands released during tissue injury or inflammation. The expression of RAGE under physiological conditions is very low, but it gets upregulated during inflammatory conditions. RAGE was found to interact with some TLRs functionally, for example, TLR4, to coordinate and regulate immune responses. Also, RAGE was found to interact with both TIRAP and MyD88 which are adaptor molecules used by TLRs to activate downstream signalling (Bertheloot et al. 2016; Sakaguchi et al. 2011). The signalling mediators activated by RAGE include MAPKs and transcription factors such as NF-κB and AP-1. Activation of NF-κB results in the transcription of various genes including cytokines, adhesion molecules and RAGE itself. Hence, activation of NF-κB provides a positive feedback loop to amplify the responses induced by ligand binding to RAGE (Chuong et al. 2009; Ibrahim et al. 2013; Kierdorf and Fritz 2013; Yamamoto et al. 2011).

18.6 Scavenger Receptors in Diseases

In view of the number of receptors that make up the scavenger receptor family, diverse ligands that the scavenger receptors recognize and the multitude of functions the SRs perform on recognition of ligands, it is expected that the scavenger receptors can be involved in the pathogenesis of various diseases. Some of the diseases in which SRs were implicated to play a major role in the disease development and progression include cardiovascular diseases such as atherosclerosis, thrombosis, Alzheimer's, type II diabetes mellitus, demyelinating disease, etc. However, the detailed mechanism and the extent of SR's involvement in the disease development still remain to be evaluated in detail.

Atherosclerosis and Alzheimer's disease are chronic inflammatory diseases involving dysregulated macrophages and characterized by a complex interaction between metabolic and immune processes leading to the formation of lesions or plaques. The key event in the development of these diseases is the inability of the macrophages to properly dispose off modified LDLs (moLDLs). These modified LDLs are highly pro-inflammatory and can induce apoptosis of cells resulting in the release of factors that further modify the LDLs and recruit more macrophages. The recruited macrophages rapidly recognize and take up the moLDLs using various SRs in order to reduce the inflammation. However, the processes distal to the SR lipid uptake that metabolize and degrade the moLDLs get overwhelmed resulting in the formation of cholesterol-laden macrophages called foam cells and the generation of atherosclerotic plaques (Kzhyshkowska et al. 2012; Moore and Freeman 2006). Similarly, in Alzheimer's disease, brains of patients contain senile plaques that contain β-amyloid fibrils, microglia and astrocytes. Ligation of AGE-modified molecules or β-amyloid proteins to the SRs present on the microglia results in macrophage activation, upregulation of adhesion molecules and generation of toxic ROS or RNS (Eugenin et al. 2016; Wilkinson and El Khoury 2012). However, these molecules significantly cause neuronal death resulting in the progression of the disease. Some of the SRs that have been extensively studied for their role in atherosclerosis and Alzheimer's disease include SR-A1, MARCO, CD36, LOX-1, SR-PSOX and RAGE.

Type II diabetes is a metabolic disorder characterized by the accumulation of fatty acids and lipids in the cells leading to changes in insulin signalling and ultimately resulting in the development of insulin resistance. Fatty acid uptake in insulin-sensitive tissue such as adipocytes, cardiac and skeletal muscles is mediated by CD36. In patients with diabetes, glucose-modified proteins such as AGEs accumulate on myelin and ECM proteins and can alter processes such as cell adhesion, growth and matrix accumulation, thus resulting in the development of peripheral neuropathy. Some of the SRs that are known to recognize AGEs include SR-PSOX, CD36, SR-A and SR-B1 (Canton et al. 2013; Goldin et al. 2006; Pickup 2004).

While the above-mentioned diseases are developed as a result of hyper activation of SRs, demyelinating disease is a result of a decrease in SR activity. Inadequate SR activity results in the accumulation of potential neurotoxic debris in the nervous system. For example, SR-A in mouse macrophages mediates removal of damaged myelin (Hendrickx et al. 2013; Popescu and Lucchinetti 2012).

References

Akira S, Takeda K (2004) Toll-like receptor signalling. Nat Rev Immunol 4:499–511 doi:10.1038/nri1391nri1391[pii]

Akira S, Takeda K, Kaisho T (2001) Toll-like receptors: critical proteins linking innate and acquired immunity. Nat Immunol 2:675–680 doi:10.1038/9060990609 [pii]

Allen IC et al (2010) The NLRP3 inflammasome functions as a negative regulator of tumorigenesis during colitis-associated cancer. J Exp Med 207:1045–1056 doi:jem.20100050 [pii]10.1084/jem.20100050

Armant MA, Fenton MJ (2002) Toll-like receptors: a family of pattern-recognition receptors in mammals. Genome Biol 3:REVIEWS3011

Arredouani MS (2014) Is the scavenger receptor MARCO a new immune checkpoint? Oncoimmunology 3:e955709. https://doi.org/10.4161/21624011.2014.955709

Baccala R, Hoebe K, Kono DH, Beutler B, Theofilopoulos AN (2007) TLR-dependent and TLR-independent pathways of type I interferon induction in systemic autoimmunity. Nat Med 13:543–551 doi:nm1590 [pii]10.1038/nm1590

Baranova IN et al (2008) Role of human CD36 in bacterial recognition, phagocytosis, and pathogen-induced JNK-mediated signaling. J Immunol 181:7147–7156

Bauernfeind FG et al (2009) Cutting edge: NF-kappaB activating pattern recognition and cytokine receptors license NLRP3 inflammasome activation by regulating NLRP3 expression. J Immunol 183:787–791 doi:jimmunol.0901363 [pii]10.4049/jimmunol.0901363

Ben J, Zhu X, Zhang H, Chen Q (2015) Class A1 scavenger receptors in cardiovascular diseases. Br J Pharmacol 172:5523–5530 doi:10.1111/bph.13105

Bertheloot D et al (2016) RAGE enhances TLR responses through binding and internalization of RNA. J Immunol 197:4118–4126 doi:10.4049/jimmunol.1502169

Bertrand MJ, Doiron K, Labbe K, Korneluk RG, Barker PA, Saleh M (2009) Cellular inhibitors of apoptosis cIAP1 and cIAP2 are required for innate immunity signaling by the pattern recognition receptors NOD1 and NOD2. Immunity 30:789–801 doi:S1074-7613(09)00203-9 [pii]10.1016/j.immuni.2009.04.011

Beutler BA (2009) TLRs and innate immunity. Blood 113:1399–1407 doi:blood-2008-07-019307 [pii] 10.1182/blood-2008-07-019307

Bierhaus A et al (2001) Diabetes-associated sustained activation of the transcription factor nuclear factor-kappaB. Diabetes 50:2792–2808

Bowdish DM et al (2009) MARCO, TLR2, and CD14 are required for macrophage cytokine responses to mycobacterial trehalose dimycolate and Mycobacterium tuberculosis. PLoS Pathog 5:e1000474. https://doi.org/10.1371/journal.ppat.1000474

Broz P, von Moltke J, Jones JW, Vance RE, Monack DM (2010) Differential requirement for Caspase-1 autoproteolysis in pathogen-induced cell death and cytokine processing. Cell Host Microbe 8:471–483 doi:S1931-3128(10)00380-X [pii]10.1016/j.chom.2010.11.007

Canton J, Neculai D, Grinstein S (2013) Scavenger receptors in homeostasis and immunity. Nat Rev Immunol 13:621–634. https://doi.org/10.1038/nri3515

Cao D, Luo J, Chen D, Xu H, Shi H, Jing X, Zang W (2016) CD36 regulates lipopolysaccharide-induced signaling pathways and mediates the internalization of Escherichia coli in cooperation with TLR4 in goat mammary gland epithelial cells. Sci Rep 6:23132. https://doi.org/10.1038/srep23132

Cervantes JL, Weinerman B, Basole C, Salazar JC (2012) TLR8: the forgotten relative revindicated. Cell Mol Immunol 9:434–438 doi:cmi201238 [pii]10.1038/cmi.2012.38

Chen ZJ (2005) Ubiquitin signalling in the NF-kappaB pathway. Nat Cell Biol 7:758–765 doi:ncb0805-758 [pii]10.1038/ncb0805-758

Chuong C, Katz J, Pauley KM, Bulosan M, Cha S (2009) RAGE expression and NF-kappaB activation attenuated by extracellular domain of RAGE in human salivary gland cell line. J Cell Physiol 221:430–434. https://doi.org/10.1002/jcp.21873

Coller SP, Paulnock DM (2001) Signaling pathways initiated in macrophages after engagement of type A scavenger receptors. J Leukocyte Biol 70:142–148

Delneste Y et al (2002) Involvement of LOX-1 in dendritic cell-mediated antigen cross-presentation. Immunity 17:353–362

Demaria O et al (2010) TLR8 deficiency leads to autoimmunity in mice. J Clin Invest 120:3651–3662 doi:42081 [pii]10.1172/JCI42081

Dostert C, Petrilli V, Van Bruggen R, Steele C, Mossman BT, Tschopp J (2008) Innate immune activation through Nalp3 inflammasome sensing of asbestos and silica. Science 320:674–677 doi:1156995 [pii] 10.1126/science.1156995

Doyle SE et al (2004) Toll-like receptors induce a phagocytic gene program through p38. J Exp Med 199:81–90. https://doi.org/10.1084/jem.20031237
Duewell P et al (2010) NLRP3 inflammasomes are required for atherogenesis and activated by cholesterol crystals. Nature 464:1357–1361 doi:nature08938 [pii] 10.1038/nature08938
Duncan JA et al (2009) Neisseria gonorrhoeae activates the proteinase cathepsin B to mediate the signaling activities of the NLRP3 and ASC-containing inflammasome. J Immunol 182:6460–6469 doi:182/10/6460 [pii]10.4049/jimmunol.0802696
Dupaul-Chicoine J et al (2010) Control of intestinal homeostasis, colitis, and colitis-associated colorectal cancer by the inflammatory caspases. Immunity 32:367–378 doi:S1074-7613(10)00082-8 [pii] 10.1016/j.immuni.2010.02.012
Etzerodt A, Moestrup SK (2013) CD163 and inflammation: biological, diagnostic, and therapeutic aspects. Antioxid Redox Signal 18:2352–2363. https://doi.org/10.1089/ars.2012.4834
Eugenin J, Vecchiola A, Murgas P, Arroyo P, Cornejo F, von Bernhardi R (2016) Expression pattern of scavenger receptors and amyloid-beta phagocytosis of astrocytes and microglia in culture are modified by acidosis: implications for Alzheimer's disease. J Alzheimer's Dis: JAD 53:857–873. https://doi.org/10.3233/JAD-160083
Fabriek BO et al (2009) The macrophage scavenger receptor CD163 functions as an innate immune sensor for bacteria. Blood 113:887–892. https://doi.org/10.1182/blood-2008-07-167064
Faustin B et al (2007) Reconstituted NALP1 inflammasome reveals two-step mechanism of caspase-1 activation. Mol Cell 25:713–724 doi: S1097-2765(07)00078-0 [pii]10.1016/j.molcel.2007.01.032
Fernandes-Alnemri T et al (2010) The AIM2 inflammasome is critical for innate immunity to Francisella tularensis. Nat Immunol 11:385–393 doi: ni.1859 [pii]10.1038/ni.1859
Ferwerda G et al (2005) NOD2 and toll-like receptors are nonredundant recognition systems of Mycobacterium tuberculosis. PLoS Pathog 1:279–285. https://doi.org/10.1371/journal.ppat.0010034
Finger JN et al (2012) Autolytic proteolysis within the function to find domain (FIIND) is required for NLRP1 inflammasome activity. J Biol Chem 287:25030–25037 doi:M112.378323 [pii]10.1074/jbc.M112.378323
Fitzgerald ML, Moore KJ, Freeman MW, Reed GL (2000) Lipopolysaccharide induces scavenger receptor A expression in mouse macrophages: a divergent response relative to human THP-1 monocyte/macrophages. J Immunol 164:2692–2700
Franchi L, Eigenbrod T, Munoz-Planillo R, Nunez G (2009) The inflammasome: a caspase-1-activation platform that regulates immune responses and disease pathogenesis. Nat Immunol 10:241–247 doi:ni.1703 [pii]10.1038/ni.1703
Frew BC, Joag VR, Mogridge J (2012) Proteolytic processing of Nlrp1b is required for inflammasome activity. PLoS Pathog 8:e1002659 doi:10.1371/journal.ppat.1002659PPATHOGENS-D-11-02650[pii]
Fukazawa A, Alonso C, Kurachi K, Gupta S, Lesser CF, McCormick BA, Reinecker HC (2008) GEF-H1 mediated control of NOD1 dependent NF-kappaB activation by Shigella effectors. PLoS Pathog 4:e1000228. https://doi.org/10.1371/journal.ppat.1000228
Furrie E, Macfarlane S, Thomson G, Macfarlane GT (2005) Toll-like receptors-2, -3 and -4 expression patterns on human colon and their regulation by mucosal-associated bacteria. Immunology 115:565–574 doi:IMM2200 [pii]10.1111/j.1365-2567.2005.02200.x
Girardin SE et al (2003) Nod1 detects a unique muropeptide from gram-negative bacterial peptidoglycan. Science 300:1584–1587. https://doi.org/10.1126/science.1084677300/5625/1584
Goldin A, Beckman JA, Schmidt AM, Creager MA (2006) Advanced glycation end products: sparking the development of diabetic vascular injury. Circulation 114:597–605. https://doi.org/10.1161/CIRCULATIONAHA.106.621854
Gomariz RP, Gutierrez-Canas I, Arranz A, Carrion M, Juarranz Y, Leceta J, Martinez C (2010) Peptides targeting Toll-like receptor signalling pathways for novel immune therapeutics. Curr Pharm Des 16:1063–1080 doi:BSP/CPD/E-Pub/00010[pii]
Gorden KB et al (2005) Synthetic TLR agonists reveal functional differences between human TLR7 and TLR8. J Immunol 174:1259–1268 doi:174/3/1259[pii]
Granucci F, Petralia F, Urbano M, Citterio S, Di Tota F, Santambrogio L, Ricciardi-Castagnoli P (2003) The scavenger receptor MARCO mediates cytoskeleton rearrangements in dendritic cells and microglia. Blood 102:2940–2947. https://doi.org/10.1182/blood-2002-12-3651
Grimes CL, Ariyananda Lde Z, Melnyk JE, O'Shea EK (2012) The innate immune protein Nod2 binds directly to MDP, a bacterial cell wall fragment. J Am Chem Soc 134:13535–13537. https://doi.org/10.1021/ja303883c
Gursel M, Gursel I, Mostowski HS, Klinman DM (2006) CXCL16 influences the nature and specificity of CpG-induced immune activation. J Immunol 177:1575–1580
Hajjar AM, Ernst RK, Tsai JH, Wilson CB, Miller SI (2002) Human Toll-like receptor 4 recognizes host-specific LPS modifications. Nat Immunol 3:354–359 doi:10.1038/ni777ni777[pii]
Hasegawa M et al (2006) Differential release and distribution of Nod1 and Nod2 immunostimulatory molecules among bacterial species and environments. J Biol Chem 281:29054–29063 doi:M602638200 [pii] 10.1074/jbc.M602638200

Hendrickx DA, Koning N, Schuurman KG, van Strien ME, van Eden CG, Hamann J, Huitinga I (2013) Selective upregulation of scavenger receptors in and around demyelinating areas in multiple sclerosis. J Neuropathol Exp Neurol 72:106–118. https://doi.org/10.1097/NEN.0b013e31827fd9e8

Hennessy EJ, Parker AE, O'Neill LA (2010) Targeting Toll-like receptors: emerging therapeutics? Nat Rev Drug Discov 9:293–307. https://doi.org/10.1038/nrd3203

Hoebe K et al (2005) CD36 is a sensor of diacylglycerides. Nature 433:523–527. https://doi.org/10.1038/nature03253

Hoffmann A, Baltimore D (2006) Circuitry of nuclear factor kappaB signaling. Immunol Rev 210:171–186 doi:IMR375 [pii]10.1111/j.0105-2896.2006.00375.x

Hornef MW, Frisan T, Vandewalle A, Normark S, Richter-Dahlfors A (2002) Toll-like receptor 4 resides in the Golgi apparatus and colocalizes with internalized lipopolysaccharide in intestinal epithelial cells. J Exp Med 195:559–570

Hsu HY, Hajjar DP, Khan KM, Falcone DJ (1998) Ligand binding to macrophage scavenger receptor-A induces urokinase-type plasminogen activator expression by a protein kinase-dependent signaling pathway. J Biol Chem 273:1240–1246

Hsu HY, Chiu SL, Wen MH, Chen KY, Hua KF (2001) Ligands of macrophage scavenger receptor induce cytokine expression via differential modulation of protein kinase signaling pathways. J Biol Chem 276:28719–28730. https://doi.org/10.1074/jbc.M011117200

Hsu YM et al (2007) The adaptor protein CARD9 is required for innate immune responses to intracellular pathogens. Nat Immunol 8:198–205 doi:ni1426 [pii] 10.1038/ni1426

Hsu LC et al (2008) A NOD2-NALP1 complex mediates caspase-1-dependent IL-1beta secretion in response to Bacillus anthracis infection and muramyl dipeptide. Proc Natl Acad Sci U S A 105:7803–7808 doi:0802726105 [pii]10.1073/pnas.0802726105

Huysamen C, Brown GD (2009) The fungal pattern recognition receptor, Dectin-1, and the associated cluster of C-type lectin-like receptors. FEMS Microbiol Lett 290:121–128. https://doi.org/10.1111/j.1574-6968.2008.01418.x

Ibrahim ZA, Armour CL, Phipps S, Sukkar MB (2013) RAGE and TLRs: relatives, friends or neighbours? Mol Immunol 56:739–744. https://doi.org/10.1016/j.molimm.2013.07.008

Imler JL, Hoffmann JA (2001) Toll receptors in innate immunity. Trends Cell Biol 11:304–311

Iwasaki A, Medzhitov R (2004) Toll-like receptor control of the adaptive immune responses. Nat Immunol 5:987–995 doi:10.1038/ni1112ni1112[pii]

Jack CS et al (2005) TLR signaling tailors innate immune responses in human microglia and astrocytes. J Immunol 175:4320–4330 doi:175/7/4320[pii]

Jeannin P et al (2005) Complexity and complementarity of outer membrane protein A recognition by cellular and humoral innate immunity receptors. Immunity 22:551–560. https://doi.org/10.1016/j.immuni.2005.03.008

Jing J et al (2013) Role of macrophage receptor with collagenous structure in innate immune tolerance. J Immunol 190:6360–6367. https://doi.org/10.4049/jimmunol.1202942

Kawai T, Akira S (2006) TLR signaling. Cell Death Differ 13:816–825 doi:4401850[pii]10.1038/sj.cdd.4401850

Kawai T, Akira S (2010) The role of pattern-recognition receptors in innate immunity: update on Toll-like receptors. Nature Immunol 11:373–384. https://doi.org/10.1038/ni.1863

Kawai T et al (2001) Lipopolysaccharide stimulates the MyD88-independent pathway and results in activation of IFN-regulatory factor 3 and the expression of a subset of lipopolysaccharide-inducible genes. J Immunol 167:5887–5894

Keestra AM et al (2013) Manipulation of small Rho GTPases is a pathogen-induced process detected by NOD1. Nature 496:233–237 doi:nature12025 [pii] 10.1038/nature12025

Kierdorf K, Fritz G (2013) RAGE regulation and signaling in inflammation and beyond. J Leukocyte Biol 94:55–68. https://doi.org/10.1189/jlb.1012519

Kim WS, Ordija CM, Freeman MW (2003) Activation of signaling pathways by putative scavenger receptor class A (SR-A) ligands requires CD14 but not SR-A. Biochem Biophys Res Commun 310:542–549

Kim S, Watarai M, Suzuki H, Makino S, Kodama T, Shirahata T (2004) Lipid raft microdomains mediate class A scavenger receptor-dependent infection of Brucella abortus. Microbial Pathog 37:11–19. https://doi.org/10.1016/j.micpath.2004.04.002

Kimbrell DA, Beutler B (2001) The evolution and genetics of innate immunity. Nat Rev Genet 2:256–267. https://doi.org/10.1038/3506600635066006 [pii]

Kirii H et al (2003) Lack of interleukin-1beta decreases the severity of atherosclerosis in ApoE-deficient mice. Arterioscler Thromb Vasc Biol 23:656–660 doi:10.1161/01.ATV.0000064374.15232.C301.ATV.0000064374.15232.C3[pii]

Kneidl J, Loffler B, Erat MC, Kalinka J, Peters G, Roth J, Barczyk K (2012) Soluble CD163 promotes recognition, phagocytosis and killing of Staphylococcus aureus via binding of specific fibronectin peptides. Cell Microbiol 14:914–936. https://doi.org/10.1111/j.1462-5822.2012.01766.x

Kobayashi K, Hernandez LD, Galan JE, Janeway CA Jr, Medzhitov R, Flavell RA (2002a) IRAK-M is a negative regulator of Toll-like receptor signaling. Cell 110:191–202

Kobayashi K et al (2002b) RICK/Rip2/CARDIAK mediates signalling for receptors of the innate and adaptive immune systems. Nature 416:194–199 doi:10.1038/416194a416194a[pii]

Kobayashi KS, Chamaillard M, Ogura Y, Henegariu O, Inohara N, Nunez G, Flavell RA (2005) Nod2-dependent regulation of innate and adaptive immunity in the intestinal tract. Science 307:731–734 doi:307/5710/731 [pii]10.1126/science.1104911

Koch M, Chitayat S, Dattilo BM, Schiefner A, Diez J, Chazin WJ, Fritz G (2010) Structural basis for ligand recognition and activation of RAGE. Structure 18:1342–1352. https://doi.org/10.1016/j.str.2010.05.017

Krieg A et al (2009) XIAP mediates NOD signaling via interaction with RIP2. Proc Natl Acad Sci U S A 106:14524–14529 doi:0907131106 [pii]10.1073/pnas.0907131106

Kzhyshkowska J, Neyen C, Gordon S (2012) Role of macrophage scavenger receptors in atherosclerosis. Immunobiology 217:492–502. https://doi.org/10.1016/j.imbio.2012.02.015

Lagathu C, Yvan-Charvet L, Bastard JP, Maachi M, Quignard-Boulange A, Capeau J, Caron M (2006) Long-term treatment with interleukin-1beta induces insulin resistance in murine and human adipocytes. Diabetologia 49:2162–2173. https://doi.org/10.1007/s00125-006-0335-z

Lamkanfi M et al (2009) Glyburide inhibits the Cryopyrin/Nalp3 inflammasome. J Cell Biol 187:61–70 doi:jcb.200903124 [pii]10.1083/jcb.200903124

Laroui H et al (2011) L-Ala-gamma-D-Glu-meso-diaminopimelic acid (DAP) interacts directly with leucine-rich region domain of nucleotide-binding oligomerization domain 1, increasing phosphorylation activity of receptor-interacting serine/threonine-protein kinase 2 and its interaction with nucleotide-binding oligomerization domain 1. J Biol Chem 286:31003–31013 doi:M111.257501 [pii]10.1074/jbc.M111.257501

Lee JG, Lim EJ, Park DW, Lee SH, Kim JR, Baek SH (2008) A combination of Lox-1 and Nox1 regulates TLR9-mediated foam cell formation. Cell Signal 20:2266–2275. https://doi.org/10.1016/j.cellsig.2008.08.022

Lee GS et al (2013) The calcium-sensing receptor regulates the NLRP3 inflammasome through Ca2+ and cAMP. Nature 492:123–127 doi:nature11588 [pii]10.1038/nature11588

Lemaitre B, Nicolas E, Michaut L, Reichhart JM, Hoffmann JA (1996) The dorsoventral regulatory gene cassette spatzle/Toll/cactus controls the potent antifungal response in Drosophila adults. Cell 86:973–983

Lightfield KL et al (2008) Critical function for Naip5 in inflammasome activation by a conserved carboxy-terminal domain of flagellin. Nat Immunol 9:1171–1178 doi:ni.1646 [pii]10.1038/ni.1646

Lightfield KL et al (2011) Differential requirements for NAIP5 in activation of the NLRC4 inflammasome. Infect Immun 79:1606–1614 doi:IAI.01187-10 [pii]10.1128/IAI.01187-10

Lubrano V, Balzan S (2014) LOX-1 and ROS, inseparable factors in the process of endothelial damage. Free radical Res 48:841–848. https://doi.org/10.3109/10715762.2014.929122

Mandrup-Poulsen T, Pickersgill L, Donath MY (2010) Blockade of interleukin 1 in type 1 diabetes mellitus. Nat Rev Endocrinol 6:158–166 doi:nrendo.2009.271 [pii]10.1038/nrendo.2009.271

Mariathasan S et al (2004) Differential activation of the inflammasome by caspase-1 adaptors ASC and Ipaf. Nature 430:213–218 doi:10.1038/nature02664nature02664[pii]

Martinon F, Burns K, Tschopp J (2002) The inflammasome: a molecular platform triggering activation of inflammatory caspases and processing of proIL-beta. Mol Cell 10:417–426 doi:S1097-2765(02)00599-3[pii]

Matloubian M, David A, Engel S, Ryan JE, Cyster JG (2000) A transmembrane CXC chemokine is a ligand for HIV-coreceptor Bonzo. Nat Immunol 1:298–304. https://doi.org/10.1038/79738

Medzhitov R (2001) Toll-like receptors and innate immunity. Nat Rev Immunol 1:135–145. https://doi.org/10.1038/35100529

Medzhitov R, Preston-Hurlburt P, Janeway CA Jr (1997) A human homologue of the Drosophila Toll protein signals activation of adaptive immunity. Nature 388:394–397. https://doi.org/10.1038/41131

Mehta JL, Chen J, Hermonat PL, Romeo F, Novelli G (2006) Lectin-like, oxidized low-density lipoprotein receptor-1 (LOX-1): a critical player in the development of atherosclerosis and related disorders. Cardiovasc Res 69:36–45. https://doi.org/10.1016/j.cardiores.2005.09.006

Miao EA et al (2010) Innate immune detection of the type III secretion apparatus through the NLRC4 inflammasome. Proc Natl Acad Sci U S A 107:3076–3080 doi:0913087107 [pii]10.1073/pnas.0913087107

Moore KJ, Freeman MW (2006) Scavenger receptors in atherosclerosis: beyond lipid uptake. Arterioscler Thromb Vasc Biol 26:1702–1711. https://doi.org/10.1161/01.ATV.0000229218.97976.43

Mukhopadhyay S, Peiser L, Gordon S (2004) Activation of murine macrophages by Neisseria meningitidis and IFN-gamma in vitro: distinct roles of class A scavenger and Toll-like pattern recognition receptors in selective modulation of surface phenotype. J Leukocyte Biol 76:577–584. https://doi.org/10.1189/jlb.0104014

Mukhopadhyay S et al (2006) MARCO, an innate activation marker of macrophages, is a class A scavenger receptor for Neisseria meningitidis. Eur J Immunol 36:940–949. https://doi.org/10.1002/eji.200535389

Mukhopadhyay S, Varin A, Chen Y, Liu B, Tryggvason K, Gordon S (2011) SR-A/MARCO-mediated ligand delivery enhances intracellular TLR and NLR function, but ligand scavenging from cell surface limits TLR4 response to pathogens. Blood

117:1319–1328. https://doi.org/10.1182/blood-2010-03-276733

Murakami T, Ockinger J, Yu J, Byles V, McColl A, Hofer AM, Horng T (2012) Critical role for calcium mobilization in activation of the NLRP3 inflammasome. Proc Natl Acad Sci U S A 109:11282–11287 doi:1117765109 [pii]10.1073/pnas.1117765109

Murshid A, Borges TJ, Calderwood SK (2015a) Emerging roles for scavenger receptor SREC-I in immunity. Cytokine 75:256–260. https://doi.org/10.1016/j.cyto.2015.02.009

Murshid A, Gong J, Prince T, Borges TJ, Calderwood SK (2015b) Scavenger receptor SREC-I mediated entry of TLR4 into lipid microdomains and triggered inflammatory cytokine release in RAW 264.7 cells upon LPS activation. PloS one 10:e0122529. https://doi.org/10.1371/journal.pone.0122529

Murshid A, Borges TJ, Lang BJ, Calderwood SK (2016) The scavenger receptor SREC-I cooperates with toll-like receptors to trigger inflammatory innate immune responses. Front Immunol 7:226. https://doi.org/10.3389/fimmu.2016.00226

Nakamura T, Suzuki H, Wada Y, Kodama T, Doi T (2006) Fucoidan induces nitric oxide production via p38 mitogen-activated protein kinase and NF-kappaB-dependent signaling pathways through macrophage scavenger receptors. Biochem Biophysl Res Commun 343:286–294. https://doi.org/10.1016/j.bbrc.2006.02.146

Ohki I et al (2005) Crystal structure of human lectin-like, oxidized low-density lipoprotein receptor 1 ligand binding domain and its ligand recognition mode to OxLDL. Structure 13:905–917. https://doi.org/10.1016/j.str.2005.03.016

O'Neill LA, Bryant CE, Doyle SL (2009) Therapeutic targeting of Toll-like receptors for infectious and inflammatory diseases and cancer. Pharmacol Rev 61:177–197 doi:pr.109.001073 [pii]10.1124/pr.109.001073

O'Neill LA, Golenbock D, Bowie AG (2013) The history of Toll-like receptors - redefining innate immunity. Nat Rev Immunol 13:453–460 doi:nri3446 [pii]10.1038/nri3446

Opitz B et al (2004) Nucleotide-binding oligomerization domain proteins are innate immune receptors for internalized Streptococcus pneumoniae. J Biol Chem 279:36426–36432 doi:10.1074/jbc.M403861200M403861200[pii]

Opitz B et al (2005) Nod1-mediated endothelial cell activation by Chlamydophila pneumoniae. Circ Res 96:319–326 doi:01.RES.0000155721.83594.2c [pii] 10.1161/01.RES.0000155721.83594.2c

Palma AS et al (2006) Ligands for the beta-glucan receptor, Dectin-1, assigned using "designer" microarrays of oligosaccharide probes (neoglycolipids) generated from glucan polysaccharides. J Biol Chem 281:5771–5779. https://doi.org/10.1074/jbc.M511461200

Park H, Adsit FG, Boyington JC (2005) The 1.4 angstrom crystal structure of the human oxidized low density lipoprotein receptor lox-1. J Biol Chem 280:13593–13599. https://doi.org/10.1074/jbc.M500768200

Park JH et al (2007) RICK/RIP2 mediates innate immune responses induced through Nod1 and Nod2 but not TLRs. J Immunol 178:2380–2386 doi:178/4/2380[pii]

Pelegrin P, Surprenant A (2006) Pannexin-1 mediates large pore formation and interleukin-1beta release by the ATP-gated P2X7 receptor. EMBO J 25:5071–5082 doi:7601378 [pii]10.1038/sj.emboj.7601378

Pepino MY, Kuda O, Samovski D, Abumrad NA (2014) Structure-function of CD36 and importance of fatty acid signal transduction in fat metabolism. Ann Rev Nutr 34:281–303. https://doi.org/10.1146/annurev-nutr-071812-161220

Pickup JC (2004) Inflammation and activated innate immunity in the pathogenesis of type 2 diabetes. Diabetes Care 27:813–823

Pikkarainen T, Brannstrom A, Tryggvason K (1999) Expression of macrophage MARCO receptor induces formation of dendritic plasma membrane processes. J Biol Chem 274:10975–10982

Pluddemann A et al (2009) SR-A, MARCO and TLRs differentially recognise selected surface proteins from Neisseria meningitidis: an example of fine specificity in microbial ligand recognition by innate immune receptors. J Innate Immun 1:153–163. https://doi.org/10.1159/000155227

Poltorak A et al (1998) Defective LPS signaling in C3H/HeJ and C57BL/10ScCr mice: mutations in Tlr4 gene. Science 282:2085–2088

Popescu BF, Lucchinetti CF (2012) Pathology of demyelinating diseases. Ann Rev Pathol 7:185–217. https://doi.org/10.1146/annurev-pathol-011811-132443

Prabhudas M et al (2014) Standardizing scavenger receptor nomenclature. J Immunol 192:1997–2006. https://doi.org/10.4049/jimmunol.1490003

Qu Y et al (2011) Pannexin-1 is required for ATP release during apoptosis but not for inflammasome activation. J Immunol 186:6553–6561 doi:jimmunol.1100478 [pii]10.4049/jimmunol.1100478

Qureshi S, Medzhitov R (2003) Toll-like receptors and their role in experimental models of microbial infection. Genes Immun 4:87–94 doi:10.1038/sj.gene.63639376363937[pii]

Resnick D, Chatterton JE, Schwartz K, Slayter H, Krieger M (1996) Structures of class A macrophage scavenger receptors. Electron microscopic study of flexible, multidomain, fibrous proteins and determination of the disulfide bond pattern of the scavenger receptor cysteine-rich domain. J Biol Chem 271:26924–26930

Sakaguchi M et al (2011) TIRAP, an adaptor protein for TLR2/4, transduces a signal from RAGE phosphorylated upon ligand binding. PloS One 6:e23132. https://doi.org/10.1371/journal.pone.0023132

Schaefer L et al (2005) The matrix component biglycan is proinflammatory and signals through Toll-like receptors 4 and 2 in macrophages. J Clin Invest 115:2223–2233. https://doi.org/10.1172/JCI23755
Schorey JS, Lawrence C (2008) The pattern recognition receptor Dectin-1: from fungi to mycobacteria. Curr Drug Targets 9:123–129
Schroder K, Tschopp J (2010) The inflammasomes. Cell 140:821–832 doi:S0092-8674(10)00075-9 [pii] 10.1016/j.cell.2010.01.040
Shimaoka T, Kume N, Minami M, Hayashida K, Sawamura T, Kita T, Yonehara S (2001) LOX-1 supports adhesion of Gram-positive and Gram-negative bacteria. J Immunol 166:5108–5114
Shimaoka T et al (2003) Cutting edge: SR-PSOX/CXC chemokine ligand 16 mediates bacterial phagocytosis by APCs through its chemokine domain. J Immunol 171:1647–1651
Silverstein RL, Febbraio M (2009) CD36, a scavenger receptor involved in immunity, metabolism, angiogenesis, and behavior. Sci Signal 2:re3. https://doi.org/10.1126/scisignal.272re3
Slack JL, Schooley K, Bonnert TP, Mitcham JL, Qwarnstrom EE, Sims JE, Dower SK (2000) Identification of two major sites in the type I interleukin-1 receptor cytoplasmic region responsible for coupling to pro-inflammatory signaling pathways. J Biol Chem 275:4670–4678
Slauenwhite D, Johnston B (2015) Regulation of NKT cell localization in homeostasis and infection. Front Immunol 6:255. https://doi.org/10.3389/fimmu.2015.00255
Stewart CR et al (2010) CD36 ligands promote sterile inflammation through assembly of a Toll-like receptor 4 and 6 heterodimer. Nat Immunol 11:155–161. https://doi.org/10.1038/ni.1836
Supajatura V, Ushio H, Nakao A, Akira S, Okumura K, Ra C, Ogawa H (2002) Differential responses of mast cell Toll-like receptors 2 and 4 in allergy and innate immunity. J Clin Invest 109:1351–1359. https://doi.org/10.1172/JCI14704
Takeda K, Akira S (2005) Toll-like receptors in innate immunity. Int Immunol 17:1–14 doi:17/1/1 [pii] 10.1093/intimm/dxh186
Tamura Y et al (2004) Scavenger receptor expressed by endothelial cells I (SREC-I) mediates the uptake of acetylated low density lipoproteins by macrophages stimulated with lipopolysaccharide. J Biol Chem 279:30938–30944. https://doi.org/10.1074/jbc.M313088200
Thakkar S, Wang X, Khaidakov M, Dai Y, Gokulan K, Mehta JL, Varughese KI (2015) Structure-based design targeted at LOX-1, a receptor for oxidized low-density lipoprotein. Sci Rep 5:16740. https://doi.org/10.1038/srep16740
Triantafilou M, Gamper FG, Haston RM, Mouratis MA, Morath S, Hartung T, Triantafilou K (2006) Membrane sorting of toll-like receptor (TLR)-2/6 and TLR2/1 heterodimers at the cell surface determines heterotypic associations with CD36 and intracellular targeting. J Biol Chem 281:31002–31011. https://doi.org/10.1074/jbc.M602794200
Viala J et al (2004) Nod1 responds to peptidoglycan delivered by the Helicobacter pylori cag pathogenicity island. Nat Immunol 5:1166–1174 doi:ni1131 [pii] 10.1038/ni1131
West AP et al (2011) TLR signalling augments macrophage bactericidal activity through mitochondrial ROS. Nature 472:476–480 doi:nature09973 [pii] 10.1038/nature09973
Wilkinson K, El Khoury J (2012) Microglial scavenger receptors and their roles in the pathogenesis of Alzheimer's disease. Int J Alzheimer's Dis 2012:489456. https://doi.org/10.1155/2012/489456
Wittel UA et al (2015) The chemokine ligand CXCL16 is an indicator of bacterial infection in necrotizing pancreatitis. Pancreatol: Off J Int Assoc Pancreatol 15:124–130. https://doi.org/10.1016/j.pan.2015.01.004
Wu Z, Sawamura T, Kurdowska AK, Ji HL, Idell S, Fu J (2011) LOX-1 deletion improves neutrophil responses, enhances bacterial clearance, and reduces lung injury in a murine polymicrobial sepsis model. Infect Immun 79:2865–2870. https://doi.org/10.1128/IAI.01317-10
Xu Y, Tao X, Shen B, Horng T, Medzhitov R, Manley JL, Tong L (2000) Structural basis for signal transduction by the Toll/interleukin-1 receptor domains. Nature 408:111–115. https://doi.org/10.1038/35040600
Xu H, Xu W, Chu Y, Gong Y, Jiang Z, Xiong S (2005) Involvement of up-regulated CXC chemokine ligand 16/scavenger receptor that binds phosphatidylserine and oxidized lipoprotein in endotoxin-induced lethal liver injury via regulation of T-cell recruitment and adhesion. Infect Immun 73:4007–4016. https://doi.org/10.1128/IAI.73.7.4007-4016.2005
Yamamoto Y et al (2011) Septic shock is associated with receptor for advanced glycation end products ligation of LPS. J Immunol 186:3248–3257. https://doi.org/10.4049/jimmunol.1002253
Yu M et al (2006) HMGB1 signals through toll-like receptor (TLR) 4 and TLR2. Shock 26:174–179 doi:10.1097/01.shk.0000225404.51320.8200024382-200608000-00011[pii]
Zaki MH, Boyd KL, Vogel P, Kastan MB, Lamkanfi M, Kanneganti TD (2010) The NLRP3 inflammasome protects against loss of epithelial integrity and mortality during experimental colitis. Immunity 32:379–391 doi:S1074-7613(10)00086-5 [pii]10.1016/j.immuni.2010.03.003
Zani IA, Stephen SL, Mughal NA, Russell D, Homer-Vanniasinkam S, Wheatcroft SB, Ponnambalam S (2015) Scavenger receptor structure and function in health and disease. Cells 4:178–201. https://doi.org/10.3390/cells4020178

Zarember KA, Godowski PJ (2002) Tissue expression of human Toll-like receptors and differential regulation of Toll-like receptor mRNAs in leukocytes in response to microbes, their products, and cytokines. J Immunol 168:554–561

Zhao W, Ma G, Chen X (2014) Lipopolysaccharide induced LOX-1 expression via TLR4/MyD88/ROS activated p38MAPK-NF-kappaB pathway. Vasc Pharmacol 63:162–172. https://doi.org/10.1016/j.vph.2014.06.008

Zhou R, Tardivel A, Thorens B, Choi I, Tschopp J (2010) Thioredoxin-interacting protein links oxidative stress to inflammasome activation. Nat Immunol 11:136–140 doi:ni.1831 [pii]10.1038/ni.1831

Zhu XD et al (2011) Caveolae-dependent endocytosis is required for class A macrophage scavenger receptor-mediated apoptosis in macrophages. J Biol chem 286:8231–8239. https://doi.org/10.1074/jbc.M110.145888

Zilbauer M et al (2007) A major role for intestinal epithelial nucleotide oligomerization domain 1 (NOD1) in eliciting host bactericidal immune responses to Campylobacter jejuni. Cell Microbiol 9:2404–2416 doi: CMI969 [pii]10.1111/j.1462-5822.2007.00969.x

Structural Basis and Functional Implications of the Membrane Pore-Formation Mechanisms of Bacterial Pore-Forming Toxins

19

Anish Kumar Mondal, Amritha Sreekumar, Nidhi Kundu, Reema Kathuria, Pratima Verma, Shraddha Gandhi, and Kausik Chattopadhyay

Abstract

Pore-forming toxins (PFTs) are a distinct class of membrane-damaging protein toxins documented in a wide array of life forms ranging from bacteria to humans. PFTs are known to act as potent virulence factors of the bacterial pathogens. Bacterial PFTs are, in general, secreted as water-soluble molecules, which upon encountering target host cells assemble into transmembrane oligomeric pores, thus leading to membrane permeabilization and cell death. Interaction of the PFTs with the target host cells can also lead to plethora of cellular responses having critical implications for the bacterial pathogenesis processes, host-pathogen interactions, and host immunity. In this review, we present an overview of our current understanding of the structural aspects of the membrane pore-formation processes employed by the bacterial PFTs. We also discuss the functional implications of the PFT mode of actions, in terms of eliciting diverse cellular responses.

Keywords

Pore-forming toxin · Cholesterol-dependent cytolysin · Membrane · Oligomerization

19.1 Introduction

Plasma membrane is one of the most critical components of cellular systems. It acts as a semipermeability barrier between the cell and the extracellular milieu. In doing so, plasma membrane compartmentalizes the cellular architecture, while at the same time allowing the cell to sense and respond to the extracellular cues. Integrity of the plasma membrane is, therefore, essential for the survival and sustenance of the living cell. Disruption of the plasma membrane is regarded as one of the ancient, yet conserved, forms of cell-killing mechanisms and is commonly observed in the diverse life forms starting from bacteria to humans. Killing of the target cells via membrane permeabilization is commonly observed in the virulence mechanisms of the wide array of pathogenic bacteria. Pathogenic bacteria employ a specialized class of membrane-damaging toxins that act by forming pores in the target cell membranes, and, accordingly, they are designated as pore-forming toxins (PFTs). PFTs represent the largest class of bacterial toxins and act as the potent virulence factors for the bacterial pathogens. Bacterial PFTs display remarkable structural and functional similarity to the PFTs

A. K. Mondal · A. Sreekumar · N. Kundu · R. Kathuria
P. Verma · S. Gandhi · K. Chattopadhyay (✉)
Centre for Protein Science, Design and Engineering, Department of Biological Sciences, Indian Institute of Science Education and Research (IISER) Mohali, Mohali, Punjab, India
e-mail: kausik@iisermohali.ac.in

K. Chattopadhyay, S. C. Basu (eds.), *Biochemical and Biophysical Roles of Cell Surface Molecules*, Advances in Experimental Medicine and Biology 1112, https://doi.org/10.1007/978-981-13-3065-0_19

found in eukaryotes, cnidarians, and higher vertebrates (Dal Peraro and van der Goot 2016; Morgan et al. 2017; Rai and Chattopadhyay 2015b). Pore formation on the membranes is a mechanism used by the pathogenic bacteria not only for the purpose of membrane permeabilization but also for various other functionalities. For example, the A-B-type toxins (like diphtheria toxin and cholera toxin) mimic PFTs in their mode of actions, where the B-components form channels across the membranes for the translocation of their catalytic A-subunits into the target cells (De Haan and Hirst 2004; Collier 1975).

The general mode of action of the bacterial PFTs involves the following steps: (i) secretion of the toxin from the bacterial pathogen, (ii) binding of the toxin to the target cell membranes, (iii) oligomerization of the toxin on the membrane, and (iv) transmembrane pore formation (Fig. 19.1). One of the most remarkable features of the bacterial PFTs is the conversion of the toxin from the water-soluble monomeric state to the oligomeric transmembrane pore form. For this, the PFT molecule undergoes array of structural/conformational changes. The most prominent structural change that occurs during the membrane pore-formation process of the PFTs is the reorganization of the pore-forming motif (s) and their insertion into the hydrophobic environment of the membrane lipid bilayer (Dal Peraro and van der Goot 2016). Thus, PFTs are the unique class of dimorphic proteins that are initially synthesized as water-soluble entities and upon interacting with the target cell membranes adopt a structure compatible with the amphipathic environment of the membrane lipid bilayer (Heuck et al. 2001). PFT structure-function paradigm, therefore, contradicts the conventional notion of the 'one-sequence-one-structure' hypothesis.

19.2 Structural Classification of the PFTs

PFTs can be classified based on the structural motifs that are employed to generate the transmembrane pore architecture. The PFTs are thus categorized either as α-PFTs that use α-helices to form the transmembrane pore or β-PFTs that generate β-barrel pores (Lesieur et al. 1997) (Fig. 19.2).

19.2.1 α-PFTs

Colicins are the typical examples of the α-PFTs. Colicins are generally secreted by *Escherichia coli* to kill bacteria of the related species by punching holes in the bacterial inner membranes (Lakey and Slatin 2001). Colicins, in their soluble state, possess hydrophobic α-helices sequestered within the amphipathic α-helical bundles (Parker et al. 1989). Upon attaching to the target

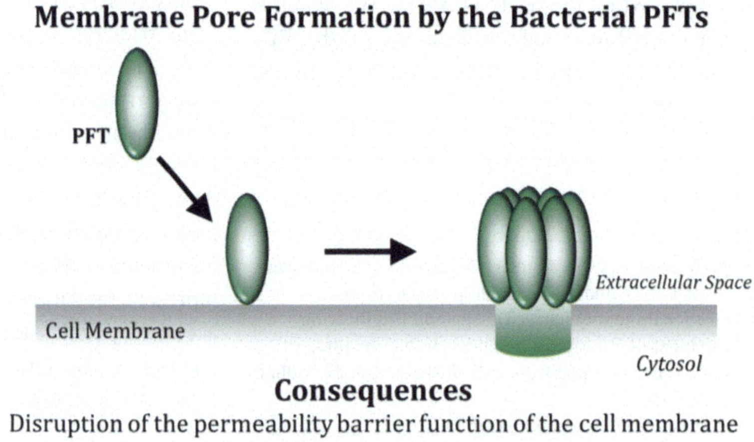

Fig. 19.1 Generalized mechanism of membrane pore formation by the bacterial PFTs

Fig. 19.2 Structural classification of the PFTs. Cytolysin A pore is shown as an example of α-PFT. *S. aureus* α-hemolysin pore is shown as an example of β-PFT. Structural co-ordinates corresponding to the transmembrane view of the PFTs were obtained from the Protein Data Bank (PDB id of Cytolysin A pore: 2WCD; PDB id of α-hemolysin pore: 7AHL). Structural models were generated using PyMOL [DeLano, W. L. (2002) The PyMOL Molecular Graphics System, found online (www.pymol.org)]

membranes, the amphipathic bundles open up allowing the hydrophobic α-helical hairpins to insert into the lipid bilayer, thus forming the transmembrane α-helical pores (Ridley et al. 2010). Colicin-fold has also been observed in other bacterial toxins as well as in eukaryotic proteins (Zakharov and Cramer 2002). The translocation domain of the diphtheria toxin possesses this colicin-fold, which mediates the pore formation in the endosomal membranes, thereby facilitating the translocation of the catalytic subunit of the toxin that causes the ADP-ribosylation of eEF2 (eukaryotic translation elongation factor 2), leading to its inactivation (Oh et al. 1999). Cry toxin, secreted by *Bacillus thuringiensis*, also exhibits colicin-fold (Xu et al. 2014). In eukaryotes, pore-forming BAX and BAK proteins of the mitochondrial apoptotic pathway display colicin-like structural folds (Westphal et al. 2011).

Cytolysin A (ClyA) and its related hemolysins secreted by certain strains of *E. coli*, *Salmonella enterica*, and *Shigella flexneri* constitute another distinct class of α-PFTs (Hunt et al. 2010). Except for a short β-tongue motif, ClyA structure displays an α-helical secondary structure disposition (Wallace et al. 2000). Upon membrane interaction, the β-tongue attaches itself to the membrane by segregating from the rest of the toxin core. This induces conformational changes in the amphipathic α-helices, which in turn approaches the membrane surface and inserts into the membrane (Mueller et al. 2009).

Actinoporins constitute another distinct family of α-PFTs. Actinoporins are the toxins secreted by the sea anemones in order to guard themselves from the predators, or to paralyse their prey, by punching holes in the target cell membranes. These proteins have a β-sandwich core structure, with the two flanking α-helical segments. Upon membrane association, the N-terminal α-helix detaches itself from the core β-sandwich domain and inserts into the membrane lipid bilayer resulting in the formation of the transmembrane pore (Rojko et al. 2016). Thermostable direct hemolysin (TDH) of *Vibrio parahaemolyticus* is the only known bacterial PFT that appears to display structural similarity to the actinoporin α-PFTs (Kundu et al. 2017; Yanagihara et al. 2010). However, it still remains unknown whether TDH indeed acts as an α-PFT.

19.2.2 β-PFTs

The β-PFT structures are generally abundant in β-strands, and the so-called 'pre-stem' motif remains packed against the core structure in the soluble state of the toxin. Upon membrane binding, the 'pre-stem' regions of the β-PFT protomers readjust themselves into antiparallel β-hairpins and insert into the membranes in a concerted manner, resulting in the formation of the amphipathic β-barrel pores (Montoya and Gouaux 2003; Heuck et al. 2001).

Staphylococcus hemolysins (α-hemolysin, γ-hemolysin, leucocidin AB, leucocidin CD Panton-Valentine leucocidin) (DuMont and Torres 2014), aerolysin from *Aeromonas hydrophila* (Podobnik et al. 2017), δ-toxins from *Clostridium perfringens* (Huyet et al. 2013), and *Vibrio cholerae* cytolysin (VCC) (De and Olson 2011) are some of the prominent members in the β-PFT family that generate small β-barrel oligomeric pores in the target membranes.

Cholesterol-dependent cytolysins (CDCs) represent a distinct subclass within the β-PFT family that form large oligomeric pore complexes (Reboul et al. 2016; Tweten et al. 2015). As the name suggests, cholesterol appears to play an obligatory role in the mode of action of these β-PFTs. CDC oligomers are composed of around 30 to 50 protomers, where each monomer contributes two β-hairpins to form the β-barrel pore. CDCs are mostly secreted by the Gram-positive bacteria; some of the prominent examples are perfringolysin O (PFO) of *Clostridium perfringens* (Johnson and Heuck 2014), listeriolysin O (LLO) from *Listeria monocytogenes* (Koster et al. 2014), intermedilysin (ILY) from *Streptococcus intermedius* (Polekhina et al. 2005), and streptolysin O (SLO) from *Streptococcus pyogenes* (Feil et al. 2014). CDCs have also been shown to possess the structural fold similar to those found in the membrane attack complexes and the perforins present in the mammalian immune system (Reboul et al. 2016).

19.3 Membrane Binding Mechanism of the PFTs

In order to exert its pore-forming activity, PFT first interacts with the target cell membranes (Gilbert et al. 2014). Membrane binding appears to increase the local concentration of the toxins onto the membrane surface which, in turn, facilitates the self-assembly of the PFT protomers, presumably due to restricted diffusion of the toxin molecules on the two-dimensional platform of the membrane surface. Binding of the PFTs may take place via interactions with the membrane lipids, as well as specific cell surface receptors, or receptor-like molecules (Geny and Popoff 2006). Analysis of the membrane interaction processes employed by the PFTs also highlights diverse structural mechanisms, where specific structural motifs within the PFT molecular architecture appear to play critical roles towards recognition of the specific cell surface molecules. Altogether, an efficient membrane interaction appears to act as the critical initial event to drive the subsequent steps of the membrane pore-formation mechanism. Interaction of the PFTs with the cell surface receptors may also trigger an array of signalling cascades, leading to the generation of a variety of cellular responses (Bischofberger et al. 2012).

Specific receptors for certain PFTs have been identified. ADAM10, a disintegrin and metalloprotease 10, appears to be the proteinaceous cellular receptor that enables high-affinity binding as well as execution of the cytotoxic activity of *Staphylococcus aureus* α-hemolysin (Wilke and Bubeck Wardenburg 2010). Intermedilysin, a pore-forming toxin from *Streptococcus intermedius*, has been shown to bind to the human cells via specific interaction with CD59, a GPI-anchored protein (Giddings et al. 2004). Cry cytotoxin produced by *Bacillus thuringiensis* acts as a biological insecticide by generating pores in the midgut epithelial cell membranes of the insects. The receptors involved in the binding of the Cry toxins include aminopeptidase N and cadherin-like proteins. Aerolysin, a β-barrel PFT secreted by the Aeromonas species, has been shown to interact with the GPI-anchored proteins (Diep et al. 1998; Hong et al. 2002).

Vibrio cholerae cytolysin (VCC), a β-barrel PFT secreted by the cholera pathogen *Vibrio cholerae*, exhibits complex multiple modes of interactions with the target cells. VCC appears to associate nonspecifically with the membrane lipid bilayer via amphipathicity-driven partitioning (Chattopadhyay et al. 2002), whereas specific interactions of the toxin with the membrane phospholipid head groups are mediated via specific structural motifs within the toxin (Rai and Chattopadhyay 2015a). In addition, VCC harbours two lectin-like domains, namely, the β-trefoil and the β-prism domains that appear to mediate interaction with the cell surface glycoconjugates (Saha and Banerjee 1997). In particular, the β-prism domain has been shown to augment binding of VCC towards the cell membranes, via specific recognition of the cell surface glycans (Rai et al. 2013; Levan et al.

2013). An earlier study has also indicated that the glycophorin B, a sialoglycoprotein, may serve as a receptor for VCC (Zhang et al. 1999). In addition to such interaction events, pore-forming activity of VCC has also been found to be regulated by the presence of specific membrane lipids, like cholesterol and sphingolipids (Zitzer et al. 2003; Paul and Chattopadhyay 2012; Zitzer et al. 2001). Exact roles of such lipids in the membrane interaction mechanism of VCC remain unclear.

A separate subclass of β-PFTs, produced by the numerous Gram-positive bacteria, is designated as the cholesterol-dependent cytolysins (CDCs) (Heuck et al. 2010). CDCs exhibit obligatory requirement of cholesterol in the target membranes to display their cytolytic activity. In most of the cases, cholesterol has been shown as the receptor for the CDCs on the target membranes. CDCs have been shown to employ specific structural motifs to recognize and bind to the membrane cholesterol (Farrand et al. 2010). In some cases, an accessory receptor has also been implicated for the CDCs. For example, human CD59 has been shown to act as an assisting molecule for the activity of the vaginolysin, a CDC that is produced by the *Gardnerella vaginalis* (Zilnyte et al. 2015).

19.4 Membrane Pore-Formation Mechanism of the PFTs

One of the most remarkable features of the PFT mode of action is their conversion from the water-soluble state to the transmembrane form. Mechanistic details of such structural/organizational changes in the PFTs during the membrane pore-formation processes still remain enigmatic. One of the most intriguing phenomena in the pore-formation mechanism of the PFTs is the formation of multimeric protein complexes on the membrane and membrane insertion of the specific pore-forming regions to generate the transmembrane pore. In many PFTs, oligomerization and insertion events are coupled and presumably synchronous. The unique characteristic of the membrane-inserted pore is that the hydrophobic residues within the pore-forming region face towards the membrane lipids, whereas hydrophilic residues line the interior lumen of the water-filled pore (Gouaux 1997). Based on the structures of the membrane-spanning regions, striking differences can be observed in the membrane insertion mechanisms of the PFTs (Iacovache et al. 2010). For α-PFTs, α-helices are used to punch holes on the membrane, whereas for the β-PFTs, β-strands of the monomeric units join hand to form a barrel-like structure that goes inside the membrane. β-PFTs can be further classified as small β-PFT or large β-PFT, based on the size of the pore they make on the membrane. *Staphylococcus aureus* α-hemolysin, aerolysin, and VCC are some of the prominent examples of the small β-PFTs that form heptameric channels, whereas CDCs make use of 30 to 50 monomeric subunits for oligomerization (Dal Peraro and van der Goot 2016).

In order to generate the transmembrane pores, a designated structural motif (pore-forming region) of the PFT must be able to insert into the core of the membrane lipid bilayer. In the water-soluble state of the protein, such a region generally remains masked from the hydrophilic aqueous environment. During the pore-formation process, this designated transmembrane region opens up and inserts into the membrane, which appears to be facilitated by partial unfolding of the protein at the tertiary structure level, upon membrane binding. This presumably generates a conformation that is highly unstable, as the exposed hydrophobic segment(s) come in contact with the aqueous environment. Thus, the interaction of the newly exposed hydrophobic regions of the transmembrane segment with the lipid membrane appears to be a spontaneous process, at least from the thermodynamic point of view. This initial process would trigger and/or guide the further steps of the membrane insertion. It is needless to say that the successful insertion step may not necessarily ensure functional pore formation. The rearrangements/reorganization of interactions within the inserted conformation is also necessary for maintaining the 'open pore' in the membranes.

19.4.1 Colicins

The structural motif crucial for membrane pore formation by colicins is composed of a bundle of α-helices. Colicin structure undergoes partial unfolding and adopts a 'molten globule' conformation, which is considered to be crucial for the subsequent membrane insertion of the pore-forming region (Wiener et al. 1997; van der Goot et al. 1991). It has been suggested that the factors triggering this unfolding could be local low pH micro-environment at the membrane interface, increased temperature, or increased negatively charged entities on the membrane. The exact stoichiometry of colicin pores remains unknown. It is proposed in an earlier study that upon insertion into the membrane, the α-helices move away from the hydrophobic core of the protein to generate an 'umbrella-like' structure, which then generates multimers of dimers or trimers to form a functional pore (Lakey et al. 1991).

19.4.2 Cytolysin A Family

The cytolysin A family of α-PFTs includes several toxins such as cytolysin A (ClyA) from *E. coli*, *S. flexneri*, and *S. enterica*, hemolysin BL (Hbl) from *B. cereus*, and non-haemolytic tripartite enterotoxin (Nhe) from *B. cereus* (Hunt et al. 2010). The structural studies on ClyA toxin show a unique pore-formation mechanism, in which a β-hairpin first interacts with the cholesterol-rich membranes; then, the distal N-terminal α-helix inserts into the membrane. Subsequently, 12 monomers come together for oligomerization and give rise to a ring-like helical barrel pore (Mueller et al. 2009; Wallace et al. 2000). Other members of the family also follow the similar pore-formation mechanism; however, the number of protomers in the final pore differs. Another unconventional way of pore formation is suggested in ClyA toxin, in which it forms pre-pore oligomers in the outer-membrane vesicles (OMVs) of the *E.coli*, and then encounters cholesterol in the target cell membranes, which finally leads to functional pore formation (Fahie et al. 2013).

19.4.3 Membrane Insertion of β-PFTs

In case of the β-PFTs, water-soluble monomers bind to the target membranes and undergo oligomerization, mostly in stoichiometric fashion, that in turn triggers/facilitates membrane insertion of the pore-forming region to create β-barrel pores (Dal Peraro and van der Goot 2016). Oligomeric transmembrane pore formation by the β-PFTs is complicated but regulated and well-orchestrated process. The factors and/or the mechanisms triggering the structural/conformational changes in the β-PFTs to form the transmembrane oligomeric pores are still unclear to a major extent. The overall understanding of the membrane insertion processes by the β-PFTs is that the β-strand pairs from the protomers of the oligomeric assembly are inserted into the membrane so that the hydrophobic residues of the membrane-spanning region are arranged outwardly from the pore lining, thereby facing and interacting with the hydrophobic fatty acyl tails of the membrane lipids. On the other hand, polar residues remain exposed towards the interior of the water-filled pore lumen, such that the passage of the polar entities through the pore is facilitated.

Leukocidins, α-haemolysin, and γ-haemolysin from the opportunistic pathogen *S. aureus*, *Vibrio cholerae* cytolysin (VCC), and δ-toxin from *C. perfringens* are some of the prominent β-PFTs that are structurally well-characterized (Dal Peraro and van der Goot 2016). Structural data have provided insights regarding their oligomerization and membrane insertion mechanisms. In general, these toxins are secreted as water-soluble monomers, and upon interacting with the membranes, they first oligomerize on the membrane surface to generate the transient, metastable pre-pore oligomeric assembly. After the oligomerization step, the so-called 'pre-stem' motif(s) (composed of a β-strand pair) from each of the protomers undergo conformational rearrangement and insert into the membrane to form the transmembrane β-barrel pore. Structural analyses have shown heptameric/octameric stoichiometric assembly of oligomerization for these types of β-PFTs. Notably, the membrane insertion step appears to be tightly correlated with the oligomerization process; however, the exact mechanism that regulates the membrane insertion step, upon formation of the 'pre-pore', remains elusive in most of the cases.

Placement of the pore-forming, amphipathic stem region into the hydrophobic interior of the membrane lipid bilayer would be thermodynamically favourable; however, traversing the long stretches of amphipathic β-strand pairs and/or synergistic insertion of the β-barrel into the membrane bilayer in the course of the pore formation would be expected to face structural/energetic constraints. Penetrating the amphipathic architecture of the membrane lipid bilayer (hydrophobic interior, with polar outer surface) by such structural motifs would require 'making and breaking' of interactions between the side chains of the 'would be' pore-forming region and membrane lipid components. It is possible that concomitant changes in the membrane lipid organization are also triggered during the pore-formation process so as to facilitate the membrane insertion step. However, such issues still remain elusive, not only in the context of the β-PFT pore-formation mechanism but also in the context of the PFT mode of actions in general.

Membrane pore-formation mechanisms of the CDC-type β-PFTs are distinct from those of the small pore-forming β-PFTs. In the water-soluble form of the CDCs, two short α-helical regions remain separated by the β-sandwich core. Upon membrane interaction and oligomerization, the α-helices show drastic conformational transition to form β-hairpins that finally insert into the membrane. Each protomer contributes two β-hairpins in the pore architecture as compared to the single β-strand pairs in the cases of the small pore-forming β-PFTs. These two β-hairpins contribute to the transmembrane β-barrel formation depending on the stoichiometry of the final pore. Structural studies have suggested that 30 to 50 monomers oligomerize by sequential addition of the protomers into the extending multimeric assembly. Conversion of the 'pre-pore' to the 'pore' state requires rotation of the core domain and tilting of the β-strands with respect to the membrane, a mechanism that is very similar to the pore-formation process of the membrane attack complex component and perforins (Reboul et al. 2016).

19.4.4 Cellular Responses against the PFTs

Membrane pore formation by the bacterial PFTs results in the breaching of the permeability barrier function of the target cell membranes that appears to be fatal in most of the cases. Exposures to PFTs often induce distinct types of cell death responses in the target cells that include apoptotic, necrotic, or pyroptotic cell death (Nelson et al. 1999; Kennedy et al. 2009). However, cells may generate multiple responses to protect themselves against the harassment caused by the PFTs as well. For example, cells may evoke pathways to repair the membrane damage upon pore formation by the PFTs (Andrews et al. 2014). Formation of the membrane blebs and outward vesiculation are also shown to be triggered due to the membrane-damaging activities of the PFTs (Babiychuk et al. 2011). Another possible mechanism to remove the PFT pores from the plasma membranes would be via endocytosis. The endocytosed pores may then be purged out into the extracellular milieu via an exosomal pathway (Idone et al. 2008; Husmann et al. 2009). Activation of distinct cellular signalling pathways in response to the bacterial PFTs has also been observed in several cases (Khilwani and Chattopadhyay 2015; Cassidy and O'Riordan 2013). Exposure to the bacterial PFTs has been shown to trigger production of pro-inflammatory and inflammatory molecules via activation of the NF-κB pathway in the target cells (Dragneva et al. 2001; Khilwani et al. 2015). PFTs also lead to modulation of the MAP kinase signal (Aguilar et al. 2009; Wiles et al. 2008). Autophagic responses are also triggered in the target cells in response to PFTs, a process necessary to interrupt the cytotoxic effects of the toxin (Gutierrez et al. 2007). The nature and the extent of such cellular responses evoked against the PFTs depend on the specific PFTs and the cell types under attack, as well as the extent of the PFT exposure.

19.5 Conclusions

PFTs and their membrane pore-formation mechanisms have been studied over past several decades. Results obtained from extensive studies provide critical insights regarding the generalized mode of actions of diverse types of PFTs including those of bacterial origin. Vast ranges of studies also highlight that, apart from the membrane-damaging direct cell-killing activity, PFTs can also elicit a plethora of cellular responses. Based on their molecular architectures, PFTs are classified into distinct structural subfamilies. At the same time, distinct structural features are acquired even within the specific PFT subfamily, so as to meet appropriate diverse functional requirements. It is also observed that PFTs from distinct subfamilies sometime display ability to mediate similar functional consequences. It still remains an enigma, how the structure-function relationships have evolved in the distinct types of protein toxins in the PFT family. More studies are required to explore the implications of the similar/distinct features in the PFT sequences and structures to understand such a notion. Structural basis of membrane pore formation by the PFTs is another issue that still remains partly understood. In depth studies, not only investigating the structural descriptions of the membrane pore complexes but also exploring the dynamics of the membrane pore-formation process would be required to address some of the unanswered questions relevant to the PFT mode of actions. Finally, it would also be important to gain more comprehensive and detailed insights regarding the diverse cellular responses generated in the target host cells against various PFTs. Such efforts are essential in understanding the implications of the PFTs for the bacterial pathogenesis as well as host-pathogen interaction processes.

Acknowledgements We acknowledge the support through funding from the Department of Biotechnology (DBT), India [DBT Grant No. BT/PR12141/BRB/10/1343/2014; DBT Grant No. BT/HRD/NBA/37/01/2014 (x)], and also through funding under the Centre of Excellence (COE) in Frontier Areas of Science and Technology (FAST) programme of the Ministry of Human Resource Development, Government of India, in the area of protein science, design, and engineering. We also thank the Indian Institute of Science Education and Research (IISER), Mohali, for the support.

References

Aguilar JL, Kulkarni R, Randis TM, Soman S, Kikuchi A, Yin Y, Ratner AJ (2009) Phosphatase-dependent regulation of epithelial mitogen-activated protein kinase responses to toxin-induced membrane pores. PLoS One 4(11):e8076. https://doi.org/10.1371/journal.pone.0008076

Andrews NW, Almeida PE, Corrotte M (2014) Damage control: cellular mechanisms of plasma membrane repair. Trends Cell Biol 24(12):734–742. https://doi.org/10.1016/j.tcb.2014.07.008

Babiychuk EB, Monastyrskaya K, Potez S, Draeger A (2011) Blebbing confers resistance against cell lysis. Cell Death Differ 18(1):80–89. https://doi.org/10.1038/cdd.2010.81

Bischofberger M, Iacovache I, van der Goot FG (2012) Pathogenic pore-forming proteins: function and host response. Cell Host Microbe 12(3):266–275. https://doi.org/10.1016/j.chom.2012.08.005

Cassidy SK, O'Riordan MX (2013) More than a pore: the cellular response to cholesterol-dependent cytolysins. Toxins 5(4):618–636. https://doi.org/10.3390/toxins5040618

Chattopadhyay K, Bhattacharyya D, Banerjee KK (2002) Vibrio cholerae hemolysin. Implication of amphiphilicity and lipid-induced conformational change for its pore-forming activity. Eur J Biochem 269(17):4351–4358

Collier RJ (1975) Diphtheria toxin: mode of action and structure. Bacteriol Rev 39(1):54–85

Dal Peraro M, van der Goot FG (2016) Pore-forming toxins: ancient, but never really out of fashion. Nat Rev Microbiol 14(2):77–92. https://doi.org/10.1038/nrmicro.2015.3

De Haan L, Hirst TR (2004) Cholera toxin: a paradigm for multi-functional engagement of cellular mechanisms (review). Mol Membr Biol 21(2):77–92. https://doi.org/10.1080/09687680410001663267

De S, Olson R (2011) Crystal structure of the Vibrio cholerae cytolysin heptamer reveals common features among disparate pore-forming toxins. Proc Natl Acad Sci U S A 108(18):7385–7390. https://doi.org/10.1073/pnas.1017442108

Diep DB, Nelson KL, Raja SM, Pleshak EN, Buckley JT (1998) Glycosylphosphatidylinositol anchors of membrane glycoproteins are binding determinants for the channel-forming toxin aerolysin. J Biol Chem 273 (4):2355–2360

Dragneva Y, Anuradha CD, Valeva A, Hoffmann A, Bhakdi S, Husmann M (2001) Subcytocidal attack by staphylococcal alpha-toxin activates NF-kappaB and induces interleukin-8 production. Infect Immun 69 (4):2630–2635. https://doi.org/10.1128/IAI.69.4.2630-2635.2001

DuMont AL, Torres VJ (2014) Cell targeting by the Staphylococcus aureus pore-forming toxins: it's not just about lipids. Trends Microbiol 22(1):21–27. https://doi.org/10.1016/j.tim.2013.10.004

Fahie M, Romano FB, Chisholm C, Heuck AP, Zbinden M, Chen M (2013) A non-classical assembly pathway of Escherichia coli pore-forming toxin cytolysin a. J Biol Chem 288(43):31042–31051. https://doi.org/10.1074/jbc.M113.475350

Farrand AJ, LaChapelle S, Hotze EM, Johnson AE, Tweten RK (2010) Only two amino acids are essential for cytolytic toxin recognition of cholesterol at the membrane surface. Proc Natl Acad Sci U S A 107 (9):4341–4346. https://doi.org/10.1073/pnas.0911581107

Feil SC, Ascher DB, Kuiper MJ, Tweten RK, Parker MW (2014) Structural studies of Streptococcus pyogenes streptolysin O provide insights into the early steps of membrane penetration. J Mol Biol 426(4):785–792. https://doi.org/10.1016/j.jmb.2013.11.020

Geny B, Popoff MR (2006) Bacterial protein toxins and lipids: pore formation or toxin entry into cells. Biol Cell 98(11):667–678. https://doi.org/10.1042/BC20050082

Giddings KS, Zhao J, Sims PJ, Tweten RK (2004) Human CD59 is a receptor for the cholesterol-dependent cytolysin intermedilysin. Nat Struct Mol Biol 11 (12):1173–1178. https://doi.org/10.1038/nsmb862

Gilbert RJ, Dalla Serra M, Froelich CJ, Wallace MI, Anderluh G (2014) Membrane pore formation at protein-lipid interfaces. Trends Biochem Sci 39 (11):510–516. https://doi.org/10.1016/j.tibs.2014.09.002

Gouaux E (1997) Channel-forming toxins: tales of transformation. Curr Opin Struct Biol 7(4):566–573

Gutierrez MG, Saka HA, Chinen I, Zoppino FC, Yoshimori T, Bocco JL, Colombo MI (2007) Protective role of autophagy against Vibrio cholerae cytolysin, a pore-forming toxin from V. cholerae. Proc Natl Acad Sci U S A 104(6):1829–1834. https://doi.org/10.1073/pnas.0601437104

Heuck AP, Moe PC, Johnson BB (2010) The cholesterol-dependent cytolysin family of gram-positive bacterial toxins. Subcell Biochem 51:551–577. https://doi.org/10.1007/978-90-481-8622-8_20

Heuck AP, Tweten RK, Johnson AE (2001) Beta-barrel pore-forming toxins: intriguing dimorphic proteins. Biochemistry 40(31):9065–9073

Hong Y, Ohishi K, Inoue N, Kang JY, Shime H, Horiguchi Y, van der Goot FG, Sugimoto N, Kinoshita T (2002) Requirement of N-glycan on GPI-anchored proteins for efficient binding of aerolysin but not Clostridium septicum alpha-toxin. EMBO J 21 (19):5047–5056

Hunt S, Green J, Artymiuk PJ (2010) Hemolysin E (HlyE, ClyA, SheA) and related toxins. Adv Exp Med Biol 677:116–126

Husmann M, Beckmann E, Boller K, Kloft N, Tenzer S, Bobkiewicz W, Neukirch C, Bayley H, Bhakdi S (2009) Elimination of a bacterial pore-forming toxin by sequential endocytosis and exocytosis. FEBS Lett 583(2):337–344. https://doi.org/10.1016/j.febslet.2008.12.028

Huyet J, Naylor CE, Savva CG, Gibert M, Popoff MR, Basak AK (2013) Structural insights into Clostridium perfringens delta toxin pore formation. PLoS One 8(6): e66673. https://doi.org/10.1371/journal.pone.0066673

Iacovache I, Bischofberger M, van der Goot FG (2010) Structure and assembly of pore-forming proteins. Curr Opin Struct Biol 20(2):241–246. https://doi.org/10.1016/j.sbi.2010.01.013

Idone V, Tam C, Goss JW, Toomre D, Pypaert M, Andrews NW (2008) Repair of injured plasma membrane by rapid Ca2+−dependent endocytosis. J Cell Biol 180(5):905–914. https://doi.org/10.1083/jcb.200708010

Johnson BB, Heuck AP (2014) Perfringolysin O structure and mechanism of pore formation as a paradigm for cholesterol-dependent cytolysins. Subcell Biochem 80:63–81. https://doi.org/10.1007/978-94-017-8881-6_5

Kennedy CL, Smith DJ, Lyras D, Chakravorty A, Rood JI (2009) Programmed cellular necrosis mediated by the pore-forming alpha-toxin from Clostridium septicum. PLoS Pathog 5(7):e1000516. https://doi.org/10.1371/journal.ppat.1000516

Khilwani B, Chattopadhyay K (2015) Signaling beyond punching holes: modulation of cellular responses by Vibrio cholerae Cytolysin. Toxins 7(8):3344–3358. https://doi.org/10.3390/toxins7083344

Khilwani B, Mukhopadhaya A, Chattopadhyay K (2015) Transmembrane oligomeric form of Vibrio cholerae cytolysin triggers TLR2/TLR6-dependent proinflammatory responses in monocytes and macrophages. Biochem J 466(1):147–161. https://doi.org/10.1042/BJ20140718

Koster S, van Pee K, Hudel M, Leustik M, Rhinow D, Kuhlbrandt W, Chakraborty T, Yildiz O (2014) Crystal structure of listeriolysin O reveals molecular details of oligomerization and pore formation. Nat Commun 5:3690. https://doi.org/10.1038/ncomms4690

Kundu N, Tichkule S, Pandit SB, Chattopadhyay K (2017) Disulphide bond restrains the C-terminal region of thermostable direct hemolysin during folding to promote oligomerization. Biochem J 474(2):317–331. https://doi.org/10.1042/BCJ20160728

Lakey JH, Massotte D, Heitz F, Dasseux JL, Faucon JF, Parker MW, Pattus F (1991) Membrane insertion of the pore-forming domain of colicin A. A spectroscopic study. Eur J Biochem 196(3):599–607
Lakey JH, Slatin SL (2001) Pore-forming colicins and their relatives. Curr Top Microbiol Immunol 257:131–161
Lesieur C, Vecsey-Semjen B, Abrami L, Fivaz M, Gisou van der Goot F (1997) Membrane insertion: the strategies of toxins (review). Mol Membr Biol 14 (2):45–64
Levan S, De S, Olson R (2013) Vibrio cholerae cytolysin recognizes the heptasaccharide core of complex N-glycans with nanomolar affinity. J Mol Biol 425 (5):944–957. https://doi.org/10.1016/j.jmb.2012.12.016
Montoya M, Gouaux E (2003) Beta-barrel membrane protein folding and structure viewed through the lens of alpha-hemolysin. Biochim Biophys Acta 1609 (1):19–27
Morgan BP, Boyd C, Bubeck D (2017) Molecular cell biology of complement membrane attack. Semin Cell Dev Biol 72:124–132. https://doi.org/10.1016/j.semcdb.2017.06.009
Mueller M, Grauschopf U, Maier T, Glockshuber R, Ban N (2009) The structure of a cytolytic alpha-helical toxin pore reveals its assembly mechanism. Nature 459(7247):726–730. https://doi.org/10.1038/nature08026
Nelson KL, Brodsky RA, Buckley JT (1999) Channels formed by subnanomolar concentrations of the toxin aerolysin trigger apoptosis of T lymphomas. Cell Microbiol 1(1):69–74
Oh KJ, Senzel L, Collier RJ, Finkelstein A (1999) Translocation of the catalytic domain of diphtheria toxin across planar phospholipid bilayers by its own T domain. Proc Natl Acad Sci U S A 96(15):8467–8470
Parker MW, Pattus F, Tucker AD, Tsernoglou D (1989) Structure of the membrane-pore-forming fragment of colicin A. Nature 337(6202):93–96. https://doi.org/10.1038/337093a0
Paul K, Chattopadhyay K (2012) Single point mutation in Vibrio cholerae cytolysin compromises the membrane pore-formation mechanism of the toxin. FEBS J 279 (21):4039–4051. https://doi.org/10.1111/j.1742-4658.2012.08809.x
Podobnik M, Kisovec M, Anderluh G (2017) Molecular mechanism of pore formation by aerolysin-like proteins. Philos Trans R Soc Lond Ser B Biol Sci 372 (1726):1. https://doi.org/10.1098/rstb.2016.0209
Polekhina G, Giddings KS, Tweten RK, Parker MW (2005) Insights into the action of the superfamily of cholesterol-dependent cytolysins from studies of intermedilysin. Proc Natl Acad Sci U S A 102 (3):600–605. https://doi.org/10.1073/pnas.0403229101
Rai AK, Chattopadhyay K (2015a) Revisiting the membrane interaction mechanism of a membrane-damaging beta-barrel pore-forming toxin Vibrio cholerae cytolysin. Mol Microbiol 97(6):1051–1062. https://doi.org/10.1111/mmi.13084
Rai AK, Chattopadhyay K (2015b) Vibrio cholerae cytolysin: structure-function mechanism of an atypical beta-barrel pore-forming toxin. Adv Exp Med Biol 842:109–125. https://doi.org/10.1007/978-3-319-11280-0_7
Rai AK, Paul K, Chattopadhyay K (2013) Functional mapping of the lectin activity site on the beta-prism domain of vibrio cholerae cytolysin: implications for the membrane pore-formation mechanism of the toxin. J Biol Chem 288(3):1665–1673. https://doi.org/10.1074/jbc.M112.430181
Reboul CF, Whisstock JC, Dunstone MA (2016) Giant MACPF/CDC pore forming toxins: a class of their own. Biochim Biophys Acta 1858(3):475–486. https://doi.org/10.1016/j.bbamem.2015.11.017
Ridley H, Johnson CL, Lakey JH (2010) Interfacial interactions of pore-forming colicins. Adv Exp Med Biol 677:81–90
Rojko N, Dalla Serra M, Macek P, Anderluh G (2016) Pore formation by actinoporins, cytolysins from sea anemones. Biochim Biophys Acta 1858(3):446–456. https://doi.org/10.1016/j.bbamem.2015.09.007
Saha N, Banerjee KK (1997) Carbohydrate-mediated regulation of interaction of Vibrio cholerae hemolysin with erythrocyte and phospholipid vesicle. J Biol Chem 272(1):162–167
Tweten RK, Hotze EM, Wade KR (2015) The unique molecular choreography of giant pore formation by the cholesterol-dependent cytolysins of Gram-positive bacteria. Annu Rev Microbiol 69:323–340. https://doi.org/10.1146/annurev-micro-091014-104233
van der Goot FG, Gonzalez-Manas JM, Lakey JH, Pattus F (1991) A ‘molten-globule’ membrane-insertion intermediate of the pore-forming domain of colicin A. Nature 354(6352):408–410. https://doi.org/10.1038/354408a0
Wallace AJ, Stillman TJ, Atkins A, Jamieson SJ, Bullough PA, Green J, Artymiuk PJ (2000) E. coli hemolysin E (HlyE, ClyA, SheA): X-ray crystal structure of the toxin and observation of membrane pores by electron microscopy. Cell 100(2):265–276
Westphal D, Dewson G, Czabotar PE, Kluck RM (2011) Molecular biology of Bax and Bak activation and action. Biochim Biophys Acta 1813(4):521–531. https://doi.org/10.1016/j.bbamcr.2010.12.019
Wiener M, Freymann D, Ghosh P, Stroud RM (1997) Crystal structure of colicin Ia. Nature 385 (6615):461–464. https://doi.org/10.1038/385461a0
Wiles TJ, Dhakal BK, Eto DS, Mulvey MA (2008) Inactivation of host Akt/protein kinase B signaling by bacterial pore-forming toxins. Mol Biol Cell 19 (4):1427–1438. https://doi.org/10.1091/mbc.E07-07-0638
Wilke GA, Bubeck Wardenburg J (2010) Role of a disintegrin and metalloprotease 10 in Staphylococcus aureus alpha-hemolysin-mediated cellular injury. Proc

Natl Acad Sci U S A 107(30):13473–13478. https://doi.org/10.1073/pnas.1001815107

Xu C, Wang BC, Yu Z, Sun M (2014) Structural insights into Bacillus thuringiensis Cry, Cyt and parasporin toxins. Toxins 6(9):2732–2770. https://doi.org/10.3390/toxins6092732

Yanagihara I, Nakahira K, Yamane T, Kaieda S, Mayanagi K, Hamada D, Fukui T, Ohnishi K, Kajiyama S, Shimizu T, Sato M, Ikegami T, Ikeguchi M, Honda T, Hashimoto H (2010) Structure and functional characterization of Vibrio parahaemolyticus thermostable direct hemolysin. J Biol Chem 285(21):16267–16274. https://doi.org/10.1074/jbc.M109.074526

Zakharov SD, Cramer WA (2002) Colicin crystal structures: pathways and mechanisms for colicin insertion into membranes. Biochim Biophys Acta 1565(2):333–346

Zhang D, Takahashi J, Seno T, Tani Y, Honda T (1999) Analysis of receptor for Vibrio cholerae El tor hemolysin with a monoclonal antibody that recognizes glycophorin B of human erythrocyte membrane. Infect Immun 67(10):5332–5337

Zilnyte M, Venclovas C, Zvirbliene A, Pleckaityte M (2015) The cytolytic activity of vaginolysin strictly depends on cholesterol and is potentiated by human CD59. Toxins 7(1):110–128. https://doi.org/10.3390/toxins7010110

Zitzer A, Bittman R, Verbicky CA, Erukulla RK, Bhakdi S, Weis S, Valeva A, Palmer M (2001) Coupling of cholesterol and cone-shaped lipids in bilayers augments membrane permeabilization by the cholesterol-specific toxins streptolysin O and Vibrio cholerae cytolysin. J Biol Chem 276(18):14628–14633. https://doi.org/10.1074/jbc.M100241200. M100241200 [pii]

Zitzer A, Westover EJ, Covey DF, Palmer M (2003) Differential interaction of the two cholesterol-dependent, membrane-damaging toxins, streptolysin O and Vibrio cholerae cytolysin, with enantiomeric cholesterol. FEBS Lett 553(3):229–231

Abiraterone and Ionizing Radiation Alter the Sphingolipid Homeostasis in Prostate Cancer Cells

20

Valentina Murdica, Giulia Mancini, Nicoletta Loberto, Rosaria Bassi, Paola Giussani, Nadia Di Muzio, Chiara Deantoni, Alessandro Prinetti, Massimo Aureli, and Sandro Sonnino

Abstract

Prostate cancer (PC) is one of the most common leading causes of cancer-related death in men. Currently, the main therapeutic approaches available for PC are based on the androgen deprivation and on radiotherapy. However, despite these treatments being initially effective in cancer remission, several patients undergo recurrence, developing a most aggressive and resistant PC.

Emerging evidence showed that abiraterone acetate drug will reduce PC recurrence by a mechanism independent of the inhibition of Cytochrome P450 17α-hydroxylase/17,20-lyase. Here we describe the involvement in the abiraterone-mediated PC cell death of a particular class of bioactive lipids called sphingolipids (SL). Sphingolipids are components of plasma membrane (PM) that organize macromolecular complexes involved in the control of several signaling pathways including the tumor cell death induced by radiotherapy. Here, we show for the first time that both in androgen-sensitive and insensitive PC cells abiraterone and ionizing radiation induce a reorganization of the plasma membrane SL composition. This event is triggered by activation of the PM-associated glycohydrolases that induce the production of cytotoxic ceramide by the in situ hydrolyses of glycosphingolipids. Taken together our data open a new scenario on the SL involvement in the therapy of PC.

Keywords

Ionizing radiation · Abiraterone · Sphingolipids · Plasma membrane · Prostate cancer · Glycohydrolases · Sialidase Neu3

20.1 Introduction

Prostate cancer (PC) is one of the most commonly diagnosed malignant tumors and the second leading cause of cancer-related death in men (Jemal et al. 2011). At early stages, PC growth and development largely depend on androgens and androgen receptor (AR)-signaling pathways and respond to androgen deprivation therapy with surgical or chemical castration. Although the use of antiandrogen, such as GnRH agonists, or AR antagonists, such as bicalutamide, is initially effective in cancer remission, by these therapies, several patients undergo recurrence developing a most aggressive castration-resistant PC (CRPC) (Kolvenbag et al. 2001; Anderson 2003). Recent

V. Murdica · G. Mancini · N. Loberto · R. Bassi
P. Giussani · A. Prinetti · M. Aureli · S. Sonnino (✉)
Department of Medical Biotechnology and Translational Medicine, University of Milano, Milan, Italy
e-mail: sandro.sonnino@unimi.it

N. Di Muzio · C. Deantoni
Department of Radiotherapy, Hospital San Raffaele, Milan, Italy

K. Chattopadhyay, S. C. Basu (eds.), *Biochemical and Biophysical Roles of Cell Surface Molecules*, Advances in Experimental Medicine and Biology 1112, https://doi.org/10.1007/978-981-13-3065-0_20

studies have shown promising data on the use of abiraterone acetate in reducing PC recurrence. It is a potent inhibitor of Cytochrome P450 17-α-hydroxylase/17,20-lyase (CYP17A1), a key enzyme in the androgen biosynthetic pathway. Both preclinical and clinical studies have shown that the antitumoral activity of abiraterone in CRPC may be independent from its effect on CYP17A1 inhibition, even if the molecular mechanisms of its off-target effects are still unclear.

Radiation therapy represents another important therapeutic approach for both primary and metastatic PC. Recently, radical radiation therapy using various hypofractionation schemes has been developed for different types of cancer diseases. On the other hand, the development of conformal/intensity-modulated techniques, in combination with precise image-guided localization of the target, allows high dose per fraction without increasing the risk of serious late injury to the normal tissues. Unfortunately, despite great efforts in the development of more advanced linear accelerators and personalized radiotherapy protocols, the occurrence of radio resistance remains one of the main drawbacks to be addressed, especially in PC. In this context, it has been demonstrated that several sphingolipids (SLs) are involved in the regulation of cell death pathways in response to various stimuli, including radiation. Therefore, they are recognized as critical targets in cancer therapy, including PC (Hajj et al. 2013). SLs are amphiphilic molecules predominantly associated with the external leaflet of the plasma membrane (PM). They are composed by a hydrophobic moiety (ceramide) inserted into the lipid bilayer along with a hydrophilic group of a different complexity protruding toward the extracellular environment (Merrill 2011). It is currently known that SLs are not only structural components (Hakomori 1990) but, together with cholesterol, saturated phospholipids, and a selected pool of proteins, constitute macromolecular complexes within the PM called lipid rafts (Kaiser et al. 2009; Sonnino et al. 2015).

To date, several studies described membrane rafts as dynamic nanoscale domains playing an important role in cell signal transduction (Simons et al. 2011). Interestingly, modifications of lipid raft SL composition are responsible for the control of several cellular processes, including cell differentiation and death (Teichgraber et al. 2008; Kabayama et al. 2007; Prinetti et al. 2009).

In particular, ceramide-enriched rafts seem to play an important role in the mechanisms involved in radiation-induced cell death in different tumor models; thus, targeting SL metabolic pathways continues to be actively investigated to improve cancer therapies. Interestingly, ionizing radiation can induce the degradation of both PM sphingomyelin and glycosphingolipids to ceramide, which in turn leads to apoptosis (Aureli et al. 2014). Besides the activation of SMase isoforms, we found that ceramide is also formed through the action of specific sphingolipid hydrolases associated with the external site of the PM (Aureli et al. 2012b). We observed in different tumor cell lines exposed to ionizing radiation that the cell death was preceded by the activation of PM-associated β-galactosidase, sialidase Neu3, β-hexosaminidase, and β-glucosidases. In particular, in irradiated breast cancer cells, we demonstrated that the increase of PM-associated glycohydrolases is responsible for the ectopic in situ production of apoptotic ceramide (Aureli et al. 2012b).

In line with these findings, here we investigated the effects of abiraterone, alone or in combination with ionizing radiation, on cell growth and sphingolipid metabolism of both androgen-sensitive and androgen-insensitive prostate cancer cells; we found that abiraterone pretreatment increases the cell death induced by ionizing radiation. In addition, both single treatments are responsible for the activation of PM glycohydrolases which, in turn, leads to altered lipid raft SL pattern.

20.2 Methods

20.2.1 Cell Cultures

Human prostate cancer cell lines LNCaP (androgen-sensitive) and DU145 (androgen-insensitive) are generous gifts of the urological research unit of San Raffaele Hospital in Milan. Cells were cultured in RPMI-1640 supplemented with 10% FCS (heat-inactivated), 1% glutamine, 1% penicillin/streptomycin, 1% sodium pyruvate, and 1% HEPES at 37 °C in a humidified, 5% CO_2 atmosphere.

20.2.2 Abiraterone and Bicalutamide Treatments

LNCaP and DU145 cell lines were treated with abiraterone at different doses in order to establish the drug toxicity curve. In the case of combined treatment with abiraterone and ionizing radiation, cells were pre-treated with the drug for 24 h (10 and 25 μM abiraterone for LNCaP and DU145, respectively) and then subjected to ionizing radiation. 20 μM Bicalutamide was administered to both LNCaP and DU145 cell lines.

20.2.3 Cell Irradiation

The most widely used technique for mathematically modeling the effect of radiation on cancer cell survival is linear quadratic (LQ) formalism. The core formula involved is an expression correlating the probability of survival of a population of mammalian cells subjected to the exposure (S) to an individual radiation dose (D) as follows:

$$S = \exp[-\alpha D - \beta D2]$$

where α and β are tumor- or normal tissue-specific parameters. Rearrangement of the equation to account for the total impact of number of individual fractions of radiation exposure (n) yields another common expression for the biologically effective dose (BED):

$$\mathrm{BED} = nD(1 + d/[\alpha/\beta])$$

The units of BED are gray (Gy), the standard unit of radiation dose. The BED is a convenient metric to compare the relative impact of a given schedule of a radiation dose on a given tumor or tissue, as long as the α/β ratio is known or closely estimated.

LNCaP and DU145 cell lines were irradiated with 6 MeV photons (direct field) with a linear accelerator, at a single dose of 9 Gy.

20.2.4 Evaluation of Cell Viability

At different time points after irradiation, abiraterone treatment, or a combination of both, cell viability was assessed by calcein staining and trypan blue exclusion assay (Aureli et al. 2012b). Briefly, for calcein staining cells were washed with PBS and then the plates were put on ice. 50 μl of a calcein-AM solution (6.25 μg/ml in PBS) were added to each well, and cells were incubated for 15 minutes at 37 °C, 5% CO_2. Calcein AM was then removed, and 100 μl of PBS and 50 μl of 1% Triton X-100 were added to each well. Plates were stirred at RT for 15 minutes in dark conditions, and then the fluorescence was detected by a microplate reader (Victor, Perkin-Elmer).

20.2.5 Plasma Membrane (PM)-Associated Glycohydrolase Assays

We recently described a simple high-throughput live cell-based assay that allows the determination of the activity of several PM glycohydrolases in living cells (Aureli et al. 2012a). PM-associated β-galactosidase (β-Gal), conduritol B epoxide (CBE)-sensitive β-glucosidase GBA1, β-glucosidase GBA2, β-hexosaminidase (β-Hex), and SMase activities were determined in control or treated living cells plated in a 96-well microplate. For GBA1 and GBA2 assays, cells were preincubated for 30 minutes at room temperature in DMEM-F12 containing 5 nM AMP-DNM (adamantane-pentyl-dNM; N-(5-adamantane-1-yl-methoxy-pentyl) deoxynojirimycin) or 1 mM CBE (Sigma), respectively. β-Gal, β-Hex, and β-Glc activities were assayed using the artificial substrates

4-methylumbelliferyl-β-D-galactopyranoside (MUB-Gal), 4-methylumbelliferil-β-N-acetylglucosaminide (MUG), and 4-methylumbelliferyl-β-D-glucopyranoside (MUB-Glc) solubilized in DMEM-F12 without phenol red at pH 6 at f 250 μM, 2 mM and 6 mM, respectively (final concentrations). At different time points (from 2 to 6 h), the quantity of 4-methylumbelliferone (MUB) produced was evaluated by a microplate reader (MUB: λex: 355 nm/λem: 460 nm) after adding 15 volumes of 0.25 M glycine pH 10.7. Standard MUB was used to establish the calibration curves in order to quantify the substrates hydrolysis and determine the enzyme activity.

20.2.6 Measurement of the PM-Associated Glycohydrolases Activity Using the Natural Radioactive Substrate [3-^{3}H](sphingosine)GM3

One day before the feeding, cells were plated in Ø 100 mm petri culture dishes at 10^4 cell/cm^2 as cell density. After 24 h, isotopically labeled [3-^{3}H](sphingosine)GM3 was solubilized in cell culture medium without FCS at the 4.5×10^{-6} M final concentration and administered to cells (Valaperta et al. 2006). Cells were first preincubated with 100 μM chloroquine for 1 h in cell medium without FCS. After medium removal and rapid washes with PBS, 6 ml of radioactive lipid-containing medium were added to each dish and followed by 4 h incubation at 37 °C in the presence of 100 μM chloroquine. At the end of pulse, cells were washed three times with complete cell culture medium and incubated in the same medium for 30 min. Then, cells were washed three times with PBS, harvested, and subjected to lipid analyses as described below.

20.2.7 Treatment of Cell Cultures with [1-^{3}H]sphingosine

[1-^{3}H]sphingosine was administered to pharmacologically treated and/or 9 Gy-irradiated cells as well as to control cells in order to metabolically label SLs at the steady state previously described (Prinetti et al. 2011). [1-^{3}H]sphingosine dissolved in methanol was transferred into a sterile glass tube and dried under a nitrogen stream; the residue was then solubilized in an appropriate volume of pre-warmed (37 °C) cell medium to obtain the desired final concentration (3×10^{-9} M). After 2 h of pulse time, the medium was removed, and cells were washed and incubated up to 72 h (chase) with fresh medium without the radioactive precursor. After chase, cells were collected, and radioactive lipids were analyzed as described below.

20.2.8 Preparation of SL-Enriched Membrane Domains by Sucrose-Gradient Centrifugation

Treated or control-cell SLs were metabolically labeled at the steady state as described above using tritiated sphingosine. Cells corresponding to 2.5 mg protein were lysed in 1.5 ml lysis buffer (1% Triton X-100, 10 mM Tris-HCl pH 7.5, 150 mM NaCl, 5 mM EDTA, 1 mM Na_3VO_4, 1 mM PMSF, and 10 μM/ml aprotinin) at 4 °C for 20 minutes and Dounce-homogenized (11 strokes tight pestle). After nuclei and cellular debris removal, the post-nuclear supernatant (PNS) was mixed with an equal volume of 85% sucrose (w/v) in 10 mM Tris-HCl buffer pH 7.5, 150 mM NaCl, 5 mM EDTA, and 1 mM Na_3VO_4 and placed at the bottom of a discontinuous sucrose concentration gradient (5–30%) in the same buffer. After ultracentrifugation for 17 h at 200,000 × g at 4 °C, 11 fractions were collected, and the radioactivity was evaluated by liquid scintillation counting. Then, the DRM fraction (5 and 6) and the high density fraction (HD) were dialyzed, lyophilized, and subjected to lipid extraction and SL analysis (Prinetti et al. 2011).

20.2.9 Radioactive Lipid Analyses

Total lipids from lyophilized cells or gradient fractions were extracted with chloroform/methanol/water 2:1:0.1 by volume, followed by a second extraction with chloroform/methanol 2:1 by volume. The total lipid extracts were subjected to a two-phase partitioning by adding 20% water to the lipid extract; the total lipid extract and the aqueous and organic phases were analyzed by HPTLC. [^{3}H]SLs of total extracts and organic phases were separated using the solvent system chloroform/methanol/water 110:40:6 by volume and those of aqueous phases with chloroform/methanol/0.2% aqueous $CaCl_2$ 50:42:11 by vol. For the specific determination of Cer content, radioactive lipids of the organic phases were separated using the solvent system hexane/chloroform/acetone/acetic acid 20:70:20:2 by volume. [^{3}H]SLs were identified by referring to the position of standards in the chromatogram and quantified by radioimaging after 48 h of acquisition (β-Imager 2000, Biospace, Paris, France) (Aureli et al. 2012b).

20.2.10 Statistics

All the experiments have been performed threefold in replicate and repeated three times. Data are presented as the mean values ± standard deviation and were tested for significance employing one- or two-way ANOVA with Bonferroni post-test analysis, as specified in figure legends. The level of significance was set at $p < 0.05$.

20.3 Results

20.3.1 Abiraterone Treatment Inhibits the Growth of Androgen-Dependent and Androgen-Independent Prostate Cancer Cell Lines

We first assessed if abiraterone exerts a direct cytostatic/cytotoxic effect on androgen-dependent (LNCaP) and androgen-independent (DU145) prostate cancer cell lines. To this purpose, both cell types were treated with increasing concentration of abiraterone (from 0.1 μM to 100 μM), and cell proliferation and viability were evaluated 24, 48, and 72 h after drug exposure. As shown in Fig. 20.1, we found that abiraterone affected the growth rate of both cell lines in the low micromolar range. As expected, LNCaP cells were more sensible to abiraterone treatment than DU145 cells as observed by the lower drug concentration as well the shorter incubation time to elicit 100% cell growth inhibition and 100% cell death. In particular, abiraterone exerted cytostatic effect at 10 μM after 24 h and at 25 μM after 48 h in LNCaP and DU145 cells, respectively, whereas 25 μM for 24 h and 100 μM for 72 h were the experimental conditions inducing the maximum cytotoxic effect in androgen-dependent and androgen-independent prostate cancer cells, respectively. These data confirm previous results revealing that abiraterone can also have a distinct activity from androgen axis (Bruno et al. 2008; Grossebrummel et al. 2016).

This finding supporting an off-target action for abiraterone was further corroborated by parallel experiments evaluating the effect of bicalutamide, a nonsteroidal competitive androgen receptor antagonist able to block androgen-regulated prostate cell growth. While the proliferation rate of the androgen-sensible LNCaP was markedly reduced after bicalutamide treatment (20 μM up to 72 h), the compound did not exert any growth inhibitory effect on DU145 cells that do not express a functional androgen receptor (data not shown) (Navone et al. 1998).

20.3.2 Abiraterone Treatment Sensitizes LNCaP and DU145 Cells to Ionizing Radiation

We next sought to investigate the possible sensitizing effect of abiraterone to ionizing radiation in the two cell types, both radioresistant prostate cancer cell lines (Chung et al. 2008). To this purpose, LNCaP and DU145 cells were pre-treated or not for 24 h with cytostatic concentrations of abiraterone followed by exposure to a single dose of ionizing radiation (9 Gy). Cell proliferation was then evaluated 72 h after irradiation in presence or in absence of

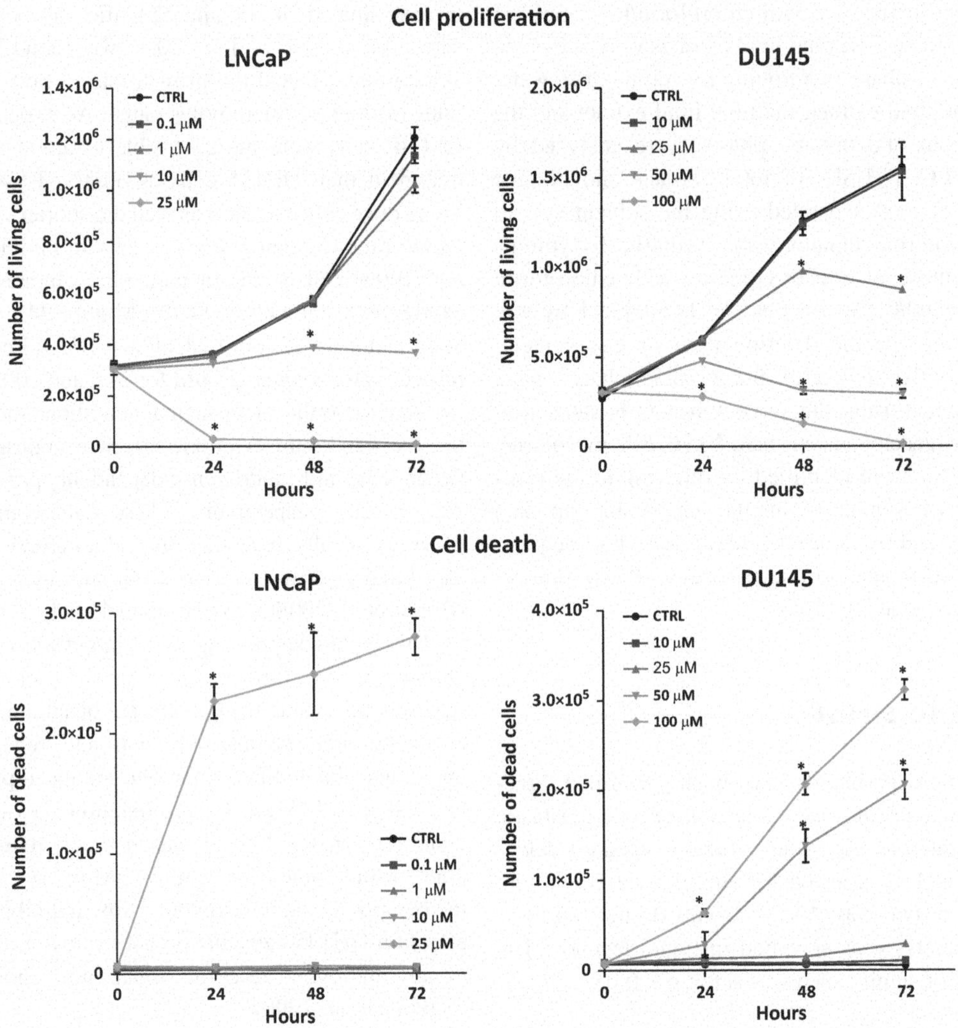

Fig. 20.1 Effect of abiraterone treatment on cell proliferation of androgen-sensitive (LNCaP) and androgen-resistant (DU145) prostate cancer cells. LNCaP and DU145 cells were incubated with different doses of abiraterone. Cell proliferation was evaluated at various time points by counting cells using trypan blue exclusion assay. Dead cells were evaluated as trypan blue-positive cells. Each value is the mean of three independent experiments performed in triplicate. $^*p < 0.03$ vs, the previous concentration

abiraterone. As shown in Fig. 20.2, in both the cell lines, abiraterone pretreatment in combination with its administration after irradiation resulted in a higher reduction of the cell survival compared to either agent alone. Interestingly, DU145 cells proved to be even more sensitive to the combined treatment of abiraterone with irradiation with respect to LNCaP cells, with a 90% and 40% reduction in cell viability.

20.3.3 Abiraterone Treatment and Ionizing Radiations Alter the Sphingolipid Pattern and Organization of Prostate Cancer Cells

Several lines of evidence have established that sphingolipid metabolism is highly deregulated in many cancer types (Huang et al. 2015). In

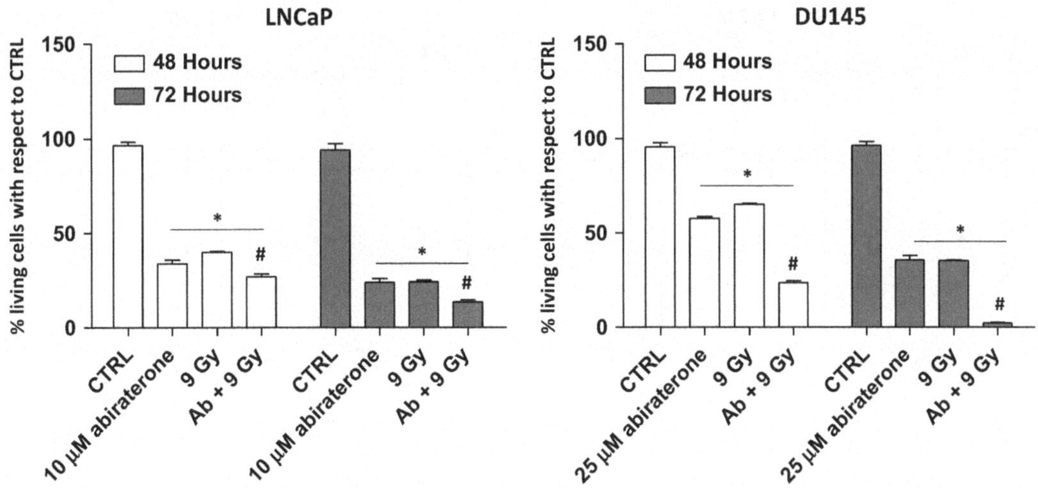

Fig. 20.2 Effect of abiraterone and ionizing radiation, alone or in combination, on LNCaP and DU145 cell viability. LNCaP and DU145 cells were exposed to a cytostatic concentration of abiraterone or vehicle at the same concentration used for drug solubilization (0.25%) for 24 h; cells were then irradiated or not at the dose indicated without or with abiraterone. Living cells were evaluated by trypan blue exclusion test 48 and 72 h after irradiation. Each value is the mean of three independent experiments performed in triplicate. *$p < 0.03$ vs CTRL, #$p < 0.001$ vs the previous treatment

particular, it is widely known in different human prostate cancer cells that ceramide, the central sphingolipid metabolite, is involved in the response to a variety of chemotherapeutic agents including ionizing radiations (Kolesnick et al. 2003; Aureli et al. 2014). However, no data on the possible role of ceramide in mediating the effect of abiraterone treatment alone or in combination with irradiation are present so far; to address this issue, we analyzed the sphingolipid content and pattern of both LNCaP and DU145 cells in the same experimental conditions reported above. To this aim, cell sphingolipids were steady-state metabolically labeled with the radioactive precursor [1-^{3}H]sphingosine and then subjected to single treatment with abiraterone, ionizing radiation, or in combination of both. Independently from either treatment alone or in combination, the total cell-incorporated radioactivity was very similar to control cells in both LNCaP and DU145 cells and mainly associated with the organic phases containing nonpolar sphingolipids (Fig. 20.3). On the other hand, we observed marked differences in the radioactivity associated with the aqueous phase-containing gangliosides between the two cell lines; in androgen-resistant DU145 cells, it was almost doubled compared to androgen-sensitive LNCaP cells (Fig. 20.3). In addition, abiraterone alone induced an increase in gangliosides content in LNCaP cells, but not in DU145 cells, whereas ionizing radiation elicited an opposite effect leading to a decrease of the radioactivity in the same lipid fraction only in androgen-sensitive cells. Conversely, in both cell lines, the combination of abiraterone plus irradiation did not affect the radioactivity distribution between the two lipid fractions (Fig. 20.3).

Thus, we further analyzed and identified the radioactive SL species by HPTLC separation followed by radiochromatoscanning. As shown in Fig. 20.4 right, we found that GM2 is the prominent ganglioside in LNCaP cells; its content increased twofold after abiraterone treatment, both as single agent and in combination with irradiation, whereas content did not significantly change after ionizing radiation exposure alone. Conversely, in the same cells, the level of phosphatidylethanolamine, the main metabolite of the organic phase, was strongly decreased,

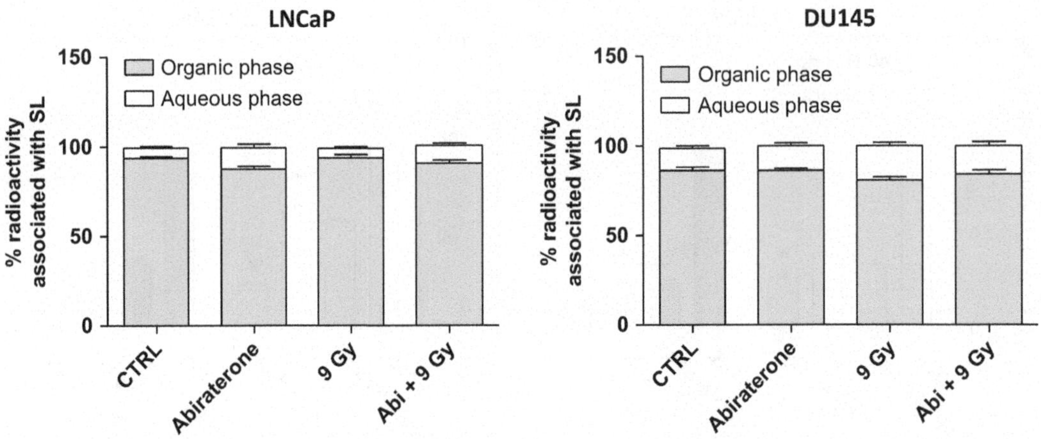

Fig. 20.3 Distribution of radioactivity between gangliosides and nonpolar lipids in LNCaP and DU145 subjected to abiraterone and ionizing radiation treatments, alone or in combination. Sphingolipids of LNCaP and DU145 were metabolically labeled at the steady state by the use of [1-^{3}H]sphingosine and exposed to abiraterone and ionizing radiation, alone or in combination, as described in Fig. 20.2. Forty-eight hours after irradiation, cells were harvested and subjected to lipid extraction. Total lipid extracts were subjected to a two-phase partitioning in order to separate gangliosides (aqueous phase) from the other lipids (organic phase). The radioactivity associated with the aqueous phases and the organic phases was evaluated by beta-counter. Data are expressed as a percentage of total incorporated radioactivity. Each value is the mean of two independent experiments performed in triplicate

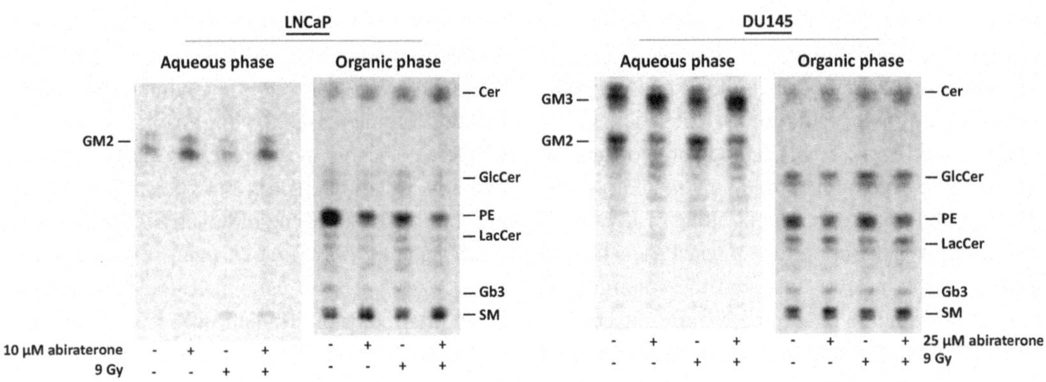

Fig. 20.4 Effects of the treatment with abiraterone and ionizing radiation, alone or in combination, on the sphingolipid pattern of LnCaP and DU145 cells. Cell sphingolipids were radiolabeled at the steady state using radioactive sphingosine and then exposed to abiraterone and ionizing radiation, alone or in combination, as described in Fig. 20.2. Lipid extraction was performed 48 h after cell irradiation. Total lipid extract was subjected to two-phase partitioning. Gangliosides were separated by HPTLC using the solvent system chloroform/methanol/0.2% $CaCl_2$ aqueous solution 50/42/11 by volume, whereas lipids contained in the organic phases were separated using the solvent system chloroform/methanol/water 110/40/6 by volume. Radioactive lipids were detected by digital autoradiography (BetaIMAGER, Biospace Lab) and identified by authentic standards

independently of the type of treatment, reaching about 70% reduction with the combined one, thus suggesting an impairment of sphingosine catabolism (Fig. 20.4 right; Table 20.1). Among the other lipid metabolites, we found that abiraterone alone and in combination with

Table 20.1 SL distribution in androgen sensitive prostate cancer cells (LNCaP) subjected to the different treatment

	Treatments			
Radioactive lipids	CTRL	10 μM abiraterone	9 Gy irradiation	Abiraterone + irradiation
Ceramide	4.9%	9.4%	11.8%	16.4%
Glucosylceramide	3.3%	2.4%	6.7%	4.2%
PE	52.8%	25.7%	32.4%	17.8%
Lactosylceramide	6.2%	4.1%	9%	6.1%
Gb3	4.7%	4.5%	6.5%	5.8%
Sphingomyelin	26.1%	49.6%	31.9%	46.4%
GM2	1,9%	4.4%	1.7%	3.3%

Table 20.2 SL distribution in androgen resistant prostate cancer cells (DU145) subjected to the different treatment

	Treatments			
Radioactive lipids	CTRL	25 μM abiraterone	9 Gy irradiation	Abiraterone + irradiation
Ceramide	3.4%	5.8%	6.2%	7.3%
Glucosylceramide	10.5%	8.8%	13.4%	9.7%
PE	17.4%	10.8%	16.0%	9.5%
Lactosylceramide	12.2%	12.4%	9.7%	10.7%
Gb3	9.1%	9.7%	8.7%	9.7%
Sphingomyelin	23.7%	26.1%	23.4%	26.7%
GM3	11.5%	18.3%	9.1%	17.9%
GM2	9.6%	3.1%	8.5%	3.7%

irradiation led to an increase in sphingomyelin content with respect to untreated cells. More interestingly, both single treatments induced a twofold increase in ceramide levels, and, of note, abiraterone treatment in combination with ionizing radiation produced further elevation in ceramide, more than 70% and 40% with respect to its level after each single treatment with abiraterone and irradiation, respectively (Fig. 20.4 right; Table 20.1).

We obtained similar results in DU145 cells, which are characterized by a ganglioside pattern with GM3 and GM2 gangliosides as major components (Fig. 20.4, left). In these cells, both treatments with abiraterone alone or in combination with irradiation led to a similar extent of increase in GM3 and reduction in GM2 content (about 1.6-fold higher and three-fold lower over control cells, respectively), whereas ionizing radiation alone did not cause any significant change in both monosialoganglioside levels (Fig. 20.4 left; Table 20.2). As observed in androgen-sensitive cells, phosphatidylethanolamine levels were also strongly decreased, although to a much lesser extent after exposure to irradiation alone. In addition, only a slight increase in sphingomyelin was elicited by the treatment with abiraterone alone or in combination, but not by radiation exposure alone. Again, all treatments caused an increase in ceramide levels, with the combined treatment eliciting the maximal elevation with respect to untreated DU145 cells (Fig. 20.4 right).

To further investigate the effect of the treatments on SL organization, we isolated detergent-resistant membrane fractions (DRM) from cells labeled at the steady state with radioactive sphingosine and subjected to single or combined treatments. As expected, radioactive SLs were highly enriched in the DRM with only a minor part found in the soluble fractions (HD). Moreover, the percentage of radioactivity associated with DRM was similar between the two cell lines with no significant differences between control and treated cells (data not shown).

The analysis of the lipid composition in DRM and HD fractions is shown in Fig. 20.5,

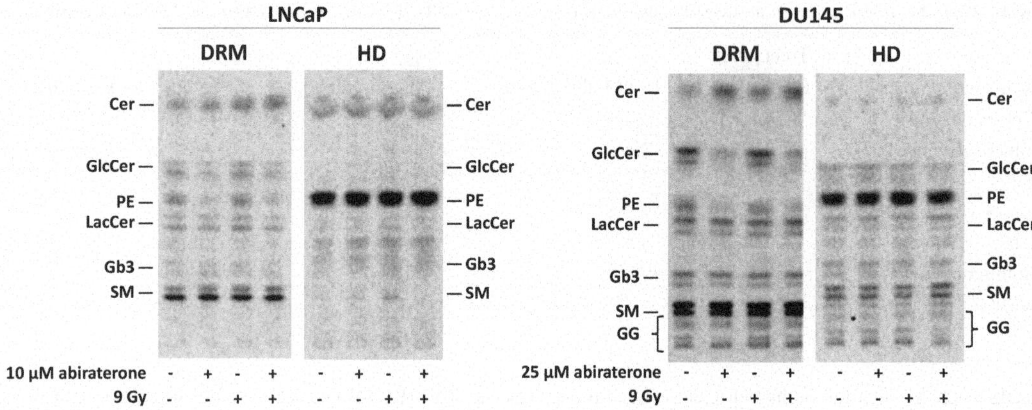

Fig. 20.5 Effects of the treatment with abiraterone and ionizing radiation, alone or in combination, on the sphingolipid pattern of LnCaP and DU145 lipids rafts. Cell sphingolipids were radiolabeled at the steady state using radioactive sphingosine and then exposed to abiraterone and ionizing radiation, alone or in combination, as described in Fig. 20.2. Forty-eight hours after irradiation, cells were harvested, and lipids rafts were isolated as detergent-resistant membrane (DRM) by cell lysis in 1% Triton X-100 and subjected to ultracentrifugation on discontinuous sucrose gradient. Radioactive lipids associated with DRM and high density fraction (HD) were separated by HPTLC using the solvent system chloroform/methanol/water 110/40/6 by volume. Radioactive lipids were detected by digital autoradiography (BetaIMAGER, Biospace Lab) and identified by authentic standards

Table 20.3 SL distribution in DRM and HD fraction obtained from androgen sensitive prostate cancer cells (LNCaP) subjected to the different treatments

	Treatments							
	CTRL		10 μM abiraterone		9 Gy irradiation		Abiraterone + irradiation	
Radioactive lipids	DRM	HD	DRM	HD	DRM	HD	DRM	HD
Ceramide	11%	13%	16%	13%	16%	13%	23%	10%
Glucosylceramide	12%	1%	6%	1%	13%	2%	9%	2%
PE	14%	76%	4%	74%	13%	67%	5%	76%
Lactosylceramide	12%	3%	12%	3%	14%	5%	8%	4%
Gb3	5%	5%	6%	6%	5%	7%	4%	6%
Sphingomyelin	46%	2%	56%	3%	39%	6%	51%	2%
GM2	ND	ND	ND	ND	ND	ND	ND	ND

independently of the type of treatment; 80–90% of radioactive PE was associated with HD fractions, thus supporting the proper DRM isolation in both the cell lines (Tables 20.3 and 20.4) (Sonnino et al. 2006). In addition, a strong enrichment of GlcCer, LacCer, and SM was observed in DRM from LNCaP compared to HD fractions (Table 20.3). In addition, in DRM derived from LNCaP cells treated with abiraterone and ionizing radiation, we observed an increase of the Cer content of about 1.5-fold followed by a parallel decrease of GlcCer with respect to those associated with CTRL DRM (Table 20.3). Importantly, in the DRM prepared from LNCaP cells subjected to the combined treatment, the Cer levels are double with respect to control DRM (Fig. 20.5 and Table 20.3). On the other hand, DRM from DU145 cells were characterized by an enrichment in Cer, GlcCer, and SM. Of note, the DRM derived from LNCaP cells treated with abiraterone both alone or in combination with ionizing radiations showed an increase in the ceramide content up to three times followed by a decrease in GlcCer content with respect to those found in DRM of CTRL cells. Whereas ionizing radiation induced only a slight increase in the Cer

Table 20.4 SL distribution in DRM and HD fraction obtained from androgen resistant prostate cancer cells (DU145) subjected to the different treatments

	Treatments							
	CTRL		10 μM abiraterone		9 Gy irradiation		Abiraterone + irradiation	
Radioactive lipids	DRM	HD	DRM	HD	DRM	HD	DRM	HD
Ceramide	5%	2%	15%	2%	8%	2%	14%	3%
Glucosylceramide	18%	7%	6%	9%	17%	7%	7%	5%
PE	10%	40%	3%	39%	6%	47%	2%	37%
Lactosylceramide	12%	8%	12%	8%	11%	8%	11%	9%
Gb3	10%	8%	9%	7%	8%	7%	7%	7%
Sphingomyelin	35%	17%	41%	17%	37%	13%	49%	23%
Gangliosides	10%	18%	11%	18%	13%	16%	10%	16%

content of DRM of treated cells (Fig. 20.5 and Table 20.4), no significant differences between control and treated cells were found in HD fractions. These data clearly indicate that the modification of the plasma membrane sphingolipids composition is mainly related to a specific membrane area involved in the control of the cell signaling.

20.3.4 Treatment with Abiraterone and Ionizing Radiation Led to Increased Activities of Plasma Membrane Glycosphingolipid Hydrolases in LNCaP and DU145 Prostate Cancer Cells

Recent lines of evidence support the involvement of PM glycosphingolipids hydrolases in the ectopic production of pro-apoptotic ceramide upon radiation treatment of breast cancer cells. We evaluated the effect of abiraterone and ionizing radiations as single treatments or in combination on the activities PM hydrolases in both LNCaP and DU145 cell lines 48 h after irradiation. In LNCaP cells, we found that abiraterone treatment induces a twofold increase in the activity of plasma membrane-associated GBA1, GBA2, β-galactosidase, and β-hexosaminidase. Similar results were obtained upon exposure to ionizing radiations except for β-galactosidase and GBA2, which increased 3- and 0.5-fold, respectively (Fig. 20.6). The combined treatment did not significantly differ from either single treatment with the exception of the β-hexosaminidase, which increased more than fivefold in comparison to untreated cells (Fig. 20.6). In DU145 cells, abiraterone treatment did not affect the activity of GBA1, β-galactosidase, and Neu3. On the contrary, the activity of GBA2 increased more than 2.5-fold and β-hexosaminidase doubled. Whereas cells exposed to ionizing radiation were characterized by twofold increase in the activity of β-galactosidase and β-hexosaminidase, we did not observe any difference in the activity of GBA1, GBA2, and Neu3 (Fig. 20.6).

Based on these results, we performed experiments aimed to prove the action of the glycosphingolipid hydrolases directly on the plasma membrane. To this purpose, we administered the natural substrate [3-^{3}H]GM3 in conditions to block lysosomal catabolism. The experiment was performed only on DU145 cells since LNCaP cells were sensitive to chloroquine treatment. As shown in Fig. 20.7, both single treatments with abiraterone and with irradiation induced an increased production of PM ceramide from GM3 catabolism (1.5- and 2.2-fold with respect to control cells, respectively); again, the combined treatment elicited the major ceramide increase being fourfold higher than control cells. These data further support that the different treatments and, in particular, the combination of abiraterone with irradiation activate the aberrant catabolism of cell surface glycosphingolipids leading to the ectopic production of cytotoxic ceramide.

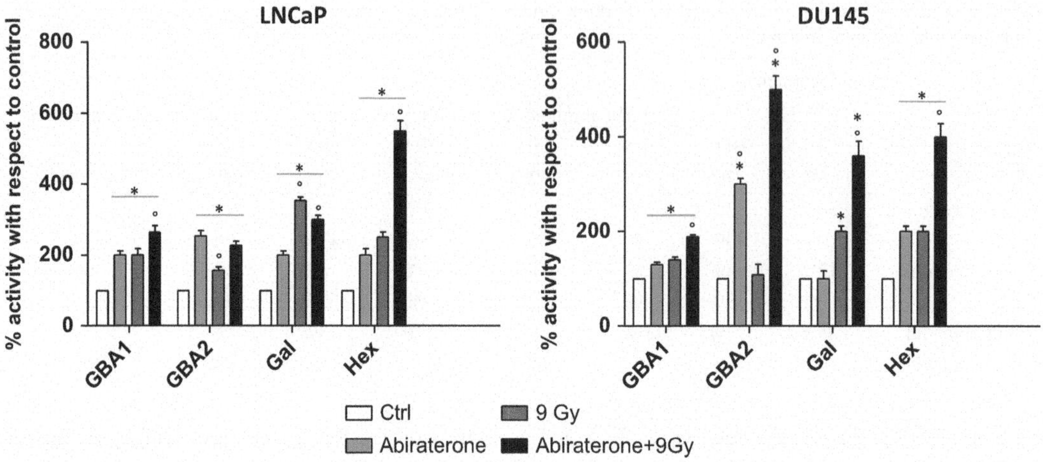

Fig. 20.6 Effects of the treatment with abiraterone and ionizing radiation, alone or in combination, on plasma membrane glycosphingolipid hydrolases activities of LNCaP and DU145 cells. The activity of β-glucocerebrosidase (GBA1), non-lysosomal β-glucosylceramidase (GBA2), β-galactosidase, and β-hexosaminidase was evaluated in living LNCaP and DU145 cells exposed to abiraterone and ionizing radiation, alone or in combination, as described in Figs. 20.2, 48 h after irradiation. The data are the average of three experiments performed in triplicate and are expressed as percentage of enzyme activity with respect to control cells. $^{*}P < 0.005$ vs ctrl; $^{\circ}P < 0.001$ vs the previous treatment

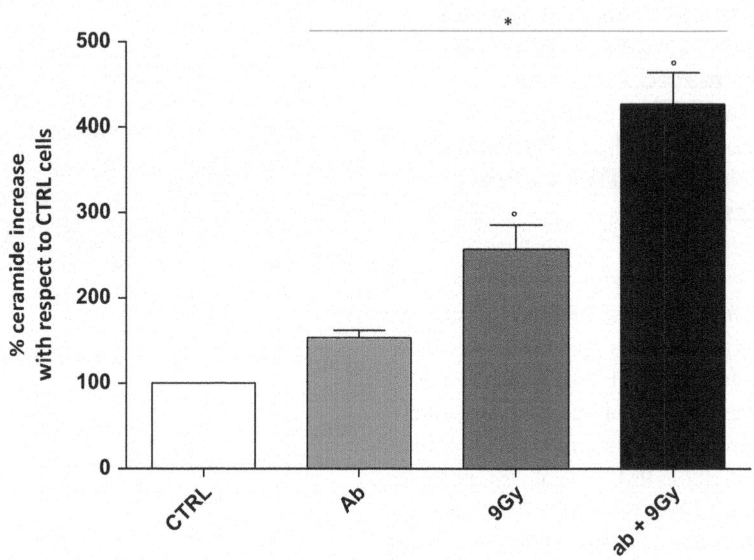

Fig. 20.7 Effects of the treatment with abiraterone and ionizing radiation, alone or in combination on GM3 ganglioside catabolism at the plasma membrane in DU145 cells. Androgen-resistant DU145 cells were exposed to abiraterone and ionizing radiation, alone or in combination, as described in Fig. 20.2. Forty-eight hours after irradiation cells were fed with [3-^{3}H]sph-GM3 in the presence of chloroquine to block lysosomal function. In the graph each bar represents the percentage increase of radioactive ceramide produced from GM3 hydrolysis at the cell surface with respect to control cells. $^{*}P < 0.0002$ vs ctrl; $^{\circ}P < 0.003$ vs the previous treatment

20.4 Summary

Our study described for the first time the effect of abiraterone alone and in combination with ionizing radiation on sphingolipid metabolism in both androgen-sensitive (LNCaP) and androgen-insensitive (DU145) prostate cancer cells. Multiple lines of evidence indicate that SLs and their metabolic enzymes are potential therapeutic targets for several tumors, including prostate cancer (Hajj et al. 2013). Among SL, ceramide has been one of the best characterized in the fate of both androgen-sensitive and androgen-insensitive prostate cancer cells; given its tumor-suppressor function, strategies aimed at enhancing its production represent promising therapeutic options (Kolesnick 1992). It is of particular interest that ceramide can be produced directly at the cell surface by the action of hydrolases able to catabolize in situ complex SLs. Among these PM enzymes, acid sphingomyelinase, β-glucocerebrosidase GBA1, β-galactosidase, and β-hexosaminidase A are the same lysosomal isoforms, whereas non-lysosomal β-glucosylceramidase GBA2 and Neu3 sialidase are mainly PM-resident enzymes. Independently from their origin, these PM enzymes are involved in the regulation of several physiological and pathophysiological cellular processes, including apoptosis (Sonnino et al. 2010). In human fibroblasts, Neu3 sialidase overexpression induces increased enzyme activity of PM-β-glucosidases and PM-β-galactosidase, leading to ectopic production of ceramide and apoptotic cell death (Valaperta et al. 2006; Aureli et al. 2009). On the other hand, ionizing radiation, via activation of PM-acid sphingomyelinase, also induces massive sphingomyelin hydrolyses to ceramide with the onset of apoptosis in several cancer cell lines (Kolesnick et al. 1994, 2003). We have previously reported that ionizing radiation-induced PM glycohydrolases activation is responsible for the production of pro-apoptotic ceramide by glycosphingolipid degradation (Aureli et al. 2011, 2012b; Valaperta et al. 2006).

Our data obtained on androgen-sensitive and androgen-insensitive prostate cancer cells are consistent with these findings. In fact, here we demonstrated that ionizing radiation induces a modification of SL pattern in both LNCaP and DU145 cell lines, resulting in increased ceramide content. Moreover, by subfractionation experiments, we found that accumulation of ceramide takes place in a specialized structure of cell membranes enriched in sphingolipids (lipids rafts).

As previously observed in breast cancer cells, we found cell exposure to ionizing radiations induces an important increase in the activity of the PM glycohydrolases responsible for the ectopic production of ceramide.

Besides ionizing radiation, several pharmacological treatments are also able to alter SL metabolism with the formation of cytotoxic ceramide. Docetaxel induces apoptosis in prostate cancer cells through increasing intracellular generation and accumulation of ceramides. Interestingly, here we described for the first time that abiraterone acetate has a similar effect to that exerted by ionizing radiation. In fact, we found that both LNCaP and DU145 cells treated with a cytostatic concentration of abiraterone are characterized by increased activity of PM glycohydrolases followed by a marked increase of ceramide in lipid rafts. More interestingly, we found that the combined treatment with abiraterone and ionizing radiation results in a very strong cytotoxic effect, reaching 100% of cell death in DU145. In these cells, the combined treatment also leads to a higher increase in the activity of PM glycohydrolases and in ceramide content if compared to each single treatment.

These data suggest that the combined treatment of abiraterone and ionizing radiation, both targeting SL metabolism leading to formation of ceramide at the cell surface, could represent intriguing and promising therapeutic strategies for prostate cancer. In addition, the development of new drugs able to enhance the activity of PM glycohydrolases, in combination with the conventional therapeutic options, could open the possibility to design future therapeutic strategies for prostate cancer, independently from their androgen sensitivity and/or radioresistance.

References

Anderson J (2003) The role of antiandrogen monotherapy in the treatment of prostate cancer. BJU Int 91 (5):455–461

Aureli M, Masilamani AP, Illuzzi G, Loberto N, Scandroglio F, Prinetti A, Chigorno V, Sonnino S (2009) Activity of plasma membrane beta-galactosidase and beta-glucosidase. FEBS Lett 583 (15):2469–2473

Aureli M, Loberto N, Lanteri P, Chigorno V, Prinetti A, Sonnino S (2011) Cell surface sphingolipid glycohydrolases in neuronal differentiation and aging in culture. J Neurochem 116(5):891–899

Aureli M, Bassi R, Loberto N, Regis S, Prinetti A, Chigorno V, Aerts JM, Boot RG, Filocamo M, Sonnino S (2012a) Cell surface associated glycohydrolases in normal and Gaucher disease fibroblasts. J Inherit Metab Dis 35(6):1081–1091

Aureli M, Bassi R, Prinetti A, Chiricozzi E, Pappalardi B, Chigorno V, Di Muzio N, Loberto N, Sonnino S (2012b) Ionizing radiations increase the activity of the cell surface glycohydrolases and the plasma membrane ceramide content. Glycoconj J 29:585

Aureli M, Murdica V, Loberto N, Samarani M, Prinetti A, Bassi R, Sonnino S (2014) Exploring the link between ceramide and ionizing radiation. Glycoconj J 31 (6–7):449–459

Bruno RD, Gover TD, Burger AM, Brodie AM, Njar VC (2008) 17alpha-Hydroxylase/17,20 lyase inhibitor VN/124-1 inhibits growth of androgen-independent prostate cancer cells via induction of the endoplasmic reticulum stress response. Mol Cancer Ther 7 (9):2828–2836

Chung JY, Cho JY, Yu KS, Kim JR, Lim KS, Sohn DR, Shin SG, Jang IJ (2008) Pharmacokinetic and pharmacodynamic interaction of lorazepam and valproic acid in relation to UGT2B7 genetic polymorphism in healthy subjects. Clin Pharmacol Ther 83(4):595–600

Grossebrummel H, Peter T, Mandelkow R, Weiss M, Muzzio D, Zimmermann U, Walther R, Jensen F, Knabbe C, Zygmunt M, Burchardt M, Stope MB (2016) Cytochrome P450 17A1 inhibitor abiraterone attenuates cellular growth of prostate cancer cells independently from androgen receptor signaling by modulation of oncogenic and apoptotic pathways. Int J Oncol 48(2):793–800

Hajj C, Haimovitz-Friedman A (2013) Sphingolipids' role in radiotherapy for prostate cancer. Handb Exp Pharmacol 216:115–130

Hakomori S (1990) Bifunctional role of glycosphingolipids. Modulators for transmembrane signaling and mediators for cellular interactions. J Biol Chem 265(31):18713–18716

Huang C, Freter C (2015) Lipid metabolism, apoptosis and cancer therapy. Int J Mol Sci 16(1):924–949

Jemal A, Bray F, Center MM, Ferlay J, Ward E, Forman D (2011) Global cancer statistics. CA Cancer J Clin 61 (2):69–90

Kabayama K, Sato T, Saito K, Loberto N, Prinetti A, Sonnino S, Kinjo M, Igarashi Y, Inokuchi J (2007) Dissociation of the insulin receptor and caveolin-1 complex by ganglioside GM3 in the state of insulin resistance. Proc Natl Acad Sci U S A 104 (34):13678–13683

Kaiser HJ, Lingwood D, Levental I, Sampaio JL, Kalvodova L, Rajendran L, Simons K (2009) Order of lipid phases in model and plasma membranes. Proc Natl Acad Sci U S A 106(39):16645–16650

Kolesnick R (1992) Ceramide: a novel second messenger. Trends Cell Biol 2(8):232–236

Kolesnick R, Fuks Z (2003) Radiation and ceramide-induced apoptosis. Oncogene 22(37):5897–5906

Kolesnick R, Golde DW (1994) The sphingomyelin pathway in tumor necrosis factor and interleukin-1 signaling. Cell 77(3):325–328

Kolvenbag GJ, Iversen P, Newling DW (2001) Antiandrogen monotherapy: a new form of treatment for patients with prostate cancer. Urology 58(2 Suppl 1):16–23

Merrill AH Jr (2011) Sphingolipid and glycosphingolipid metabolic pathways in the era of sphingolipidomics. Chem Rev 111(10):6387–6422

Navone NM, Logothetis CJ, von Eschenbach AC, Troncoso P (1998) Model systems of prostate cancer: uses and limitations. Cancer Metastasis Rev 17 (4):361–371

Prinetti A, Loberto N, Chigorno V, Sonnino S (2009) Glycosphingolipid behaviour in complex membranes. Biochim Biophys Acta 1788(1):184–193

Prinetti A, Cao T, Illuzzi G, Prioni S, Aureli M, Gagliano N, Tredici G, Rodriguez-Menendez V, Chigorno V, Sonnino S (2011) A glycosphingolipid/caveolin-1 signaling complex inhibits motility of human ovarian carcinoma cells. J Biol Chem 286 (47):40900–40910

Simons K, Sampaio JL (2011) Membrane organization and lipid rafts. Cold Spring Harb Perspect Biol 3(10): a004697

Sonnino S, Prinetti A, Mauri L, Chigorno V, Tettamanti G (2006) Dynamic and structural properties of sphingolipids as driving forces for the formation of membrane domains. Chem Rev 106(6):2111–2125

Sonnino S, Aureli M, Loberto N, Chigorno V, Prinetti A (2010) Fine tuning of cell functions through remodeling of glycosphingolipids by plasma membrane-associated glycohydrolases. FEBS Lett 584(9):1914–1922

Sonnino S, Aureli M, Mauri L, Ciampa MG, Prinetti A (2015) Membrane lipid domains in the nervous system. Front Biosci 20:280–302

Teichgraber V, Ulrich M, Endlich N, Riethmuller J, Wilker B, De Oliveira-Munding CC, van Heeckeren AM, Barr ML, von Kurthy G, Schmid KW, Weller M, Tummler B, Lang F, Grassme H, Doring G, Gulbins E (2008) Ceramide accumulation mediates inflammation, cell death and infection susceptibility in cystic fibrosis. Nat Med 14(4):382–391

Valaperta R, Chigorno V, Basso L, Prinetti A, Bresciani R, Preti A, Miyagi T, Sonnino S (2006) Plasma membrane production of ceramide from ganglioside GM3 in human fibroblasts. FASEB J 20(8):1227–1229

A Glycomic Approach Towards Identification of Signature Molecules in CD34$^+$ Haematopoietic Stem Cells from Umbilical Cord Blood

21

Suchandra Chowdhury, Kaushik Bhattacharya, Chandan Mandal, Susmita Mondal, Sayantani Sarkar, Sarmila Chandra, Subir Banerjee, and Chitra Mandal

Abstract

Umbilical cord blood (UCB) is a powerful storehouse for normal CD34$^+$ haematopoietic stem cells (HSCs), often used for allogeneic bone marrow (BM) transplantation in malignant and non-malignant diseases. The glycomic especially the sialoglycomic aspect of these HSCs has been unravelled in this study. Cell surface expression of the glycans with the related enzymatic activities has been compared with the BM of childhood acute lymphoblastic leukaemia, a common BM-associated malignancy. An enhanced cell surface expression of α2,3-linked sialic acid, P- and E-selectins, and intercellular adhesion molecule along with reduced expression of L-selectin distinguishes CD34$^+$ HSCs of UCB from leukaemic samples. More importantly, high expression of *O*-acetylated sialoglycoproteins, a hallmark of lymphoblasts, is drastically reduced in the CD34$^+$ HSCs of UCB and is substantiated by the low activity of sialylate-*O*-acetyltransferase and high sialidase activity. In contrast, a significant variation is evident in the expression of sialic acid, α2,6-linked sialic acids, and the sialyltransferase activity. Taken together, these studies indicate a few signature molecules, forming a unique glycomic template, which may be a potential indicator, reassuring the normal profile of these stem cells, to be used for future transplantation.

Keywords

CD34$^+$ haematopoietic stem cell · Glycomics · Sialic acid · Cell adhesion molecules · Sialylate-*O*-acetyltransferase · Sialidase

21.1 Introduction

Allogeneic haematopoietic stem cell (HSC) transplantation has provided curative therapy for severe bone marrow (BM)-related malignant and non-malignant diseases in patients. However, only a few of them have the option of a human leukocyte antigen (HLA) identical sibling donor. Substantial advances in the use of alternative donors like unrelated volunteer donors and umbilical cord blood (UCB) donors enable almost all patients to benefit from this life-saving treatment. Nevertheless, each donor type has its own demerits (Chowdhury et al. 2014; Hough et al. 2009).

S. Chowdhury · K. Bhattacharya · C. Mandal · S. Mondal · S. Sarkar · C. Mandal (✉)
Cancer Biology and Inflammatory Disorder Division, Council of Scientific and Industrial Research (CSIR)-Indian Institute of Chemical Biology, Kolkata, India
e-mail: cmandal@iicb.res.in

S. Chandra
Clinical Haematology Service, Kolkata, India

S. Banerjee
Department of Anaesthesiology and critical care, Medical College and Hospital, Kolkata, India

K. Chattopadhyay, S. C. Basu (eds.), *Biochemical and Biophysical Roles of Cell Surface Molecules*, Advances in Experimental Medicine and Biology 1112, https://doi.org/10.1007/978-981-13-3065-0_21

The problems inhibiting the wider use of alternative donors are the probability of identifying a 'suitable' donor, the speed with which a graft can be acquired, and the risk of graft-versus-host disease. Since UCB stem cells are the simplest to collect, allowing greatest flexibility in HLA matching (Chowdhury et al. 2014; Hough et al. 2009), therefore, at present, the strategy of expanding the potential donor pool involves transplantation with UCB, which maintains an acceptable level of treatment-related complications (Brown and Boussiotis 2008).

Following this success, UCB banks have been established to facilitate UCB transplants (Hough et al. 2009) that offer parents the opportunity to store UCB for possible future use for their child or other family members. These banks have grown rapidly across the world with a cumulative repository (Chowdhury et al. 2014; Hollands and McCauley 2009).

A plethora of studies has been conducted to investigate the in vitro and in vivo characterization of UCB cells, with the ultimate goal of optimizing and increasing their clinical use. Several approaches have been initiated to understand the phenotypic, biochemical, and biological properties of these cells (Satomaa et al. 2009). However, the profile of these HSCs from the glycomic especially sialoglycomic aspect has never been satisfactorily explored till date.

Carbohydrates are essential constituents of glycoproteins, glycolipids, and proteoglycans (Satomaa et al. 2009). Among them, sialic acids, a family of 9-carbon carboxylated monosaccharides, have diverse modifications (Schauer 2000; Mandal et al. 2000). Commonly modified derivatives like *O*-acetylated sialoglycoproteins (Neu5,9Ac_2-GPs) (Sinha et al. 1999a, 1999b; Pal et al. 2004; Chowdhury et al. 2007, 2008; Chowdhury and Mandal 2009) and *O*-acetylated sialoglycolipids (9-*O*AcGD3) (Mukherjee et al. 2007) are well-established cancer-associated antigens. They influence numerous biological reactions by reacting with specific surface receptors or via masking of carbohydrate recognition sites (Schauer 2000; Mandal et al. 2000). The cellular sialic acid content is metabolically regulated and controlled by several enzymes, main contributory ones being sialate-*O*-acetyltransferase (SOAT), sialyltransferase, and sialidase (Mandal et al. 2000, 2010; Mondal et al. 2009).

In addition, selectins (L, E, and P; CD62) along with some cell surface adhesion molecules (CAMs), namely, intercellular adhesion molecule (ICAM-1) and vascular cell adhesion molecule (VCAM-1), play fundamental roles in numerous cellular processes like growth, differentiation, embryogenesis, immune cell transmigration, and cancer metastasis (Chowdhury et al. 2007).

With this background information, the main aim of this study has been to explore the status of sialic acids and CAMs, including selectins, on the HSCs isolated from UCB. In parallel, the orchestration of the three main enzymes regulating the overall sialylation of these HSCs has also been investigated. A major characteristic of these HSCs is the phenotypic expression of CD34 antigen (Storms et al. 2005). Therefore, for this investigation, the haematopoietic cells expressing CD34 have been selected as the study population. Since childhood acute lymphoblastic leukaemia (ALL) is an impairment of HSC maturation, the BM of these patients at diagnosis, i.e. before treatment and in clinical remission (CR), has been used for comparison throughout (George et al. 2001).

To the best of our knowledge, this sialoglycomic approach reports for the first time a pattern of signature molecules on the $CD34^+$ HSCs of UCB. The lower expression of *O*-acetylated sialoglycoproteins along with the higher expression of α2,3-linked sialic acids, on these cells, gives an extra level of restriction, forming the blueprint, distinguishing them from the leukaemic cells. This finding has been substantiated by the respective enzymatic activities. Additionally, an enhanced cell surface expression of P- and E-selectins and ICAM-1 along with a diminished expression of L-selectin is observed. Interestingly, these parameters are completely different in diagnostic ALL samples. Taken together, this glycomic template can serve as a potential indicator, reassuring the normal profile of these stem cells, to be used for future transplantation.

21.2 Materials and Methods

21.2.1 Study Subject

Human UCB samples (~20 ml, n = 15) were collected from healthy full-term neonates in Alsevers' solution, immediately after delivery by gravitation. Additionally, BM samples from clinically confirmed childhood ALL patients at diagnosis (BM_{ALL}, n = 15) and in CR (BM_{CR}) were used for comparison. The Institutional Human Ethical Committee as per the protocol of the Indian Council of Medical Research approved the study. Samples were collected with informed consent from parents/guardians.

21.2.2 Flow Cytometric Detection of Sialic Acids and Different Adhesion Molecules in $CD34^+$ Cells of UCB

Mononuclear cells (MNCs) were isolated from all different clinical samples separately and processed as described earlier (Chowdhury et al. 2008). Briefly, cells ($1x10^6$) were incubated with PE-CD34 and FITC conjugated/biotinylated/unconjugated analytical probes comprising a lectin/monoclonal antibody (mAb), at 4 °C for 1 h in dark. Using appropriate FITC secondary antibodies or streptavidin, stained cells were analysed in FACSCalibur using CellQuest Pro software (BD Biosciences, USA). Fifty thousand events were analysed by gating the nucleated cells (R1, Fig. 21.1a).

FITC conjugated lectin, Achatinin-H having a preferential affinity towards Neu5,9Ac_2α2-6β-D-GalNAc and recognizing Neu5,9Ac_2-GPs, was used (Mandal et al. 2000; Sinha et al. 1999a, 1999b; Pal et al. 2004; Chowdhury et al. 2007, 2008; Chowdhury and Mandal 2009). Additionally, FITC-*Limulus polyphemus* agglutinin (LPA) and two biotinylated plant lectins, namely, *Maackia amurensis* agglutinin (MAA) and *Sambucus nigra* agglutinin (SNA) having specificity towards Neu5Ac, Neu5Acα2-3Gal, and Neu5Acα2-6Gal/GalNAc, respectively, were used (Vector Laboratories). The selectins (L, E, and P), ICAM-1, and VCAM-1 were probed with anti-CD62L, anti-CD62E, anti-CD62P, anti-CD54, and anti-CD106 mAbs, respectively. Anti-CD62E and anti-CD54 were detected using FITC-streptavidin (BD Biosciences, USA). Purified anti-CD62P, anti-CD62L, and anti-CD106 were detected using FITC-anti-murine IgG (Chowdhury et al. 2007).

21.2.3 Functional Characterization of the $CD34^+$ Cells of UCB Using Aldehyde Dehydrogenase (ALDH)

Viable $CD34^+$ cells expressing cytosolic ALDH were identified by flow cytometry using Aldefluor reagent (Stemcell Technologies Inc., Canada) as

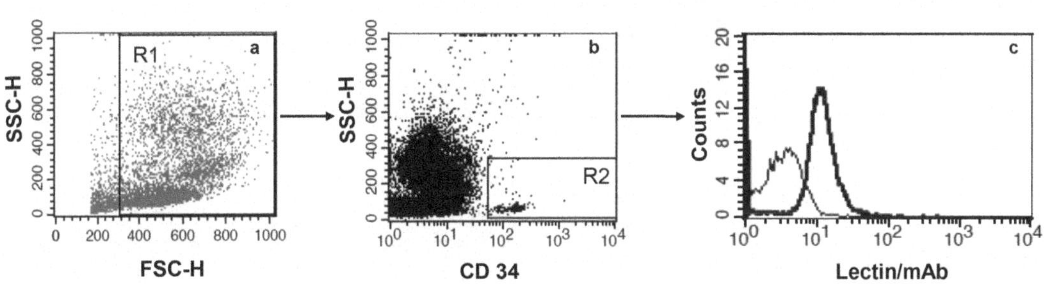

Fig. 21.1 Phenotypic identification of the study population in umbilical cord blood (UCB) showing status of a specific antigen. A representative dot plot with forward (FSC-H) versus side (SSC-H) scatter, showing the region gated for the haematopoietic cells (HSCs), designated as R1 (**a**). Dot plot showing cells in R1, gated for positive expression of CD34 antigen (R2, **b**). Cells in R2 assessed in a representative histogram plot presenting the log fluorescence intensity of a representative probe used for the study (**c**)

per the manufacturer's instruction. Briefly, MNCs ($1x10^6$) were suspended in ALDEFLUOR assay buffer. The activated Aldefluor substrate was added, and half of this cell suspension was transferred to a tube containing diethylaminobenzaldehyde (DEAB), a specific inhibitor of ALDH that serves as control. Samples were incubated for 30 min at 37 °C for optimum enzyme activity and analysed in a flow cytometer.

21.2.4 Detection of the Activity of Enzymes Responsible for Sialylation in $CD34^+$ Populations of UCB

The activities of the three main enzymes regulating sialylation were determined by isolating the CD34-enriched population from MNCs of UCB samples using magnetic columns (Stemcell Technologies Inc., Canada). For comparison, these assays were repeated in parallel with the $CD34^+$ cells of BM_{ALL} and BM_{CR}. The activity was determined using cell lysate prepared by sonication of $1–2x10^6$ cells in the water on ice (Mandal et al. 2009).

21.2.5 Sialate-O-Acetyltransferase (SOAT)

Bovine submaxillary mucin known to contain Neu5,9Ac2α2-6β-D-GalNAc was de-O-acetylated and used as the exogenous acceptor. Cell lysate and [^{3}H]AcCoA were used as enzyme source and donor, respectively. Briefly, cell lysate (60 μg), de-O-AcBSM, and [3H]AcCoA (10 μl, 0.05 μCi, 2.4 μM) in 10 μl buffer containing Tris-HCl (10 mM, pH 7.0), KCl (150 mM), and $MgCl_2$ (1 mM) in a total volume of 100 μl were incubated at 37 °C for 1 h. The reaction was stopped with ice-cold perchloric acid (PCA, 10%). After centrifugation the precipitated protein was washed with ice-cold PCA (4%). The washed protein was dissolved in NaOH and incubated overnight at 25 °C. The protein solution was then neutralized, and radioactivity was measured in a scintillation counter (Packard Bioscience Company, USA). In the control tube, the buffer was used instead of the enzyme source (Mandal et al. 2009).

21.2.6 Sialyltransferase Activity

In this assay (Mondal et al. 2009), the reaction mixture consisting of cell lysate (125 μg, enzyme source), asialofetuin (acceptor), 0.25 mCi CMP-^{14}Sia (donor, Amersham Bioscience, Germany), 50 nmol CMP-Sia, 1 mM DANA (2,3-dihydro-2-deoxy-NeuAc, an inhibitor of sialidase, Calbiochem), and 50 mM cacodylate buffer (0.10% triton X-100, 10 mM $MnCl_2$, 150 mM NaCl, pH 6.5) in a total final volume 100 μl was incubated at 37 °C for 1 h and processed as above.

21.2.7 Sialidase Activity

The total sialidase activity was determined by using 4-MU-Neu5Ac as the substrate. The reaction mixture of the assay contained cell lysate (100 μg), 4-MU-Neu5Ac (30 nmole), and BSA (15 μl, 40 μg/ml) in sodium acetate buffer (50 mM, pH 4.6). After incubation at 37 °C for 1 h, the reaction was terminated with glycin/NaOH buffer (1.5 ml, 0.2 M, pH 10.8). Assays were performed in triplicate in a final volume of 100 μl. Fluorescence emission was measured on fluorimeter (PerkinElmer) with excitation at 365 nm and emission at 450 nm as discussed earlier (Mandal et al. 2010).

21.2.8 Statistical Analysis

Statistical analysis was performed using the GraphPad Prism software program (GraphPad Software Inc., CA). Results were expressed as mean ± SD for each individual set of experiment. Two-tailed P value for unpaired t test $(P) < 0.05$ was considered significant.

21.3 Results and Discussion

21.3.1 Physical and Functional Parameters Defining the HSCs of UCB

The CD34 antigen is a 110–115 kDa transmembrane sialoglycoprotein, expressed by early haematopoietic progenitor cells with the highest expression on the most immature stem cells (George et al. 2001). Incidentally, a large percentage of acute leukaemias are $CD34^+$ (Krause and Van Etten 2007).

Since cell surface CD34 is a well-established marker to identify the HSC population, we have studied exclusively those cells with phenotypic expression of CD34 antigen. The nucleated cells have been gated as R1 (Fig. 21.1a) and analysed in a dot plot where the $CD34^+$ cells are designated as R2 (Fig. 21.1b). Flow cytometric analysis of the percentages of $CD34^+$ cells revealed considerable statistical differences among the three sets of samples studied. $CD34^+$ HSCs of UCB were significantly low (4–10%) as compared to BM_{ALL} (~75%), whereas $CD34^+$ cells of BM_{CR} samples were 1–5% as expected.

The phenotypic expression of CD34 has been supported by the co-expression of ALDH in the $CD34^+$ cells (Fig. 21.2). Since the activity of intracellular ALDH is highest in the HSCs, in contrast to lymphocytes, the former can be easily identified based on this differential expression (Storms et al. 2005). Cells in the $ALDH^+$ region (R2, Fig. 21.2b) had more than 90% expression of CD34 (Fig. 21.2c), attesting to the functional activity and viability of the selected study population.

21.3.2 Differential Pattern of Sialylation on the $CD34^+$ HSCs of UCB as Compared to $CD34^+$ Cells from BM_{ALL} and BM_{CR}

Following the characterization of the study population ($CD34^+ALDH^+$ cells, Fig. 21.1c), we investigated the diversity of sialic acid derivatives in these cells (Fig. 21.3; Table 21.1). BM_{ALL} represent the diseased population with which comparison has been drawn. In contrast, the patients in CR are declared to be clinically asymptomatic and have been used in this study, as a substitute for the normal BM.

A close association exists between childhood ALL and Neu5,9Ac_2-GPs. The latter is overexpressed at the onset of the disease, followed by reduced expression with treatment in CR and reappearance with relapse (Mandal et al. 2000; Sinha et al. 1999a, 1999b; Pal et al. 2004; Chowdhury et al. 2007; Chowdhury et al. 2008; Chowdhury and Mandal 2009). In this study, cell surface expression of linkage-specific Neu5,9Ac_2-GPs on $CD34^+$ HSCs of UCB is

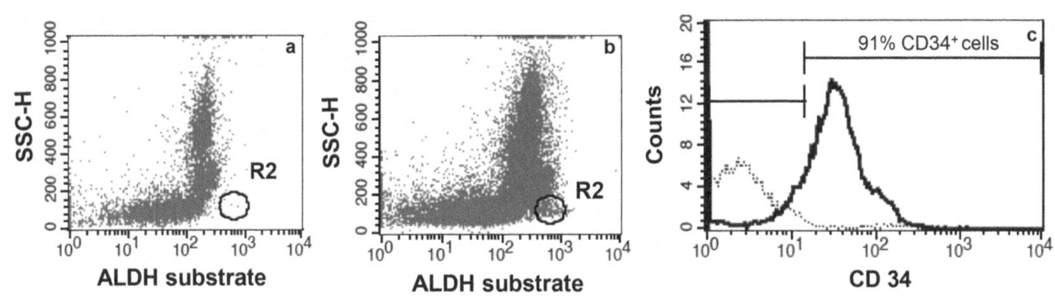

Fig. 21.2 Functional identification of $CD34^+$ HSCs of UCB with aldehyde dehydrogenase (ALDH) activity. Cells in R1 (Fig. 21.1a) were analysed in a dot plot with FL1-H versus SSC for control setting of Aldefluor assay, as per manufacturer's instruction. FL1 contains the DAAA, a fluorescent substrate for ALDH, where the reaction has been blocked by DEAB to adjust R2 such that no events appear in the R2 region. This setting served as a control for subsequent analyses (**a**). In UCB sample, the R2 region was positive for cells with high ALDH activity ($ALDH^+$, **b**). $ALDH^+$ cells in R2 plotted in a histogram, with CD34, revealed >90% $CD34^+$ cells (**c**)

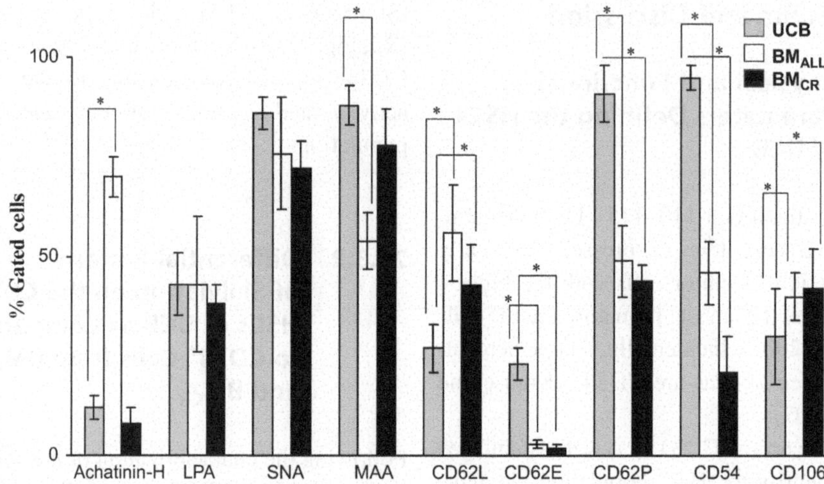

Fig. 21.3 **Expression of sialic acid and its derivatives along with cell adhesion molecules on CD34$^+$ cells of UCB as compared to BM_{ALL} at diagnosis and BM_{CR} in clinical remission**. CD34$^+$ cells (gated as shown in Fig. 21.1a–b) from the aforesaid samples were analysed by two-colour flow cytometric analysis. Bar depicts the mean ± SD of the percentage of cells expressing sialoglycoproteins and adhesion molecules as evidenced by binding with fluorescence-tagged lectins (Achatinin-H, LPA, SNA, and MAA) and specific mAbs against adhesion molecules (CD62L, CD62E, CD62P, CD54, and CD106). *$P < 0.05$ indicates a significant difference between two test groups

minimal as detected by FITC-Achatinin-H. In contrast, the percentage of Neu5,9Ac_2-GPs$^+$ cells was very high in BM_{ALL} (~sixfold more than the former, Table 21.1), while it is significantly low in BM_{CR} as reported earlier [11–14]. Since this expression is comparable to that of BM_{CR}, therefore, the lower expression of this disease-associated antigen is a hallmark describing the CD34$^+$ HSCs of UCB.

In contrast to the expression of Neu5,9Ac_2-GPs, the α2,3-linked sialic acid content (Neu5Acα2-3Gal$^+$ cells), as identified by MAA, is significantly more (~1.6-fold more) in CD34$^+$ cells of UCB samples, as compared to the BM_{ALL} samples, confirming it as a marker with down-regulated expression in BM-associated malignancy, in this case childhood ALL (Table 21.1). Therefore, an elevated level of α2,3-linked sialic acid can also serve as a potent indicator of the normal CD34$^+$ HSCs of UCB.

Contrary to the differential expression of Neu5,9Ac_2-GPs and Neu5Acα2-3Gal, the percentage of cells having cell surface sialoglycoproteins with terminal Neu5Acα2-6Gal/GalNAc, as detected using SNA, had statistically insignificant variation between UCB and diseased samples (Table 21.1). Similarly, the total sialic acid content (ranging between 35 and 51%, Table 21.1) in the CD34$^+$ cells of UCB/BM_{ALL}/BM_{CR} is not significantly different from each other as detected by FITC-LPA. Therefore, total Neu5Ac or α2,6-linked sialic acid content does not hold promise as a potential biomarker for normal HSCs.

21.3.3 Differential Expression of CAMs Along with Selectins on the CD34$^+$ Cells of UCB, BM_{ALL}, and BM_{CR}

CAMs play a vital role in the initial steps of leukocyte trafficking and can be grouped into four major families including the immunoglobulin (Ig) superfamily CAMs, integrins, cadherins, and selectins. Among these, selectins mainly leukocyte endothelial CAMs (LECAMs) are a family of carbohydrate-binding proteins which bind fucosylated carbohydrates, especially sialylated Lewis^x and mucins. The three selectins are

Table 21.1 Cell surface molecular profile and enzymatic activities of CD34$^+$ HSCs of UCB compared to CD34$^+$ cells of BM_{ALL} and BM_{CR}

Parameters		CD34$^+$ cells UCB	BM_{ALL}	BM_{CR}
[a]Sialic acids and derivatives	Neu5,9Ac$_2$α2-6β-D-GalNAc	12 ± 3	70 ± 5	8 ± 4
	Neu5Ac	43 ± 8	43 ± 17	38 ± 5
	Neu5Acα2-6Gal/GalNAc	86 ± 5	76 ± 7	72 ± 9
	Neu5Acα2-3Gal	88 ± 4	54 ± 14	78 ± 7
[a]Cell adhesion molecules	L-selectin/LECAM-1/CD62L	27 ± 6	56 ± 12	43 ± 10
	E-selectin/LECAM-2/CD62E	23 ± 4	3 ± 1	2 ± 1
	P-selectin/LECAM-3/CD62P	91 ± 7	49 ± 9	44 ± 4
	ICAM-1/CD54	95 ± 3	46 ± 8	21 ± 9
	VCAM-1/CD106	30 ± 12	40 ± 6	42 ± 10
[b]Enzymes	Sialate-*O*-acetyltransferase (SOAT) (*p*mole/min × mg protein)	0.55 ± 0.02	0.84 ± 0.03	0.46 ± 0.02
	α2,6 sialyltransferase (ST6Gal I) (*p*mole/min × mg protein)	0.48 ± 0.03	0.57 ± 0.04	0.40 ± 0.04
	Sialidase (*n*mole/min × mg protein)	2.7 ± 0.1	0.3 ± 0.08	2.81 ± 0.09

[a]Expression in terms of percent positive cells; [b]enzymes regulating the overall sialylation of cells; *UCB* umbilical cord blood, BM_{ALL} bone marrow (BM) at diagnosis of childhood acute lymphoblastic leukaemia (ALL), BM_{CR} BM of childhood ALL patients in clinical remission

leukocyte (L)-, endothelial (E)-, and platelet (P)-selectins (Chowdhury et al. 2007).

In addition to the linkage-specific sialic acids, the selectins also provide a different sketch on the CD34$^+$ HSCs of UCB (Fig. 21.3, Table 21.1). Among the existing selectins studied here, E-selectin is exclusively expressed on the CD34$^+$ HSCs of UCB (~eightfold more) in contrast to negligible expression in BM_{ALL} and BM_{CR} samples. This significantly higher manifestation of E-selectin, therefore, can be utilized as a steadily expressing stem cell antigen.

L-selectin$^+$ cells showed considerable difference between all the three sets of samples studied. CD34$^+$ HSCs of UCB have the least amount of L-selectin$^+$ cells, significantly less (~twofold less) than BM_{ALL} and BM_{CR} samples. Consequently, this basal level of expression of L-selectin on CD34$^+$ HSCs of UCB, in contrast to over expression in diseased samples, can serve as an emblem for normal HSCs.

Analogous to the previous selectins, P-selectin does not exhibit a significant variation between BM_{ALL} and BM_{CR} samples. However, this expression becomes twofold more in the CD34$^+$ HSCs of UCB rendering it to be another indicator of these cells.

The Ig superfamily CAMs (ICAM-1 and VCAM-1) are calcium-independent transmembrane glycoproteins (Chowdhury et al. 2007). Interestingly, the ICAM-1 expression is extremely high in the CD34$^+$ cells from UCB, being twofold and fourfold more than that of BM_{ALL} and BM_{CR}, respectively (Table 21.1). Owing to this extremely significant variation in this expression in UCB, the potential of ICAM-1 as a biomarker of normal HSCs can be ensured. The CD34$^+$ cells of UCB had the least expression of VCAM-1 as compared to more CD34$^+$ cells in BM_{ALL}, and BM_{CR}, but statistically, insignificant difference rules out this antigen to be used as a marker of normal stem cells.

21.3.4 Lower Sialate-O-Acetyltransferase (SOAT) in CD34$^+$ Cells of UCB Compared to BM_{ALL}

The nascent sialoglycoconjugates are modified by SOAT which transfer the acetyl group from acetyl-coenzyme A onto sialoglycoconjugates. We have demonstrated that the enhanced activity of SOAT is responsible for the transfer of the *O*-

acetyl group to the sialic acid of glycoproteins in leukaemic cells (Mandal et al. 2009). In this study, the status of the enzymes regulating *O*-acetylated sialic acid was evaluated in all three populations (Fig. 21.4a, Table 21.1). In the CD34$^+$ cells of UCB, the SOAT activity was significantly less (~twofold) than that of CD34$^+$ cells of BM_{ALL} while almost similar to that of BM_{CR}. This finding supports the previously assured lower expression of cell surface Neu5,9Ac_2-GPs in UCB. Thus both the enzyme activity and the antigen expression show a trait opposite to the leukaemic blasts and hence can be used for the nomenclature of normal CD34$^+$ HSCs from UCB.

21.3.5 α2,6-Sialyltransferase Activity (ST6Gal I) Remains Almost Unaltered in the CD34$^+$ Cells

Sialyltransferase belongs to the enzyme class which adds sialic acid residues to sialoglycoproteins and sialoglycolipids during their biosynthesis. β-Galactoside α2,6-sialyltransferase (ST6Gal I) is the only sialyltransferase having the capability of adding sialic acid in α2,6-linkage to Galβ1,4GlcNAc, a sequence commonly found in *N*-linked chains of glycoproteins (Mondal et al. 2009). ST6Gal I activity in CD34$^+$ HSCs of UCB did not exhibit significant variation from the CD34$^+$ cells of BM_{ALL} and BM_{CR} samples. Thus, like the total α2,6-linked sialic acid content, the ST6Gal I activity cannot be used as an indicator of normal HSCs.

21.3.6 Higher Sialidase Activity in CD34$^+$ Cells from UCB Compared to BM_{ALL}

Sialidase removes sialic acids during sialoglycoconjugate degradation (Mandal et al. 2010). In the CD34$^+$ cells of UCB, the sialidase activity is significantly higher than that of the BM_{ALL} samples while almost comparable with that of BM_{CR}-derived CD34$^+$ cells (Fig. 21.4b, Table 21.1). Hence this high activity of the sialidase can be used as a specific indicator of normal HSCs.

Thus, the low expression of Neu5,9Ac_2-GPs is corroborated by the low activity of SOAT in CD34$^+$ HSCs of UCB. Similar expression of α2,6-linked sialic acids in UCB has been supported by analogous sialyltransferase activity while the high activity of sialidase accounts for the comparable sialic acid expression on CD34$^+$ HSCs of UCB. Hence, the synchronized activities of the aforesaid enzymes maintain the cell surface

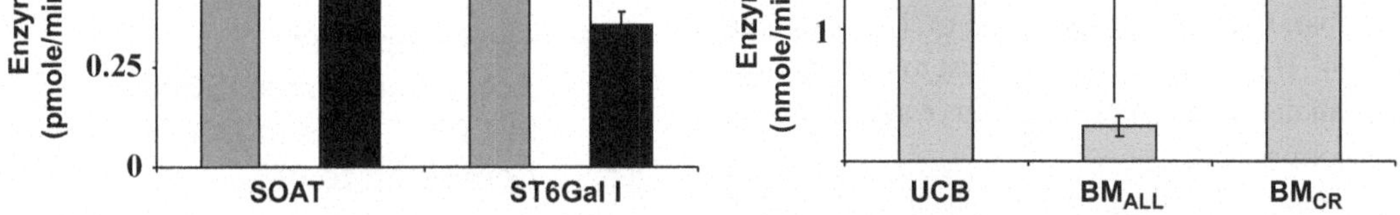

Fig. 21.4 Modulation of activities of sialate-*O*-acetyltransferase (SOAT), α2,6-sialyltransferase (ST6Gal I), and sialidase in CD34$^+$ cells of UCB as compared to BM_{ALL} and in BM_{CR}. Enzyme activities were determined using cell lysates as enzyme sources, and the reactions were performed in duplicate. *$P < 0.05$, a significant difference between two test groups. (**a**) The enzyme activities were evaluated in pmole/min × mg protein and represented as mean ± SEM. (**b**) The enzyme activity was evaluated in nmole/min × mg protein and represented as mean ± SEM

expression of the respective sialic acid derivatives.

21.4 Conclusion

O-Acetylated sialic acids play a multifaceted role in childhood ALL (Pal et al. 2004; Chowdhury et al. 2007; Mukherjee et al. 2008, 2009; Ghosh et al. 2005a, 2005b, 2007). The exclusive expression of two *O*-acetylated sialoglycoproteins (120 and 90 kDa) and their impaired immunological functions reflect their disease association. In contrast, normal lymphocytes have a minimal expression of 144 and 36 kDa Neu5,9Ac2-GPs with the low affinity towards Achatinin-H (Pal et al. 2004). The splendour of the present study is the lower expression of these *O*-acetylated sialoglycoproteins confirmed by the low SOAT activity in the $CD34^+$ HSCs of UCB attesting to their normal characteristics. Therefore, the profile of *O*-acetylated sialoglycoproteins in UCB is a unique outcome in this investigation. Additionally, the status of other molecules along with linkage-specific sialic acids on $CD34^+$ HSCs of UCB, in contrast to $CD34^+$ BM_{ALL}, also reassures their normal profile. Based on these observations, the established sialoglycopattern of normal $CD34^+$ HSCs of UCB opens up an unconventional avenue to monitor these cells before transplantation, thereby deserving special attention in clinical therapeutics.

Acknowledgement This work received financial support from the CSIR, New Delhi, Govt. of India. CM sincerely acknowledges the financial support from JC Bose National Fellowship (DST) and Distinguished Biotechnology Research Professor (DBT).

References

Brown JA, Boussiotis VA (2008) Umbilical cord blood transplantation: basic biology and clinical challenges to immune reconstitution. Clin Immunol 127:286–297

Chowdhury S, Mandal C (2009) *O*-acetylated sialic acids: multifaceted role in childhood acute lymphoblastic leukaemia. Biotechnol J 4:361–374

Chowdhury S, Bandyopadhyay S, Chandra S, Mandal C (2007) Comparative analysis of differential expression of sialic acids and adhesion molecules on mononuclear cells of bone marrow and peripheral blood in childhood acute lymphoblastic leukaemia at diagnosis and clinical remission. Indian J Biochem Biophys 44:357–365

Chowdhury S, Bandyopadhyay S, Mandal C, Chandra S, Mandal C (2008) Flow-cytometric monitoring of disease-associated expression of 9-O-acetylated sialoglycoproteins in combination with known CD antigens, as an index for MRD in children with acute lymphoblastic leukaemia: a two-year longitudinal follow-up study. BMC Cancer 8:40

Chowdhury S, Chandra S, Mandal C (2014) 9-O-acetylated sialic acids differentiating normal haematopoietic precursors from leukemic stem cells with high aldehyde dehydrogenase activity in children with acute lymphoblastic leukaemia. Glycoconj J 31 (6–7):523–535

George AA, Franklin J, Kerkof K, Shah AJ, Price M, Tsark E, Bockstoce D, Yao D, Hart N, Carcich S, Parkman R, Crooks GM, Weinberg K (2001) Detection of leukemic cells in the CD34(+)CD38(−) bone marrow progenitor population in children with acute lymphoblastic leukemia. Blood 97:3925–3930

Ghosh S, Bandyopadhyay S, Pal S, Das B, Bhattacharya DK, Mandal C (2005a) Increased interferon gamma production by peripheral blood mononuclear cells in response to stimulation of overexpressed disease-specific 9-O-acetylated sialoglycoconjugates in children suffering from acute lymphoblastic leukaemia. Br J Haematol 128:35–41

Ghosh S, Bandyopadhyay S, Mallick A, Pal S, Vlasak R, Bhattacharya DK, Mandal C (2005b) Interferon gamma promotes survival of lymphoblasts overexpressing 9-O-acetylated sialoglycoconjugates in childhood acute lymphoblastic leukaemia (ALL). J Cell Biochem 95:206–216

Ghosh S, Bandyopadhyay S, Mukherjee K, Mallick A, Pal S, Mandal C, Bhattacharya DK, Mandal C (2007) O-acetylation of sialic acids is required for the survival of lymphoblasts in childhood acute lymphoblastic leukemia (ALL). Glycoconj J 24:17–24

Hollands P, McCauley C (2009) Private cord blood banking: current use and clinical future. Stem Cell Rev Rep 5:195–203

Hough R, Cooper N, Veys P (2009) Allogeneic haemopoietic stem cell transplantation in children: what alternative donor should we choose when no matched sibling is available? Br J Haematol 147 (5):593–613

Krause DS, Van Etten RA (2007) Right on target: eradicating leukemic stem cells. Trends Mol Med 13:470–481

Mandal C, Chatterjee M, Sinha D (2000) Investigation of 9-O-acetylated sialoglycoconjugates in childhood acute lymphoblastic leukaemia. Br J Haematol 110:801–812

Mandal C, Srinivasan GV, Chowdhury S, Chandra S, Mandal C, Schauer R, Mandal C (2009) High level of sialate-*O*-acetyltransferase activity in lymphoblasts of childhood acute lymphoblastic leukaemia (ALL): enzyme characterization and correlation with disease status. Glycoconj J 26:57–73 and references therein

Mandal C, Tringali C, Mondal S, Anastasia L, Chandra S, Venerando B, Mandal C (2010) Down regulation of membrane-bound Neu3 constitutes a new potential marker for childhood acute lymphoblastic leukemia and induces apoptosis suppression of neoplastic cells. Int J Cancer 126(2):337–349

Mondal S, Chandra S, Mandal C (2009) Elevated mRNA level of hST6Gal I and hST3Gal V positively correlates with the high risk of pediatric acute leukemia. Leuk Res 34(4):463–470

Mukherjee K, Chowdhury S, Mondal S, Mandal C, Chandra S, Bhadra RK, Mandal C (2007) 9-O-acetylated GD3 triggers programmed cell death in mature erythrocytes. Biochem Biophys Res Commun 362:651–657

Mukherjee K, Chava AK, Mandal C, Dey SN, Kniep B, Chandra S, Mandal C (2008) O-acetylation of GD3 prevents its apoptotic effect and promotes survival of lymphoblasts in childhood acute lymphoblastic leukaemia. J Cell Biochem 105:724–734

Mukherjee K, Chava AK, Bandyopadhyay S, Mallick A, Chandra S, Mandal C (2009) Co-expression of 9-O-acetylated sialoglycoproteins and their binding proteins on lymphoblasts of childhood acute lymphoblastic leukemia: an anti-apoptotic role. Biol Chem 390:325–335

Pal S, Ghosh S, Bandyopadhyay S, Mandal C, Bandhyopadhyay S, Bhattacharya DK, Mandal C (2004) Differential expression of 9-O-acetylated sialoglycoconjugates on leukemic blasts: a potential tool for long-term monitoring of children with acute lymphoblastic leukemia. Int J Cancer 111:270–277

Satomaa T, Heiskanen A, Mikkola M, Olsson C, Blomqvist M, Tiittanen M, Jaatinen T, Aitio O, Olonen A, Helin J, Hiltunen J, Natunen J, Tuuri T, Otonkoski T, Saarinen J, Laine J (2009) The N-glycome of human embryonic stem cells. BMC Cell Biol 10:42 and references therein

Schauer R (2000) Achievements and challenges of sialic acid research. Glycoconj J 17:485–499

Sinha D, Mandal C, Bhattacharya DK (1999a) Identification of 9-O acetyl sialoglycoconjugates (9-OAcSGs) as biomarkers in childhood acute lymphoblastic leukaemia using a lectin, AchatininH, as a probe. Leukemia 13:119–125

Sinha D, Mandal C, Bhattacharya DK (1999b) A novel method for prognostic evaluation of childhood acute lymphoblastic leukemia. Leukemia 13:309–312

Storms RW, Green PD, Safford KM, Niedzwiecki D, Cogle CR, Colvin OM, Chao NJ, Rice HE, Smith CA (2005) Distinct hematopoietic progenitor compartments are delineated by the expression of aldehyde dehydrogenase and CD34. Blood 106:95–102

Part VI

Microbial Cell Surface

Vitamin C: A Natural Inhibitor of Cell Wall Functions and Stress Response in Mycobacteria

22

Kirtimaan Syal and Dipankar Chatterji

Abstract

Tuberculosis, caused by *Mycobacterium tuberculosis*, has re-emerged as a threat to human race. Conventional antibiotic treatments are failing due to different stress response strategies adopted by bacterial pathogens. Since time immemorial, Vitamin C is known to protect against pathogens by boosting immunity in humans. Recently, Vitamin C has been shown to directly kill *M. tuberculosis* including multiple drug-resistant strains by generation of oxidative radicals through Fenton's reaction. Concurrently, it inhibits (p)ppGpp-mediated stringent response thus effectively shutting down long-term survival and persistence in mycobacteria. Here, we have discussed historical perspective and recent evidences on Vitamin C-mediated inhibition of several key pathways of *M. tuberculosis* such as (p)ppGpp synthesis and mycobacterial cell wall function. Several cell wall components including mycolic acids are critical for mycobacterial virulence. We observed downregulation of various mycolic acids in *M. smegmatis* upon treatment with Vitamin C, and data have been presented here. Vitamin C has been shown to inhibit the biofilm growth as well as disrupt the formed biofilm in mycobacteria. Additionally, Vitamin C role in cell-mediated and humoral immunity has been elucidated. Vitamin C is toxic at high concentration; therefore we have proposed the idea of derivatizing Vitamin C in order to lower the minimal inhibition concentration (MIC) necessary to target *M. tuberculosis*.

Keywords

Vitamin C · (p)ppGpp · Biofilm formation · Immune response · Oxidative radicals · Mycobacteria · Mycolic acid

Abbreviations

(p)ppGpp	Guanosine tetraphosphate or guanosine pentaphosphate
C1q	Complement component 1, q subcomponent
MDR	Multidrug resistant
NK cells	Natural killer cells
RF3	Release factor 3
RNAP or RNA Pol	RNA polymerase
RNS	Reactive nitrogen species
ROS	Reactive oxygen species
TLC	Thin-layer chromatography
XDR	Extensively drug resistant

K. Syal · D. Chatterji (✉)
Molecular Biophysics Unit, Indian Institute of Science, Bangalore, India
e-mail: dipankar@iisc.ac.in

K. Chattopadhyay, S. C. Basu (eds.), *Biochemical and Biophysical Roles of Cell Surface Molecules*, Advances in Experimental Medicine and Biology 1112, https://doi.org/10.1007/978-981-13-3065-0_22

22.1 Introduction

Vitamins are organic compounds essential to humans that must be obtained from food except Vitamin D, which can be made inside human body provided there is enough exposure to sunlight. 'Vitamin' word has been derived by the combination of 'vita' (in Latin, vita means life) and chemical group 'amine'. Vitamin is a misnomer as most vitamins do not contain amine group. Like all other vitamins, humans must obtain Vitamin C from their diet. Most animals can synthesize Vitamin C in the body from D-glucose or D-galactose except for primates, humans, guinea pigs and fruit bats. The human body lacks D-gulonolactone oxidase enzyme necessary for Vitamin C synthesis (Lehninger et al. 2008). The deficiency of Vitamin C caused devastation across the world until it was isolated and purified by Albert Szent-Gyorgyi (Nobel Laureate, Physiology or Medicine, 1937) (Lehninger et al. 2008). The recommended daily intake of Vitamin C has long been disputed. As per US Food and Nutrition Board of the Institute of Medicine, an adult man and woman need 90 milligrams/day and 75 milligrams/day of Vitamin C, respectively (Institute of Medicine (US) Panel on Dietary Antioxidants and Related Compounds. Dietary Reference Intakes for Vitamin C, Vitamin E, Selenium, and Carotenoids. Washington (DC) 2000). However, Vitamin C could be administered up to 2 g/day by oral or injection route for therapeutic purposes like treating scurvy (Institute of Medicine (US) Panel on Dietary Antioxidants and Related Compounds. Dietary Reference Intakes for Vitamin C, Vitamin E, Selenium, and Carotenoids. Washington (DC) 2000). Linus Pauling (Nobel Laureate, Chemistry, 1954) advocated the administration of Vitamin C for treating common cold, flu and cancer (Hemila 1997). At that time, it was contested that, nevertheless in the light of recent evidences, interest in Vitamin C-based therapy has again rejuvenated (Berger and Oudemans-van Straaten 2015). Earlier, Vitamin C was shown to enhance the function of the immune system. It specifically upregulates the T-lymphocyte proliferation and increases the levels of complement protein C1q thus strengthening the cell-mediated immunity and humoral immunity, respectively (Badr et al. 2012; Johnston et al. 1987). More recently, it has been discovered that Vitamin C can directly act and kill dreadful pathogen *Mycobacterium tuberculosis* by Fenton's reaction (Vilcheze et al. 2013). Additionally, it targets (p)ppGpp synthesis thus bringing down the stringent response in mycobacteria (Syal et al. 2017a). Tough mycobacterial cell wall with defined architecture is essential for its virulence and survival. It protects mycobacteria from host immune response. One of its constituents, namely, mycolic acids, is critical virulence factors, and any inhibition in mycolic acid synthesis compromises mycobacterial survival inside the host. Vitamin C has been shown to inhibit mycobacterial cell wall function and biofilm formation (Syal et al. 2017a). On one side, Vitamin C enhances the immune system, and on the other side, it directly targets the pathogen such as *M. tuberculosis*, which is incredible and unique. It has been shown to be effective against drug-susceptible, multidrug-resistant (MDR) and extensively drug-resistant (XDR) *M. tuberculosis* strains supporting the contention of the incorporation of Vitamin C in anti-mycobacterial therapy (Vilcheze et al. 2013). However, necessity of high concentration to directly target *M. tuberculosis* is a limiting factor. It has been shown that tuberculosis patients have lower blood concentration of Vitamin C thus associating its deficiency with the higher incidence of tuberculosis (Awotedu et al. 1984). Vitamin C is believed to protect from infectious diseases. Vitamin C also helps in prognosis of tuberculosis by promoting healing of lesions (Soh et al. 2017; Chakraborty et al. 2014). Most importantly, bacteria can evolve into resistant strains due to its ability to persist attributed to (p)ppGpp synthesis which is considered as the master regulator of stress response and persistence (Syal et al. 2015c). Inhibition of (p)ppGpp function compromises bacterial survival in hostile conditions. Still, many key molecular mechanisms involved in assimilation of Vitamin C by immune cells, cell protection, regulation of oxidation and reduction are largely unknown.

22.2 Vitamin C and Historical Perspective

Many ancient excerpts described large-scale devastation inflicted by the deficiency of Vitamin C. For example, Vasco da Gama lost two-third of his crew during exploration of trade routes to India (Lehninger et al. 2008) due to scurvy. In the early seventeenth century, James Lind (surgeon from the Royal Army) carried out first control study where he observed quick recovery of sailors from scurvy on inclusion of lemons and oranges in their diet (Lehninger et al. 2008). Almost after 40 years, the British admiral made concentrated lemon juices mandatory for all British sailors. Later, Vitamin C was discovered as a cure for scurvy. It was reported to be essential for the development of connective tissues, bone formation and healthy gums (Chambial et al. 2013). Multiple autopsy reports showed complication of pneumonia in Vitamin C-deficient patients (Hemila and Louhiala 2007). However, it was contested as a secondary complication of scurvy.

The Vitamin C levels were reported to be lower in comparison to the healthy controls in infectious diseases including tuberculosis (Dunn 1945). Linus Pauling advocated the administration of high concentration of Vitamin C in cases of pathogenic infections. Though several studies concluded and agreed upon the positive correlation between Vitamin C deficiency and infections, it was not clear whether lower levels of Vitamin C result in infection or infection leads to the lower levels of Vitamin C. It needs further investigation (Hemila 2017).

Vitamin C is a well-known antioxidant. The pathogenic infestation leads to high levels of reactive oxidant species (ROS) in immune cells, and Vitamin C helps in containing the harmful effects of such oxidative radicals in the host cells. Phagocytes transport Vitamin C into the cell to prevent the toxic effects of ROS. This transport reduces the plasma concentration of Vitamin C which is needed to be replenished. Thus, Vitamin C helps in the regulation of immune response to pathogens. This topic has been further explained in the next section. Recently, Harri Hemilä elegantly summarized the outcome of several studies carried out on guinea pigs infected with bacterial pathogens and fed with Vitamin C-rich diet or pure Vitamin C (Hemila 2017). In most of these studies, guinea pigs have been used as a model organism as like humans, they cannot synthesize Vitamin C and are dependent on dietary intake of Vitamin C.

Vitamin C was shown to increase the survival of terminally ill cancer patients. Pauling and Cameron reported that the supplementation of Vitamin C (10 g/day) to the conventional cancer therapy helped in saving 10.3% of the terminally ill cases, whereas none of the patients administered with conventional therapy alone survived (Padayatty et al. 2006). Another study reported a sixfold increase in survival of cancer patients on supplementation of therapy with Vitamin C. Though there are statistical evidences, comparing one cancer patient with another may not be correct as many parameters like patient condition, doses of medicines, case history or response to medicines/Vitamin C could be different (Padayatty et al. 2006).

Most importantly, Vitamin C is necessary for the hydroxylation of collagen. Collagen is a crucial component of connective tissue. It has a triple helix structure with very high hydroxyproline (unusual amino acid) content (Lehninger et al. 2008).

22.3 Vitamin C and Fenton's Reaction

It is evident that many bactericidal antibiotics corrupt iron metabolism leading to the synthesis of highly reactive hydroxyl radicals by Fenton's reaction. In 2013, Jacob Jr. et al. showed that Vitamin C through Fenton's reaction kills drug-susceptible and drug-resistant *M. tuberculosis* (Vilcheze et al. 2013). Interestingly, *M. tuberculosis* was shown to be highly susceptible to the high concentration of Vitamin C, unlike most Gram-negative and Gram-positive bacteria (Vilcheze et al. 2013). Susceptibility of mycobacteria to Vitamin C was dependent on high ferrous ion levels in its cytoplasm. It is worth mentioning that hydroxyl radicals are generated by the combination of Haber-Weiss

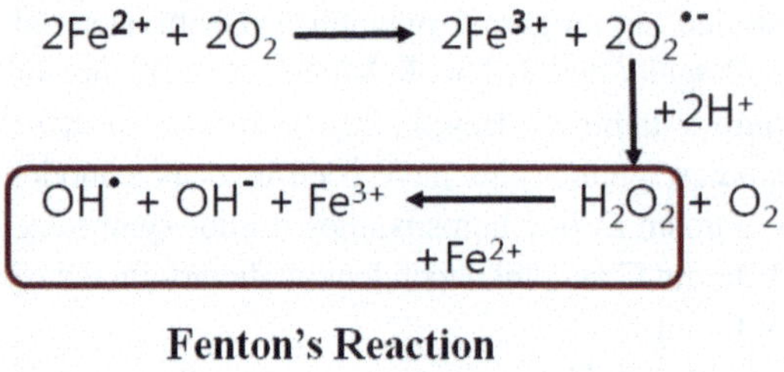

Fig. 22.1 Generation of hydroxyl radicals through Fenton's reaction. (Vilcheze et al. 2013; Aisen et al. 2001; Williams et al. 2014)

cycle and Fenton's reaction (Vilcheze et al. 2013) (Fig. 22.1).

Vitamin C functions as a pro-oxidant in Fenton's reaction. Iron is essential for growth of *M. tuberculosis* and development of tuberculosis disease (Sargazi et al. 2017). *M. tuberculosis* acquires iron from the host with the help of siderophores and mycobactin (Sritharan 2016). It has been shown that mutation in mycobactin can be detrimental for survival of *M. tuberculosis* inside the host (Reddy et al. 2013). Hydroxyl radicals produced from Fenton's reaction induce cell death by DNA damage which is partly due to the oxidation of guanine nucleotide pool (Vilcheze et al. 2013). It has been shown that Vitamin C can sterilize *M. tuberculosis* cultures in vitro as well. It increases iron levels inside mycobacterial cells. The conversion of ferrous ion to ferric ion leads to the production of ROS (Vilcheze et al. 2013). On one hand, Vitamin C acts as a pro-oxidant in Fenton's reaction, but on the other hand, it quenches ROS and reactive nitrogen species (RNS) in immune cells thus regulating immune function (Carr and Maggini 2017).

22.4 (p)ppGpp Master Regulator of Stress Response and Persistence

In 1969, Cashel and Gallant reported first the appearance of a spot on thin-layer chromatographic (TLC) analysis of nucleotides in lane corresponding to the cell extract from the starved *Escherichia coli* cells in comparison to that of the control (Cashel and Gallant 1969). In the 1980s, Bremer's group from the University of Texas (Little et al. 1983) and Ishihama's group

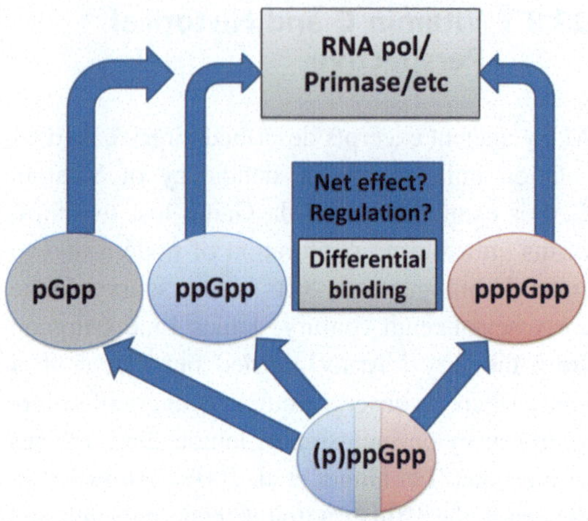

Fig. 22.2 Schematic view: plausible differential regulatory roles of pGpp, ppGpp and ppGpp (Syal and Chatterji 2015)

from Kyoto University observed that ppGpp regulates transcription but the mechanism was not clear (Glass et al. 1986). In 1994, one of us showed, for the first time, that (p)ppGpp directly binds to RNA polymerase at the C-terminal domain of beta subunit of RNA polymerase (Reddy et al. 1995). Later, Hernandez group from State University of New York confirmed the binding pocket at the interface of beta-beta' subunit of RNA polymerase (Toulokhonov et al. 2001). The binding sites have been long contested (Perederina et al. 2004; Vrentas et al. 2008), but soon with the availability of RNAP-ppGpp crystal structure from Steitz's group (Zuo et al. 2013; Ross et al. 2013), it is taken as resolved. The latter crystallization experiment was conducted by soaking and not co-crystallization, so ppGpp binding with different conformations of RNA polymerase might be different. Further, ppGpp and pppGpp were considered as the same molecule, but studies showed their differential binding to RNA polymerase explaining different effects on transcription (Mechold et al. 2013; Syal and Chatterji 2015). Still, many questions remain open like what is the net regulation and effect? Is there a special function associated with each of them? How are levels of pGpp, ppGpp and pppGpp determined? Is there any mechanism or mode of ppGpp and pppGpp interconversion (Fig. 22.2)? The primary target of (p)ppGpp is RNA polymerase in Gram-negative bacteria.

Table 22.1 (p)ppGpp binds to several partners, and the list indicates natural function of (p)ppGpp target along with the effect of (p)ppGpp binding

(p)ppGpp binding partners	Function	Effect of (p)ppGpp on their function
RNA polymerase	**Transcription**	**Inhibits or activates depending on the AT and GC content of the promoters (in Gram-negative bacteria) (Jores and Wagner** 2003)
RF3 (release factor)	Release of RF1 and RF2 from ribosome complex. Role in translation	Inhibits RF3 function and translation (Kihira et al. 2012)
Guanylate kinase (GMK)	**Guanosine metabolism**	**Inhibits (critical in Gram-positive bacteria as they lack transcription-mediated regulation) (Liu et al.** 2015)
ppGpp synthetase – SAS	Allosterically activates synthesis of (p)ppGpp	Stress response (Steinchen et al. 2015)
Rel from *M. smegmatis*	(p)ppGpp synthesis and hydrolysis	Binds C-terminal domain at high concentration. High concentration switches from synthesis to hydrolysis. Also, enzyme kinetics illustrated activation of enzyme at low concentration of (p)ppGpp. Model needs more investigation (Syal et al. 2015b)
Primase DnaG	DNA replication	Inhibits its function (Steinchen et al. 2015)
Ldc1, lysine decarboxylase	Acid resistance	Inhibits in a pH-dependent manner. Needs more validation (Kanjee et al. 2011)
BipA	Critical for the biogenesis of 70S ribosomes	Inhibits ribosome assembly (Kumar et al. 2015)
Obg GTPase	Implicated in ribosome biogenesis, stress response and DNA replication	Prevents assembly of 50S subunit of ribosome with 30S. Needs more validation (Verstraeten et al. 2015)

Most important partners are in bold font

However, it has been shown to regulate various other processes including translation and DNA replication in *E. coli* (Table 22.1).

(p)ppGpp is synthesized by RelA/SpoT enzyme and hydrolysed by SpoT enzyme in *E. coli* (Braedt and Gallant 1977; Weyer et al. 1976). (p)ppGpp synthetase (RSH) domain is largely conserved in bacteria, algae and plants (van der Biezen et al. 2000). (p)ppGpp synthesis involves transfer of pyrophosphate from 5'OH position of ATP to the 3'OH group of GDP or GTP leading to the synthesis of ppGpp or pppGpp (Syal et al. 2015c). In mycobacteria, (p)ppGpp is synthesized and hydrolysed by a bifunctional Rel enzyme (Jain et al. 2007). Both synthesis and hydrolysis activities are concentrated at the N-terminal domain of the enzyme (Jain et al. 2007). Earlier, we showed that ppGpp at high concentration binds to C-terminal domain of bifunctional Rel enzyme from *M. smegmatis* switching its activity from synthesis to hydrolysis (Syal et al. 2015b). Structure of N-terminal domain of Rel enzyme from *Streptococcus equisimilis* revealed two conformations: (a) open confirmation, synthesis off, hydrolysis on, and (b) close confirmation, synthesis on, hydrolysis off. Both synthesis and hydrolysis activities were shown to be mutually exclusive (Hogg et al. 2004). The C-terminal domain has been shown to play a key role in regulation of (p)ppGpp synthesis and hydrolysis (Syal et al. 2015b) (Fig. 22.3).

Earlier, it has been shown that ppGpp analogue can inhibit (p)ppGpp synthesis in *Bacillus subtilis* (Wexselblatt et al. 2012). On an important note, survival of (p)ppGpp null strain of *M. tuberculosis* was also shown to be compromised inside the host (Klinkenberg et al. 2010). Therefore, we decided to synthesize artificial analogues as well as look for the natural analogues for inhibiting (p)ppGpp synthesis in mycobacteria. We observed that (p)ppGpp

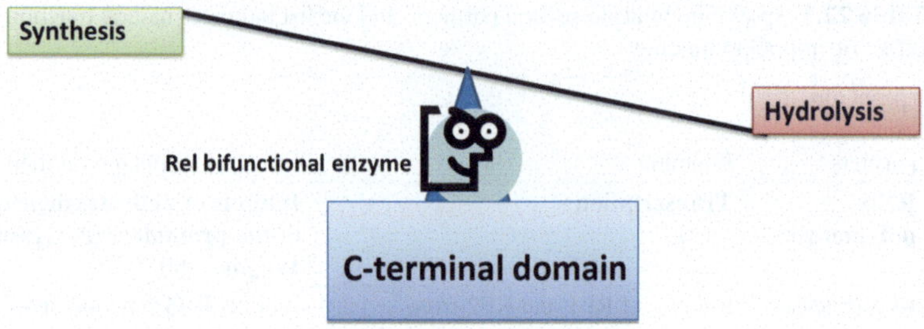

Fig. 22.3 Schematic view of C-terminal domain-mediated regulation of (p)ppGpp synthesis and hydrolysis by N-terminal domain of Rel enzyme from *M. smegmatis*

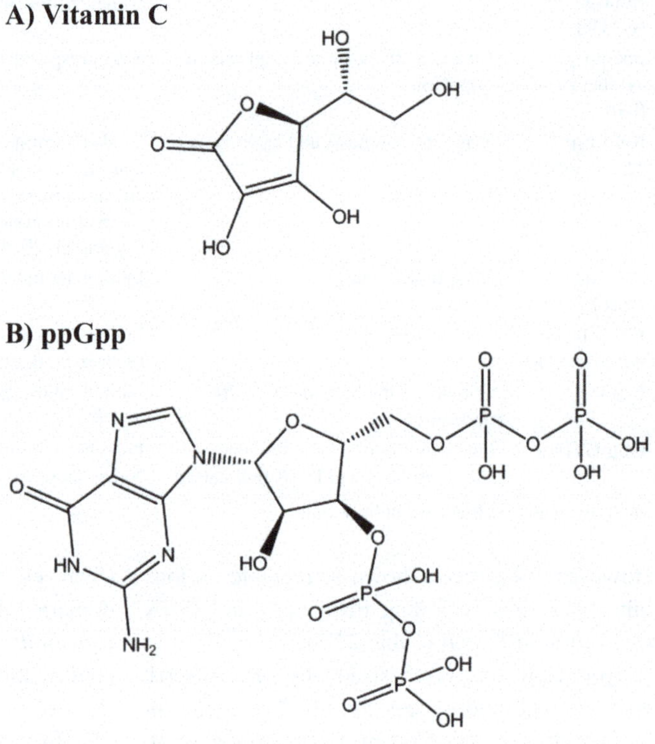

Fig. 22.4 Chemical Structure of Vitamin C and ppGpp molecule

analogue can inhibit (p)ppGpp synthesis in mycobacteria (Syal et al. 2017b). Also, we showed that targeting (p)ppGpp synthesis leads to the inhibition of long-term survival and biofilm formation in mycobacteria (Syal et al. 2017b). Vitamin C was shown to be directly involved in regulation of mycobacterial persistence, but the mechanism was largely unknown (Sikri et al. 2015). We tested the effect of Vitamin C on the formation of (p)ppGpp. Vitamin C inhibited both in vitro and in vivo synthesis of (p)ppGpp in *M. smegmatis* in a dose-dependent manner (Syal et al. 2017a). The inhibition of (p)ppGpp synthesis was observed to be specific. We determined the binding of Vitamin C to bifunctional Rel from *M. smegmatis* by isothermal titration calorimetry (Syal et al. 2017a). Further, enzyme kinetics was followed where $K_{0.5}$ was calculated to be increased with the corresponding reduction of Vmax value. It suggested mixed inhibition. As expected, Vitamin C also inhibited long-term survival and biofilm formation in *M. smegmatis* (Syal et al. 2017a). Vitamin C and ppGpp structures are presented in Fig. 22.4.

22.5 Vitamin C and Mycobacterial Cell Wall

Mycobacterial cell envelope is unique and critical for virulence of pathogenic strains. It is hydrophobic in nature and waxy coated rich in mycolic acids (Syal et al. 2015c). The mycobacterial cell wall has several key signatures such as arabinogalactan and oligosaccharides. A representation of mycobacterial cell membrane is shown in Fig. 22.4. The three layers, the plasma membrane (PM), the cell wall and the outermost layer, constitute the cell envelope (Syal et al. 2015c). Vitamin C is a small molecule and thus permeable to cell membrane.

Proteomic analysis revealed downregulation of Wag31, cell shape protein, by threefolds in Vitamin C-treated mycobacterial cells (Mishra and Sarkar 2015). Further, expression of FtsZ, key cell division protein, was also found to be suppressed upon treatment with Vitamin C in *M. smegmatis* (Mishra and Sarkar 2015). Interestingly, lipid biosynthesis *psd* gene coding for phosphatidylserine decarboxylase (catalyses the reaction of phosphatidylserine to phosphatidylethanolamine) was reported to be upregulated. In other organisms, *psd* gene has been linked with the ROS generation. Lipids are extraordinarily sensitive to oxidative damage which can impair their function. In alignment to the previous studies (Vilcheze et al. 2013), we also observed overall downregulation of lipids upon Vitamin C treatment in *M. smegmatis* (data not shown) (Syal et al. 2016). Further, we observed downregulation of mycolic acids upon Vitamin C treatment in *M. smegmatis* (Fig. 22.5), as determined by densitometric analysis of TLC sheet by a method described elsewhere (Syal et al. 2017a).

22.6 Vitamin C and Biofilms

Many bacteria exist in biofilm community in addition to the planktonic single-cell state. Biofilm formation involves aggregation of cells in a defined manner (Basaraba and Ojha 2017). Microbial cells in biofilms are predominantly organized into outer layer, matrix and attachment layer. Outer layer protects bacterial cells residing below in inside layers from hostile conditions like temperature or antibiotics. Outer layer cells restrict the permeability of antibiotics at the cost of their own survival for the greater benefit of the biofilm community thus increasing the minimum inhibition concentration of antibiotics to the thousandfold higher value (Basaraba and Ojha 2017; Syal et al. 2015c). Therefore, pathways necessary for biofilm formation constitute a fascinating target (Maiti et al. 2017; Naresh et al. 2012). The middle matrix layer secretes the necessary buffer of polysaccharides and metabolites that keep the biofilm layer together and intact. The attachment layer keeps the biofilm layer attached to substratum. Molecular cascades responsible for switching planktonic to biofilm form are not understood. Secondary messenger (p)ppGpp has been shown to play a key role in activation of biofilms (Liu et al. 2017; Nunes-Alves 2014). In fact, we reported that inhibition of (p)ppGpp synthesis compromises the ability of *M. smegmatis* and *M. tuberculosis* to form biofilm (Syal et al. 2017b). Additionally, it disrupts the formed biofilm suggesting that (p)ppGpp is not only involved in the formation of biofilms but also critical for its maintenance (Syal et al. 2017b). As described before, we noticed downregulation of (p)ppGpp levels in a concentration-dependent manner. As expected, Vitamin C inhibited biofilm formation in *M. smegmatis*, which could be predominantly attributed to the inhibition of (p)ppGpp synthesis (Syal et al. 2017a).

22.7 Vitamin C Boosts Immunity

Evidently, Vitamin C plays a key role in the immune system. It is known for its immunostimulatory and anti-inflammatory properties (Sorice et al. 2014). Many reports suggest that both cell-mediated and humoral immune response is improved by Vitamin C supplementation. Vitamin C enhanced the natural killer (NK) cell activity, and it led to the higher levels of C1q protein and antibodies in blood plasma (Heuser and Vojdani 1997; Chambial et al. 2013). Free

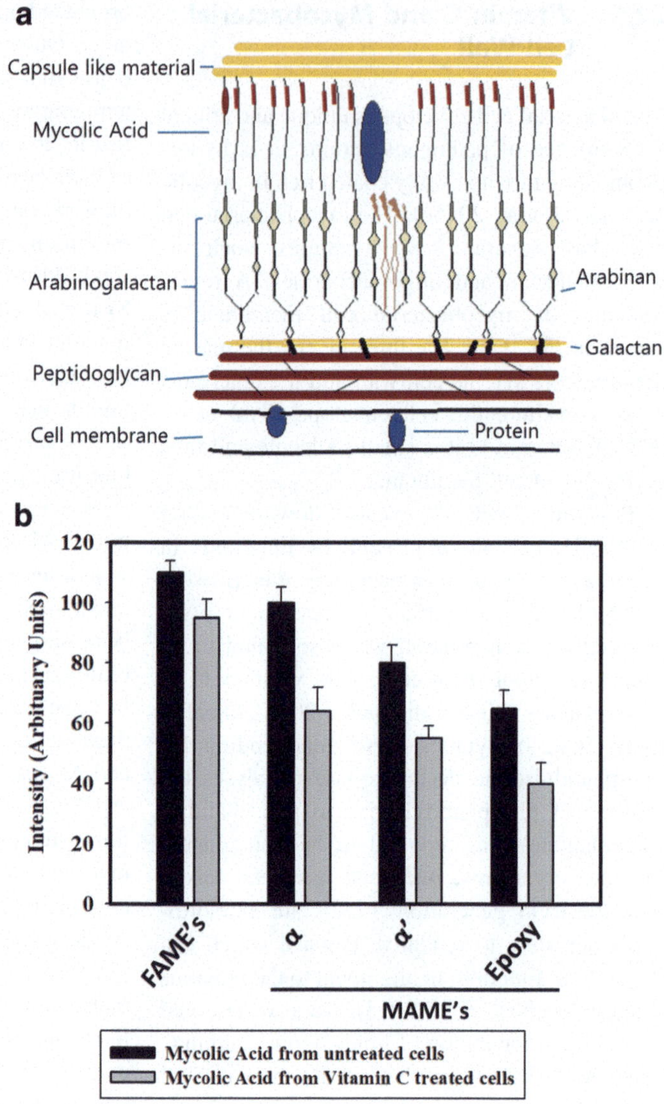

Fig. 22.5 (**a**) Schematic view of mycobacterial cell envelope. (**b**) Densitometric analysis of mycolic acids by thin-layer chromatography extracted from *M. smegmatis* cells treated with Vitamin C in comparison to that of the control

radicals like ROS are continuously generated in immune cells. Free radicals are important for functioning of the immune system but in a regulated manner. At high levels, free radicals may damage the essential biomolecules including DNA and proteins thus inhibiting immune cell function and causing cell death. At moderate levels, both ROS and RNS serve as the key regulatory molecules. ROS participate in defence against bacterial infections (Bhattacharya 2015). In activated macrophages, ROS target and digest the engulfed bacteria. Similarly, RNS play a critical role in targeting the engulfed pathogen (Bhattacharya 2015). Additionally, RNS act as a key signalling molecule responsible for growth and coordination of various immune cells including macrophages, T lymphocytes, neutrophils and NK cells (Birben et al. 2012). Vitamin C quenches the free radicals whenever required thus regulating the ROS function in immune cells. On the other hand, Vitamin C acts as a pro-oxidant leading to the production of free radicals mediated by Fenton's reaction, as described before.

22.8 Discussion

Vitamin C has been shown to be critical for the maintenance of immune system. Many pathogenic infestations, common cold and cancer have been associated with the deficiency of Vitamin C. As mentioned before, whether the deficiency of Vitamin C is the cause of susceptibility to pathogenic infestation or it is the symptom due to consumption of Vitamin C in immune cells is not clear (Fig. 22.6). Recent evidences suggest direct effects of Vitamin C manifested via Fenton's reaction or ppGpp synthesis in mycobacteria.

Additionally, we observed inhibition of mycolic acid synthesis (Fig. 22.5) and change in lipid profile (data not shown) in *M. smegmatis* upon Vitamin C treatment. Many beneficial effects (Fig. 22.7) were observed at higher concentration of Vitamin C, and such levels of Vitamin C could be toxic to humans. Excessive Vitamin C levels can lead to generation of free radicals through iron metabolism which in contrast are damaging to cells. Besides, Vitamin C catabolism by-product oxalic acid is a known risk factor for kidney stone. Excess of Vitamin C may also lead to erosion of tooth enamel, diarrhoea, gastrointestinal problems, Vitamin B_{12} deficiency, higher iron levels and higher uric acid (Institute of Medicine (US) Panel on Dietary Antioxidants and Related Compounds. Dietary Reference Intakes for Vitamin C, Vitamin E,

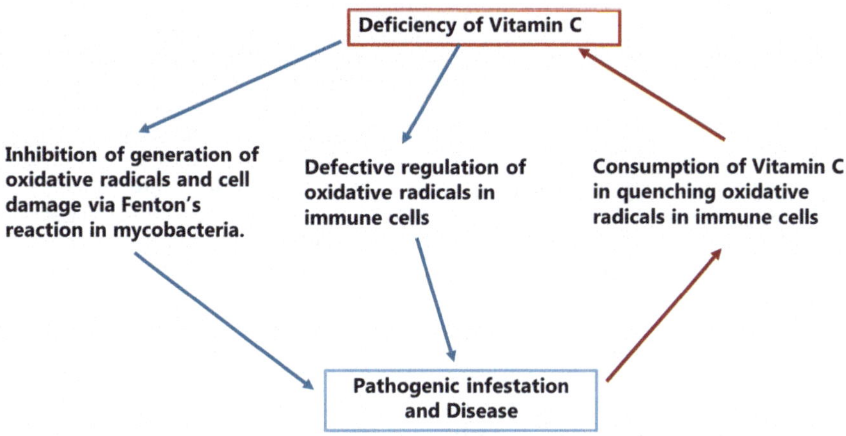

Fig. 22.6 Deficiency of Vitamin C and infection

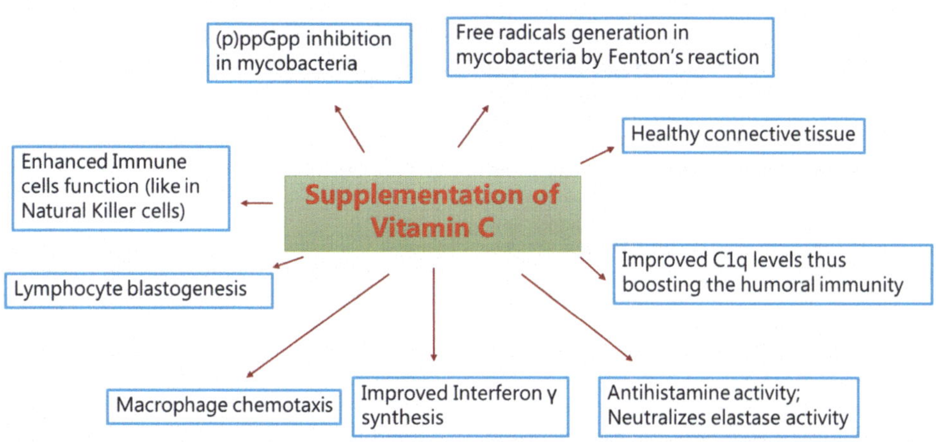

Fig. 22.7 Supplementation of Vitamin C and its potential effects

Selenium, and Carotenoids. Washington (DC) 2000).

In order to overcome the adverse effects, Vitamin C can be administered directly at the site of action like in case of tuberculosis; it can be administered through inhalers for the local higher concentration thus preventing the accumulation of Vitamin C or its by-product in plasma or any other organ (Syal et al. 2015a). Further, we propose that the modification of Vitamin C so that it cannot degrade into oxalic acid would be an effective way, without compromising its natural function.

Acknowledgement Authors acknowledge the Department of Science and Technology, Government of India, and Department of Biotechnology, Government of India, for funding the laboratory.

References

Aisen P, Enns C, Wessling-Resnick M (2001) Chemistry and biology of eukaryotic iron metabolism. Int J Biochem Cell Biol 33 (10):940–959. doi: S1357-2725(01)00063-2 [pii]

Awotedu AA, Sofowora EO, Ette SI (1984) Ascorbic acid deficiency in pulmonary tuberculosis. East Afr Med J 61(4):283–287

Badr G, Bashandy S, Ebaid H, Mohany M, Sayed D (2012) Vitamin C supplementation reconstitutes polyfunctional T cells in streptozotocin-induced diabetic rats. Eur J Nutr 51(5):623–633. https://doi.org/10.1007/s00394-011-0176-5

Basaraba RJ, Ojha AK (2017) Mycobacterial biofilms: revisiting tuberculosis bacilli in extracellular necrotizing lesions. Microbiol Spectr 5(3). https://doi.org/10.1128/microbiolspec.TBTB2-0024-2016

Berger MM, Oudemans-van Straaten HM (2015) Vitamin C supplementation in the critically ill patient. Curr Opin Clin Nutr Metab Care 18(2):193–201. https://doi.org/10.1097/MCO.0000000000000148

Bhattacharya S (2015) Reactive oxygen species and cellular defense system. In: Rani V, Yadav UCS (eds) Free radicals in human health and disease. Springer, New Delhi, pp 17–29. https://doi.org/10.1007/978-81-322-2035-0_2

Birben E, Sahiner UM, Sackesen C, Erzurum S, Kalayci O (2012) Oxidative stress and antioxidant defense. World Allergy Organ J 5(1):9–19. https://doi.org/10.1097/WOX.0b013e3182439613

Braedt G, Gallant J (1977) Role of the rel gene product in the control of cyclic adenosine 3′,5′-monophosphate accumulation. J Bacteriol 129(1):564–566

Carr AC, Maggini S (2017) Vitamin C and immune function. Nutrients 9(11). doi: nu9111211 [pii]10.3390/nu9111211

Cashel M, Gallant J (1969) Two compounds implicated in the function of the RC gene of Escherichia coli. Nature 221(5183):838–841

Chakraborty S, Syal K, Bhattacharyya R, Banerjee D (2014) Vitamin deficiency and tuberculosis: need for urgent clinical trial for management of tuberculosis. J Nutr Health Food Sci 2(2):1–6

Chambial S, Dwivedi S, Shukla KK, John PJ, Sharma P (2013) Vitamin C in disease prevention and cure: an overview. Indian J Clin Biochem 28(4):314–328. https://doi.org/10.1007/s12291-013-0375-3375 pii

Dunn ME (1945) A Study of Vitamin C Deficiency in Patients with Pulmonary Tuberculosis. Ulster Med J 14(1):17–33

Glass RE, Jones ST, Ishihama A (1986) Genetic studies on the beta subunit of Escherichia coli RNA polymerase. VII. RNA polymerase is a target for ppGpp. Mol Gen Genet 203(2):265–268

Hemila H (1997) Vitamin C supplementation and the common cold--was Linus Pauling right or wrong? Int J Vitam Nutr Res 67(5):329–335

Hemila H (2017) Vitamin C and infections. Nutrients 9(4): 339. doi: nu9040339 [pii]10.3390/nu9040339

Hemila H, Louhiala P (2007) Vitamin C may affect lung infections. J R Soc Med 100(11):495–498. doi: 100/11/495 [pii]10.1177/014107680710001109

Heuser G, Vojdani A (1997) Enhancement of natural killer cell activity and T and B cell function by buffered vitamin C in patients exposed to toxic chemicals: the role of protein kinase-C. Immunopharmacol Immunotoxicol 19(3):291–312. https://doi.org/10.3109/08923979709046977

Hogg T, Mechold U, Malke H, Cashel M, Hilgenfeld R (2004) Conformational antagonism between opposing active sites in a bifunctional RelA/SpoT homolog modulates (p)ppGpp metabolism during the stringent response [corrected]. Cell 117(1):57–68. doi: S0092867404002600 [pii]

Institute of Medicine (US) Panel on Dietary Antioxidants and Related Compounds (2000) Dietary reference intakes for vitamin C, vitamin E, selenium, and carotenoids. Washington, DC

Jain V, Saleem-Batcha R, Chatterji D (2007) Synthesis and hydrolysis of pppGpp in mycobacteria: A ligand mediated conformational switch in Rel. Biophys Chem 127(1–2):41–50. https://doi.org/10.1016/j.bpc.2006.12.003

Johnston CS, Kolb WP, Haskell BE (1987) The effect of vitamin C nutriture on complement component C1q concentrations in guinea pig plasma. J Nutr 117 (4):764–768

Jores L, Wagner R (2003) Essential steps in the ppGpp-dependent regulation of bacterial ribosomal RNA promoters can be explained by substrate competition. J Biol Chem 278(19):16834–16843. doi:10.1074/jbc.M300196200M300196200 [pii]

Kanjee U, Gutsche I, Alexopoulos E, Zhao B, El Bakkouri M, Thibault G, Liu K, Ramachandran S, Snider J, Pai EF, Houry WA (2011) Linkage between the bacterial acid stress and stringent responses: the structure of the inducible lysine decarboxylase. EMBO J 30(5):931–944. doi: emboj20115 [pii] 10.1038/emboj.2011.5

Kihira K, Shimizu Y, Shomura Y, Shibata N, Kitamura M, Nakagawa A, Ueda T, Ochi K, Higuchi Y (2012) Crystal structure analysis of the translation factor RF3 (release factor 3). FEBS Lett 586(20):3705–3709. doi: S0014-5793(12)00686-2 [pii]10.1016/j.febslet.2012.08.029

Klinkenberg LG, Lee JH, Bishai WR, Karakousis PC (2010) The stringent response is required for full virulence of Mycobacterium tuberculosis in guinea pigs. J Infect Dis 202(9):1397–1404. https://doi.org/10.1086/656524

Kumar V, Chen Y, Ero R, Ahmed T, Tan J, Li Z, Wong AS, Bhushan S, Gao YG (2015) Structure of BipA in GTP form bound to the ratcheted ribosome. Proc Natl Acad Sci U S A 112(35):10944–10949. doi: 1513216112 [pii]10.1073/pnas.1513216112

Lehninger A, Nelson D, Cox M (2008) Lehninger principles of biochemistry. W. H. Freeman. doi: citeulike-article-id:3823091

Little R, Ryals J, Bremer H (1983) rpoB mutation in Escherichia coli alters control of ribosome synthesis by guanosine tetraphosphate. J Bacteriol 154 (2):787–792

Liu K, Myers AR, Pisithkul T, Claas KR, Satyshur KA, Amador-Noguez D, Keck JL, Wang JD (2015) Molecular mechanism and evolution of guanylate kinase regulation by (p)ppGpp. Mol Cell 57 (4):735–749. doi: S1097-2765(14)01015-6 [pii]10.1016/j.molcel.2014.12.037

Liu H, Xiao Y, Nie H, Huang Q, Chen W (2017) Influence of (p)ppGpp on biofilm regulation in Pseudomonas putida KT2440. Microbiol Res 204:1–8. doi: S0944-5013(16)30672-3 [pii]10.1016/j.micres.2017.07.003

Maiti K, Syal K, Chatterji D, Jayaraman N (2017) Synthetic Arabinomannan Heptasaccharide Glycolipids Inhibit Biofilm Growth and Augment Isoniazid Effects in Mycobacterium smegmatis. Chembiochem 18 (19):1959–1970. https://doi.org/10.1002/cbic.201700247

Mechold U, Potrykus K, Murphy H, Murakami KS, Cashel M (2013) Differential regulation by ppGpp versus pppGpp in Escherichia coli. Nucleic Acids Res 41 (12):6175–6189. doi: gkt302 [pii]10.1093/nar/gkt302

Mishra A, Sarkar D (2015) Qualitative and quantitative proteomic analysis of Vitamin C induced changes in Mycobacterium smegmatis. Front Microbiol 6:451. https://doi.org/10.3389/fmicb.2015.00451

Naresh K, Avaji PG, Maiti K, Bharati BK, Syal K, Chatterji D, Jayaraman N (2012) Synthesis of beta-arabinofuranoside glycolipids, studies of their binding to surfactant protein-A and effect on sliding motilities of M. smegmatis. Glycoconj J 29(2–3):107–118. https://doi.org/10.1007/s10719-012-9369-2

Nunes-Alves C (2014) Targeting (p)ppGpp disrupts biofilms. Nat Rev Microbiol 12:461. https://doi.org/10.1038/nrmicro3302

Padayatty SJ, Riordan HD, Hewitt SM, Katz A, Hoffer LJ, Levine M (2006) Intravenously administered vitamin C as cancer therapy: three cases. CMAJ 174 (7):937–942. doi: 174/7/937 [pii]10.1503/cmaj.050346

Perederina A, Svetlov V, Vassylyeva MN, Tahirov TH, Yokoyama S, Artsimovitch I, Vassylyev DG (2004) Regulation through the secondary channel--structural framework for ppGpp-DksA synergism during transcription. Cell 118(3):297–309. doi: 10.1016/j.cell.2004.06.030S0092867404006270 [pii]

Reddy PS, Raghavan A, Chatterji D (1995) Evidence for a ppGpp-binding site on Escherichia coli RNA polymerase: proximity relationship with the rifampicin-binding domain. Mol Microbiol 15(2):255–265

Reddy PV, Puri RV, Chauhan P, Kar R, Rohilla A, Khera A, Tyagi AK (2013) Disruption of Mycobactin Biosynthesis Leads to Attenuation of Mycobacterium tuberculosis for Growth and Virulence. J Infect Dis 208(8):1255–1265

Ross W, Vrentas CE, Sanchez-Vazquez P, Gaal T, Gourse RL (2013) The magic spot: a ppGpp binding site on E. coli RNA polymerase responsible for regulation of transcription initiation. Mol Cell 50(3):420–429. doi: S1097–2765(13)00249–9 [pii]10.1016/j.molcel.2013.03.021

Sargazi A, Gharebagh RA, Aali H, Oskoee HO, Sepehri Z (2017) Role of essential trace elements in tuberculosis infection: A review article. Indian J Tuberc 64 (4):246–251. doi: S00195707(16)30296-7 [pii] 10.1016/j.ijtb.2017.03.003

Sikri K, Batra SD, Nandi M, Kumari P, Taneja NK, Tyagi JS (2015) The pleiotropic transcriptional response of Mycobacterium tuberculosis to vitamin C is robust and overlaps with the bacterial response to multiple intracellular stresses. Microbiology 161(Pt 4):739–753. doi: mic.0.000049 [pii]10.1099/mic.0.000049

Soh AZ, Chee CBE, Wang YT, Yuan JM, Koh WP (2017) Dietary Intake of Antioxidant Vitamins and Carotenoids and Risk of Developing Active Tuberculosis in a Prospective Population-Based Cohort Study. Am J Epidemiol 186(4):491–500. doi: 3831256 [pii] 10.1093/aje/kwx132

Sorice A, Guerriero E, Capone F, Colonna G, Castello G, Costantini S (2014) Ascorbic acid: its role in immune system and chronic inflammation diseases. Mini Rev Med Chem 14(5):444–452. doi: MRMC-EPUB-60293 [pii]

Sritharan M (2016) Iron Homeostasis in Mycobacterium tuberculosis: Mechanistic Insights into Siderophore-Mediated Iron Uptake. J Bacteriol 198 (18):2399–2409. doi: JB.00359-16 [pii]10.1128/JB.00359-16

Steinchen W, Schuhmacher JS, Altegoer F, Fage CD, Srinivasan V, Linne U, Marahiel MA, Bange G (2015) Catalytic mechanism and allosteric regulation of an oligomeric (p)ppGpp synthetase by an alarmone. Proc Natl Acad Sci U S A 112(43):13348–13353. doi: 1505271112 [pii]10.1073/pnas.1505271112

Syal K, Chatterji D (2015) Differential binding of ppGpp and pppGpp to E. coli RNA polymerase: photo-labeling and mass spectral studies. Genes Cells 20 (12):1006–1016. https://doi.org/10.1111/gtc.12304

Syal K, Chakraborty S, Bhattacharyya R, Banerjee D (2015a) Combined inhalation and oral supplementation of Vitamin A and Vitamin D: a possible prevention and therapy for tuberculosis. Med Hypotheses 84 (3):199–203. doi: S0306-9877(15)00005-5 [pii] 10.1016/j.mehy.2014.12.022

Syal K, Joshi H, Chatterji D, Jain V (2015b) Novel pppGpp binding site at the C-terminal region of the Rel enzyme from Mycobacterium smegmatis. FEBS J 282(19):3773–3785. https://doi.org/10.1111/febs.13373

Syal K, Maiti K, Naresh K, Chatterji D, Jayaraman N (2015c) Synthetic glycolipids and (p)ppGpp analogs: development of inhibitors for mycobacterial growth, biofilm and stringent response. Adv Exp Med Biol 842:309–327. https://doi.org/10.1007/978-3-319-11280-0_20

Syal K, Maiti K, Naresh K, Avaji PG, Chatterji D, Jayaraman N (2016) Synthetic arabinomannan glycolipids impede mycobacterial growth, sliding motility and biofilm structure. Glycoconj J 33 (5):763–777. doi: 10.1007/s10719-016-9670-610.1007/s10719-016-9670-6 [pii]

Syal K, Bhardwaj N, Chatterji D (2017a) Vitamin C targets (p)ppGpp synthesis leading to stalling of long-term survival and biofilm formation in Mycobacterium smegmatis. FEMS Microbiol Lett 364(1). doi: fnw282 [pii]10.1093/femsle/fnw282

Syal K, Flentie K, Bhardwaj N, Maiti K, Jayaraman N, Stallings CL, Chatterji D (2017b) Synthetic (p)ppGpp Analogue Is an Inhibitor of Stringent Response in Mycobacteria. Antimicrob Agents Chemother 61(6). doi: AAC.00443-17 [pii]10.1128/AAC.00443-17

Toulokhonov, II, Shulgina I, Hernandez VJ (2001) Binding of the transcription effector ppGpp to Escherichia coli RNA polymerase is allosteric, modular, and occurs near the N terminus of the beta′-subunit. J Biol Chem 276(2):1220–1225. doi:10.1074/jbc.M007184200M007184200 [pii]

van der Biezen EA, Sun J, Coleman MJ, Bibb MJ, Jones JD (2000) Arabidopsis RelA/SpoT homologs implicate (p)ppGpp in plant signaling. Proc Natl Acad Sci U S A 97(7):3747–3752. doi:10.1073/pnas.060392397060392397 [pii]

Verstraeten N, Knapen WJ, Kint CI, Liebens V, Van den Bergh B, Dewachter L, Michiels JE, Fu Q, David CC, Fierro AC, Marchal K, Beirlant J, Versees W, Hofkens J, Jansen M, Fauvart M, Michiels J (2015) Obg and Membrane Depolarization Are Part of a Microbial Bet-Hedging Strategy that Leads to Antibiotic Tolerance. Mol Cell 59(1):9–21. doi: S1097-2765(15)00347-0 [pii]10.1016/j.molcel.2015.05.011

Vilcheze C, Hartman T, Weinrick B, Jacobs WR, Jr. (2013) Mycobacterium tuberculosis is extraordinarily sensitive to killing by a vitamin C-induced Fenton reaction. Nat Commun 4:1881. doi: ncomms2898 [pii]10.1038/ncomms2898

Vrentas CE, Gaal T, Berkmen MB, Rutherford ST, Haugen SP, Vassylyev DG, Ross W, Gourse RL (2008) Still looking for the magic spot: the crystallographically defined binding site for ppGpp on RNA polymerase is unlikely to be responsible for rRNA transcription regulation. J Mol Biol 377(2):551–564. doi: S0022-2836(08)00075-2 [pii]10.1016/j.jmb.2008.01.042

Wexselblatt E, Oppenheimer-Shaanan Y, Kaspy I, London N, Schueler-Furman O, Yavin E, Glaser G, Katzhendler J, Ben-Yehuda S (2012) Relacin, a novel antibacterial agent targeting the Stringent Response. PLoS Pathog 8(9):e1002925. doi: 10.1371/journal.ppat.1002925PPATHOGENS-D-12-00921 [pii]

Weyer WJ, de Boer HA, de Boer JG, Gruber M (1976) The sequence of ppGpp and pppGpp in the reaction scheme for magic spot synthesis. Biochim Biophys Acta 442 (1):123–127. doi: 0005-2787(76)90183-0 [pii]

Williams SM, Chandran AV, Vijayabaskar MS, Roy S, Balaram H, Vishveshwara S, Vijayan M, Chatterji D (2014) A histidine aspartate ionic lock gates the iron passage in miniferritins from Mycobacterium smegmatis. J Biol Chem 289(16):11042–11058. doi: M113.524421 [pii]10.1074/jbc.M113.524421

Zuo Y, Wang Y, Steitz TA (2013) The mechanism of E. coli RNA polymerase regulation by ppGpp is suggested by the structure of their complex. Mol Cell 50(3):430–436. doi: S1097-2765(13)00248-7 [pii] 10.1016/j.molcel.2013.03.020

The Wrappers of the 1,2-Propanediol Utilization Bacterial Microcompartments

23

Naimat K. Bari, Gaurav Kumar, and Sharmistha Sinha

Abstract

The propanediol utilization bacterial microcompartments are specialized protein-based organelles in *Salmonella* that facilitate the catabolism of 1,2-propanediol when available as the sole carbon source. This smart prokaryotic cell organelle compartmentalizes essential enzymes and substrates in a volume of a few attoliters compared to the femtoliter volume of a bacterial cell thereby enhancing the enzyme kinetics and properly orchestrating the downstream pathways. A shell or coat, which is composed of a few thousand protein subunits, wraps a chain of consecutively acting enzymes and serves as ducts for the diffusion of substrates, cofactors, and products into and out of the core of the microcompartment. In this article we bring together the properties of the wrappers of the propanediol utilization bacterial microcompartments to update our understanding on the mechanism of the formation of these unique wraps, their assembly, and interaction with the encapsulated enzymes.

Keywords

Bacterial microcompartments · BMC domain fold · Shell proteins · Propanediol utilization

N. K. Bari · G. Kumar · S. Sinha (✉)
Institute of Nano Science and Technology, Mohali, Punjab, India
e-mail: naimat.ph14210@inst.ac.in; gaurav.ph17202@inst.ac.in; sinhas@inst.ac.in

23.1 Introduction

Bacterial microcompartments (MCPs) are interesting genre of prokaryotic inclusions which were discovered 75 years ago by electron microscopy, and initially these structures were thought to be viruses inside bacteria. Till date, however, the MCPs have not shown any resemblance to any known viral capsid protein. Several genetic and molecular biology studies have confirmed that these structures are specialized chambers inside bacteria that aid metabolism by concentrating substrates and enzymes in a confined space (Cheng et al. 2011; Kerfeld et al. 2005; Shively et al. 1973a; Yeates et al. 2011, 2013). MCPs were first discovered in cyanobacteria and were named as carboxysomes because of their role in CO_2 fixation by the Calvin cycle (G Drews 1956; Shively et al. 1973b). Later on, homologues of these carboxysomes were discovered in several bacteria, and the literature witnessed the reports on the characterization of at least seven different genres of MCPs in 16% of bacteria across 23 different phyla in 30 loci, and more are being discovered (Axen et al. 2014; Beeby et al. 2009; Farah Abdul-Rahman and Jeffrey 2013; Jorda et al. 2013; Sutter et al. 2017). All these different MCPs have similar structure which comprises of a protein shell that encapsulates a cluster of related enzymes for a particular function. Genomic and structural data indicate that the shell proteins are present in a wide range of bacteria,

K. Chattopadhyay, S. C. Basu (eds.), *Biochemical and Biophysical Roles of Cell Surface Molecules*, Advances in Experimental Medicine and Biology 1112, https://doi.org/10.1007/978-981-13-3065-0_23

and there exists a high level of structural sophistication leading to the development of these wrappers of a definite size and shape in vivo. Genetic, biochemical, and microbiological studies on the MCPs indicate that the general function of these prokaryotic organelles is to optimize metabolic pathways having intermediates that are toxic to the cell or are volatile which can diffuse out freely through the cell membrane (Jorda et al. 2013, 2015; Kerfeld et al. 2010; Yeates et al. 2011, 2013). Recent biophysical studies have shed light on the topological and mechanical features of these organelles. Using the β-carboxysome as a paradigm, it has been shown that these organelles have a flexible organization and soft mechanical properties as compared to rigid viruses (Faulkner et al. 2017).

The general scheme for a MCP action involves movement of a metabolic substrate from the cytoplasm to the MCP core where it is acted upon by a series of enzymes along with the formation of a toxic intermediate (Havemann et al. 2002; Penrod and Roth 2006; Rondon et al. 1995). The outer wrapper proteins of the MCPs play a crucial role in trapping this intermediate inside and channelizing it to downstream enzymes for further processing (Dou et al. 2008; Havemann et al. 2002; Price and Badger 1989; Rondon et al. 1995). As and when required, the product (s) move out of the MCP, where they enter the central metabolism of the cell thereby providing energy for cell growth. The basic function of the MCPs thus is to segregate molecular processes leading to proper channelization of the nutrients in an energy-efficient approach. Among the common MCPs reported till date, the carboxysomes play a major role in carbon fixation, while the Eut (ethanolamine utilization) and the Pdu (1,2-propanediol utilization) MCPs act as metabolosomes for the utilization of specific substrates and are crucial for development and dissemination of enteric pathogens. The carboxysomes account for close to 25% of the carbon fixation of the nature and are present in almost all bacteria using the Calvin cycle for carbon utilization (Kerfeld et al. 2010; Rae et al. 2013). Carboxysomes encapsulate two enzymes: carbonic anhydrase and RuBisCO. The carbonic anhydrase transforms bicarbonate to produce CO_2 which is then fixed to ribulose-1,5-bisphosphate by the enzyme RuBisCO (Chen et al. 2013). Sequestration by the carboxysomes allows the concentration of CO_2 to be elevated in the immediate vicinity of RuBisCO, improving its catalytic efficiency (Dou et al. 2008; Price et al. 2008; Rae et al. 2013). The other main MCPs, Pdu-MCP and Eut-MCP found in the enteric pathogen *Salmonella*, help the organism to grow in the infected intestine leading to their propagation in new hosts (Thiennimitr et al. 2011; Winter and Baumler 2011; Winter et al. 2010). Both 1,2-propanediol and ethanolamine are the by-products of the incomplete utilization of fucose or rhamnose in the intestine (Badía et al. 1985; Baldoma and Aguilar 1988). The Pdu- and Eut-MCPs operate by sequestering toxic/volatile catabolic intermediate like propionaldehyde during 1,2-propanediol utilization (in Pdu-MCPs) or formaldehyde during ethanolamine utilization (in Eut-MCPs). A schematic representation of the action of the Pdu-MCPs is shown in Fig. 23.1. Apart from enhancing the enzyme activity by compartmentalization, the MCPs also help to reduce cytotoxicity, DNA damage, and the loss of valuable nutrient source.

All the MCPs reported till date have shown an icosahedral structure, though deviations from a perfect icosahedra have been observed in some cases (Bobik 2006; Cannon et al. 2001; Cheng et al. 2008; Shively et al. 1973a). The core elements in assembly of a MCP include the formation of the external shell or wrap with the concurrent encapsulation of the enzyme cluster inside. The wrap or the envelope is made up of two different kinds of proteins: hexagonal proteins that make up the faces of the polyhedral structure and pentagonal proteins that make up the vertices of the polyhedra. The hexagonal shell proteins belong to the family of bacterial microcompartment domain (BMC) which is typically about 90 amino acids long and have an α/β fold, while the vertex proteins belong to the BMV (bacterial microcompartment vertex) family of proteins (Kerfeld et al. 2010; Yeates et al. 2011). The BMC domain hexagonal proteins self-assemble into a flat disc-shaped hexagonal

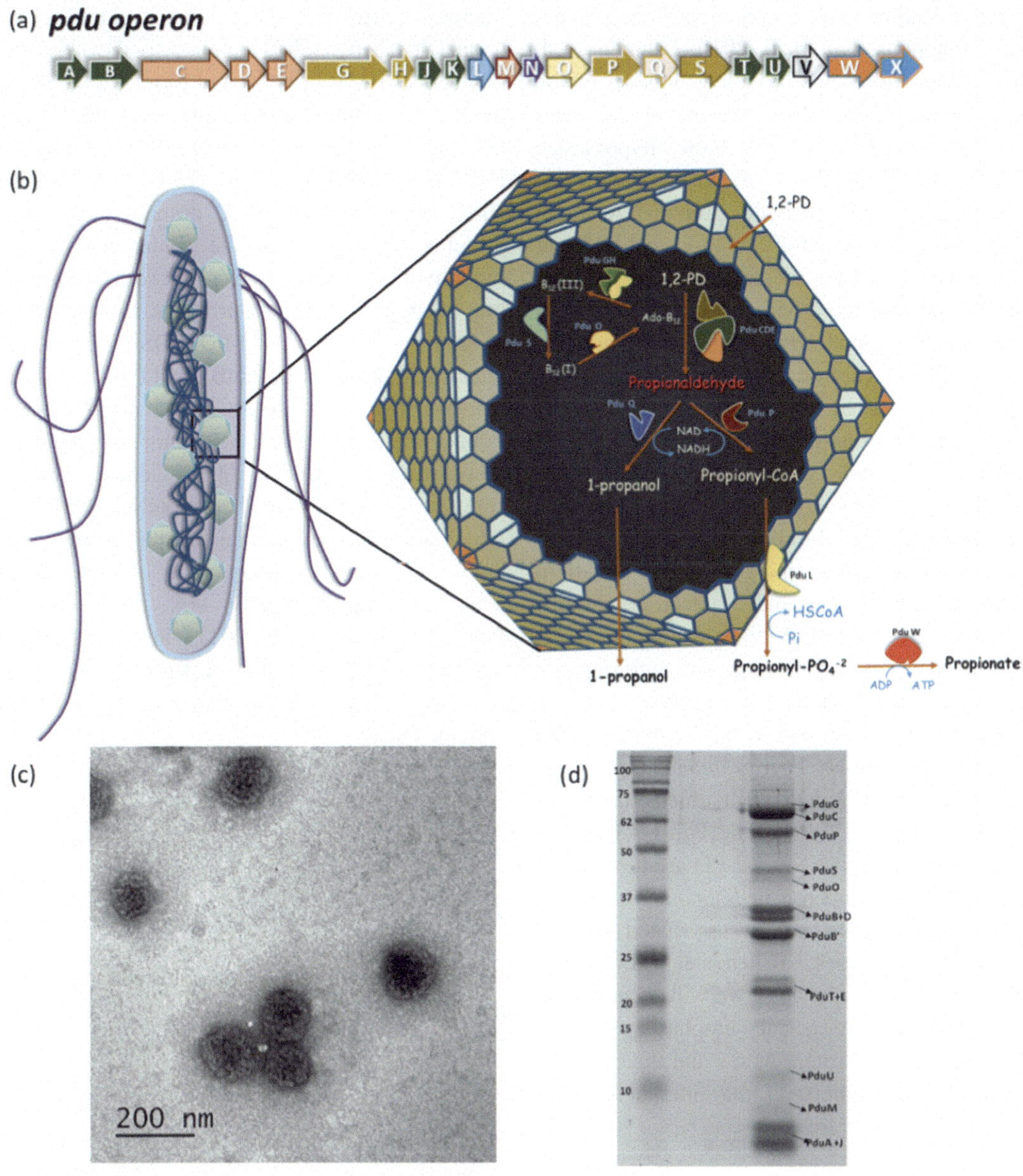

Fig. 23.1 Pdu-MCP. (**a**) Schematic representation of the pdu operon in *Salmonella enterica* Serovar Typhimurium LT2. (**b**) Schematic representation of the Pdu-MCP structure and functional pathway indicating the role of the enzymes. (**c**) Electron micrographs and (**d**) SDS-PAGE of purified Pdu-MCP from LT2 strain [Sinha Lab]

structure that forms the basic building block of the external wrap. In most of the shell proteins, the two sides of the disc have different properties with respect to their hydrophobicity and curvature. These discs typically have a concave hydrophobic surface which homes the C-termini and the N-termini of the proteins and has been suggested to be facing the MCP lumen (Tanaka et al. 2009). These C-termini and N-termini are involved in the interactions with the luminal enzymes leading to the proper orchestration of the MCP functions (Fan et al. 2010; Lehman et al. 2017). Although there is no direct evidence on the arrangement of the shell proteins of the outer wrap of the MCPs, several studies hypothesize that the wrap is made up of mixed sheets composed of the multiple BMC proteins. A single shell thus is composed of few thousand different proteins of 4–8 types (depending on the type of MCP), and each type bears a special functional role. The pentagonal vertex proteins belong to the family of BMV (bacterial microcompartment vertex) proteins and are crucial for the closure of the polyhedra. This unique architecture of the MCPs helps in efficient encapsulation of enzymes, substrates, and cofactors in a volume as low as an attoliter compared to the femtoliter volume of the cell and offers an energy-efficient paradigm for bacteria to survive under special conditions. Lessons from the MCP structure-function relationship have of late become very intriguing in developing novel synthetic bioreactors.

Followed by carboxysomes, the most studied MCPs are the Pdu-MCPs in *Salmonella* which aid the vitamin B12-dependent metabolism of 1,2-propanediol (1,2-PD). The Pdu-MCPs grow only in the presence of 1,2-PD as the carbon source. The *Pdu* operon in *Salmonella* encodes all the shell proteins and the associated 1,2-PD-degrading enzymes (Chen et al. 1994) (Fig. 23.1a). In this chapter we will focus on the Pdu-MCPs with special emphasis on the properties of the protein wraps and their mechanism of interaction with the internal enzyme cluster.

23.2 Pdu-MCP Envelope

The present understanding of the Pdu-MCP is that the outer wrap is composed of eight structural proteins, PduA, PduB, PduB', PduJ, PduK, PduT, PduU, and PduN. PduM has also been suggested to be a part of the shell, although its role is unclear (Sinha et al. 2012). The shell proteins PduA, PduB, PduB', PduJ, PduK, PduT, and PduU belong to the family of bacterial microcompartment (BMC) domain and are hexagonal in shape. PduN belongs to the class of pentagonal BMV proteins and forms the vertices of MCPs. While PduA, PduJ, and PduU are hexamers of six monomeric BMC domains, PduB/B' and PduT are trimers of two tandem BMC domains. The BMC domain hexagonal proteins have pores at the center which are thought to play crucial roles in the selective transport of enzyme substrates, products, and cofactors (Kerfeld et al. 2005). 2D-IEF-SDS-PAGE have identified PduA, PduB, PduB', and PduJ as major components of the Pdu-MCP shell while PduK, PduN, PduT, and PduU as minor components (Havemann and Bobik 2003). Genetic studies suggest that among the shell proteins, PduB/B', PduJ, and PduN are crucial for the assembly of the Pdu-MCP, and deletion of any of these genes leads to the formation of nonfunctional Pdu-MCPs (Cheng et al. 2011). In the following few paragraphs, we will discuss each of the shell proteins and its role in wrapping the internal enzyme cluster.

23.3 Shell Proteins of Pdu-MCP

23.3.1 PduA

PduA protein comprises 16% of the entire shell of the Pdu-MCP and has been the most studied shell protein in the Pdu-MCP system (Havemann and Bobik 2003; Havemann et al. 2002). This protein is 94 amino acid long and is encoded by the *pduA* gene of the *pdu* operon. It is a hexamer with the monomers containing only one BMC domain. The crystal structure of PduA shows adjacent hexamers interacting together to form an extended two-dimensional molecular sheet. These sheets are stabilized by specific interactions of a number of conserved edge amino acids (K26, N29, and R79) and are crucial for the architecture of the whole MCP envelope (Pang et al. 2014; Sinha et al. 2014). The mutation K26A in PduA leads to the complete abolition of the Pdu-MCP (Fig. 23.2), while the mutations N29A and R79A lead to the formation of enlarged and porous

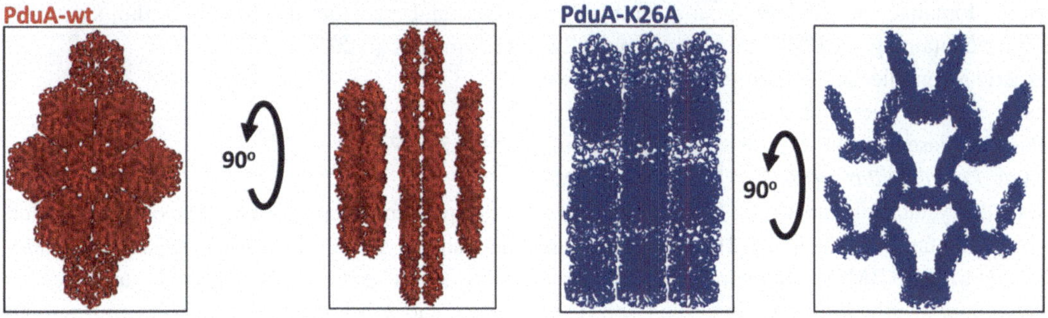

Fig. 23.2 Crystal structure of PduA (PDB ID-3NGK) showing interactions of adjacent hexamers to form an extended two-dimensional molecular sheet. The mutation K26A (PDB ID-4PPD) disrupts these edge-to-edge interactions, preventing the formation of an extended sheet

Pdu-MCPs. The crystal structure of the PduA-K26A mutant protein hexamer assemblies displays two crystal forms 1 and 2. These crystal forms when superimposed on the wild-type PduA hexamer structure showed RMSDs of 1.0 Å and 0.9–1.2 Å, respectively, suggesting that PduA-K26A protein had no major conformational differences compared to wild-type PduA. However, the PduA-K26A mutant did not form extended protein sheets in crystals in contrast to the wild-type PduA (Fig. 23.2). One crystal form of PduA-K26A forms one-dimensional sheets leading to the formation of a three-dimensional shell protein assembly that cannot form the MCP facet. Hence, the mutant proteins cannot be incorporated in the MCPs leading to the non-formation of the polyhedral structures. It is also envisaged that PduA-K26A protein may interact with the internal enzyme cluster similar to the wild-type protein leading to non-closure of the shell and eventually a damaged architecture of the MCPs. Stability studies done at the individual protein level also demonstrate lower transition temperature for the K26A mutant compared to the wild type (**Sinha Lab, unpublished results**). It has also been shown that these K26 and R79 residues stabilize the PduA assembly in bacterial cytoplasm as well (Pang et al. 2014). PduA protein presents a central pore of 0.6 nm diameter which is lined by the backbone amides of the residues G39 and S40 and side chain of S40 from each of the six monomers. This pore has been shown to be conduit for transport of 1,2-PD across the MCP envelope (Chowdhury et al. 2015). Molecular dynamics simulations study has shown that the pore of PduA hexamer has a lower-energy barrier for the movement of 1,2-PD, compared to the movement of toxic metabolite propionaldehyde (Park et al. 2017). The anchoring of 1,2-PD in the pore occurs by the formation of two hydrogen bonds between the 1,2-PD and the pore-lining atoms, which is less probable for propionaldehyde. A *ΔpduA* developed in *Salmonella* shows enlarged Pdu-MCPs compared to the wild type along with growth arrest due to propionaldehyde toxicity and an increased growth rate at limiting vitamin B12 concentrations. These studies suggest the importance of PduA protein in the integrity and function of the Pdu-MCPs (Cheng et al. 2011; Havemann et al. 2002; Sinha et al. 2014).

23.3.2 PduB/PduB'

Close to one-half of the Pdu shell is made up of PduB (28%) and PduB' (25%). These two proteins are formed by a pair of overlapping genes *pduB/B'* in the *pdu* operon with their start sites 111 base pairs apart, resulting in extra 37 amino acids at the N-terminal of PduB'. The role of this extra 37 amino acid is still unclear. It has been suggested by Lehman et al. that these 37 amino acids play a role in the association of the shell with the internal enzyme cluster (Lehman et al. 2017). The PduB/B' proteins have tandem

BMC domains, and the biological assembly is a trimer containing six BMC domains. These oligomeric assemblies in the shell proteins are known as pseudohexamers. Our understanding of the PduB is based on the crystal structure of PduB from *Lactobacillus reuteri*, which is the only solved crystal structure (PDB ID-4FAY) of the protein so far (Pang et al. 2012). PduB protein is related to those BMC domain proteins that have a gated pore like the EutL protein or the CsoS1D in Eut-MCP and α-carboxysome, respectively (Klein et al. 2009; Sagermann et al. 2009; Takenoya et al. 2010). Both the EutL and CsoS1D proteins display two conformations in the crystal structures: open central pore and closed central pore which implicates crucial roles of these tandem BMC repeats in gated transport of materials in and out of the MCPs. The domain duplication allows greater flexibility involved in the dynamics of pore activity. The crystal structure of PduB from *L. reuteri*, however, shows only a closed pore, and each unit of the trimer binds to one molecule of glycerol suggesting an involvement in substrate transport. PduB from the *Lactobacillus reuteri* also forms extended two-dimensional sheets in crystals, though the intersheet contact is reported to be out of plane, indicating a lower stability. Modelling studies have confirmed that PduB and PduA form mixed sheets where the edges completely align suggesting better packing of the proteins as expected in the physiological shell. *Salmonella* strains with deleted *pduB* genes from the *pdu* operon are unable to form the MCPs. Interestingly, the electron micrographs of *pduB*-deleted *Salmonella* cells show protein deposits without the presence of polyhedral MCPs. This suggests that in the absence of the PduB/B' proteins, the shell cannot be formed, and enzyme clusters with fragments of the shell remain as protein inclusions inside the cell. These deletion mutants also show the phenotype of porous shell with aldehyde toxicity and retarded growth, indicating a nonfunctional shell. A recent study shows that N-terminal 37 amino acids of the PduB shell protein are crucial in binding the central enzyme core to the shell and provide insights into the sidedness (Lehman et al. 2017).

23.3.3 PduJ

PduJ protein covers close to 22% of the Pdu shell and shares greater than 80% sequence identity with PduA protein. The crystal structures of PduA and PduJ overlap with same pore-lining and edge contact residues. Interestingly, however, it has been demonstrated that PduJ has no role in the transport of 1,2-PD across the shell. By elegant gene replacement studies in the chromosome locus, Choudhury et al. have shown that when *pduA* is replaced by *pduJ*, the latter acquires the function of the former (Chowdhury et al. 2016). These studies suggest that the locus of the gene on the *pdu* operon decides the function of the gene under question. This study is one of the first studies in the field that relates the function of a gene with its location. The *pduJ* deletion in the MCPs leads to the formation of elongated MCPs in contrast to the porous MCPs in case of *pduA* deletion. The elongated MCP advocates have faulty closure of faces of the MCP polyhedra, and hence it is hypothesized that PduJ forms the edges that join the facets of the Pdu-MCPs. However, studies on the effect of the replacement of the *pduJ* gene by *pduA* gene on the joining of facets of Pdu-MCPs have not yet been reported.

23.3.4 PduK

PduK is an interesting shell protein with an N-terminal (~90 amino acids) consisting of the MCP domain fold and a C-terminal (~70 amino acids) of unknown function. This protein is known to bind iron (Cheng et al. 2011). The C-terminal has a Fe–S-binding motif (residues 130–152) which suggests that this protein might be involved in some redox or electron transfer reactions at the Pdu-MCP surface (Crowley et al. 2010). However, there are no experimental evidences to support this hypothesis. Unlike the *pduB/B'* deletion which leads to a nonfunctional strain, the *pduK* deletion leads to a functionally active MCP with an aggregated morphology (Cheng et al. 2011). Based on these observations, it has been inferred that PduK is involved in the segregation of MCPs in vivo especially during cell division. However, more experimental evidences are required to strengthen this hypothesis.

23.3.5 PduT

The PduT has two interesting characteristics: it is a pseudohexameric protein consisting of 184-residue-long tandem domains of the canonical (i.e., non-permuted) variety of the α/β BMC fold connected by a short linker sequence. The first domain is made up of residues 1–88, while the second domain comprises of residues 94–184 (Pang et al. 2011). The two tandem domains of PduT although conform closely to the canonical BMC fold and align with each other with RMSD = 1.14, their sequences are only 24% identical to each other. PduT forms only 3% of the entire shell, and unlike the other shell proteins, it does not form an extended sheet as observed in the crystal structures. Another interesting feature of the PduT protein is the presence of 4[Fe–S] cluster at the central pore (Crowley et al. 2010). The other proteins in the Pdu-MCP having the [Fe–S] centers are PduS and PduK (Chowdhury et al. 2014). EPR studies on a PduT homologue from *C. freundii* have confirmed the presence of a [4Fe–4S] cluster. The pore region of PduT has three cysteine residues C38, one from each subunit. The [4Fe–4S] cluster has an essentially cubic structure, and the modeled placement of the cluster orients it with its cubic body diagonal along the threefold axis of symmetry down the center of the pore in the PduT trimer. The metal cluster and its three cysteine ligands obey this symmetry. In addition, the [4Fe–4S] cluster calls for a fourth S-ligand to an iron atom lying on the threefold symmetry. This fourth S-atom in the 4[Fe–S] center has not yet been identified. The[4Fe–4S] cluster in PduT is reported to be accessible from both the luminal and the cytoplasmic side of the MCP and suggests a possible role in assisting redox chemistry. In the crystal structure of PduT, C-108 and C-136 are located close enough to form a disulfide bond. However the side chains are positioned in such a way that the disulfide bond is not formed. It is suggested that the formation of this disulfide bond due to rotamer conformational changes might be resulting in a redox-sensitive change in the pore properties. These cysteine positions are conserved in a sequence alignment of PduT homologues, while they are absent in other Pdu shell proteins. As of now the role of redox coupling in the Pdu-MCP function is not obvious; however, meticulous studies on the structure-activity relationship of PduT might suggest crucial phenomenon. Mutations in PduT have been reported to marginally weaken the growth of *Salmonella* on 1,2-PD and do not significantly affect MCP structure or function. All these taken together suggest that PduT might be important for *Salmonella* under different physiochemical conditions. Like PduB, PduT is also a protein with tandem BMC domains; however, unlike PduB, PduT does not form extended sheets in crystal structures (Crowley et al. 2010).

23.3.6 PduU

PduU is a minor component of the Pdu shell. This hexameric protein displays a circularly permuted BMC domain fold and fails to form hexagonal layers in crystal structures. PduU forms the blocked pore, bendable type of the shell protein (Crowley et al. 2008) and bears structural similarity with the EutS protein of the Eut-MCP (Tanaka et al. 2010). On the convex side of the central pore, PduU is capped by a parallel six-stranded β-barrel structure which is composed of one β-sheet from each of the monomers. Due to the presence of the β-barrel, the central pore of PduU is predicted to be not involved in any transport mechanism. However, Jorda et al. (2015) have shown that PduV, a Pdu enzyme with unknown function, binds to this β-barrel structure of PduU through its N-terminal region. This binding interaction might play a role in targeting PduV to Pdu-MCPs (Jorda et al. 2015). The concave face of PduU has a deep cavity lined by several hydrophobic residues. These residues are suggested to interact with luminal proteins, though it is not certain if this side of the PduU protein faces the lumen of the Pdu-MCP.

23.3.7 PduN

PduN is a bacterial microcompartment vertex protein which is a pentamer, and the closest analogues are the CcmL protein in carboxysomes, the EutN protein in the Eut-MCPs, or the GrpN protein in Grp-MCP (Wheatley et al. 2013). These vertex proteins form the vertices of the MCPs and help to form a closed polyhedral structure. PduN is the least abundant protein in the Pdu-MCP and is detected only by Western blotting or in vivo fluorescent tagging. This less abundance of PduN is expected, as a perfect icosahedra will have only 12 vertices. It has been reported that the copy numbers of the vertex proteins in a MCP is ~60 which is very less as compared to a few thousand copies of flat facet proteins. Although less abundant, PduN plays a crucial role in delivering the closed structure of the MCPs, and the deletion of *pduN* gene results in the formation of totally abnormal Pdu-MCPs with completely altered functions during growth on 1,2-PD (Cheng et al. 2011). This observation holds true for related vertex proteins in the β-carboxysomes as well (Cai et al. 2009).

23.3.8 PduM

PduM is the only shell protein that is devoid of a BMC or a BMV domain and yet is crucial for the assembly of the Pdu-MCPs. This protein does not have any analogue in the carboxysome of the Eut-MCP system. It is a 163-amino acid-long protein of unknown function, the deletion of which leads to the formation of abnormal nonfunctional Pdu-MCPs. Electron micrographs demonstrate abnormal protein aggregates in a *pduM*-deleted *Salmonella* strain, while biochemical assays suggest a nonfunctional MCP with phenotype of propionaldehyde toxicity (Sinha et al. 2012).

23.4 Interactions Between Pdu Shell Proteins and Core Enzymes

The Pdu-MCPs encapsulate at least five different enzymes inside the lumen (PduCDE, PduP, PduO, PduGH, and PduS) (Crowley et al. 2008; Sinha et al. 2012). The last decade has shed a lot of light on the underlying principles for the self-assembly of MCPs especially on how the enzymes are encapsulated inside the protein shell. The role of N-terminal extensions or the encapsulation peptides of the core enzymes has been shown to be very significant in this regard (Aussignargues et al. 2015). For the Pdu-MCP, the first report demonstrated that the N-terminal of PduP (propionaldehyde dehydrogenase) protein is crucial for its encapsulation within the MCP. The N-terminal extension of PduP when tagged to heterologous proteins such as green fluorescent protein (GFP) resulted in their successful entrapment within Pdu microcompartment (Fan et al. 2010). Scanning mutagenesis studies have revealed that the residues E7, I10, and L14 in the N-terminal of PduP are essential for the encapsulation of PduP. Further, in silico modelling has shown that the N-terminal of PduP interacts with the C-terminal of PduA, a shell protein (Fan et al. 2012). In another study, it has been shown that the first 18 amino acids in the N-terminal of the medium subunit (PduD) are necessary and sufficient for the encapsulation of PduCDE (diol dehydratase) complex (Chowdhury et al. 2014). Impaired encapsulation of this enzyme has been observed upon deletion of N-terminal extension of PduD subunit (Fan and Bobik 2011). However, specific binding interaction between PduCDE and any of the shell proteins needs more experimental evidences.

Beside molecular biology approach, computational and bioinformatics approaches have enabled us to look into the depth of the protein assembly in microcompartments. Computational modelling has predicted various interactions between shell proteins and core enzymes of Pdu microcompartment. The Pdu interactome model prepared using coevolution-based methods suggests that the shell protein PduA acts as a "universal hub" for directing the core enzymes PduC, PduD, PduE, PduL, and PduP into the microcompartment. These enzymes have N-terminal extensions which bind to a cleft on the concave face of PduA hexamer (Jorda et al. 2015). Three important predictions have been made in this model that needs to be analyzed carefully. First, the enzyme PduP binds to only two of the shell proteins, namely, PduA and PduJ. This has been confirmed earlier by the co-elution experiments conducted by Fan and co-workers, where only PduA and PduJ co-eluted with His-tagged PduP (Fan et al. 2012). Second, all the three subunits of PduCDE diol dehydratase binds to PduA hexamer through their N-terminal extensions. This explains the impaired encapsulation of PduCDE within the microcompartment upon deletion of N-terminal of PduD subunit (Fan and Bobik 2011). However, it has also been observed that the chromosomal deletion of PduA or PduJ has no effect on the diol dehydratase activity of the purified microcompartments (Sinha et al. 2014), which suggests an association of PduCDE with the Pdu microcompartment even in the absence of PduA or PduJ. This can be explained by the fact that PduA is 80% identical to PduJ in terms of amino acid sequence (Crowley et al. 2010). It is plausible that the void created by the deletion of PduA is taken up by architecturally similar PduJ; but there is a lack of evidence in this regard. Also, deletion of PduJ results in the formation of an impaired and elongated microcompartment (Cheng et al. 2011), and therefore, increased diol dehydratase activity of purified microcompartments in case of chromosomal deletion of PduJ must be studied in this context. Third, the shell protein PduB doesn't bind to any of the core enzymes. This is indeed thought-provoking, as a recent study has shown the importance of N-terminal of PduB' in binding shell protein to the core enzymes. Substituting isoleucine with a hydrophilic amino acid, threonine ($I\Delta T$), at the tenth position in the N-terminal of PduB' has a deleterious effect on the encapsulation of core enzymes, including PduCDE and PduP (Lehman et al. 2017). This severe depletion in the concentration of core enzyme within microcompartments in $I\Delta T$ mutants indicates the importance of hydrophobic residue isoleucine in core–shell interaction. It can be hypothesized that the shell protein PduB' has no specific interaction with any of the core enzymes but has an overall global impact in retaining the core enzymes within Pdu microcompartment through hydrophobic interactions. This hypothesis is based on the fact that PduB and PduB' constitute 50% of the Pdu shell proteins (Havemann and Bobik 2003). Therefore, deletion of isoleucine at the tenth residue of PduB' may critically reduce the hydrophobic environment within the microcompartment as PduB accounts for almost 50% of the shell. This may result in a diminished retention of core enzymes within the microcompartment. However, experimental evidences are needed to support this hypothesis. Although it appears that the encapsulation strategies for the enzymes inside the Pdu-MCPs follow a common mechanism, apart from PduCDE and PduP, there are no reports on the encapsulation of other enzymes like PduG, PduH, PduO, PduS, and PduQ. An interesting area of exploration would be identification of the interacting shell protein-enzyme pair. This would lead to a better understanding of the packaging principles of the megadalton-sized MCPs.

23.5 Conclusions and Perspectives

The literature till date has a virtuous amount of information regarding the overall shape, size, and structure of the bacterial microcompartments. Genetic and biochemical studies on the Pdu-MCPs have enhanced our understanding regarding the interaction of the shell proteins and internal enzyme clusters. However, these understandings can be reinforced using co-crystallization experiments with the shell proteins and the enzymes. Further, in vitro binding studies between the shell proteins and the enzymes will shed light on the intensity and dynamics of these interactions. Strategies to mimic the in vivo shell protein assemblies in vitro also require some attention. At present only one substrate-shell protein pair (1,2-propanediol-PduA) has been identified for the Pdu-MCPs. The entry routes for vitamin B12 and other chemicals like NADH and ATP still remain unexplored. Computational studies with respect to modelling and docking the shell protein and cofactors may provide useful insights into the structure-function activity of the individual shell proteins. With respect to translational approaches, these genre of organelles have a vast prospective to be applied as nanoreactors for biofuel or pharmaceuticals production or as containers for storage of useful biomaterials once the evolution, transport, and dynamics are comprehensively demonstrated.

References

Aussignargues C, Paasch BC, Gonzalez-Esquer R, Erbilgin O, Kerfeld CA (2015) Bacterial microcompartment assembly: the key role of encapsulation peptides. Commun Integr Biol 8:e1039755

Axen SD, Erbilgin O, Kerfeld CA (2014) A taxonomy of bacterial microcompartment loci constructed by a novel scoring method. PLoS Comput Biol 10: e1003898

Badía J, Ros J, Aguilar J (1985) Fermentation mechanism of fucose and rhamnose in Salmonella typhimurium and Klebsiella pneumoniae. J Bacteriol 161:435–437

Baldoma L, Aguilar J (1988) Metabolism of L-fucose and L-rhamnose in Escherichia coli: aerobic-anaerobic regulation of L-lactaldehyde dissimilation. J Bacteriol 170:416–421

Beeby M, Bobik TA, Yeates TO (2009) Exploiting genomic patterns to discover new supramolecular protein assemblies. Protein Sci 18:69–79. https://doi.org/10.1002/pro.1

Bobik TA (2006) Polyhedral organelles compartmenting bacterial metabolic processes. Appl Microbiol Biotechnol 70:517–525. https://doi.org/10.1007/s00253-005-0295-0

Cai F, Menon BB, Cannon GC, Curry KJ, Shively JM, Heinhorst S (2009) The pentameric vertex proteins are necessary for the icosahedral carboxysome shell to function as a CO2 leakage barrier. PLoS One 4:e7521

Cannon GC, Bradburne CE, Aldrich HC, Baker SH, Heinhorst S, Shively JM (2001) Microcompartments in prokaryotes: carboxysomes and related polyhedra. Appl Environ Microbiol 67:5351–5361

Chen P, Andersson DI, Roth JR (1994) The control region of the pdu/cob regulon in Salmonella typhimurium. J Bacteriol 176:5474–5482

Chen AH, Robinson-Mosher A, Savage DF, Silver PA, Polka JK (2013) The bacterial carbon-fixing organelle is formed by shell envelopment of preassembled cargo. PLoS One 8:e76127

Cheng S, Liu Y, Crowley CS, Yeates TO, Bobik TA (2008) Bacterial microcompartments: their properties and paradoxes. BioEssays 30:1084–1095

Cheng S, Sinha S, Fan C, Liu Y, Bobik TA (2011) Genetic analysis of the protein shell of the microcompartments involved in coenzyme B12-dependent 1,2-propanediol degradation by Salmonella. J Bacteriol 193:1385–1392

Chowdhury C, Sinha S, Chun S, Yeates TO, Bobik TA (2014) Diverse bacterial microcompartment organelles. Microbiol Mol Biol Rev 78:438–468

Chowdhury C, Chun S, Pang A, Sawaya MR, Sinha S, Yeates TO, Bobik TA (2015) Selective molecular transport through the protein shell of a bacterial microcompartment organelle. Proc Natl Acad Sci U S A 112:2990–2995. https://doi.org/10.1111/mmi.13423

Chowdhury C, Chun S, Sawaya MR, Yeates TO, Bobik TA (2016) The function of the PduJ microcompartment shell protein is determined by the genomic position of its encoding gene. Mol Microbiol 101:770–783. https://doi.org/10.1128/jb.00785-16

Crowley CS, Sawaya MR, Bobik TA, Yeates TO (2008) Structure of the PduU shell protein from the Pdu microcompartment of Salmonella. Structure 16:1324–1332

Crowley CS, Cascio D, Sawaya MR, Kopstein JS, Bobik TA, Yeates TO (2010) Structural insight into the mechanisms of transport across the Salmonella enterica Pdu microcompartment shell. J Biol Chem 285:37838–37846

Dou Z, Heinhorst S, Williams EB, Murin CD, Shively JM, Cannon GC (2008) CO_2 fixation kinetics of Halothiobacillus neapolitanus mutant carboxysomes lacking carbonic anhydrase suggest the shell acts as a diffusional barrier for CO_2. J Biol Chem 283:10377–10384. https://doi.org/10.1074/jbc.M709285200

Fan C, Bobik TA (2011) The N-terminal region of the medium subunit (PduD) packages adenosylcobalamin-dependent diol dehydratase (PduCDE) into the Pdu microcompartment. J Bacteriol 193:5623–5628

Fan C et al (2010) Short N-terminal sequences package proteins into bacterial microcompartments. Proc Natl Acad Sci 107:7509–7514

Fan C, Cheng S, Sinha S, Bobik TA (2012) Interactions between the termini of lumen enzymes and shell proteins mediate enzyme encapsulation into bacterial microcompartments. Proc Natl Acad Sci 109:14995–15000

Farah Abdul-Rahman EP, Jeffrey LB (2013) The distribution of polyhedral bacterial microcompartments suggests frequent horizontal transfer and operon reassembly. J Phylogenet Evol Biol 1:1–7. https://doi.org/10.4172/2329-9002.1000118

Faulkner M et al (2017) Direct characterization of the native structure and mechanics of cyanobacterial carboxysomes. Nanoscale 9:10662–10673. https://doi.org/10.1039/c7nr02524f

Drews G, Niklowitz W (1956) Beiträge zur Cytologie der Blaualgen. II. Zentroplasma und granulare Einschlüsse von Phormidium uncinatum. Arch Mikrobiol 24 (2):147–162

Havemann GD, Bobik TA (2003) Protein content of polyhedral organelles involved in coenzyme B12-dependent degradation of 1,2-propanediol in Salmonella enterica serovar typhimurium LT2. J Bacteriol 185:5086–5095

Havemann GD, Sampson EM, Bobik TA (2002) PduA is a shell protein of polyhedral organelles involved in coenzyme B(12)-dependent degradation of 1,2-propanediol in Salmonella enterica serovar typhimurium LT2. J Bacteriol 184:1253–1261

Jorda J, Lopez D, Wheatley NM, Yeates TO (2013) Using comparative genomics to uncover new kinds of protein-based metabolic organelles in bacteria. Protein Sci 22:179–195. https://doi.org/10.1002/pro.2196

Jorda J, Liu Y, Bobik TA, Yeates TO (2015) Exploring bacterial organelle interactomes: a model of the protein-protein interaction network in the Pdu microcompartment. PLoS Comput Biol 11:e1004067

Kerfeld CA, Sawaya MR, Tanaka S, Nguyen CV, Phillips M, Beeby M, Yeates TO (2005) Protein structures forming the shell of primitive bacterial organelles. Science 309:936–938. https://doi.org/10.1126/science.1113397

Kerfeld CA, Heinhorst S, Cannon GC (2010) Annu Rev Microbiol 64:391–408. https://doi.org/10.1146/annurev.micro.112408.134211

Klein MG et al (2009) Identification and structural analysis of a novel carboxysome shell protein with implications for metabolite transport. J Mol Biol 392:319–333. https://doi.org/10.1016/j.jmb.2009.03.056

Lehman BP, Chowdhury C, Bobik TA (2017) The N terminus of the PduB protein binds the protein shell of the Pdu microcompartment to its enzymatic core. J Bacteriol 199:e00785–e00716

Pang A, Warren MJ, Pickersgill RW (2011) Structure of PduT, a trimeric bacterial microcompartment protein with a 4Fe–4S cluster-binding site. Acta Crystallogr D Biol Crystallogr 67:91–96

Pang A, Liang M, Prentice MB, Pickersgill RW (2012) Substrate channels revealed in the trimeric lactobacillus reuteri bacterial microcompartment shell protein PduB. Acta Crystallogr D Biol Crystallogr 68:1642–1652. https://doi.org/10.1107/s0907444912039315

Pang A, Frank S, Brown I, Warren MJ, Pickersgill RW (2014) Structural insights into higher order assembly and function of the bacterial microcompartment protein PduA. J Biol Chem 289:22377–22384. https://doi.org/10.1074/jbc.M114.569285

Park J, Chun S, Bobik TA, Houk KN, Yeates TO (2017) Molecular dynamics simulations of selective metabolite transport across the propanediol bacterial microcompartment shell. J Phys Chem B 121:8149–8154. https://doi.org/10.1021/acs.jpcb.7b07232

Penrod JT, Roth JR (2006) Conserving a volatile metabolite: a role for carboxysome-like organelles in Salmonella enterica. J Bacteriol 188:2865–2874. https://doi.org/10.1128/JB.188.8.2865-2874.2006

Price GD, Badger MR (1989) Expression of human carbonic anhydrase in the cyanobacterium Synechococcus PCC7942 creates a high CO(2)-requiring phenotype : evidence for a central role for Carboxysomes in the CO (2) concentrating mechanism. Plant Physiol 91:505–513

Price GD, Badger MR, Woodger FJ, Long BM (2008) Advances in understanding the cyanobacterial CO2-concentrating-mechanism (CCM): functional components, ci transporters, diversity, genetic regulation and prospects for engineering into plants. J Exp Bot 59:1441–1461. https://doi.org/10.1093/jxb/erm112

Rae BD, Long BM, Badger MR, Price GD (2013) Functions, compositions, and evolution of the two types of carboxysomes: polyhedral microcompartments that facilitate CO2 fixation in cyanobacteria and some proteobacteria. Microbiol Mol Biol Rev 77:357–379

Rondon MR, Horswill AR, Escalante-Semerena JC (1995) DNA polymerase I function is required for the utilization of ethanolamine, 1,2-propanediol, and propionate by Salmonella typhimurium LT2. J Bacteriol 177:7119–7124

Sagermann M, Ohtaki A, Nikolakakis K (2009) Crystal structure of the EutL shell protein of the ethanolamine ammonia lyase microcompartment. Proc Natl Acad Sci U S A 106:8883–8887

Shively JM, Ball F, Brown DH, Saunders RE (1973a) Functional organelles in prokaryotes: polyhedral inclusions (carboxysomes) of Thiobacillus neapolitanus. Science 182:584–586

Shively JM, Ball FL, Kline BW (1973b) Electron microscopy of the carboxysomes (polyhedral bodies) of Thiobacillus neapolitanus. J Bacteriol 116:1405–1411

Sinha S, Cheng S, Fan C, Bobik TA (2012) The PduM protein is a structural component of the microcompartments involved in coenzyme B(12)-dependent

1,2-propanediol degradation by Salmonella enterica. J Bacteriol 194:1912–1918

Sinha S, Cheng S, Sung YW, McNamara DE, Sawaya MR, Yeates TO, Bobik TA (2014) Alanine scanning mutagenesis identifies an asparagine–arginine–lysine triad essential to assembly of the shell of the Pdu microcompartment. J Mol Biol 426:2328–2345

Sutter M, Greber B, Aussignargues C, Kerfeld CA (2017) Assembly principles and structure of a 6.5-MDa bacterial microcompartment shell. Science 356:1293–1297. https://doi.org/10.1126/science.aan3289

Takenoya M, Nikolakakis K, Sagermann M (2010) Crystallographic insights into the pore structures and mechanisms of the EutL and EutM shell proteins of the ethanolamine-utilizing microcompartment of Escherichia coli. J Bacteriol 192:6056–6063

Tanaka S, Sawaya MR, Phillips M, Yeates TO (2009) Insights from multiple structures of the shell proteins from the beta-carboxysome. Protein Sci 18:108–120

Tanaka S, Sawaya MR, Yeates TO (2010) Structure and mechanisms of a protein-based organelle in. E coli Sci 327:81–84. https://doi.org/10.1126/science.1179513

Thiennimitr P et al (2011) Intestinal inflammation allows Salmonella to use ethanolamine to compete with the microbiota. Proc Natl Acad Sci U S A 108:17480–17485

Wheatley NM, Gidaniyan SD, Liu Y, Cascio D, Yeates TO (2013) Bacterial microcompartment shells of diverse functional types possess pentameric vertex proteins. Protein Sci 22:660–665

Winter SE, Baumler AJ (2011) A breathtaking feat: to compete with the gut microbiota, Salmonella drives its host to provide a respiratory electron acceptor. Gut Microbes 2:58–60

Winter SE et al (2010) Gut inflammation provides a respiratory electron acceptor for Salmonella. Nature 467:426–429

Yeates TO, Thompson MC, Bobik TA (2011) The protein shells of bacterial microcompartment organelles. Curr Opin Struct Biol 21:223–231. https://doi.org/10.1016/j.sbi.2011.01.006

Yeates TO, Jorda J, Bobik TA (2013) The shells of BMC-type microcompartment organelles in bacteria. J Mol Microbiol Biotechnol 23:290–299. https://doi.org/10.1159/000351347

F-type Lectin Domains: Provenance, Prevalence, Properties, Peculiarities, and Potential

24

Sonal Mahajan and T. N. C. Ramya

Abstract

F-type lectins are phylogenetically widespread albeit selectively distributed lectins with an L-fucose-binding sequence motif and an F-type lectin fold. Several F-type lectins from fishes have been extensively studied, and structural information is available for F-type lectin domains from fish and bacterial proteins. F-type lectins have been demonstrated to be involved in self−/nonself-recognition and therefore have an important role in pathogen defense in many metazoan animals. F-type lectin domains also have been implicated in functions related to fertilization, protoplast regeneration, and bacterial virulence. We have recently analyzed and reported the taxonomic spread, phylogenetic distribution, architectural contexts, and sequence characteristics of prokaryotic and eukaryotic F-type lectin domains. Interestingly, while eukaryotic F-type lectin domains were frequently present as stand-alone domains, bacterial F-type lectin domains were mostly found co-occurring with enzymatic or nonenzymatic domains in diverse domain architectures, suggesting that the F-type lectin domain might be involved in targeting enzyme activities or directing other biological functions to distinct glycosylated niches in bacteria. We and others have probed the fine oligosaccharide-binding specificity of several F-type lectin domains. The currently available wealth of sequence, structural, and biochemical information about F-type lectin domains provides opportunities for the generation of designer lectins with improved binding strength and altered binding specificities. We discuss the prevalence, provenance, properties, peculiarities, and potential of F-type lectin domains for future applications in this review.

Keywords

F-type lectin domain · L-fucose · Motif · Domain architectures · Structural features

S. Mahajan · T. N. C. Ramya (✉)
Institute of Microbial Technology, Chandigarh, India
e-mail: sonalmahajan@imtech.res.in;
ramya@imtech.res.in

24.1 Introduction to the F-type Lectin Domain

The F-type lectin domain (FLD) is a fairly short, ~140 residue domain containing a conserved L-fucose-binding sequence motif, HX(26)RXDX(4) R/K (where X refers to any residue), and a calcium-binding motif, h2DGx (where h refers to a small hydrophobic residue such as Val, Ala, or Ile and x refers to a small hydrophilic residue such as Asn, Asp, or Ser) (Vasta et al. 2008). Several eukaryotic F-type lectins have been well studied biochemically and structurally. Among them is the European eel lectin, *Anguilla anguilla*

K. Chattopadhyay, S. C. Basu (eds.), *Biochemical and Biophysical Roles of Cell Surface Molecules*, Advances in Experimental Medicine and Biology 1112, https://doi.org/10.1007/978-981-13-3065-0_24

agglutinin (AAA), which is well known for its use in blood typing (Baldus et al. 1996). F-type lectins from several other animals have also been biochemically characterized (Bianchet et al. 2010; Cammarata et al. 2001, 2007; Fleming et al. 2009; Cassels et al. 1994; Saito et al. 1997). Besides these animal lectins, biochemical studies have also been conducted on Bryohealin, an F-type lectin from the marine alga, *Bryopsis plumosa* (Kim et al. 2006), and researchers including us have studied the saccharide-binding property of FLDs from microbial proteins (Farrand et al. 2008; Feil et al. 2012; Mahajan et al. 2017).

F-type lectins from fishes have an important role as molecular pattern receptors, recognizing glycan epitopes on the surface of pathogens and facilitating opsonization (Vasta et al. 2017). Concordant with their role in innate immunity, fish F-type lectins are expressed in the liver as well as in the gills and intestine, organs that are exposed to the outer environment, and the expression level of F-type lectins in certain fishes is increased upon pathogen or LPS challenge (Vasta et al. 2017; Salerno et al. 2009; Odom and Vasta 2006; Honda et al. 2000). F-type lectins from sea bass, *Dicentrarchus labrax*, and gilt head, *Sparus aurata*, have been demonstrated to opsonize *Escherichia coli*; exposure of formalin-killed *E. coli* to these fucolectins enhanced their phagocytosis by peritoneal macrophages when compared to unexposed controls (Cammarata et al. 2012; Salerno et al. 2009). The F-type lectin from rock bream, *Oplegnathus fasciatus*, has been shown to limit infection by viral hemorrhagic septicemia virus (Cho et al. 2014). The multivalent nature of the F-type lectin (as in AAA) and the opposite arrangement of tandem domains (as in *Morone saxatilis* agglutinin, MsaFBP32) facilitate binding to glycans on the microbial surface with high avidity and cross-linking of nonself and self glycans on microbial and host phagocytic cells, respectively (Odom and Vasta 2006; Vasta et al. 2017).

The highly diversified repertoire of FLDs in the sperm bindin proteins of the Pacific oyster, *Crassostrea gigas*, plays a role in prevention of polyspermy during fertilization (Springer et al. 2008), and an F-type lectin in the pearl oyster, *Pinctada martensii*, has been demonstrated to protect against infectious challenge by the pathogen, *Vibrio alginolyticus* (Chen et al. 2011). The horseshoe crab lectin, tachylectin-4, also plays a role in the recognition of Gram-negative bacteria by recognizing S-type lipopolysaccharides through O-antigens (Saito et al. 1997).

Bryohealin, an F-type lectin from the green alga, *Bryopsis plumosa*, is involved in agglutination of subcellular organelles – an essential step in protoplast formation and regeneration of new living cells (Kim et al. 2006; Yoon et al. 2008). The bacterial F-type lectin domain in *Streptococcus mitis* lectinolysin has been demonstrated to have a role in virulence. It enables recognition of and adhesion to difucosylated glycans on host platelets and the subsequent disruption of the host cell membrane via the action of its co-associated domain, a cholesterol-dependent cytolysin domain. Although the functions of many other biochemically characterized FLDs are unknown, possible functions have been inferred from their biochemical properties and/or tissue expression data. For instance, the FLD in the *furrowed* gene of *Drosophila* is thought to play a role in cell adhesion (Odom and Vasta 2006; Vasta et al. 2017); the FLD in the purple sea urchin, *Strongylocentrotus purpuratus*, might have a function in complement regulation (Multerer and Smith 2004); the FLDs in *Streptococcus pneumoniae* SP2159 might have a role in virulence (Boraston et al. 2006); and F-type lectins in the tungara frog (*Engystomops pustulosus*), common periwinkle (*Littorina littorea*), and blunt-gaper clam (*Mya truncata*) are thought to have a role in innate immune defense against pathogens (Fleming et al. 2009; Arivalagan et al. 2016; Gorbushin and Borisova 2015; Vasta et al. 2017).

In this mini-review, we will summarize the existing knowledge about the taxonomic spread, architectural contexts, sequence characteristics, and saccharide-binding specificity of biochemically characterized FLDs and deliberate the possible potential of FLD sequence diversity, structure, and function on glycobiology applications.

24.2 Prevalence of FLDs in Various Life Forms

FLDs are taxonomically widespread, occurring in viruses, prokaryotes, and eukaryotes (Bishnoi et al. 2015). However, their distribution is very selective in all these life forms. Figure 24.1 shows the distribution of FLDs in the tree of life.

Among viruses, FLDs occur only in a family of dsDNA viruses called *Phycodnaviridae*; in the viruses *Emiliania huxleyi virus 203* and *Ostreococcus tauri* virus strains 1, 2, and OsV5; and in *Ostreococcus lucimarinus* virus strains Olv1, Olv3, and Olv6 (Bishnoi et al. 2015). The hosts of these viruses are freshwater and marine algae, which also have FLDs, and it is likely that the prevalence of viral FLDs is due to lateral transfer from these algal hosts (Bishnoi et al. 2015). Indeed, phylogenetic analysis indicating clustering of the viral FLDs not only with those of algae, *Thalassiosira oceanica*, *Volvox carteri* f. nagariensis, and *Chlorella variabilis* but also with those of bivalve mollusks, *Crassostrea* spp., and *Mytilus edulis*, which are filter feeders that take up microalgae (together with their associated viruses), hints at possible lateral transfer events between mollusk, algal, and viral FLDs (Bishnoi et al. 2015; Vasta et al. 2017).

Among prokaryotes, FLDs do not occur in Archaea but are found in a few eubacterial phyla – *Acidobacteria*, *Actinobacteria*, *Bacteroidetes*, *Cyanobacteria*, *Firmicutes*, *Planctomycetes*, *Proteobacteria*, and *Verrucomicrobia* (Bishnoi et al. 2015). The selective distribution of FLDs is again apparent from their patchy pattern of occurrence in the organisms within these phyla. For instance, FLDs are found only in a few of the many *Campylobacter* and *Geobacter* species whose genomes are sequenced and available in the public database (Bishnoi et al. 2015). Bacteria with FLDs belong to a wide range of ecological niches – some are free living and are found in soil, glacial ice, freshwater, or marine water, and others are commensals of algae, higher plants, or animals (including human beings) (Bishnoi et al. 2015).

Among eukaryotes, FLDs occur in a coccolithophore (orphan order Isochrysidales), in diatoms (phylum Bacillariophyta), in a cryptomonad (orphan class Cryptophyta), in a brown alga (phylum Phaeophyceae), in green algae (phylum Chlorophyta), in fungi (orphan order Peronosporales), and in the metazoan phyla, Cnidaria, Arthropoda, Annelida, Mollusca, Hemichordata, and Chordata (Bishnoi et al. 2015). FLDs are especially abundantly represented in the cephalochordate, *Branchiostoma floridae*, in mollusks, bony fishes (teleosts), amphibians, and insects (Bishnoi et al. 2015).

24.3 Provenance of the FLD

The selective and discontinuous distribution of FLDs in eukaryotes and prokaryotes suggests that the FLDs were likely lost from several taxa due to lack of fitness value (Odom and Vasta 2006; Bishnoi et al. 2015). It is also possible that the FLDs diverged beyond recognition (by sequence similarity) to non-FLD FA58C or to other domains in certain taxa or that the presence of the FLDs in certain taxa is due to lateral transfer (Bishnoi et al. 2015). The relatively greater abundance and more extensive representation of FLDs in eukaryotes makes it likely that the FLD first originated in eukaryotes and was then laterally transferred to bacteria and to viruses (Odom and Vasta 2006; Bishnoi et al. 2015). In fact, the phylogenetic tree of all FLDs shows eukaryotic FLDs interspersed among separate clusters of bacterial FLDs, and although the percent sequence similarity and domain organization pattern are similar among FLDs from related organisms, there are several atypical phylogenetic associations that hint at lateral transfers (Bishnoi et al. 2015).

Structurally fold-wise, the FLD (classified under the family fucose-binding lectins) is similar to the discoidin domain and galactose-binding domains classified together with FLD in the Pfam database under the coagulation factor 5/8 C-terminal (FA58C)/discoidin (PF00754) family (Holm and Sander 1993; Bishnoi et al. 2015). Structural homology is shared with domains of proteins from varied sources and with diverse functions such as discoidins, a single-strand

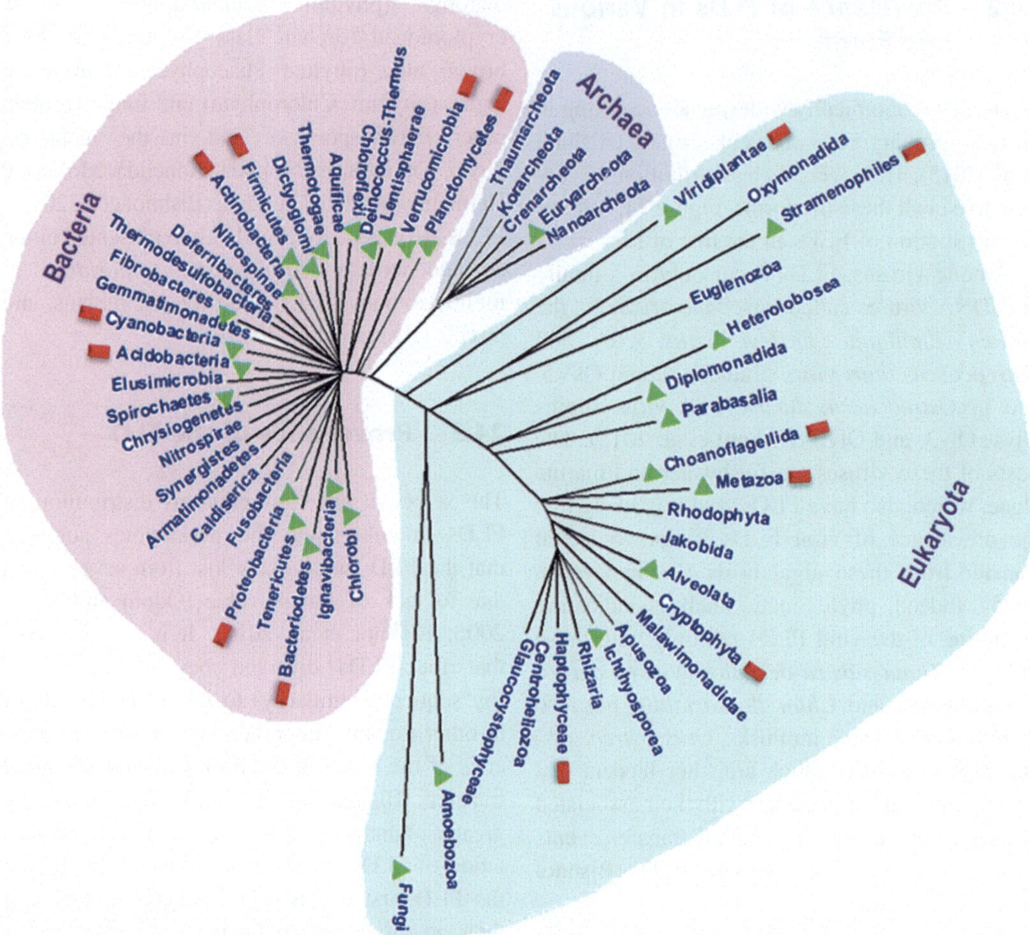

Fig. 24.1 Phylogenetic tree representing distribution of FLDs in various life forms. Green triangular marker corresponds to taxa containing FA58C domain according to Pfam database (Finn et al. 2014). Red bars point to taxa containing FLDs identified by our different search strategies in different life forms. (Figure adapted from Bishnoi et al. 2015)

DNA break repair complex, neuropilin, human clotting factors V and VIII, bacterial sialidases, a yeast allantoicase, a fungal galactose oxidase, and a human ubiquitin ligase subunit (Vasta et al. 2004). The FLD even shares a similar placement of the binding site with these domains (Vasta et al. 2004). Thus it is possible that the FLD and these other domains either share a similar carbohydrate-binding domain as their ancestor or that the FLD has been evolutionarily co-opted into these other domains (Vasta et al. 2004).

24.4 Domain Context of FLDs

Viral proteins have stand-alone FLDs or FLDs associated with a pentraxin domain (Bishnoi et al. 2015). Significantly, they are found in different architectures from their host algal counterparts, perhaps due to the lateral transfer of just the FLD into different contexts (Bishnoi et al. 2015).

With the exception of a few FLDs in *S. mitis*, *S. pneumoniae*, and *Myxococcus hansupus*, which exist as tandem repeats, bacterial FLDs occur as a single copy in tandem with one or more other protein domains (Bishnoi et al. 2015). The co-occurring protein domains include carbohydrate-binding modules (CBMs) belonging to CBM6 agarase, CBM6 cellulase, CBM6 xylanase, ricin-B lectin, and NPCBM (novel putative carbohydrate-binding module)-associated domain. They also include carbohydrate-active enzyme domains such as alpha-L-fucosidase; glycosyl hydrolases (GHs) belonging to GH16, GH20, and GH98; alginate lyase; beta-*N*-acetylglucosaminidase; glycosyltransferases (GTs) belonging to GT2/GT92; and GT17, fucosyltransferase, and LicD (lipopolysaccharide cholinephosphotransferase family of proteins involved in glycan modification). Other enzyme domains such as methyltransferases 21 and 23, lipase, thiol cytolysin, and membrane-bound dehydrogenase and nonenzymatic domains such as Ig, cytochrome C, tetratricopeptide (TPR) repeats, and HEAT repeats also co-occur with FLDs (Bishnoi et al. 2015). The domain organization is largely similar in related organisms (Bishnoi et al. 2015). For instance, ricin-B lectin, GH16, and Por secretion system C-terminal sorting domain are co-occurring domains in several members of *Flavobacteriia*, GH98 is a co-occurring domain in *Bacilli*, and alpha-L-fucosidase, Ig-domain, and glucan-binding domain (YG repeat/cell wall-binding repeat) are co-occurring domains in *Actinobacteria* (Bishnoi et al. 2015). Figure 24.2 shows some of the common domain architectures in bacterial FLD-containing proteins.

The co-occurring domains of eukaryotic FLD-containing proteins are listed in Table 24.1. Unlike their bacterial counterparts, vertebrate FLDs mostly occur stand-alone in a single copy or in multiple tandem repeats (Bishnoi et al. 2015). In a few eukaryotic classes, FLDs are also frequently found paired with C-type lectins (Bishnoi et al. 2015). In amphibians, the FLD is also often found in tandem with a pentraxin domain, and in insects, FLDs typically have a C-type lectin domain as well as Sushi or complement control protein (CCP) domains in tandem (Bishnoi et al. 2015). Algae, fungi, cnidarians, arachnids, hemichordates, and mollusks, however, have FLDs occurring in tandem with other domains, too, and the cephalochordate, *Branchiostoma,* is remarkable for its profusion of FLD-containing proteins containing diverse co-occurring domains in different architectures (Bishnoi et al. 2015).

24.5 Glycan Binding by FLDs

Several characterized FLDs have been demonstrated to bind to L-fucose and fucosylated oligosaccharides and display variations in their fine oligosaccharide-binding specificities (Table 24.2). For instance, *Anguilla anguilla* agglutinin (AAA) binds to L-fucose and to fucosylated oligosaccharides such as the blood group antigen H type 1 and the Le^a antigen, but does not bind to the Le^x antigen (Bianchet et al. 2002). The F-type lectin from *M. saxatilis*, MsaFBP32, demonstrates specific binding to L-fucose; it does not bind to other sugars such as D-fucose, D-galactose, L-galactose, D-glucosamine, D-mannosamine, and D-arabinose (Odom and Vasta 2006). It also binds to the fucosylated oligosaccharides, H-disaccharide, 3-fucosyllactose, and Fucα1-6GlcNAc. It shows similar binding to these fucosylated oligosaccharides, with no preference for α1-2- or α1-3-linked L-fucose, whereas AAA binds preferentially to α1-2-linked L-fucose (Odom and Vasta 2006; Bianchet et al. 2002). MsaFBP32 has also been shown to bind to fucosylated gastric mucins (Odom and Vasta 2006). Docking studies of *M. saxatilis* F-type lectin, MsaFBP32, which has two tandem FLDs suggest that the N-terminal FLD may bind to Lewis and H-trisaccharides and the C-terminal FLD to simple L-fucose (Bianchet et al. 2010). *S. pneumoniae* SP2159 binds to L-fucose, blood group H-disaccharide, and A and B trisaccharides but shows maximal binding preference for Le^y antigen, type 2 blood group H-trisaccharide, and 2′-fucosyllactose, and it does not bind to Le^a antigen and Le^x antigen (Boraston et al. 2006). The F-type lectin domain of *S. mitis* lectinolysin displays binding preference for the difucosylated glycans, $Lewis^b$ and $Lewis^y$ antigens

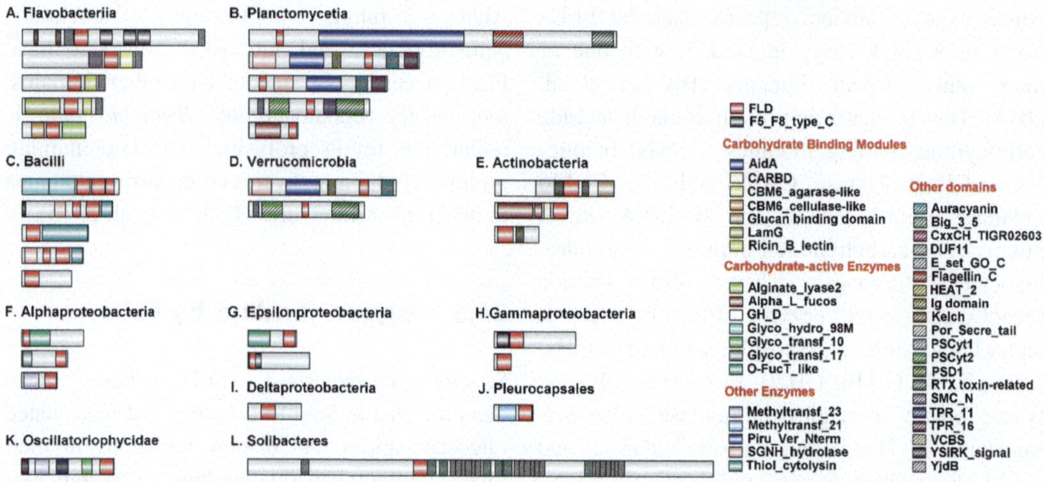

Fig. 24.2 Representative domain architectures of FLD-containing polypeptides in Eubacteria. Different taxonomic classes of Eubacteria are represented, except for *Oscillatoriophycideae* (subclass) and *Pleurocapsales* (order) where no class designation is available. Both these taxa belong to the phylum *Cyanobacteria*. The domain organizations are represented using software called illustrator of biological sequences (IBS) (Liu et al. 2015). Domain abbreviations are according to Conserved Domain Database (CDD). (Figure adapted from Bishnoi et al. 2015)

(Farrand et al. 2008). We recently characterized the glycan-binding specificities of three bacterial FLDs (Mahajan et al. 2017). The FLD from *Cyanobium* sp. PCC 7001, *Cy*FkbMFLD, demonstrates a narrow glycan-binding specificity as it binds only to a subset of complex fucosylated glycans with an extended type 2 blood group H motif, Fucα1-2Galβ1-4GlcNAcβ1-3Galβ1-4GlcNAcβ (Mahajan et al. 2017). The FLD from *Myxococcus hansupus, Mh*FLD1, shows specificity toward blood group B, H type 1, and Lewis^{b} motifs, and the FLD from *Leucothrix mucor, Lm*FLD, binds to a wide range of fucosylated glycans, including A, B, H, and Lewis antigens (Mahajan et al. 2017).

Despite the evidence of numerous FLDs binding to L-fucose and fucosylated oligosaccharides, it is also apparent that the FLD has nominal but not absolute specificity for L-fucose. There are examples of atypical F-type lectins which show binding to non-fucosylated glycans. For instance, besides binding to L-fucose and fucosylated oligosaccharides, AAA can bind to the galactose derivatives, 3-O-methyl-D-galactose and 3-O-methyl-D-fucose, as well as to colitose (3-deoxy-L-fucose) (Springer and Desai 1970; Vasta et al. 2004). Bryohealin, an F-type lectin from *Bryopsis plumosa*, does not bind to L-fucose, but instead recognizes *N*-acetylglucosamine and *N*-acetylgalactosamine (Kim et al. 2006). Ranaspumin-4, an F-type lectin and foam nest component of tungara frog (*Engystomops Pustulosus*), does not bind to L-fucose but instead binds to D-galactose, lactose, and melibiose (Fleming et al. 2009). Tachylectin-4, an F-type lectin from horseshoe crab (*Tachypleus tridentatus*), binds to terminal L-fucose residues and to bacterial lipopolysaccharide, likely, through the L-fucose-like colitose and abequose sugars present in it (Saito et al. 1997). A *Dicentrarchus labrax* fucose-binding lectin demonstrates binding to L-fucose as well as D-galactose and also to the disaccharide D-galactosides, melibiose and lactulose (Cammarata et al. 2001). The *Sparus aurata* serum lectin, SauFBP32, preferentially recognizes fucosyl residues, albeit it binds weakly to galactose and oligosaccharides such as melibiose and lactulose which have galactose residues at the nonreducing end (Cammarata et al. 2007).

Table 24.1 Representative domain architectures of FLD-containing polypeptides in different taxonomic classes of Eukaryotes

Phylum	Class	Organism	Protein ID	Domain organization
A. Domain architectures of FLD-containing polypeptides in algae and invertebrates				
Haptophyta	Haptophyceae-Isochrysidales	*Emiliania huxleyi CCMP1516*	485628619	FLD, MAM
Bacillariophyta	Coscinodiscophyceae	*Thalassiosira pseudonana CCMP1335*	224012533	DUF1800 (2), DUF1501, FLD
		Thalassiosira oceanica	397570057	FLD(3), PT
Chlorophyta	Ulvophyceae	*Bryopsis maxima*	292657239	FLD
	Trebouxiophyceae	*Chlorella variabilis*	307104012	Gal_Lectin, FLD
	Chlorophyceae	*Volvox carteri f. nagariensis*	302853363	F5_F8_type_C, PAN_1, FLD, PAN_APPLE
Heterokontophyta	Oomycetes-Peronosporales	*Phytophthora infestans T30-4*	301107029	OmpH, RCC1, FLD
Cnidaria	Anthozoa	*Nematostella vectensis*	156347710	FA58C, FLD
Annelida	Polychaeta	*Capitella teleta*	443734247	FLD, CLECT
Arthropoda	Arachnida	*Ixodes scapularis*	240979804	Commd8, FLD
			241783054	CUB (2), FLD (2), ApeC
	Malacostraca	*Macrobrachium rosenbergii*	397881228	FLD
	Branchiopoda	*Daphnia pulex*	321474073	FLD, CLECT, CCP (4)
	Insecta	*Drosophila melanogaster*	161077826	CCP, FLD, CLECT, Atrophin-1
		Nasonia vitripennis	345482661	CCP, FLD, CLECT, CCP (4)
	Merostomata	*Tachypleus tridentatus*	156713139	FLD
Mollusca	Bivalvia	*Crassostrea gigas*	146285283	F5_F8_type_C (2), FLD, F5_F8_type_C (2)
			405953631	FLD, NADAR
			405955840	PAN_1, CCP, FLD
			405960138	EGF_CA, FLD
Hemichordata	Enteropneusta	*Saccoglossus kowalevskii*	585709924	FLD (7)
			291220830	ASC, LDLa (3), FLD (2), PAN_AP_HGF, ASC
			291225543	LamG (3), FLD
			291242391	LamG (31), Herpes_BLLF1, FLD (5)
B. Domain architectures of FLD-containing polypeptides in chordates				
Chordata	Cephalochordata-Branchiostomidae	*Branchiostoma floridae*	229286868	FLD
			260810536	FLD (3)
			260812465	FLD, F5_F8_type_C (2), FLD, F5_F8_type_C
			260831061	HEAT_2, F8_type_C
			260789746	F5_F8_type_C, WSC
			261289281	An_peroxidase, FLD
			260794384	Gal_Lectin, FLD, WSC, Gal_Lectin, F5_F8_type_C

(continued)

Table 24.1 (continued)

Phylum	Class	Organism	Protein ID	Domain organization
			229292595	FA58C (2), CCP, F5_F8_type_C (2)
			260810530	FLD (2), CLECT, KR
			229273176	MACPF, Furin-like, ChtBD1_GH18_2, F5_F8_type_C, FLD
			229293864	LamG, FLD, LamG, FLD, LamG, FLD (2)
			229290813	FLD, CLECT, KR, FA58C, FLD, PAN_APPLE, Atrophin-1, CLECT, F5_F8_type_C, PAN_APPLE, lamp, CLECT
			260786006	FLD, CLECT (2), FA58C, GON, WSC, KR, CLECT, Ig
			260831928	GMP_PDE_delta, FLD (2), CLECT, TSP1, CPW_WPC, CUB, TSP1, GPS, 7tm_GPCRs
			260831065	CLECT, TFIIFa, FLD, GON, FLD, CLECT, CUB, SR (2), SerH, CLECT, LDLa, CRD_FZ, KR, CCP
			229290364	FA58C, FLD, FN3 (2), FXa_inhibition, FN3
			260787565	LRRNT, LRR, LRRCT, FLD, CCP (4), EGF_3, CCP (2)
			260787625	NEL, LRR, PCC, FLD, CCP
			260793715	FA58C, LINK, FLD, Laminin_G_3, PAN_AP_HGF, Laminin_G_3, FA58C, FN3 (2), GPS, PLAT, PKD_channel, NT_Pol-beta-like, TPR
			260794149	FLD, EGF_CA, REJ
			260794525	LCCL, CLECT, link, trefoil, FLD, CLECT, fn3 (2)
			260795378	TMEM108, F5_F8_type_C, FLD, CLECT
			260800881	KR, FLD, UDPGT
			260800893	KR, FLD, GON
			260806424	NHL, Gal_lectin, FLD, ChtBD1_GH18_2
			260815028	Laminin_G_3, FLD, collagen (2)
			260825869	P-loop_NTPase, FLD, CLECT (2)
			260824649	WSC, FLD (2) Sulfotransfer_3
			260783098	MACPF, CLECT, FLD, CLECT, SMC_N
			260799503	FLD, G8, PANDER_like_TMEM2, FLD, PANDER_like
			260794963	P-loop_NTPase, COG5635, LRR_RI, FLD, Gal_Lectin, FLD
			260826486	An_peroxidase_like, FLD, CLECT, peroxinectin_like (2)
			260820309	FLD, CLECT, KR, FA58C (2), SRCR

(continued)

Table 24.1 (continued)

Phylum	Class	Organism	Protein ID	Domain organization
			260841546	FLD, Gal_lectin, F5_F8_type_C, PAN_APPLE, TSP1 (8), Neur_chan_LBD
			260820305	FLD, KR (2), FLD (4), KR, SR, ChtBD1_GH18_2, CLECT (2)
			260806492	PAN_1, SR, ChtBD1_GH18_2, SR, TSP1, SR, TSP1, SR (2), TSP1 (7), FLD, SR (2), TSP1 (2), CLECT (2), CUB (2), SR, FLD, CUB, FLD, FA58C, SR (2), TSP1 (7), SR (2), TSP1 (8)
			260831067	TM2, Sulfotransfer_2, Mucin2_WxxW,WSC, FLD
			260835699	FLD (12), EGF_CA (5), GLTP
			260791880	WSC, F5_F8_type_C (2), WSC, TECPR, Hyd_WA, TECPR, Hyd_WA (2), FLD, F5_F8_type_C (2), CLECT
			260785974	CUB, LamG, SR, Endomucin, Mucin2_WxxW, Methyltransf_FA, CRD_FZ, ChtBD1, Methyltransf_FA, CRD_FZ, ChtBD1, Gal_Lectin, FLD, LamG, FLD
	Chondrichthyes	*Callorhinchus milii*	387915284	FLD
	Coelacanthimorpha-Coelacanthiformes	*Latimeria chalumnae*	942209564	FLD
	Actinopteri	*Anguilla japonica*	9651031	FLD
		Tetraodon nigroviridis	47210978	FLD (2)
		Oncorhynchus mykiss	185134285	FLD (4)
	Amphibia	*Engystomops pustulosus*	37906080	FLD
		Xenopus laevis	187608266	FLD (2)
			115528319	FLD (3)
			78191615	FLD (4)
			213627704	FLD (5), PTX
	Sauria-Testudines	*Chelonia mydas*	465970729	FLD
	Mammalia	*Ornithorhynchus anatinus*	620970996	FLD (5), DNA_pol3_gamma3

Domain abbreviations are according to Conserved Domain Database (CDD). Domain organization is shown according to CDD representation of domains from N- to C-terminus in each polypeptide. Number written in brackets next to some of the domains is representing the number of repeats of that particular domain. Table adapted from (Bishnoi et al. 2015)

24.6 Structural Features of FLD

Experimentally determined atomic-resolution three-dimensional structures are available for the F-type lectins, *A. anguilla agglutinin* (AAA, PDB ID: 1 K12), *M. saxatilis* MsaFBP32 (PDB ID: 3CQO), *S. pneumoniae* SP2159 (PDB ID: 2J1S), and *S. mitis* lectinolysin (PDB ID: 3LEI). The FLD adopts a beta-barrel structure having jelly roll topology with a unique fold called the

Table 24.2 Glycan-binding specificities of characterized FLDs. (A) Eukaryotic and (B) prokaryotic FLDs

Name of lectin	Glycan-binding specificity		References
	Binders	Non-binders	
A			
Anguilla anguilla agglutinin (AAA)	L-fucose, H type-1 antigen, Le^a antigen, 3-O-methyl-D-galactose, 3-O-methyl-D-fucose, colitose	Le^x antigen	Bianchet et al. (2002)
Morone saxatilis fucose-binding lectin (MsaFBP32)	L-fucose, H-disaccharide, 3-fucosyllactose, Fucα6GlcNAc, mucin	D-fucose, D-galactose, L-galactose, D-glucosamine, D-mannosamine HCl, D-arabinose	Odom and Vasta (2006)
Tachypleus tridentatus fucose-binding lectin (Tachylectin-4)	L-fucose, bacterial lipopolysaccharide		Saito et al. (1997)
Bryopsis plumosa fucose-binding lectin (Bryohealin)	*N*-acetylglucosamine, *N*-acetylgalactosamine	L-fucose	Yoon et al. (2008)
Engystomops pustulosus fucose-binding lectin (Ranaspumin-4)	D-galactose, lactose and melibiose	L-fucose	Fleming et al. (2009)
Dicentrarchus labrax fucose-binding lectin	L-fucose, D-galactose, melibiose, lactulose		Cammarata et al. (2001)
Sparus aurata fucose-binding lectin (SauFBP32)	L-fucose, D-galactose, melibiose, lactulose		Cammarata et al. (2007)
B			
Streptococcus pneumoniae SP2159	L-fucose, $Lewis^y$ antigen, H type 2 di- and trisaccharide, blood group A and B trisaccharides, 2′-fucosyllactose	$Lewis^a$, $Lewis^x$ antigens	Boraston et al. (2006)
Streptococcus mitis lectinolysin	$Lewis^b$, $Lewis^y$ antigens		Farrand et al. (2008)
Cyanobium sp. PCC 7001 fucose-binding lectin (*Cy*FkbMFLD)	Extended H type 2 saccharides, Fucα1-2Galβ1-4GlcNAcβ1-3Galβ1-4GlcNAcβ		Mahajan et al. (2017)
Myxococcus hansupus fucose-binding lectin (*Mh*FLD1)	Blood group B, H type 1, and $Lewis^b$	$Lewis^x$, $Lewis^a$, and blood group A	Mahajan et al. (2017)
Leucothrix mucor fucose-binding lectin (*Lm*FLD)	Blood group A, B, H, and Lewis antigens		Mahajan et al. (2017)

F-type lectin fold (Bianchet et al. 2002). Analysis of the crystal structure of AAA reveals that the structure comprises two sheets of five (β2, β3, β6, β7, and β10) and three (β5, β8, and β11) antiparallel beta-strands (Bianchet et al. 2002). One end of the barrel is closed by the strands β4 and β9 (Bianchet et al. 2002). At the other end is an L-fucose-binding cleft, bounded by five loops or complementarity-determining regions (CDR1–5) that connect the β-strands (Bianchet et al. 2002). The side chains of His52 and two Arg residues, Arg79 and Arg86, protrude out of this cleft and make hydrogen bonding contacts with the ring O5, the axial 4-OH, and the equatorial 3-OH oxygen atoms of L-fucose (Bianchet et al. 2002). His52, Arg79, and Arg86 are held in position by hydrogen bonds with Ser53, Asp81, and His144. The C1 and C2 atoms of L-fucose rest on the cysteine disulfide bridge formed by Cys82 and Cys83 (Bianchet et al. 2002). Amino acids of the CDR1 and CDR2 loops, Leu23, His27, Phe45, and Tyr46, form a hydrophobic pocket which ensconce the C6 methyl group of L-fucose (Bianchet et al. 2002) (Fig. 24.3a).

A 3_{10} helix-rich substructure including helices, ⍰1, ⍰2, and ⍰3, and loops CDR1 and CDR2 makes a Ca^{2+}-binding motif. In AAA, the Ca^{2+} is tightly coordinated by seven oxygen atoms from Asn35, Asp38, Asn40, Ser49, Cys146, and Glu147, which includes side chain and main chain oxygen

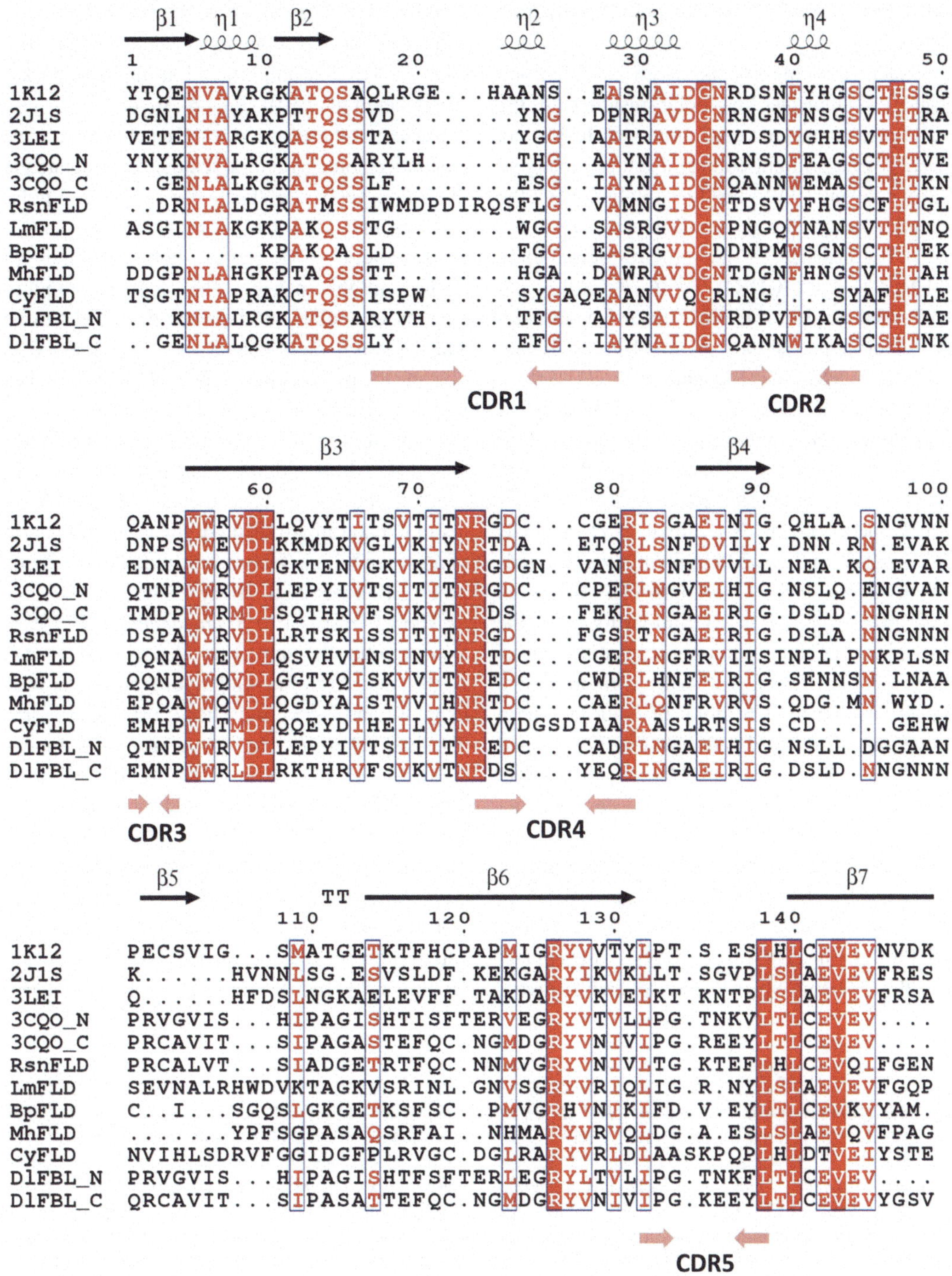

Fig. 24.3 Surface representations of the fucose-binding cleft of FLDs. (**a**) 1K12 (*A. anguilla* agglutinin with fucose), (**b**) 3CQO_N (*M.* agglutinin N-terminus FLD with fucose), (**c**) 3CQO_C (*M. saxatilis* C-terminus FLD

atoms from Ser49, side chain oxygen atoms from Asp38 and Glu147, and main chain oxygen atoms from the remaining residues. Two disulfide bridges (Cys50 to Cys146 and Cys108 to Cys124) and two salt bridges (Arg41 to Glu149 and Asp64 to Arg131) help to hold the structure together.

Modeling experiments with AAA and MsaFBP32 indicate that polar residues of the CDR1 also contact the third moiety of the glycan, CDR2 residues might contact a fourth moiety of the glycan, and CDR4 residues might contact the second moiety of the glycan (Bianchet et al. 2002, 2010). For instance, in AAA, the residues, Glu26 and His27 of the CDR1, Tyr46 of CDR2, and Arg79 and Asp81 of CDR4, make contact with the subterminal residues of blood group H and Lewis[a] trisaccharides (Bianchet et al. 2002).

Carbohydrate docking studies in MsaFBP32 show that similar interactions are involved in binding to L-fucose (Bianchet et al. 2010). His44, Arg71, and Arg78 make hydrogen bonds with L-fucose in the N-terminal FLD, and His 190, Arg217, and Arg223 make hydrogen bonds with L-fucose in the C-terminal FLD (Bianchet et al. 2010) (Fig. 24.3b, c). Similar to AAA, the N-terminal FLD has hydrophobic residues, Phe37 and Tyr26, in CDR1 that make a hydrophobic pocket nestling the C6 group of L-fucose and a contiguous disulfide bond between Cys74 and Cys75, which makes van Der Waals contact with the C1 and C2 atoms of L-fucose (Bianchet et al. 2010) (Fig. 24.3b). However, in the C-terminal FLD, the shape of the binding pocket is altered by the bulkier residues, Trp183 in CDR1 that contacts the C6 group of L-fucose and Phe220 in CDR4 that makes close contact with the 2-OH group of L-fucose, and disfavors substitution at the 2-OH group of L-fucose (Bianchet et al. 2010) (Fig. 24.3c). Also, while residues Tyr18, His20, His22, Asp36, and Glu38 make up a continuous polar ridge in the N-terminal FLD, the binding pocket in the C-terminal FLD is more shallow with a broader and more apolar ridge with residues Phe166, Glu167, Trp183, and Glu184 around the pocket (Bianchet et al. 2010) (Fig. 24.3b, c).

Based on these structures, Bianchet et al. have postulated some structural features that dictate the fine oligosaccharide specificity of FLDs – polar residues in CDR1 facilitate interactions with a third moiety of a glycan, bulky residues in CDR2 might narrow the L-fucose-binding pocket, an apolar area between the CDR2 ridge and L-fucose might enable a fourth moiety of a glycan to be accommodated, CDR4 residues might interact with second moiety of the glycan, and substitution of the Cys residues in the CDR4 that make the contiguous disulfide bond might result in the flat apolar ridgelike structure being disrupted and might disfavor the accommodation of a glycan with L-fucose substituted at the 2-OH group (Bianchet et al. 2010).

The FLD SpX-1 in *S. pneumoniae* SP2159 is also structurally similar to AAA with His54, Arg81, and Arg88 making hydrogen bonds with L-fucose and Tyr31 and Phe47 forming a hydrophobic pocket for the C6 group of L-fucose (Boraston et al. 2006) (Fig. 24.3d). Minimal direct and some water-mediated hydrogen bonds are observed between the protein and the subterminal moieties of blood group A tetrasaccharide, type 2 trisaccharide, and Lewis[y] antigen, and Boraston et al. suggest that binding is principally determined by specific interactions between the protein and L-fucose and shape complementarity

Fig. 24.3 (continued) with fucose), (**d**) 2J1U (*S. pneumoniae* FLD with A tetrasaccharide), (**e**) 3LEG (*S. mitis* FLD with Lewis Y Antigen), (**f**) *L. mucor* FLD, (**g**) *M. hanspus* FLD, (**h**) *Cyanobium* sp. FLD, (**i**) Ranaspumin FLD, (**j**) *B. plumosa* FLD, (**k**) *D. labrax* N-terminus FLD, and (**l**) *D. labrax* C-terminus FLD. Structural models of *Lm*FLD, *Mh*FLD1, RsnFLD, *Bp*FLD, and *Dl*FBL were modeled using 1 k12 as template using Fold and Function Assignment System (FFAS) server (Jaroszewski et al. 2005), and structural model of *Cy*FLD was made in I-TASSER server (Roy et al. 2010; Yang et al. 2015; Zhang 2008) using 3LEI as template. All the structures were aligned with 1 K12 complexed with fucose. The surface charge and hydrophobicity were visualized in PyMOL (PyMOL Molecular Graphics System, Version 2.0 Schrödinger, LLC) by coloring Lys, Arg, and His in blue; Asp and Glu in red; Trp, Try, and Phe in yellow; Cys, Met, Ser, Thr, Asn, and Gln in green; and Ile, Leu, Val, Gly Pro, and Ala in white. Fucose is represented in stick model with elemental coloring. (Figure adapted from Mahajan et al. 2017) between the binding cleft and the saccharides attached to L-fucose (Boraston et al. 2006). Different substituents on the saccharides might result in intramolecular restraints and thereby alter conformation of the glycan, thus affecting shape complementarity (Boraston et al. 2006).

In *S. mitis* lectinolysin, the L-fucose-binding pocket comprises the residues His85, Arg112, and Arg120 that directly interact with L-fucose through hydrogen bonds, Gly115 that makes a water-mediated hydrogen bond with L-fucose, and Tyr62, Tyr78, Phe88, and Val117 that make hydrophobic interactions with L-fucose (Feil et al. 2012) (Fig. 24.3e). Water-mediated hydrogen bonds and van der Waals interactions are present between the protein and the other saccharide moieties of the fucosylated glycans, Lewis^{b} and Lewis^{y} (Feil et al. 2012).

Structural models of *Lm*FLD and *Mh*FLD indicate similar positively charged pockets as AAA (Mahajan et al. 2017). The predicted fucose-binding pocket of *Lm*FLD displays the basic residues, His48, Arg75, and Arg82, for hydrogen bonding with L-fucose, and Asp77 to orient Arg75 (Fig. 24.3f). The bulky residues, Trp25 and Tyr41, which seem to constrict the binding site, offer a hydrophobic ridge for interacting with the C6 group of L-fucose (Fig. 24.3f). The polar residues, Asn42 and Gln40, extend from the CDR1 ridge and might interact with subterminal moieties (Fig. 24.3f). A disulfide bridge formed by Cys78 and Cys79 offers a flat ridge that might make van der Waals contact with the carbon atoms of L-fucose (Fig. 24.3f). An additional apolar ridge formed by Tyr141 is observed near the His48, further constricting the shallow pocket (Fig. 24.3f). The predicted L-fucose-binding pocket of *Mh*FLD is relatively broad and contains the basic residues, His67, Arg94, and Arg101, for hydrogen bonding with L-fucose, Asp96 to orient Arg94, the hydrophobic residue, Phe60 to accommodate the C6 group of L-fucose, polar residues, His44 and His61 extending from the CDR1 ridge to interact with subterminal moieties, a disulfide bridge formed by Cys97 and Cys98 to make van der Waals contact with the carbon atoms of L-fucose, and a protuberance formed by His70 (Fig. 24.3g).

The structural model of *Cy*FLD also displays the conserved His and Arg residues, His48, Arg75, and Arg82 in a cleft; however the cleft is quite distinct from the others, being deeper and surrounded by a high ridge with the hydrophobic residues, Trp26 and Tyr45 of the CDR1 and Asp51 on one side (Mahajan et al. 2017) (Fig. 24.3h). A narrow polar ridge formed by Asp78, Ser80, and Asp81 is adjacent to the cleft, replacing the flat Cys disulfide bridge (Fig. 24.3h). The unusually long CDR1 and short CDR2 could be a reason for this unusual cleft and for the ability of *Cy*FLD to bind to glycans with extended type 2 H motif (Mahajan et al. 2017) (Fig. 24.3h).

The structural model of Ranaspumin-4 reveals the residues His67, Arg94, and Arg100 forming a positively charged cleft bounded by Tyr76, Phe66, Phe61, Asp96, and Phe97 (Mahajan et al. unpublished data) (Fig. 24.3i). Tyr60 is present in a hydrophobic hollow adjacent to the positively charged cleft, and Phe97 forms a conspicuous protrusion instead of the flat ridge formed by the contiguous Cys disulfide bridge in AAA and other FLDs (Fig. 24.3i). CDR1 forms a distinct ridge and has the polar residues, Asp39 and Gln42 (Fig. 24.3i). These unusual features likely determine the binding of Ranaspumin-4 to D-galactose rather than L-fucose.

The structural model of Bryohealin, which binds to *N*-acetylglucosamine and *N*-acetylgalactosamine rather than to L-fucose, shows a positively charged cleft formed by the residues, His169, Arg196, and Arg203, bounded by a hydrophobic ridge formed by Phe146 and Trp162, a hydrophobic protrusion formed by Tyr255, and a flat apolar ridge formed by the Cys199 and Cys200 disulfide bridge (Mahajan et al. unpublished observation) (Fig. 24.3j). Glu197 and Asp198 form a negatively charged ridge extending the ridge formed by the contiguous Cys residues, and Asp145 forms an unusual negatively charged area just adjacent to the positively charged cleft (Fig. 24.3j).

The structural models of the N-terminal and C-terminal FLDs of the fucose- and galactose-binding lectin from sea bass, *Dicentrarchus labrax*, show interesting differences. The N-terminal FLD has a relatively broad, positively charged cleft with the basic residues, His62, Arg89, and Arg96, bounded by a flat apolar ridge formed by the contiguous Cys92 and Cys93 on one side and short ridges formed by Phe153, Glu65, Tyr36, Phe55, and Asp91 (Mahajan et al. unpublished observation) (Fig. 24.3k). The C-terminal FLD has a small positively charged hollow with the basic residues, His 208, Arg235, and Arg241 bounded mainly by the bulky, hydrophobic residues, Tyr184, Trp201, Tyr238, and Tyr297 (Fig. 24.3l).

24.7 Sequence Characteristics of FLD

The FLD sequence has two sequence motifs – an L-fucose-binding FLD motif and a Ca^{2+}-binding motif (Fig. 24.4). The L-fucose-binding motif, conventionally regarded as HX(26)RXDX(4)R/K (Vasta et al. 2008), includes the His and the two Arg residues that are involved in making hydrogen bond contacts with the ring O5, 4-OH, and 3-OH atoms of the L-fucose. Upon analysis of the many more FLD sequences available in the public database, we found that this motif can be regarded as HX(26-31)RX(0-2)DX(1-10)R/K for prokaryotic FLDs, HX(26-29)RXD/N/SX(4)R/K for viral FLDs, and HX(26-36)RX(0-2)DX(4-14)R/K for eukaryotic FLDs (Bishnoi et al. 2015).

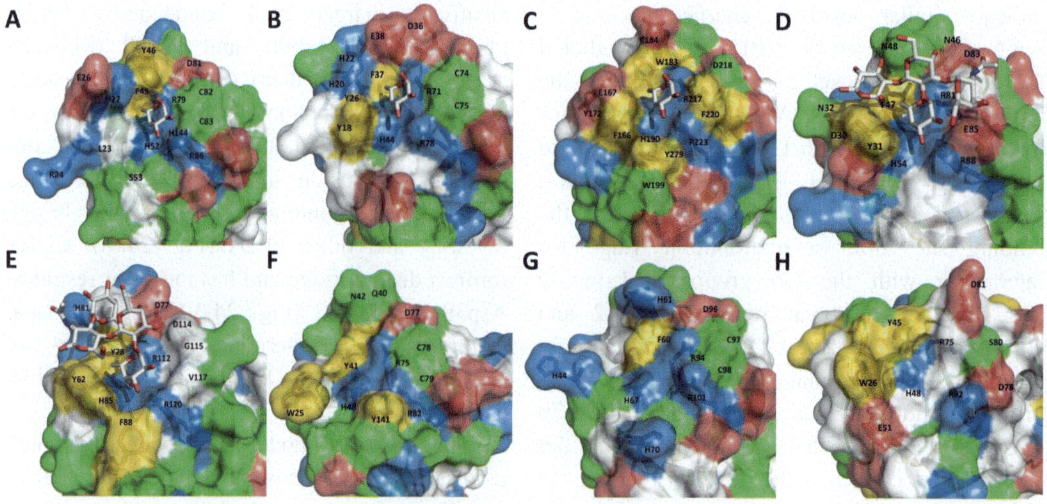

Fig. 24.4 Sequence alignment of FLDs. 1K12 (*Anguilla anguilla* agglutinin), 2J1S (*S. pneumoniae* FLD), 3LEI (*S. mitis* FLD), 3CQO_N (*Morone saxatilis* agglutinin N-terminus FLD), 3CQO_C (*Morone saxatilis* C-terminus FLD), *L. mucor* FLD, *M. hanspus* FLD, *Cyanobium* sp. FLD, Ranaspumin FLD, *B. plumosa* FLD, *D. labrax*_N terminus FLD, and *D. labrax*_N terminus FLD. Alignment was performed with T-Coffee multiple sequence alignment server (Notredame et al. 2000) and is represented using ESPript program (Gouet et al. 2003). (Figure adapted from Bishnoi et al. 2015)

However, approximately one third of the FLD sequences lack one or more of the conserved His, Asp, and/or Arg residues (Bishnoi et al. 2015). Polar as well as apolar residues are found to substitute the His and the Arg residues at these positions in these FLD sequences (Bishnoi et al. 2015). The Asp residue in the FLD sequence motif is mostly substituted by the physicochemically similar amino acid, Glu, or by Asn, Ser, or Gly in these sequences (Bishnoi et al. 2015).

The His/Arg/Arg residues are thought to be critical for L-fucose binding, and mutagenesis of the first Arg residue has been demonstrated to lead to loss of lectin activity in *S. mitis* lectinolysin (Farrand et al. 2008). In the FLDs that occur as plural tandem repeats on a single polypeptide, mutations of the His/Arg/Arg might be tolerated due to redundancy. However, naturally occurring variations of the His/Arg/Arg are observed in FLDs that occur singly, too (Bishnoi et al. 2015). Our studies on a few bacterial FLDs with His substitution indicate that certain polar amino acids can effectively replace His without loss of lectin activity (Khairnar et al. unpublished data). In other FLDs with apolar residues substituting the His, polar residues in the close vicinity might instead be involved in hydrogen bonding with L-fucose. It is also possible that certain FLDs with apolar residues substituting the His/Arg/Arg do not possess L-fucose-binding lectin activity, but have evolved to serve some other function such as binding to some other biological macromolecule.

There are two contiguous Cys residues that form a disulfide bridge (Fig. 24.4) that is involved in making contact with the ring atoms C1 and C2 of L-fucose in AAA. This unusual disulfide bridge is conserved in >50% eukaryotic FLDs and in all viral FLDs but present in only ~20% of all prokaryotic FLDs. It is absent in FLDs belonging to the classes *Alphaproteobacteria*, *Epsilonproteobacteria*, *Planctomycetia*, and *Verrucomicrobia* and phylum *Cyanobacteria* and present in FLDs from *Solibacteres* and *Deltaproteobacteria*, most FLDs from *Flavobacteriia*, and in a few FLDs from *Gammaproteobacteria*, *Actinobacteria*, and *Bacilli* (Bishnoi et al. 2015). Its selective presence suggests that the FLD tolerates its absence. Unlike the characterized eukaryotic FLDs, which are secreted proteins and have signal peptides, more than 50% of the bacterial FLDs do not have a signal peptide (Bishnoi et al. 2015), suggesting that they are either secreted through noncanonical pathways or that they are indeed located in the cytoplasm. Besides being a constituent of the cell wall and signaling factors, L-fucose may also be present within bacterial cells, for instance, as a component of carotenoids or secondary metabolites (Maki and Renkonen 2004; Riely et al. 2004; Samuel and Reeves 2003; Takaichi et al. 2001), and it is possible that FLDs are involved in processes related to their function (Bishnoi et al. 2015). It is therefore likely that the FLD first originated in a nonreducing environment in the eukaryotes, but later the disulfide bridge and the contiguous Cys residues were lost following lateral transfer to prokaryotes, loss of signal peptide, and change of locale to a reducing environment (Bishnoi et al. 2015).

In AAA, besides this unusual disulfide bridge, there are Cys residues forming two other disulfide bridges that stabilize the structure (Bianchet et al. 2002). These Cys residues are absent in the prokaryotic and viral FLDs (Bishnoi et al. 2015). Among the eukaryotic FLDs too, they are absent in FLDs from diatoms and green algae (Bishnoi et al. 2015). There are also a few peculiarities associated with these disulfide bridges in a few eukaryotic FLDs. In certain proteins with two FLD repeats from bony fishes, one FLD has the contiguous Cys residues but lacks the two Cys residues involved in the internal of the two other disulfide bridges, while the other FLD possesses both the internal and external disulfide bridges but lacks the contiguous Cys residues involved in the unusual disulfide bridge (Odom and Vasta 2006; Vasta et al. 2008; Bishnoi et al. 2015). Further, a couple of FLDs from *Branchiostoma floridae* have four contiguous Cys residues instead of just two (Bishnoi et al. 2015).

The Ca^{2+}-binding site in FLDs is located in the region of the FLD including the loops CDR1 and CDR2, and although the Ca^{2+} is not involved directly in ligand binding, it does help to stabilize the overall structure of the protein and in particular contour the CDR1 and CDR2 loops (Bianchet et al. 2002). In AAA, the Ca^{2+}-binding site comprises the amino acids, Asn35, Asp38, Asn40, Ser49, Cys146, and Glu147 (Bianchet et al. 2002). Of these, Asp38, Ser49, and Glu147 interact with Ca^{2+} through their side chains, and Ser49 seems especially important as it interacts both through its side chain and main chain oxygen atoms (Bianchet et al. 2002). Asp38 is conserved in the majority of prokaryotic, eukaryotic, and viral FLDs and in a few cases substituted with Asn, Glu, or Gln. The amino acids, Ser49 and Glu147, are conserved in >50% of eukaryotic and prokaryotic FLDs but absent in viral FLDs (Bishnoi et al. 2015). Glu147 is substituted by Asp, Gln, Asn, Ser, or Lys in some FLDs and Ser49 substituted by Thr in some proteins (Bishnoi et al. 2015). In yet others lacking an amino acid residue with a side chain oxygen atom in this position, a water molecule might contribute toward Ca^{2+} coordination, or the coordination geometry might be altered (Vasta et al. 2008).

Besides these motifs, a characteristic feature of the FLD is the presence of hypervariable regions termed the complementarity-determining regions (CDRs) (Bianchet et al. 2002) (Fig. 24.4). These CDRs are located in the loops that frame the fucose-binding region and play an important role in shaping the fucose-binding pocket. The length and sequence of CDR1 especially has been suggested to have an important role in determining the fine oligosaccharide specificity of the FLD (Bianchet et al. 2002). It varies from 5 to 15 residues in bacterial sequences and is 9 residues long in ~50% of the prokaryotic FLDs (Bishnoi et al. 2015). In eukaryotic FLDs the CDR1 is 7 to 23 residues long, and in viral FLDs, 9 to 11 residues long (Bishnoi et al. 2015).

24.8 Future of FLDs

Historically, eel serum F-type lectins have been widely used for blood typing as anti-H hemagglutinins (Judd 1980) and in lectin histochemistry studies aimed at differentiating cells according to their developmental stages, differentiation, and/or malignancy, based on the expression of several glycans (Wagner 1988; Danguy et al. 1988). With the substantial body of research conducted on F-type lectins and FLDs from diverse organisms, structural and biochemical aspects of FLD binding to fucose and various fucosylated ligands are currently well understood. Additionally, the small size of the FLD (only ~140 residues) and the β-barrel fold make it amenable for engineering applications. Further, the multivalent nature of most FLDs either by oligomerization or by tandem duplication within a polypeptide potentiates binding to and/or cross-linking of cell surface glycans. For instance, in the *S. pneumoniae* protein, SP2159, the presence of three FLDs in the polypeptide contributes to its high apparent affinity for immobilized fucose due to an avidity effect (Boraston et al. 2006). On the other hand, AAA and MsaFBP32 achieve a similar avidity effect via the formation of a non-covalent trimer with the three fucose-binding sites oriented in the same direction (Bianchet et al. 2002, 2010). The structure of MsaFBP32 presents another potential application of FLDs in cross-linking two different surfaces with different glycan epitopes by means of two tandem FLDs oriented in two opposite directions and with distinct specificities (Bianchet et al. 2010). Therefore, it is clear that FLDs have great potential in applications involving the recognition of various fucosylated glycoconjugates, such as in diagnosis of cancer cells with altered fucosylation.

Our extensive survey of FLDs in different life forms, both prokaryotic and eukaryotic, drew our attention toward FLDs co-occurring with a range of other domains including carbohydrate-binding modules, carbohydrate-active enzymes, glycosyl transferases, and other proteins of immune origin (Bishnoi et al. 2015). This remarkable diversity of domain architectures is not only of great interest from an evolutionary point of view, but it also suggests that FLDs are likely being recruited to target diverse functions to specific fucosylated niches in these organisms.

One such example of an FLD directing the biological activity of its co-occurring domain has been previously demonstrated by Farrand et al. with the *S. mitis* protein, lectinolysin, wherein FLD was shown to enhance the pore-forming activity of its co-occurring thio-cytolysin domain by directing it to difucosylated glycans present on the cell membrane of platelets, thereby causing platelet aggregation (Farrand et al. 2008). Another study hinting at the role of FLD in directing activity of its co-occurring domain is of *S. pneumoniae* SP2159, which has three tandem FLDs in coassociation with a glycosyl hydrolase domain (Boraston et al. 2006). The FLD shows specificity toward the fucosylated oligosaccharides of blood antigens present in the lungs and likely directs the glycosyl hydrolase domain, a known virulence factor, to these host tissues bearing these antigens (Boraston et al. 2006). Our recently published study suggests that the FLD might not only work by targeting the co-occurring domain to a specific niche but also modulate the activity of the co-occurring glycoenzymatic domain. Our studies on *Streptosporangium roseum* F-type lectin have demonstrated that a *cis*-positioned FLD can enhance alpha-L-fucosidase activity for aqueous, soluble fucosylated oligosaccharides (Bishnoi et al. 2018). These studies point to the potential of FLDs in applications involving targeting, such as in targeting therapeutics to cancer cells with altered fucosylation, or involving improved or designer glycoenzyme activity.

These applications of FLDs are limited by a few factors. One, currently characterized FLDs recognize only a limited subset of glycan epitopes. Two, there is a paucity of studies demonstrating the functional roles of FLDs that co-occur with other domains. Three, FLDs, like most lectins, have relatively low binding affinity for their ligands. These problems can be alleviated by the study of more diverse naturally occurring FLDs from different life forms and by rational engineering approaches, utilizing the wealth of available sequence and structural FLD space. In research aimed toward this goal, we have found that by drawing from the diversity contained within the sequence space of naturally occurring F-type lectin domains, we can indeed generate engineered F-type lectin domains with improved binding strength and altered ligand-binding specificity (Mahajan et al. 2018; Sharma et al. unpublished data).

To conclude, FLDs with their small size and fold amenable to protein engineering, well-characterized sequence motifs, structural mode of fucose binding, and diverse domain architectures offer tremendous potential for use in applications involving directing, modulating, or targeting distinct biological activities or functions to selected glycosylated niches with efficacy. Further research on diverse naturally occurring FLDs in different domain architectures and protein engineering experiments with the FLD fold can help realize these applications.

Acknowledgments The authors' research described in this review was enabled by a research grant from the Department of Science and Technology, Government of India, to RTNC (FAST-TRACK grant no. SR/FT/LS-87/2012 to RTNC) and infrastructure and research facilities provided by the CSIR-Institute of Microbial Technology, Chandigarh (manuscript number 05/2018). SM is a DBT Senior Research Fellow.

References

Arivalagan J, Marie B, Sleight VA, Clark MS, Berland S, Marie A (2016) Shell matrix proteins of the clam, Mya truncata: roles beyond shell formation through proteomic study. Mar Genomics 27:69–74. https://doi.org/10.1016/j.margen.2016.03.005

Baldus SE, Thiele J, Park YO, Hanisch FG, Bara J, Fischer R (1996) Characterization of the binding specificity of Anguilla anguilla agglutinin (AAA) in comparison to Ulex europaeus agglutinin I (UEA-I). Glycoconj J 13 (4):585–590

Bianchet MA, Odom EW, Vasta GR, Amzel LM (2002) A novel fucose recognition fold involved in innate immunity. Nat Struct Biol 9(8):628–634. https://doi.org/10.1038/nsb817

Bianchet MA, Odom EW, Vasta GR, Amzel LM (2010) Structure and specificity of a binary tandem domain F-lectin from striped bass (*Morone saxatilis*). J Mol Biol 401(2):239–252. https://doi.org/10.1016/j.jmb.2010.06.018

Bishnoi R, Khatri I, Subramanian S, Ramya TN (2015) Prevalence of the F-type lectin domain. Glycobiology 25(8):888–901. https://doi.org/10.1093/glycob/cwv029

Bishnoi R, Mahajan S, Ramya TNC (2018) An F-type lectin domain directs the activity of Streptosporangium roseum alpha-L-fucosidase. Glycobiology 28 (11):860–875

Boraston AB, Wang D, Burke RD (2006) Blood group antigen recognition by a *Streptococcus pneumoniae* virulence factor. J Biol Chem 281(46):35263–35271. https://doi.org/10.1074/jbc.M607620200

Cammarata M, Vazzana M, Chinnici C, Parrinello N (2001) A serum fucolectin isolated and characterized from sea bass Dicentrarchus labrax. BBA-Gen Subjects 1528(2-3):196–202

Cammarata M, Benenati G, Odom EW, Salerno G, Vizzini A, Vasta GR, Parrinello N (2007) Isolation and characterization of a fish F-type lectin from gilt head bream (*Sparus aurata*) serum. Biochim Biophys Acta 1770(1):150–155. https://doi.org/10.1016/j.bbagen.2006.09.015

Cammarata M, Salerno G, Parisi MG, Benenati G, Vizzini A, Vasta GR, Parrinello N (2012) Primary structure and opsonic activity of an F-lectin from serum of the gilt head bream Sparus aurata (Pisces, Sparidae). Ital J Zool 79(1):34–43. https://doi.org/10.1080/11250003.2011.596167

Cassels FJ, Odom EW, Vasta GR (1994) Hemolymph lectins of the blue crab, Callinectes sapidus, recognize selected serotypes of its pathogen Vibrio parahaemolyticus. Ann N Y Acad Sci 712:324–326

Chen J, Xiao S, Yu Z (2011) F-type lectin involved in defense against bacterial infection in the pearl oyster (Pinctada martensii). Fish Shellfish Immunol 30(2):750–754. https://doi.org/10.1016/j.fsi.2010.12.025

Cho SY, Kwon J, Vaidya B, Kim JO, Lee S, Jeong EH, Baik KS, Choi JS, Bae HJ, Oh MJ, Kim D (2014) Modulation of proteome expression by F-type lectin during viral hemorrhagic septicemia virus infection in fathead minnow cells. Fish Shellfish Immunol 39(2):464–474. https://doi.org/10.1016/j.fsi.2014.05.042

Danguy A, Kiss R, Pasteels JL (1988) Lectins in histochemistry. A survey. Biol Struct Morphog 1(3):93–106

Farrand S, Hotze E, Friese P, Hollingshead SK, Smith DF, Cummings RD, Dale GL, Tweten RK (2008) Characterization of a streptococcal cholesterol-dependent cytolysin with a Lewis y and b specific lectin domain. Biochemistry 47(27):7097–7107. https://doi.org/10.1021/Bi8005835

Feil SC, Lawrence S, Mulhern TD, Holien JK, Hotze EM, Farrand S, Tweten RK, Parker MW (2012) Structure of the lectin regulatory domain of the cholesterol-dependent Cytolysin Lectinolysin reveals the basis for its Lewis antigen specificity. Structure 20(2):248–258. https://doi.org/10.1016/j.str.2011.11.017

Finn RD, Bateman A, Clements J, Coggill P, Eberhardt RY, Eddy SR, Heger A, Hetherington K, Holm L, Mistry J, Sonnhammer EL, Tate J, Punta M (2014) Pfam: the protein families database. Nucleic Acids Res 42(Database issue):D222–D230. https://doi.org/10.1093/nar/gkt1223

Fleming RI, Mackenzie CD, Cooper A, Kennedy MW (2009) Foam nest components of the tungara frog: a cocktail of proteins conferring physical and biological resilience. Proc Biol Sci 276(1663):1787–1795. https://doi.org/10.1098/rspb.2008.1939

Gorbushin AM, Borisova EA (2015) Lectin-like molecules in transcriptome of Littorina littorea hemocytes. Dev Comp Immunol 48(1):210–220. https://doi.org/10.1016/j.dci.2014.10.007

Gouet P, Robert X, Courcelle E (2003) ESPript/ENDscript: extracting and rendering sequence and 3D information from atomic structures of proteins. Nucleic Acids Res 31(13):3320–3323

Holm L, Sander C (1993) Protein structure comparison by alignment of distance matrices. J Mol Biol 233(1):123–138. https://doi.org/10.1006/jmbi.1993.1489

Honda S, Kashiwagi M, Miyamoto K, Takei Y, Hirose S (2000) Multiplicity, structures, and endocrine and exocrine natures of eel fucose-binding lectins. J Biol Chem 275(42):33151–33157. https://doi.org/10.1074/jbc.M002337200

Jaroszewski L, Rychlewski L, Li Z, Li W, Godzik A (2005) FFAS03: a server for profile-profile sequence alignments. Nucleic Acids Res 33(Web Server issue):W284–W288. https://doi.org/10.1093/nar/gki418

Judd JW, Issitt PD (1980) The role of lectins in blood group serology. CRC Crit Rev Clin Lab Sci 12(3):171–214

Kim GH, Klochkova TA, Yoon KS, Song YS, Lee KP (2006) Purification and characterization of a lectin, bryohealin, involved in the protoplast formation of a marine green alga *Bryopsis plumosa* (Chlorophyta). J Phycol 42(1):86–95. https://doi.org/10.1111/j.1529-8817.2005.00162.x

Liu W, Xie Y, Ma J, Luo X, Nie P, Zuo Z, Lahrmann U, Zhao Q, Zheng Y, Zhao Y, Xue Y, Ren J (2015) IBS: an illustrator for the presentation and visualization of biological sequences. Bioinformatics 31(20):3359–3361. https://doi.org/10.1093/bioinformatics/btv362

Mahajan S, Khairnar A, Bishnoi R, Ramya TNC (2017) Microbial F-type lectin domains with affinity for blood group antigens. Biochem Biophys Res Commun 491(3):708–713. https://doi.org/10.1016/j.bbrc.2017.07.125

Mahajan S, Ramya TNC (2018) Nature-inspired engineering of an F-type lectin for increased binding strength. Glycobiology:cwy082

Maki M, Renkonen R (2004) Biosynthesis of 6-deoxyhexose glycans in bacteria. Glycobiology 14(3):1R–15R. https://doi.org/10.1093/glycob/cwh040

Multerer KA, Smith LC (2004) Two cDNAs from the purple sea urchin, Strongylocentrotus purpuratus, encoding mosaic proteins with domains found in factor H, factor I, and complement components C6 and C7. Immunogenetics 56(2):89–106. https://doi.org/10.1007/s00251-004-0665-2

Notredame C, Higgins DG, Heringa J (2000) T-coffee: a novel method for fast and accurate multiple sequence alignment. J Mol Biol 302(1):205–217. https://doi.org/10.1006/jmbi.2000.4042

Odom EW, Vasta GR (2006) Characterization of a binary tandem domain F-type lectin from striped bass (*Morone saxatilis*). J Biol Chem 281(3):1698–1713. https://doi.org/10.1074/jbc.M507652200

Riely BK, Ane JM, Penmetsa RV, Cook DR (2004) Genetic and genomic analysis in model legumes bring Nod-factor signaling to center stage. Curr Opin Plant Biol 7(4):408–413. https://doi.org/10.1016/j.pbi.2004.04.005

Roy A, Kucukural A, Zhang Y (2010) I-TASSER: a unified platform for automated protein structure and function prediction. Nat Protoc 5(4):725–738. https://doi.org/10.1038/nprot.2010.5

Saito T, Hatada M, Iwanaga S, Kawabata S (1997) A newly identified horseshoe crab lectin with binding specificity to O-antigen of bacterial lipopolysaccharides. J Biol Chem 272(49):30703–30708

Salerno G, Parisi MG, Parrinello D, Benenati G, Vizzini A, Vazzana M, Vasta GR, Cammarata M (2009) F-type lectin from the sea bass (Dicentrarchus labrax): purification, cDNA cloning, tissue expression and localization, and opsonic activity. Fish Shellfish Immun 27(2):143–153. https://doi.org/10.1016/j.fsi.2009.01.004

Samuel G, Reeves P (2003) Biosynthesis of O-antigens: genes and pathways involved in nucleotide sugar precursor synthesis and O-antigen assembly. Carbohydr Res 338(23):2503–2519. https://doi.org/10.1016/j.carres.2003.07.009

Springer GF, Desai PR (1970) The immunochemical requirements for specific activity and the physiochemical properties of eel anti-human blood-group H(O) 7 S globulin. Vox Sang 18(6):551–554

Springer SA, Moy GW, Friend DS, Swanson WJ, Vacquier VD (2008) Oyster sperm bindin is a combinatorial fucose lectin with remarkable intra-species diversity. Int J Dev Biol 52(5-6):759–768. https://doi.org/10.1387/ijdb.082581ss

Takaichi S, Maoka T, Masamoto K (2001) Myxoxanthophyll in *Synechocystis sp* PCC 6803 is myxol 2′-dimethyl-fucoside, (3R,2′S)-myxol 2′-(2,4-di-O-methyl-alpha-L-fucoside), not rhamnoside. Plant Cell Physiol 42(7):756–762. https://doi.org/10.1093/Pcp/Pce098

Vasta GR, Ahmed H, Odom EW (2004) Structural and functional diversity of lectin repertoires in invertebrates, protochordates and ectothermic vertebrates. Curr Opin Struct Biol 14(5):617–630. https://doi.org/10.1016/j.sbi.2004.09.008

Vasta GR, Odom EW, Bianchet MA, Amzel LM, Saito K, Ahmed H (2008) F-type lectins: a new family of recognition factors. In: Vasta GR, Ahmed H (eds) Animal lectins: a functional view. CRC Press, London

Vasta GR, Amzel LM, Bianchet MA, Cammarata M, Feng C, Saito K (2017) F-type lectins: a highly diversified family of Fucose-binding proteins with a unique sequence motif and structural fold, involved in self/non-self-recognition. Front Immunol 8:1648. https://doi.org/10.3389/fimmu.2017.01648

Wagner M (1988) Light and electron microscopic lectin histochemistry using fluorochromes and ferritin as labels. Acta Histochem Suppl 36:115–123

Yang J, Yan R, Roy A, Xu D, Poisson J, Zhang Y (2015) The I-TASSER suite: protein structure and function prediction. Nat Methods 12(1):7–8. https://doi.org/10.1038/nmeth.3213

Yoon KS, Lee KP, Klochkova TA, Kim GH (2008) Molecular characterization of the lectin, bryohealin, involved in protoplast regeneration of the marine alga *Bryopsis plumosa* (Chlorophyta). J Phycol 44(1):103–112. https://doi.org/10.1111/j.1529-8817.2007.00457.x

Zhang Y (2008) I-TASSER server for protein 3D structure prediction. BMC Bioinformatics 9:40. https://doi.org/10.1186/1471-2105-9-40

The manufacturer's authorised representative in the EU is Springer Nature Customer Service Centre GmbH, Europaplatz 3, 69115 Heidelberg, Germany. If you have any concerns regarding our products, please contact ProductSafety@springernature.com

Printed and bound by CPI Group (UK) Ltd, Croydon, CR0 4YY
25/09/2026
02222900-0020